T0321808

Superconductivity

Superconductivity is among the most exciting of quantum phenomena in condensed matter physics, and has important applications across science and technology, from fusion reactors to particle accelerators. This self-contained text provides a comprehensive account of the physical foundations of superconductivity and related recent developments in the field. Beginning with a detailed description of the BCS theory of superconductivity, the book then describes the subsequent successes of this landmark theory and proceeds to more advanced topics such as Josephson effects and vortices. The strong coupling theory of superconductivity is introduced in later chapters, providing a springboard to important current research on hydride superconductors, which have displayed very high critical temperatures. Recent manifestations of superfluidity in ultracold atoms physics are also described. This book will give readers a solid grounding in the theory and applications of superconductivity, and an appreciation of its broader importance in the field of modern condensed matter physics.

Roland Combescot is Professor Emeritus in Physics at École Normale Supérieure (ENS) and Sorbonne Université (formerly Université Pierre et Marie Curie) in Paris, and has been a member of l'Institut Universitaire de France (IUF) since 2005. He has an outstanding research and publication record in the area of superconductivity and superfluidity, including nearly thirty articles published in the influential journal, *Physical Review Letters*. He has taught a course on the theory of superconductivity to MSc students at ENS for over twenty years.

Superconductivity

An Introduction

ROLAND COMBESCOT

Ecole Normale Supérieure, Paris

CAMBRIDGE
UNIVERSITY PRESS

University Printing House, Cambridge CB2 8BS, United Kingdom

One Liberty Plaza, 20th Floor, New York, NY 10006, USA

477 Williamstown Road, Port Melbourne, VIC 3207, Australia

314–321, 3rd Floor, Plot 3, Splendor Forum, Jasola District Centre, New Delhi – 110025, India

103 Penang Road, #05–06/07, Visioncrest Commercial, Singapore 238467

Cambridge University Press is part of the University of Cambridge.

It furthers the University's mission by disseminating knowledge in the pursuit of education, learning, and research at the highest international levels of excellence.

www.cambridge.org
Information on this title: www.cambridge.org/9781108428415
DOI: 10.1017/9781108560184

First published 2022

A catalogue record for this publication is available from the British Library.

ISBN 978-1-108-42841-5 Hardback

Contents

Preface

As is often the case, this book is an outgrowth of lecture notes for a course on superconductivity I had the pleasure to teach for several years. Teaching implies naturally severe limitations of time for the course duration, which leads to strong constraints on the subject matter that is taught. These boundary conditions disappear in principle for a book, and I have taken this opportunity to include a number of points I could not address at all during my lectures, for lack of time.

However, this book has not been written with the a priori intention of extending the scope of what I taught. My teaching was an introduction to superconductivity, and my aim in this book has been to stay at the same introductory level. I have rather taken the extended space one enjoys in a book to include subject matter that in my mind should have been included logically and coherently in the course but could not be included for lack of time. I have been able in this way to include most of the points I regretted omitting from the course. This description actually only corresponds to the first six chapters of this book. Indeed, I have taken this opportunity to include some more recent subject matter from the field of cold atoms physics, which I feel is quite relevant for our understanding of superconductivity, as I describe in more detail below.

Coherently with this introductory spirit, I have tried to be quite explicit in my writing, providing all the necessary details for understanding by the same kind of students as the ones I was teaching. On balance between going into detail or skipping the "obvious," I have chosen the former. Naturally, there are limits to this, as providing too much detail makes the text burdensome. This choice has also been mostly valid at the beginning of chapters, with the finishing parts being usually being devoted to more advanced matter and going accordingly at a somewhat more accelerated pace.

In the same spirit, I have done my best to make this book as self-contained as possible. The understanding requires only basic knowledge of electromagnetism, quantum mechanics, statistical physics and little of solid state physics (and some mathematics). But otherwise I have rather chosen to start from first principles, without for example referring to linear response or scattering theory.

Regarding the organization of the chapters, my aim has been to go directly at the microscopic understanding of superconductivity provided by BCS theory. Whether accepted or rejected, it is the reference frame of our present-day comprehension. Another ingredient that I feel is quite important when teaching science nowadays, particularly when exploring such a strange phenomenon as superconductivity, is to describe the evolution of ideas that has led to the present understanding. This is an essential point for explaining how the present scientific knowledge has been built, which allows for its future evolution on a sound basis. Ideally, this should be done by following the historical evolution of the field, but this comes rapidly in contradiction with good teaching practices because the real history is often quite long and somewhat tortuous. Hence one needs to identify shortcuts to obtain a clear logical presentation. Nevertheless, I have kept in mind this spirit, trying to explain wherever possible where the matter comes from.

This approach appears particularly in the first chapter, which presents the early ingredients of superconductivity and closes with a section summarizing the ideas leading to BCS theory. The second chapter deals with the basics of BCS theory. The last sections of this chapter are more advanced and are addressed to readers who are somewhat dissatisfied with the BCS wave function, which is fairly frequent. As such, they may be skipped upon first reading. The third chapter extends BCS theory to the nonzero temperature situation. The last sections of this chapter are an appropriate place to deal in depth with the microscopics of the pairing interaction, and then to open the door leading to symmetries and mechanisms other than the standard BCS ones, which are present in unconventional superconductors. Again, these sections may be skipped at first reading. The portion of the manuscript devoted to BCS closes with the fourth chapter, which is mostly devoted to the response of a superconductor to an electromagnetic perturbation, according to BCS theory. One aim of this chapter is to show that BCS theory indeed describes a system that is superconducting. This is a worthwhile purpose, although the road is fairly long.

The next chapter deals with the fascinating manifestations of quantum mechanics at the macroscopic scale that appear in superconductors, through flux quantization and the Josephson effects. These give rise in particular to remarkable applications. The sixth chapter is devoted to the beautiful Ginzburg–Landau theory, which leads to the introduction of vortices, so crucial for strong current applications of superconductivity. It does not come in the proper historical order, but I feel that this is pedagogically the appropriate position.

The following chapters arise from the opportunity presented by the recent remarkable progress in the physics of ultracold atoms, allowing one to display the BEC–BCS crossover. This establishes a direct physical link between these two related aspects of superfluidity and enlightens our understanding of pairing in superconductivity. Hence, all the more since I have had a direct interest in this field, it has been quite tempting to include a chapter on the BEC–BCS crossover, both for logical reasons and also because this is beautiful physics. This has not been such an easy matter, since it implied, at a simple level, the introduction of several new concepts. In particular, this has inspired first a complete chapter on Bose–Einstein condensation, a necessary ingredient to fully understand the BEC–BCS crossover. This chapter is also welcome for enhancing one's understanding of superconductivity, since it allows one to present in detail the physics of superfluidity that underlies superconductivity.

The final chapter is by far the most difficult matter of this book and also was the most difficult to write. Hence its position at the end of this book is quite appropriate. Nevertheless, its presence is logically necessary. Indeed, it serves to complete the unsatisfactory handling of BCS theory regarding the time-dependent nature of the pairing interaction. This completion is rewarded by some remarkable agreements between experiment and theory, confirming the validity of our understanding of superconductivity in the corresponding compounds. On the other hand, this chapter is also justified by the very recent discovery of hydride superconductors with very high critical temperatures, for which the formalism presented in this chapter seems to be the appropriate one. However, although I have done my best to stay in the spirit of the preceding chapters, this one is not as self-contained as the preceding ones. A proper complete presentation of the matter would have led to far too technical and far too long explanations for this book. Hence some stages require a leap of faith, which I have tried to patch as well as I could.

On a final note, I would like to thank Xavier Leyronas for all the pleasant time I had sharing with him teaching superconductivity. And I want to take this opportunity to express my gratitude and thoughts to my friend, Dierk Rainer, from whom I learned so much.

1 Phenomenology

1.1 Basic Properties

1.1.1 Infinite Conductivity

Superconductivity was discovered in 1911 by H. Kamerlingh Onnes [1] in Leiden. Kamerlingh Onnes was a pioneer in reaching experimentally low temperatures and exploring physics in this by then new realm. He was the first in 1908 to liquify helium gas (he got the Nobel Prize for this in 1913), which provided him a powerful cooling agent at these temperatures.

Kamerlingh Onnes was interested in the variation of the conductivity of metals at low temperature, which was at the time a controversial matter. He chose to work on mercury because it is much easier to purify by distillation than other metals, since it is liquid at room temperature. In this way, he could get rid of impurities, which contribute to the electrical resistivity of metals, and study the intrinsic low-temperature behavior. Upon cooling his solid mercury wire, he observed a slow decrease of its resistance, corresponding to his expectations, and then around the temperature of 4.2 K, a sudden drop to a value so low that he could not actually measure it, as seen in Fig. (1.1). This experimental disappearance of the resistance implies a vanishing resistivity ρ for mercury below the "critical temperature" of $T_c = 4.2$ K,

$$\rho = 0 \tag{1.1}$$

Kamerlingh Onnes called "superconductivity" this property of a metal to have in effect an infinite conductivity $\sigma = 1/\rho$. The year after his discovery, he found that tin and lead were superconductors with critical temperatures of 3.7 K and 7.2 K, respectively.

Naturally it is meaningless to claim that Eq. (1.1) is an experimental result; it is only a logical extrapolation of the experimental result. An experiment always has a limited accuracy and can only claim that the resistivity is extremely small. Nevertheless experimentalists have pushed as far as they could to determine how small the resistivity of a superconductor is. A clever and striking way to do this indirectly is to observe persistent currents. Indeed one can generate currents by induction in a metal having the shape of a ring. For a standard metal with a nonzero resistivity, these currents decay rapidly by dissipation due to Joule heating. However, for a superconductor, Eq. (1.1) holds and no dissipation occurs, so the induced currents can persist indefinitely. These currents can be observed by the magnetic field they generate. Kamerlingh Onnes performed first such an

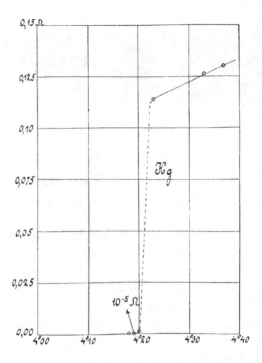

Fig. 1.1 The resistance of mercury as a function of temperature as measured by H. Kamerlingh Onnes [1].

experiment in 1914, and observed persistent currents for hours. This kind of experiment has been repeated; persistent currents have been observed for several years, and the decay time for the persistent current has been evaluated to $\sim 10^5$ years. This comes quite close to an experimental proof of Eq. (1.1).

1.1.2 Critical Temperature

Kammerlingh Onnes realized naturally that in principle, superconductivity could allow one to generate very large electric currents and, as a result, very large magnetic fields. Obviously the very low temperature at which the phenomenon occurs makes it in practice quite inconvenient to set up such a device. Hence the value of the critical temperature is not only an important quantity to characterize superconductivity, but it is also of utmost practical interest. This has led to the exploration of a number of materials for their possible superconducting properties. Among the elements of the periodic table, 33 are superconductors at atmospheric pressure (with an additional 24 which become superconductors under pressure). The one with the highest critical temperature is niobium, with $T_c = 9.26$ K.

More generally, it has been progressively realized that, far from being an exceptional phenomenon, superconductivity appears quite frequently at low temperature. Its absence may be due to the competing appearance of another kind of transition, for example toward magnetic order. It should be noted that superconducting transitions may be found at very low temperature, which while not being practically useful may be of fundamental interest. There is no lower bound for T_c. In particular copper, the best standard conductor, does not display a superconducting transition at the lowest temperatures presently reached. Similarly gold and silver are not superconductors.

Coming back to the more practically interesting purpose of finding high T_c, a number of metallic alloys have also been explored. In this search, the one with highest critical temperature has been Nb_3Ge with $T_c = 23.2$ K. This is already above the boiling point of liquid hydrogen, which is at 20 K under atmospheric pressure, but not enough to be of practical interest. Hence, although it has a slightly lower $T_c = 18.3$ K, Nb_3Sn is rather used industrially because it can withstand high currents and magnetic fields. Nevertheless it is NbTi that is presently the industrially preferred compound for practical reasons, although its T_c is only 10 K. It is this alloy that is mostly used for the production of the high magnetic fields required in magnetic resonance imaging (MRI) in standard medical devices. Similarly this is the alloy used in large high-energy particle accelerators, although the need for higher fields induces a switch to Nb_3Sn. In all cases, the required low temperature is obtained through liquid Helium cryogenics, which has seen much development to large scale for this purpose.

Despite many efforts, progress in increasing T_c had become so slow in the fifties and sixties that researchers in the field of superconductivity tended to believe that there was some kind of intrinsic upper bound for T_c and that in practice its increase was near saturation. Hence it has been a great shock to this community when in 1986 Georg Bednorz and Alex Müller found that a perovskite[1] in "the La-Ba-Cu-O system" (more precisely with chemical composition $Ba_xLa_{1-x}CuO_{3-y}$) is a superconductor with T_c around 35 K. They received the Nobel Prize the next year for this breakthrough. Then things progressed very rapidly, essentially guided by chemical reasoning that leads one to replace an element with a chemically similar one. The following year, $T_c = 93$ K was reached in $YBa_2Cu_3O_7$. More generally, the critical temperature of $YBa_2Cu_3O_{7-x}$ is very sensitive to the "doping" x since $YBa_2Cu_3O_6$ is an antiferromagnetic insulator. It reaches $T_c = 95$ K for $x \simeq 0.07$. Such critical temperatures have represented an essential step in the increase of T_c since they are above the boiling point of liquid nitrogen, which is at 77 K under atmospheric pressure. Liquid nitrogen is routinely obtained in the gas industry, with a typical annual world production of 8 million tons. Hence cooling such a superconductor below its critical temperature is a considerably simpler matter than when helium has to be used.

Further progress with these kinds of compounds has led to the discovery of bismuth compounds of general formula $Bi_2Sr_2Ca_nCu_{n+1}O_{2n+6-\delta}$ with T_c ranging from 95 K to 107 K depending on n, thallium compounds $Tl_mBa_2Ca_{n-1}Cu_nO_{2n+m+2+\delta}$ with a highest T_c reaching 127 K, and finally mercury compounds $HgBa_2Ca_2Cu_3O_{8+\delta}$ with a highest critical temperature of 135 K found in 1993. The highest critical temperature reported to date in these cuprates superconductors has been 166 K in $HgBa_2Ca_2Cu_3O_{8+\delta}$ at a pressure of 23 GPa. A remarkable feature of all these compounds is that above the critical temperature, they are fairly bad metals with poor conductivity, in contrast with the earlier superconductors discussed at the beginning of this section. They are also quasi-bidimensional materials since they are essentially stacks of CuO_2 planes, which are their conducting part, with a fairly weak electronic coupling between the planes.

---

[1] This is a material with the same crystal structure as $CaTiO_3$.

The cuprate discovery clearly showed that there was no barrier around 20 K for T_c. Hence this produced an incentive to check the low-temperature properties of various materials. As early as 1988, superconductivity was found in $Ba_{0.6}K_{0.4}BiO_3$ around 30 K. Like the cuprates, this compound is a perovskite, but it does not contain CuO_2 planes and it is tridimensional. Hence, it is in a somewhat different class of materials.

Markedly different are the "doped" fullerenes, with the main ingredient being the fullerene molecule C_{60}, which has the shape of a soccer ball. It can be doped with various alkali, with Cs_3C_{60} reaching in 1995 a surprising $T_c = 40$ K under pressure. These materials could almost be considered as organic superconductors, with other organic compounds having much lower T_c.

A surprising result was then found in 2001, where MgB_2 was discovered to have a critical temperature of 39 K. This compound is similar to the alloys investigated earlier with the hope of finding higher critical temperature, and its late discovery looks like a miss of earlier searches.

More interesting is the discovery of superconductivity in iron-based compounds. Indeed, starting in 2006, superconductivity in these compounds was investigated because they have fairly high T_c, considering that the magnetic properties of Fe were believed to be detrimental to superconductivity. In 2008, a T_c of 26 K was reported in $LaO_{1-x}F_xFeAs$ (called an "oxypnictide"[2]) with $x = 0.05 - 0.12$, followed the same year by the finding of $T_c = 55$ K in $SmO_{1-x}F_xFeAs$. These iron-based materials form a rich family with several parent compounds.

Finally the quite recent last step in this progress in T_c has been the evidence for superconductivity in various hydrides around 200 K, under high pressure. A first result of $T_c = 203$ K in H_3S at 155 GPa has appeared in 2015. Then, in 2018, $T_c = 215$ K was reached in LaH_{10} (although the stoichiometry in H may be somewhat uncertain in these compounds), followed by a claim for superconductivity at $T_c = 260$ K around 200 GPa in the same compound. Clearly the search for superconductivity in these hydrides is not over. It is already quite close to the long-lasting dream of finding superconductivity at room temperature.

1.1.3 Meissner Effect

We now come to the second defining property of a superconductor. Although it has been found by Meissner and Ochsenfeld [2] in 1933, a fairly long time after the discovery by Kamerlingh Onnes, it turned out to be a fundamental feature of the superconducting state. Meissner and Ochsenfeld cooled a sample of tin in the presence of an applied magnetic field **H**. They expected no change of the field when the temperature was going below the critical temperature T_c for tin. Instead, when measuring the field in the vicinity of the superconducting tin, they found strong changes, as if tin behaved as a magnetic material. These modifications were consistent with the magnetic induction **B** going to zero inside the superconducting tin sample. As a result the field lines are pushed away from the superconductor, as shown in Fig. (1.2). The field is "expelled" from the superconductor.

---

[2] A pnictide element is an element belonging to the nitrogen column in the periodic table: N, P, As, Sb, Bi.

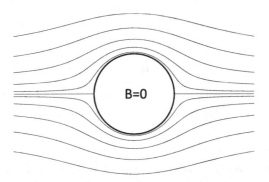

Fig. 1.2 Schematic view of the magnetic field lines for a spherical superconductor, in the presence of an applied magnetic field. Without the superconductor or for a normal metal, the magnetic field lines would just be horizontal parallel lines.

If the temperature was first lowered below the critical temperature T_c with a zero applied magnetic field (zero-field-cooling), and then the magnetic field would be raised at fixed temperature, this experimental result could easily be understood as resulting from the infinite conductivity of the superconductor. Indeed, in this case, raising the magnetic field gives rise to induced currents in the superconductor, and from Lenz's law, they oppose the variation of the induction inside the superconductor. For a standard metal, these induced currents decay by dissipation due to the metal resistivity. However, with the infinite conductivity of the superconductor there is no such dissipation; these currents run forever, and Lenz's law can reach its full effect of maintaining the magnetic induction at its initial value $\mathbf{B} = \mathbf{0}$.

This "freezing" of the induction lines is, for example, well known in plasma physics, where very large conductivity (although not infinite) can be found. Basically the infinite conductivity σ forces the electric field $\mathbf{E} = \mathbf{j}/\sigma$ to be zero inside the superconductor regardless of the current $\mathbf{j}$. Then for $\mathbf{E} = \mathbf{0}$, Maxwell's equation $\mathbf{curl}\,\mathbf{E} = -\partial\mathbf{B}/\partial t$ implies that the magnetic induction cannot change. However, in this zero-field-cooled case, we would have reached an out-of-equilibrium situation lasting forever.

By contrast, the Meissner effect cannot be explained by Lenz's law, since it is obtained by merely changing the temperature at a fixed field, so that no induced currents can arise. Rather, one comes to the conclusion that the situation depicted by Fig. (1.2) corresponds to a thermodynamical equilibrium situation for the superconductor since, for given temperature and field, it is the one that is found regardless of the order in which the temperature is lowered and the magnetic field is raised.

Nevertheless, although infinite conductivity cannot fully explain by itself the Meissner effect, it is an essential ingredient of the effect. Indeed the fact that the induction is zero inside the superconductor is physically due to the existence of permanent currents in the superconductor which screen the external field, and they can persist only because conductivity is infinite.

If one does not look for a microscopic understanding of the superconductor and stays only at a macroscopic level, one can summarize the Meissner effect by the fact that it is a magnetic material with the property $\mathbf{B} = \mathbf{0}$ in the superconductor in the presence of an applied magnetic field **H**. By definition of the magnetic susceptibility χ of the material, we have

$$\mathbf{B} = \mu_0(\mathbf{H} + \mathbf{M}) = \mu_0(1 + \chi)\mathbf{H} \tag{1.2}$$

since the magnetization **M** is related to the field by $\mathbf{M} = \chi\mathbf{H}$. Here μ_0 is the vacuum permeability. Hence we have for a superconductor

$$\chi = -1 \tag{1.3}$$

In other words, a superconductor is perfect diamagnet. In general, diamagnets tend to weaken the external magnetic field. A superconductor does it perfectly by reducing the magnetic induction to zero.

Let us conclude by stressing that the standard procedure for identifying a metallic compound as a superconductor is to show experimentally that it has zero resistivity and displays the Meissner effect. "Zero" resistivity only may just correspond to a very good conductor with a resistivity below the experimental resolution. The Meissner effect is usually checked both when the compound is cooled in a zero field (and then a field is applied) and when it is cooled through the critical temperature in the presence of a magnetic field, to check that the field is properly expelled from the superconductor below the critical temperature. A practical complication is that the Meissner effect may not be complete: some parts of the superconductor may actually stay in the normal state due to inhomogeneities, impurities, and other kinds of defects. As a result, the measured susceptibility may not be as strong as it should be, which may bring some ambiguity in the identification of the compound as a superconductor.

1.1.4 Critical Field

The Meissner effect makes it clear physically that there must be a critical field, beyond which the superconductor no longer exists at a given temperature. Indeed there is clearly a magnetic energy cost to the distortion of the field from its value in the absence of the superconductor. This will be quantified below in the next subsection. This is compensated by the lowering in energy associated with the spontaneous transition from the normal to the superconducting state. However, this gain in energy is a fixed amount, independent of the field, whereas the magnetic energy cost increases indefinitely with the field. Obviously, if the field is too high, the total energy for going into the superconducting state will be positive and this state will no longer be stable with respect to the normal state of the metal. Hence, for a given temperature T, there is a critical field $H_c(T)$ beyond which superconductivity disappears.

The existence of this critical field was actually discovered experimentally by Kamerlingh Onnes in 1914, not much after his initial discovery of superconductivity, and much earlier than the Meissner effect. Indeed, soon after his discovery, Kamerlingh Onnes was interested in the possibility of producing high magnetic fields by electromagnets with superconducting wires, in which huge currents could in principle be fed. The existence of a critical field puts a fundamental limit to the production of such high currents, since

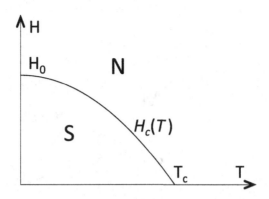

Phase diagram of a type I superconducting metal in the presence of a magnetic field H. In the domain below the critical field curve $H_c(T)$, the metal is in the superconducting state (S), while above this curve it is in the normal state (N).

the magnetic field produced by the current itself destroys superconductivity when it goes beyond the critical field. Hence there is an upper limit for the supercurrent carried by the superconductor, which is called the critical current.

Experimentally the critical field $H_c(T)$ decreases with increasing temperature and goes smoothly to zero at the critical temperature. The resulting phase diagram is pictured in Fig. (1.3). The experimental results turn out to be very close to a parabolic law:

$$H_c(T) = H_0 \left[1 - \left(\frac{T}{T_c} \right)^2 \right] \tag{1.4}$$

Actually this simple situation holds only for one class of superconductors, called type I superconductors, which have been the first to be discovered and studied. A second class of superconductors, called type II superconductors, was discovered by Shubnikov in 1933. In these type II superconductors, instead of having a sudden transition from the super-conducting state to the normal state upon increasing the magnetic field, the transition is progressive. More precisely, below a first critical field $H_{c1}(T)$ the situation is exactly the same as in type I superconductors and one has a complete exclusion of the field from the bulk of the superconductor. However, above $H_{c1}(T)$, there is a progressive admission of the field in the bulk of the superconductor, until an upper critical field $H_{c2}(T)$ is reached where superconductivity disappears completely and the metal is in the normal state. The state between $H_{c1}(T)$ and $H_{c2}(T)$ is called the "mixed state." This more complex situation is depicted in Fig. (1.4).

Since there is a partial admission of the field in the mixed state, it is intuitively clear that the magnetic energy cost is lowered, compared to the situation of full exclusion we have seen for type I superconductors. Hence, in this case, superconductivity can survive to higher magnetic fields than for type I. Indeed type II superconductors are the only ones of interest for applications where superconductivity has to survive very high fields or very high currents.

Finally it is important to stress that if we want to directly apply the above considerations to a real superconducting sample, we have to take it with a shape infinitely elongated in the

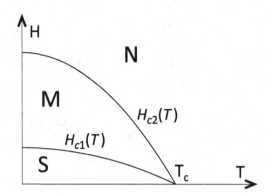

Fig. 1.4   Phase diagram of a type II superconducting metal in the presence of a magnetic field H. In the domain below the critical field curve $H_{c1}(T)$, the metal is in the superconducting state (S); between $H_{c1}(T)$ and $H_{c2}(T)$, it is in the mixed state (M); and above $H_{c2}(T)$, it is in the normal state (N).

direction of the field – for example, a very long cylinder parallel to the field. Otherwise, as for any magnetic material, we have to take into account the demagnetizing field created by the sample itself, which is naturally inhomogeneous and depends on the sample shape. As a result, the strength of the actual field depends on the spatial position. For example, in the situation depicted qualitatively in Fig. (1.2), the field at the equator of the sphere will be stronger than at the poles.

If we take the case of a type I superconductor, this may result in the fact that if the applied field is strong enough, some parts of the metal will have a field larger than the critical field and will accordingly be in the normal state, while some other parts will have a field smaller than the critical field and be superconducting. Naturally the field distribution itself depends on which parts of the sample are normal and which parts are superconducting. Hence one finds a situation where there is a mixture of normal and superconducting domains in the metallic sample. This is called the "intermediate state." Finding the distribution of domains that minimizes the energy is quite a complicated problem, and the result may be a fairly complex structure. Correspondingly, it is also quite difficult to experimentally determine this structure with good precision.

1.1.5 Thermodynamics

The interpretation of the Meissner effect in magnetic terms, leading to the conclusion that a superconductor can be considered as a perfect diamagnet, allows one to relate in a simple way the thermodynamic properties of a superconductor and its magnetic properties. Actually this holds only for type I superconductors, to which we here restrict ourselves, the case of type II superconductors being somewhat more involved.

Thermodynamics tells us that the appropriate thermodynamical potential, to investigate the properties of a system as a function of its temperature T and its volume V, is its free energy $F(T, V) = E - TS$, rather than its energy E. Indeed the differential of the free energy, in terms of the entropy S and the pressure p of the system, is

$$dF = -SdT - p\,dV \tag{1.5}$$

whereas we have for the energy $dE = TdS - pdV$. Note that here, since we deal with a system with a fixed number of particles N, we do not have to include a contribution $\mu\,dN$, where μ is the chemical potential, in contrast with situations we will deal with later on.

When a magnetic field is present, this has to be generalized to take into account the magnetic variables. In terms of the magnetic field H and the magnetic induction B, electromagnetism tells us that the energy increase dE, due to a change dB in magnetic induction, is given by $dE = HdB$. Actually, in standard electromagnetism, one deals usually with space-dependent field $H(\mathbf{r})$ and induction $B(\mathbf{r})$, and the local change in energy $H(\mathbf{r})dB(\mathbf{r})$ has to be summed over all space to give the total energy change $\int d\mathbf{r}\,H(\mathbf{r})dB(\mathbf{r})$. However, we consider here a homogeneous system, so we do not need to take into account space dependence.

To take into account the change in energy due to magnetic variables, this electromagnetic contribution has to be added to the above energy variation, and similarly to the free energy variation, to give

$$dF = -SdT - pdV + HdB \tag{1.6}$$

However, by analogy with the case of a temperature T imposed by an external source, we rather want to consider the superconductor submitted to an external magnetic field H, produced, for example, by external currents in a coil. For this purpose, it is more convenient to perform a Legendre transform and consider the thermodynamical potential $G = F - HB$, which has a differential

$$dG = -SdT - pdV - BdH \tag{1.7}$$

Actually the effects associated with the volume variation of a superconductor, at the low temperature we are interested in, are in practice extremely small, and hence we will disregard them. We will assume that the superconductor has a fixed volume, which we take for convenience equal to the unit volume. Hence, we may omit the pdV term in Eq. (1.7), which reduces merely to

$$dG = -SdT - BdH \tag{1.8}$$

Consider first the compound in its normal state, for which we assume that there are no magnetic properties at all, so that $B = \mu_0 H$ where μ_0 is again the vacuum permeability. Hence $BdH = \mu_0 HdH$. Integrating Eq. (1.8) at fixed temperature with the magnetic field going from 0 to H, we obtain the normal state potential $G_n(T, H)$ in terms of $G_n(T, 0)$ as

$$G_n(T, H) - G_n(T, 0) = \int_0^H dG = -\mu_0 \int_0^H HdH = -\frac{1}{2}\mu_0 H^2 \tag{1.9}$$

In the superconducting state, we may perform the same calculation, which gets even simpler since from the Meissner effect we merely have $B = 0$ in the superconductor. This leads to

$$G_s(T, H) - G_s(T, 0) = \int_0^H dG = -\int_0^H BdH = 0 \tag{1.10}$$

Let us now, for a given temperature T, take the magnetic field equal to $H_c(T)$, which is the field corresponding to the transition line from the superconducting to the normal state. On this line, there is a thermodynamical equilibrium between the superconducting and the normal states, which implies that the corresponding potential G for these two phases are equal:

$$G_s(T, H_c(T)) = G_n(T, H_c(T)) \tag{1.11}$$

Making use of Eq. (1.9) and Eq. (1.10), this leads to

$$G_n(T, 0) - G_s(T, 0) = \frac{1}{2}\mu_0 H_c^2(T) \tag{1.12}$$

Hence we have just to measure the magnetic field corresponding to the normal-superconducting phase transition, and we know the thermodynamic potential $G_s(T, 0)$ in the zero field in the superconducting state as soon as we know the corresponding thermodynamic potential $G_n(T, 0)$ in the normal state. And from Eq. (1.9) and Eq. (1.10), we have the same information for any magnetic field.

We make use of Eq. (1.12) to investigate the order of the normal-superconducting phase transition. From Eq. (1.8), the entropy of any phase is given by $S = -\partial G/\partial T|_H$. Having the entropy of the two phases, the latent heat L is obtained from $L = T(S_n - S_s)$, to be evaluated on the transition line. From Eq. (1.9), Eq. (1.10), and Eq. (1.12), we obtain

$$\begin{aligned} S_n - S_s &= -\frac{\partial(G_n(T, H) - G_s(T, H))}{\partial T}\bigg|_H \\ &= -\frac{\partial(G_n(T, 0) - G_s(T, 0))}{\partial T} = -\mu_0 H_c(T)\frac{dH_c(T)}{dT} \end{aligned} \tag{1.13}$$

and the latent heat is given by

$$L = -\mu_0 T H_c(T)\frac{dH_c(T)}{dT} \tag{1.14}$$

We see from Fig. (1.3) that experimentally the transition field $H_c(T)$ decreases with increasing temperature, so that $dH_c(T)/dT < 0$ and the latent heat Eq. (1.14) is positive. This means that the transition from superconducting to normal state is first order. The only exception is at the zero field, where $H_c(T) = 0$, so that $L = 0$ and the transition is second order at the standard critical temperature T_c of the superconductor.

For this second-order phase transition at zero field, we can obtain the specific heat jump $C_s - C_n$ from the experimental knowledge of $H_c(T)$. Since $C = T \partial S/\partial T$, we have at temperature T

$$C_s - C_n = T \frac{d(S_s - S_n)}{dT} = \frac{1}{2}\mu_0 T \frac{d^2 H_c^2(T)}{dT^2}\bigg|_{T_c} \tag{1.15}$$

At $T = T_c$ where $H_c(T) = 0$, this quantity reduces to $\mu_0 T_c(dH_c(T)/dT)^2$, which is positive. But without further microscopic knowledge, there is nothing more to be said about this. However, it is interesting to try to go further by introducing some phenomenological considerations.

We first take the very good parabolic approximation for the critical field $H_c(T) = H_0[1 - (T/T_c)^2]$ and insert it in Eq. (1.15). Furthermore, we assume the temperature to be low enough for the standard linear dependence of the normal state specific heat $C_n(T)$ to be valid. Hence $C_n(T) = \gamma T$, where γ is the Sommerfeld constant. We extrapolate this expression below T_c, corresponding to a possible metastable normal state for the metal. All this gives

$$C_s(T) = \gamma T - 2\mu_0 H_0^2 \frac{T}{T_c^2}\left[1 - 3\left(\frac{T}{T_c}\right)^2\right] \tag{1.16}$$

Hence the temperature dependence of $C_s(T)$ is a combination of a linear law and a cubic law. However, for most standard low-temperature superconductors, there is experimentally no hint of a linear dependence, whereas a cubic dependence describes fairly well the general behavior of $C_s(T)$ (although such a law is by no means exact, as we will later see). In order to eliminate the linear contribution, this leads us to conclude that

$$\frac{\gamma T_c^2}{\mu_0 H_0^2} = 2 \tag{1.17}$$

Here we have found quite easily a very simple relation linking a thermal energy γT_c^2 to a magnetic energy $\mu_0 H_0^2$. It will be much more difficult to obtain this dimensionless ratio from BCS theory.

Taking into account this relation, the superconducting specific heat Eq. (1.16) reads

$$C_s(T) = 3\gamma T\left(\frac{T}{T_c}\right)^2 \tag{1.18}$$

which implies $C_s(T_c) = 3\gamma T_c$, to be compared with the normal state specific heat $C_n(T_c) = \gamma T_c$ at the same temperature. This leads for the specific heat jump to

$$\left.\frac{C_s - C_n}{C_n}\right|_{T_c} = 2 \tag{1.19}$$

This will again have to be compared to the BCS theory result. A schematic view of the normal state and superconducting state specific heat is displayed in Fig. (1.5).

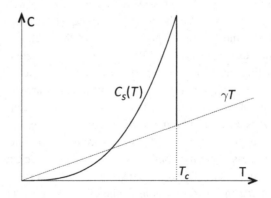

Fig. 1.5 Schematic view of the normal state $C_n(T) = \gamma T$ and superconducting state specific heat $C_s(T)$, together with the specific heat jump at T_c.

1.2 London Theory

The equation proposed by brothers Fritz and Heinz London in 1935 is the first major step in understanding superconductivity. It is in essence a phenomenological equation, without a well-structured demonstration. Hence we first present this equation, discuss it, and consider its consequences before later on coming back to the justification given by London.

The London equation aims at describing the reaction of the superconductor to the application of a magnetic field, as it occurs for example in the Meissner effect. The magnetic field $\mathbf{B}(\mathbf{r})$ is characterized by the vector potential $\mathbf{A}(\mathbf{r})$, related to the field by $\mathbf{B}(\mathbf{r}) = \mathrm{curl}\,\mathbf{A}(\mathbf{r})$. The superconductor reacts by the creation of currents, characterized by the current density $\mathbf{j}_s(\mathbf{r})$. The London equation reads

$$\mathbf{j}_s(\mathbf{r}) = -\frac{n_s e^2}{m}\mathbf{A}(\mathbf{r}) \tag{1.20}$$

relating at equilibrium the current density $\mathbf{j}_s(\mathbf{r})$ to the vector potential $\mathbf{A}(\mathbf{r})$ at the same location $\mathbf{r}$ in the superconductor. This equation may be seen as a case of "linear response": when a physical system is subject to a small perturbation, it is usually slightly modified. The modification (the "response") is quite often proportional to the perturbation; hence the response is linear. In the present case, the perturbation is characterized by $\mathbf{A}(\mathbf{r})$ and the superconductor response by $\mathbf{j}_s(\mathbf{r})$; hence the linear relation between these two quantities given by Eq. (1.20).

In the coefficient entering Eq. (1.20), e is the electronic charge, m is the electronic mass, and n_s is some electronic density. This expression for the coefficient can mostly be seen as resulting from dimensional considerations. This can be seen by looking at the units corresponding to the various physical quantities entering the equation. But it is easier to realize that, since $\mathbf{p} - e\mathbf{A}$ is entering the kinetic energy in the expression of the Hamiltonian of the electron in the presence of the magnetic field, the momentum $\mathbf{p} = m\mathbf{v}$, where $\mathbf{v}$ is the electronic velocity, has the same dimension as $e\mathbf{A}$. Since we have $\mathbf{j}_s = n_s e\mathbf{v}$ for the current density, this shows that the coefficient in Eq. (1.20) has indeed the correct dimension.

On the other hand, the precise value of the coefficient is unknown since we have not specified exactly what is n_s. We expect it to be related to the electronic density n of the conduction electrons, which participate in the conduction process in the normal metal (the electrons making up the ions of the crystal lattice are expected to stay out of the superconductivity phenomenon). And indeed when we will need to evaluate n_s below, we will take $n_s = n$ as a reasonable guess. But we will see later on in the text, where we will obtain specific expressions for n_s, that things are more complicated and that most of the time this simple guess is not the appropriate one, although we expect it to give the correct order of magnitude. Finally this uncertainty on n_s makes it so that we do not need to worry about introducing the (bare) mass m of the electron in Eq. (1.20). Indeed we could wonder if some other electronic mass, related to the band structure of the metal, would not be the correct one. But we may just consider at this stage that all these uncertainties are included in n_s.

Even if we have argued that Eq. (1.20) describes a linear response, this equation is very strange since it relates a physical quantity, namely the current density $\mathbf{j}_s(\mathbf{r})$, to an unphysical one, the vector potential $\mathbf{A}(\mathbf{r})$. From gauge invariance, we could add to $\mathbf{A}(\mathbf{r})$ the gradient $\mathrm{grad}\,\chi(\mathbf{r})$ of any scalar function $\chi(\mathbf{r})$ without modifying the field $\mathbf{B}(\mathbf{r})$. From Eq. (1.20) this would modify the current density without changing the field, which is physically absurd. Hence, as such, Eq. (1.20) cannot be correct. What is missing becomes clear when one realizes that the continuity equation $e\,\partial n(\mathbf{r})/\partial t + \mathrm{div}\,\mathbf{j}_s(\mathbf{r}) = 0$ (which expresses electronic charge conservation) implies $\mathrm{div}\,\mathbf{j}_s(\mathbf{r}) = 0$, since we are at equilibrium so that $\partial n(\mathbf{r})/\partial t = 0$. From Eq. (1.20) this implies $\mathrm{div}\,\mathbf{A}(\mathbf{r}) = 0$. Hence the gauge is not free. In addition to Eq. (1.20), the vector potential must satisfy

$$\mathrm{div}\,\mathbf{A}(\mathbf{r}) = 0 \tag{1.21}$$

The vector potential is said to be in the London gauge when this condition is satisfied. One can see that, with this condition, together with appropriate boundary conditions, the vector potential $\mathbf{A}(\mathbf{r})$ is uniquely determined from the magnetic field $\mathbf{B}(\mathbf{r})$ (if one starts with a vector potential which is not in the London gauge, the above scalar function $\chi(\mathbf{r})$ to obtain a vector potential satisfying Eq. (1.21) is uniquely determined). Hence $\mathbf{A}(\mathbf{r})$ becomes a physical quantity directly related to $\mathbf{B}(\mathbf{r})$, and the above physical problem disappears.

Another way to have an equation relating only physical quantities is to get rid of the vector potential by merely taking the curl of Eq. (1.20). This leads to

$$\mathrm{curl}\,\mathbf{j}_s(\mathbf{r}) = -\frac{n_s e^2}{m}\mathbf{B}(\mathbf{r}) \tag{1.22}$$

This tells us which currents $\mathbf{j}_s(\mathbf{r})$ the field $\mathbf{B}(\mathbf{r})$ is generating in the superconductor. On the other hand, these currents will produce themselves magnetic fields, the precise relation between them being given by the Maxwell–Ampère equation

$$\mathrm{curl}\,\mathbf{B}(\mathbf{r}) = \mu_0\,\mathbf{j}_s(\mathbf{r}) \tag{1.23}$$

where μ_0 is the vacuum permeability (we assume that the superconducting metal is non-magnetic, as is very often the case). Taken together, Eq. (1.22) and Eq. (1.23) give us the space dependence of $\mathbf{j}_s(\mathbf{r})$ and $\mathbf{B}(\mathbf{r})$. If we want to obtain an equation for $\mathbf{B}(\mathbf{r})$, we take the curl of Eq. (1.23) and make use of Eq. (1.22) for $\mathrm{curl}\,\mathbf{j}_s(\mathbf{r})$. This leads us to

$$\mathrm{curl}\,\mathrm{curl}\,\mathbf{B}(\mathbf{r}) = \mathrm{grad}\,\mathrm{div}\mathbf{B}(\mathbf{r}) - \Delta\mathbf{B}(\mathbf{r}) = -\frac{\mu_0 n_s e^2}{m}\mathbf{B}(\mathbf{r}) \tag{1.24}$$

where the first equality is just an identity from vector calculus, valid for any vector $\mathbf{B}(\mathbf{r})$, which can be easily checked by taking the Cartesian components of both sides of the equality. Making use of Maxwell's equation $\mathrm{div}\mathbf{B}(\mathbf{r}) = 0$, we end up with

$$\Delta\mathbf{B}(\mathbf{r}) = \frac{\mu_0 n_s e^2}{m}\mathbf{B}(\mathbf{r}) \tag{1.25}$$

The physical implication of Eq. (1.25) is most easily seen by considering a one-dimensional situation. Specifically we take the case where the superconductor occupies the whole $x > 0$ half-space while there is a vacuum in the $x < 0$ half-space, as seen in Fig. (1.6)a. Clearly in this case all physical quantities depend only on the x variable. In

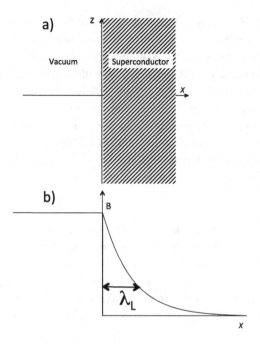

Fig. 1.6 (a) Sketch of a simple one-dimensional situation: superconductor for $x > 0$ and vacuum for $x < 0$. (b) Exponential decrease of the magnetic field inside the superconductor.

order to have the electronic charges staying in the superconductor, the currents have to flow parallel to the vacuum-superconductor interface. If the y-axis is along these currents, the magnetic field is along the z-axis from Eq. (1.22). Then Eq. (1.25) reduces to

$$\frac{d^2}{dx^2}B_z(x) = \frac{\mu_0 n_s e^2}{m}B_z(x) \tag{1.26}$$

The solution of this differential equation is a simple exponential. Comparing the two sides of the equation, it is clear that the coefficient on the right-hand side has the dimension of the inverse of a length squared, since it compares to a second derivative with respect to length on the left-hand side. Hence we introduce the London penetration depth λ_L defined as

$$\lambda_L = \left(\frac{m}{\mu_0 n_s e^2}\right)^{1/2} \tag{1.27}$$

Then the solution of Eq. (1.26) is

$$B_z(x) = B_0 \exp(-\frac{x}{\lambda_L}) \tag{1.28}$$

where B_0 is the value of the magnetic field at the superconductor-vacuum interface $x = 0$. This is depicted in Fig. (1.6)b). We have discarded the exponentially growing solution of

Eq. (1.26) as unphysical since one would get a very large energy for the superconductor, due to the very large magnetic field for large x. Numerical (or analytical) solutions of the three-dimensional equation Eq. (1.25) give the same physical behavior of a magnetic field decreasing essentially exponentially over a typical length λ_L when one goes from the surface into the superconductor. Hence the magnetic field $\mathbf{B}(\mathbf{r})$ is vanishingly small inside the superconductor, which is just the Meissner effect. Therefore the London equation leads immediately to the Meissner effect, obviously a remarkable success of this theory.

Nevertheless the magnetic field penetrates the superconductor over a typical length λ_L, and it is of interest to evaluate it quantitatively. This can be done by plugging into the definition in Eq. (1.27) values of the various physical quantities typical of a standard metal. But it is more interesting to notice that it is directly related to an important physical characteristic of the metal, namely its plasma frequency[3] ω_p, defined by

$$\omega_p^2 = \frac{ne^2}{m\epsilon_0} \tag{1.29}$$

where ϵ_0 is the vacuum permittivity. From standard electromagnetism $\epsilon_0 \mu_0 c^2 = 1$, where c is the speed of light, so from Eqs. (1.27) and (1.29) we have merely

$$\lambda_L = \frac{c}{\omega_p} \tag{1.30}$$

if we take $n_s = n$.

Let us recall that ω_p is the frequency of spontaneous small oscillations of the homogeneous electron gas in the metal, with equilibrium density n, under the restoring force due to the electric field created by the electronic charge oscillation itself. This plasma frequency is linked to the phenomenon of "ultraviolet transparency": the metal is transparent to electromagnetic waves with a frequency beyond the plasma frequency, while electromagnetic waves with a frequency below the plasma frequency are absorbed. Standard metals are opaque to visible light; therefore the plasma frequency is located beyond this optical domain, indeed in the ultraviolet range. From Eq. (1.30) the London penetration depth is just the wavelength of an electromagnetic wave with a frequency equal to the plasma frequency. Hence we have typically $\lambda_L \sim 100\,\text{nm}$ in standard metals.

This typical value for λ_L is quite interesting. Indeed it is quite large compared to atomic sizes. This justifies the above large-scale treatment, where the physical quantities correspond to averages over many elementary cells of the metal. On the other hand, λ_L is quite small compared to a macroscopic scale, so for a standard size sample, the magnetic field is zero over essentially the whole sample, which corresponds to the experimental observation in the Meissner effect. Accordingly, London theory provides a fully coherent account of the Meissner effect.

Although quite small, the value of the penetration depth given by the London theory provided for experiment a definite goal that was not out of reach. Indeed experimentalists have succeeded in showing that the magnetic field indeed penetrates slightly in a superconductor. They were able to measure λ_L. The results were in rough agreement with the estimate from the London theory. However, discrepancies were found,

[3] This plasma frequency is derived in Section 3.7.1.

which are in part at the origin of the generalization of London theory proposed by Pippard, the subject of the next section.

Let us finally come to the justification[4] given by London of Eq. (1.20). London was deeply convinced that quantum mechanics is required to explain superconductivity, that superconductivity is a macroscopic manifestation of quantum mechanics. As we will see, this remarkable physical intuition has been fully vindicated by the following developments in our understanding of superconductivity. This is all the more impressive from London because at the time of the paper publication, a coherent formulation of quantum mechanics itself had been achieved only very recently, in 1927 at the Solvay conference, with the Copenhagen interpretation.

Accordingly, London envisioned a superconductor as being described by a single multi-electronic wave function. Then the current is given by its quantum mechanical expression. In the current $e\mathbf{v}$ carried by a single electron, we have to express the velocity in terms of the electron momentum $\mathbf{p}$ by $\mathbf{v} = \mathbf{p}/m$, which becomes in the presence of a magnetic field $\mathbf{v} = (\mathbf{p} - e\mathbf{A})/m$ from classical electrodynamics. In quantum mechanics, the momentum $\mathbf{p}$ becomes an operator, and to obtain the current, we need to calculate the following average in the electronic wavefunction, described by $|\psi\rangle$:

$$\mathbf{j}_s = \frac{e}{m}\langle\psi|\mathbf{p} - e\mathbf{A}|\psi\rangle = \frac{e}{m}\langle\psi|\mathbf{p}|\psi\rangle - \frac{e^2}{m}\langle\psi|\psi\rangle\mathbf{A} \tag{1.31}$$

where a summation over all the electrons is implicitly understood.

In the absence of a magnetic field $\mathbf{A} = \mathbf{0}$, we expect, at least for symmetry reasons, that the electronic system in its ground state has a zero average momentum $\langle\psi|\mathbf{p}|\psi\rangle = \mathbf{0}$. This leads to $\mathbf{j}_s = \mathbf{0}$, as physically expected. In the presence of a magnetic field $\mathbf{A} \neq \mathbf{0}$, London argued that the electronic ground state could display some "rigidity" so that $|\psi\rangle$ would be unchanged and one would still have $\langle\psi|\mathbf{p}|\psi\rangle = \mathbf{0}$. This sounds strange, but it is actually similar to what happens in an insulating compound: in the absence of electric field $\mathbf{E} = \mathbf{0}$, all the electronic states of the valence band are full, those of the conduction band are empty, and there is no net current. When a small electric field $\mathbf{E} \neq \mathbf{0}$ is applied, the situation is unchanged because there is a finite gap separating the top of the valence band from the bottom of the conduction band. A weak electric field is unable to transfer an electron from the valence to the conduction band, in much the same way as an isolated atom cannot be ionized by a small electric field. London thought that, in a superconductor, a gap could similarly exist in the energy spectrum, between the ground state and the lowest energy excited states, which would account for the rigidity of the wave function. Finally, for a single electron, we would have from normalization $\langle\psi|\psi\rangle = 1$. Hence, taking into

---

[4] In their original work [3], F. and H. London obtained Eq. (1.20) by an argument equivalent to the following. The equation of motion for a free electron $m\,d\mathbf{v}/dt = e\mathbf{E}$ leads for the electronic current $\mathbf{j} = ne\mathbf{v}$ to $m\,d\mathbf{j}/dt = ne^2\mathbf{E}$, in the linear regime where the departure of the electronic density from its equilibrium value n is small enough to be negligible. Assuming that the electric field derives entirely from a time-dependent vector potential, $\mathbf{E} = -d\mathbf{A}/dt$ leads to $m\,d\mathbf{j}/dt = -ne^2d\mathbf{A}/dt$. Integrating with respect to time leads to Eq. (1.20), assuming the integration constant to be zero. The London brothers realized that Eq. (1.20) leads to a natural explanation of the Meissner effect. On the other hand, they knew that the above argument is by no means a derivation of Eq. (1.20), since Eq. (1.20) contains only time-independent physical quantities for which all time derivatives are zero. They understood that Eq. (1.20) should be considered as a constitutive equation for a superconductor, requiring a specific justification which they sketched at the end of their paper.

account all the involved electrons in the wave function, it is natural to take $\langle \psi | \psi \rangle = n_s$ for the properly normalized wave function. All this makes it so that Eq. (1.31) indeed reduces to Eq. (1.20).

This justification is somewhat imprecise, partly because we did not want to go into too much detail. Nevertheless we will see that a number of its ingredients have been validated later on. This makes the Londons' intuition quite impressive.

1.3 Electromagnetic Response

1.3.1 Pippard Theory

We now come to an extension of the London equation, which in addition to providing a better agreement between experiment and theory, introduces important physical ideas. It was proposed by Pippard [4] in 1953 while he was experimentally studying the effect of impurities on the penetration depth. Pippard was also motivated theoretically by his knowledge of the physics of normal metals. Indeed electrons located at $\mathbf{r}$ in a metal are influenced by electrons located nearby at $\mathbf{r}'$; their physical properties are correlated. In this respect, the London equation Eq. (1.20) is somewhat peculiar, since the current at $\mathbf{r}$ depends only on the vector potential $\mathbf{A}$ at $\mathbf{r}$ and not at nearby location $\mathbf{r}'$. In other words, it is a purely local relation.

If $\mathbf{r}'$ is too far away from $\mathbf{r}$, we naturally do not expect that physical properties at $\mathbf{r}$ are influenced by what happens at $\mathbf{r}'$; there is surely some limit to the range of correlation for the physical properties. A fairly obvious origin for the limitation of this range, well known in normal metals, is the existence of imperfections, which is a departure from the perfect arrangement of ions or atoms in the ideal crystal. A frequent source of imperfection is the existence of atoms foreign to the metal, which are called impurities. But a departure from the perfect crystalline arrangement of atoms belonging to the metal has the same kind of effect. In the following, we use the term "impurity" as a generic term for any kind of imperfection.

The existence of impurities provides a clear origin to the limitation of the correlation range. A classical view is to see the electron bouncing back and forth on all the impurities around it, and as a result unable to go far away to carry information. In the corresponding quantum view, the electron in the perfect metal is a propagating wave extending throughout the crystal. Impurities in the metal scatter this wave, which introduces dephasing. The destructive interferences between all the possible paths from these scattering processes make it impossible for an electron to carry information far away with a sizeable probability. Whatever the physical picture, the net result is the existence of a mean free path ℓ for the electron, which is the typical distance on which it can propagate before having a first encounter or scattering with an impurity. Clearly this mean free path sets the order of magnitude for the electronic correlation range.

Pippard proposed to replace the local London equation Eq. (1.20) with the following non-local relation

$$\mathbf{j}_s(\mathbf{r}) = -\frac{3}{4\pi\xi_0}\frac{n_s e^2}{m}\int d\mathbf{r}' \frac{\mathbf{R}[\mathbf{R}\cdot\mathbf{A}(\mathbf{r}')]}{R^4}\, e^{-R/\xi} \tag{1.32}$$

where $\mathbf{R} = \mathbf{r} - \mathbf{r}'$ and ξ_0 is by definition the value of ξ in the pure superconductor. Let us consider the various ingredients in this formula. The essential physical point is that the current $\mathbf{j}_s$ at $\mathbf{r}$ is now determined by the value of the vector potential $\mathbf{A}$ in a region of typical size ξ around $\mathbf{r}$, since R cannot be much larger than ξ due to the factor $e^{-R/\xi}$. The choice of the exponential as a cut-off function is made for simplicity, since this formula is anyway a phenomenology.

Eq. (1.32) is designed to reduce to the London equation in the "local" limit, where ξ is very small. Indeed, in this case, since $\mathbf{r}'$ must be very close to $\mathbf{r}$ to give a sizeable contribution, we have $\mathbf{A}(\mathbf{r}') \simeq \mathbf{A}(\mathbf{r})$. Hence $\mathbf{A}(\mathbf{r}')$ can be taken outside the integral, and we are left with the calculation of $\int d\mathbf{R}(R_i R_j/R^4)e^{-R/\xi}$, where we have changed the integration variable from $\mathbf{r}'$ to $\mathbf{R}$, and we have introduced the Cartesian component R_i of $\mathbf{R}$. For symmetry reasons, this integral is zero when $i \neq j$, and when $i = j$ the result is the same for $i = x, y$, and z. Adding these three results gives $\int d\mathbf{R}(R_x^2 + R_y^2 + R_z^2)e^{-R/\xi}/R^4 = 4\pi \int_0^\infty dR\, e^{-R/\xi} = 4\pi\xi$. So $\int d\mathbf{R}(R_i R_j/R^4)e^{-R/\xi} = 4\pi\delta_{ij}\xi/3$, where δ_{ij} is the Kronecker delta and Eq. (1.32) becomes

$$\mathbf{j}_s(\mathbf{r}) = -\frac{\xi}{\xi_0}\frac{n_s e^2}{m}\mathbf{A}(\mathbf{r}) \tag{1.33}$$

Hence, in the case of a pure superconductor where $\xi = \xi_0$, Eq. (1.32) reduces to the London equation, Eq. (1.20). Note that a simpler kernel in the integral Eq. (1.32), such as $e^{-R/\xi}/R^2$, could lead to the same result. Pippard made his specific choice Eq. (1.32) by analogy with the anomalous skin effect in normal metals.

In the presence of impurities, one expects the range ξ to be smaller since, as we have seen, impurities reduce the correlation range. When the mean free path ℓ is very short, a reasonable assumption is that the correlation range ξ cannot extend much beyond ℓ, so $\xi \simeq \ell$. To evaluate ξ between the two limits of very pure ($\ell \to \infty$) and very dirty ($\ell \ll \xi_0$) superconductor, Pippard proposed the simple interpolation formula

$$\frac{1}{\xi} = \frac{1}{\xi_0} + \frac{1}{\ell} \tag{1.34}$$

which indeed gives the expected result in the two above limits, and that agreed reasonably well with his experimental results. Hence Eq. (1.32) together with Eq. (1.34) accounted for the increase in penetration depth Pippard observed when the impurity content was increased.

A very important physical point in the Pippard theory is the introduction of the length ξ, which characterizes the non-locality of the superconductor response and is called "coherence length" in superconductivity. However, this concept is the same as the "correlation length," which comes in for phase transitions. Pippard understood that ξ_0 has to be related to a characteristic length emerging from the microscopic theory of superconductivity, but it will be easier to be more specific when we will deal with BCS theory.

The Pippard formula has the form of a convolution product, namely the three-dimensional equivalent of a one-dimensional integral expression $h(y) = \int dx\, f(y - x)g(x)$.

It is well known that if one goes to the Fourier transforms of the various functions, $f_F(q) = \int dx\, e^{-iqx} f(x)$ with $f(x) = (1/2\pi) \int dq\, e^{iqx} f_F(q)$, and so on, the Fourier transform of $h(y)$ is merely the simple product of the Fourier transforms of $f(x)$ and $g(x)$: $h_F(q) = f_F(q) g_F(q)$. The same result holds for three-dimensional integrals. We introduce for example the Fourier transform of the vector potential $\mathbf{A}(\mathbf{q}) = \int d\mathbf{r}\, e^{-i\mathbf{q}\cdot\mathbf{r}} \mathbf{A}(\mathbf{r})$. Here, for simplicity, we drop the index F and do not write explicitly $\mathbf{A}_F(\mathbf{q})$ for the Fourier transform, relying on the standard variable name $\mathbf{q}$ to make it clear that we deal with Fourier transforms.

In this way, introducing the Cartesian components, the Fourier transform of Eq. (1.32) can be written as

$$\mathbf{j}_{s,i}(\mathbf{q}) = -K_{ij}(\mathbf{q}) A_j(\mathbf{q}) \tag{1.35}$$

where a summation over j is understood and

$$K_{ij}(\mathbf{q}) = \frac{3}{4\pi\xi_0} \frac{n_s e^2}{m} \int d\mathbf{R}\, \frac{R_i R_j}{R^4} e^{-i\mathbf{q}\cdot\mathbf{R}} e^{-R/\xi} \tag{1.36}$$

Here the direction of $\mathbf{q}$ provides a specific direction, and we take the z-axis along this direction. Just as above, for symmetry reasons, $K_{ij}(\mathbf{q})$ is zero when $i \neq j$. When $i = j$ the result is the same for $i = x, y$, and we call it $K_\perp(q)$, with $q = |\mathbf{q}|$, but it is different from the result for $i = z$, which we call $K_\parallel(q)$. On the other hand, just as for the London equation, we have to satisfy the continuity equation $\operatorname{div} \mathbf{j}_s(\mathbf{r}) = 0$, which implies for the Fourier transform $\mathbf{q} \cdot \mathbf{j}_s(\mathbf{q}) = 0$. From Eq. (1.35) and the above symmetries, this implies $\mathbf{q} \cdot \mathbf{A}(\mathbf{q}) = 0$. This is the London gauge condition, which must be satisfied by the Fourier transform $\mathbf{A}(\mathbf{q})$ of the vector potential for Eq. (1.35) to be valid. But this condition implies that we do not need to know $K_\parallel(q)$; it is irrelevant for our purpose.

The expression for $K_\perp(q)$ is not difficult to obtain from Eq. (1.36). We just give the result without the explicit calculation, in order to discuss the physical implications,

$$\mu_0 K_\perp(q) = \frac{1}{\lambda_L^2} \frac{\xi}{\xi_0} \frac{3}{2(q\xi)^2} \left[\left(1 + (q\xi)^2\right) \frac{\arctan(q\xi)}{q\xi} - 1 \right] \tag{1.37}$$

where the London penetration depth λ_L is given by Eq. (1.27). One checks that for $q \to 0$, $\mu_0 K_\perp(q)$ goes to $(1/\lambda_L^2)(\xi/\xi_0)$, which is the London result (for $\xi = \xi_0$). This is actually obvious from Eq. (1.36). In the limit of large q, $K_\perp(q)$ decreases as $1/q$. This makes sense physically, since one expects the superconductor response to be weaker for a rapidly oscillating magnetic field than for a constant one. This is also clear directly from Eq. (1.36), where for large q the exponential $e^{-i\mathbf{q}\cdot\mathbf{R}}$ is rapidly oscillating as a function of R, which makes the overall integral small by cancellation of positive and negative contributions.

Hence Pippard theory explicitly displays the physically reasonable fact that the superconductor response depends on the wavevector q, in contrast with London theory, which provides a response independent of the wavevector. The characteristic length for this dependence is the coherence length ξ. This length comes in addition to the London penetration depth λ_L, which is a typical length scale for the variation of the vector potential and the supercurrents in the superconductor.

However, if this penetration depth is much larger than the coherence length, this means that the vector potential is essentially constant on the scale of the coherence length. In this case, as we have seen, the Pippard equation, Eq. (1.32), reduces to the London equation, Eq. (1.20). This physical situation turns out, as we will see later, to be the one found in type II superconductors, as introduced in Section 1.1.4 (they are sometimes called London superconductors).

On the other hand, in the opposite physical situation, where the coherence length is much larger than the penetration depth, we naturally need to apply the Pippard equation. This physical situation is the one found in type I superconductors (sometimes called Pippard superconductors). The transition from type I to type II superconductors will be discussed in more detail within the discussion of the Ginzburg–Landau theory in Chapter 6.

We finally note that for type I superconductors, the calculation of the actual penetration depth is more subtle since we have to use a complicated response function like Eq. (1.37) instead of a simple constant, and the simple exponential decay found in Eq. (1.28) is no longer valid. This point is discussed more explicitly in the Further Reading section at the end of this chapter.

1.3.2 Response Function, the Link Between Infinite Conductivity and the Meissner Effect

In the preceding subsection we have seen, as an outcome of the Pippard theory, that the response function $K(q)$ has a natural dependence on the wavevector q (from now on, we drop the subscript $\perp$, which is unimportant once we know that $\mathbf{j}_s(\mathbf{q})$ and $\mathbf{A}(\mathbf{q})$ must be orthogonal to $\mathbf{q}$). We have considered only time-independent situations corresponding physically to the superconductor in its ground state, in the presence of a static magnetic field. However, we may generalize the concept of response function to the case where the potential vector is time-dependent. Going again to the Fourier transform, we consider a vector potential $\mathbf{A}(\mathbf{q}, \omega)$ having a given wave vector $\mathbf{q}$ and frequency ω. Generalizing Eq. (1.35), the supercurrent $\mathbf{j}_s$ produced by the superconductor is

$$\mathbf{j}_s(\mathbf{q}, \omega) = -K(\mathbf{q}, \omega)\mathbf{A}(\mathbf{q}, \omega) \tag{1.38}$$

which serves as a definition for the response function $K(\mathbf{q}, \omega)$. This response function provides a full knowledge of the reaction of a superconductor to an electromagnetic perturbation. In particular, we may use it to specify when the response corresponds to an infinite conductivity or to a Meissner effect.

Indeed the conductivity characterizes the response to an electric field, but we can produce a time-dependent electric field $\mathbf{E}(t) = -\partial \mathbf{A}(t)/\partial t$ by applying a time-dependent vector potential $\mathbf{A}(t)$. For a vector potential $\mathbf{A}(t) = e^{-i\omega t}\mathbf{A}(\omega)$ with given frequency ω, we will have a corresponding electric field $\mathbf{E}(t) = e^{-i\omega t}\mathbf{E}(\omega)$, given by $\mathbf{E}(\omega) = i\,\omega\mathbf{A}(\omega)$. Here again $\mathbf{E}(\omega)$ and $\mathbf{A}(\omega)$ are shorthands for $\mathbf{E}_F(\omega)$ and $\mathbf{A}_F(\omega)$, and so on. On the other hand, for the conductivity, we have in mind the response to a uniform electric field with no space variation. In terms of the Fourier transform, this corresponds to the case where the wave vector is zero $\mathbf{q} = \mathbf{0}$. From the response, Eq. (1.38), $\mathbf{j}_s(\mathbf{0}, \omega) = -K(\mathbf{0}, \omega)\mathbf{A}(\mathbf{0}, \omega)$ and the definition $\mathbf{j}_s(\omega) = \sigma(\omega)\mathbf{E}(\omega)$ of the conductivity at frequency ω, we obtain

$$\sigma(\omega) = i\,\frac{K(\mathbf{0}, \omega)}{\omega} \tag{1.39}$$

Finally when we say that the conductivity is infinite, this is only under static conditions, which means in the limit of zero frequency. Hence we have infinite conductivity provided $K(\mathbf{0}, \omega)/\omega$ goes to infinity when the frequency goes to zero. To obtain this result, it is enough that $K(\mathbf{0}, \omega)$ has a nonzero limit when ω goes to zero. This is what will happen in regular cases. Hence, barring more exotic situations, we have infinite conductivity provided the response function satisfies

$$\lim_{\omega \to 0} K(\mathbf{0}, \omega) \equiv K_0 \neq 0 \tag{1.40}$$

On the other hand, we have seen that the Meissner effect occurs under static conditions, which means at zero frequency $\omega = 0$. When London theory applies, we have a simple exponential decay of the magnetic field upon penetration in the superconductor. However, the decay is not that simple when we have to use Pippard theory, but in order to obtain the physics of the Meissner effect, we merely want a rapid decay (in practice exponential) of the field when, starting from the surface, we go deep enough in the superconductor. When we go in Fourier transform, a large distance from the surface translates into a small wave vector. Hence only in the limit $q \to 0$ do we want to recover the behavior resulting from the London theory (this is discussed in more detail in the "Further Reading" section). This is obtained as soon as the response is nonzero, just as in the London theory. Accordingly, we obtain the Meissner effect when

$$\lim_{q \to 0} K(\mathbf{q}, 0) \equiv K_0' \neq 0 \tag{1.41}$$

Comparing conditions Eq. (1.40) and Eq. (1.41), one sees that both involve $\lim_{q \to 0} \lim_{\omega \to 0} K(\mathbf{q}, \omega)$. The only difference is that the respective orders of the two limits, $\lim_{q \to 0}$ and $\lim_{\omega \to 0}$, are exchanged when we go from one condition to the other. It is very tempting to believe that there is no singularity at $q = 0$ and $\omega = 0$, that the order of these two limits is unimportant, and that $K_0 = K_0'$, which would imply that infinite conductivity and the Meissner effect are automatically linked. However, this is incorrect, as it is easily recognized by considering the case of free electrons: they display infinite conductivity but no Meissner effect, as is well known. We will see this explicitly in Chapter 4. Hence there is in general a singularity at $q = 0$ and $\omega = 0$ for the response function, so infinite conductivity and the Meissner effect are not linked in principle. However, we will see that this singularity is removed in superconductivity, in the framework of BCS theory.

1.4 Physical Ideas, Microscopic Origin of Superconductivity

This last section is an introduction to the various physical ideas leading to the microscopic theory of superconductivity, which will be considered in the following chapters. It is often said that the lapse of time between the discovery of superconductivity in 1911 and the publication of the microscopic theory of Bardeen, Cooper, and Schrieffer (BCS) in 1957 has been quite long. Considering the difficulty of the matter, together with the ten years of world wars that to some extent kept people away from basic research, one could easily argue contrarily that the pace of progress has been remarkable. For example, the matter of high T_c superconductivity in cuprates is still quite controversial, even though the discovery goes back to 1986.

One main difficulty in elaborating a microscopic theory has been the proper treatment of electronic interactions in metals. Indeed superconductivity is clearly the result of inter-electronic interactions that cooperate to produce a state in which dissipation has disappeared. However, it is a difficult matter to properly handle the strong Coulomb interaction between electrons in metals. The elaborations of the proper theoretical frameworks, including the random phase approximation (RPA) and the Landau Fermi liquid theory [5, 6, 7], appeared essentially at the same time as the BCS theory.

A main ingredient in the progress of ideas has been the strong similarity between superconductivity and the superfluidity of liquid ^{4}He, discovered in 1937 by Kapitza [8] and by Allen and Misener [9]. At atmospheric pressure, ^{4}He goes from a gas to a liquid phase at 4.2 K. A remarkable property of this liquid is that for low enough pressure, it is still present down to the lowest temperature, that is $T = 0$, instead of going to a solid phase, which happens for all other known liquids. This is due to the very light mass of the helium atom and the very weak interactions between helium atoms (except for the hard core repulsion, which comes in at a short distance). This liquid ^{4}He behaves as an ordinary "normal" liquid until, when temperature is lowered, its viscosity suddenly vanishes below 2.17 K.

Viscosity is a universal property of liquids corresponding physically to the transfer of momentum from some part of the liquid to the neighboring ones. This transfer occurs as the result of collisions between atoms or molecules making up the liquid. Qualitatively it describes the rubbing of different parts of the liquid against each other. As a result of viscosity, different parts of a liquid tend to flow at the same velocity. Similarly scattering of the particles making up the liquid against the wall of the container makes the liquid have a vanishing velocity near an immobile wall. In particular, in capillaries, a liquid has a hard time moving at all since it stands still against the wall, and viscosity makes its velocity to be barely different from zero in the middle of the capillary. For example, we are used to the fact that, in ordinary life, water does not flow in very small apertures. As a result, a very small hole in a pipe does not give rise to a leak. This is the physical principle behind standard waterproof plumbing.

In contrast, quite the opposite occurs with liquid ^{4}He when, by going below 2.17 K, it loses its viscosity and becomes superfluid. As a result, superfluid ^{4}He can run in extremely small pores and capillaries. This is immediately noticed by an experimentalist who sees many leaks appearing in the vessel containing liquid ^{4}He when it goes below 2.17 K, even though it was a perfectly fine tight vessel above 2.17 K. Similarly superfluid ^{4}He can run in the extremely thin ^{4}He films covering the walls of a vessel and leak out efficiently in this way. There are many other spectacular experimental manifestations[5] of the superfluidity of ^{4}He.

A simple qualitative way to characterize this superfluid behavior is to say that ^{4}He behaves as a perfect fluid. Its flow is fully coherent, with all the atoms moving in accordance with its single velocity field and no apparent role for inter-atomic collisions. This is to be contrasted with the behavior of a standard fluid, where collisions occur all over the

---

[5] Many videos of such experiments can be found by searching the Internet.

fluid, which results in its viscous properties, with no relation between the atomic collisions occurring in various places of the fluid. These processes are not directly related, and they are fully incoherent.

This contrast between the coherent behavior of the superfluid and the incoherence of the standard fluid was rapidly ascribed to a direct manifestation of quantum mechanics [10]. Naturally such an interpretation of superfluidity was by no means obvious, and there were competing proposals, but it is crucial that it turned out to be basically correct. The link between coherence and quantum mechanics is clear if we consider the naive picture of the electron orbiting around the proton in the ground state of the hydrogen atom. The "motion" of the electron is described by its wave function, which is a coherent field, and there is naturally no dissipation in this motion, which is everlasting. If we could pile up a large number of electrons in the state described by this wave function, we would have something very much like a superfluid current of electrons. Naturally such a simple picture does not work, first of all because electrons are fermions and it is forbidden to put more than a single fermion in a specific state.

However, ^{4}He atoms are bosons, and for particles following Bose statistics [42], there is no such exclusion principle, so in principle one can put any number of bosons in a specific state. Actually, not so long before the discovery of superfluidity in ^{4}He, Einstein [11] had studied the peculiar condensation of a gas of non-interacting particles following this bosonic statistics. In particular, in the ground state of this system obtained at zero temperature, all the particles are indeed in the same one-particle state, corresponding to the lowest kinetic energy. However, the link with experimental reality was quite unclear. London went on to suggest [10], with caution due to the interactions between ^{4}He atoms, that the transition of liquid ^{4}He from its normal to its superfluid phase could be related to this Bose–Einstein condensation. Nevertheless ^{4}He atoms have a very strong hard-core repulsion, which makes liquid ^{4}He quite different from an ideal non-interacting gas. Hence the link with the Bose–Einstein condensation was rejected by some physicists. Landau [12], for example, went on to build a theory for the hydrodynamics of superfluid ^{4}He which, while relying directly on quantum mechanics, had nothing to do with Bose–Einstein condensation. We will return to Landau's theory later on in Chapter 7.

Unexpectedly, a side effect of the Second World War brought a clear answer to the question of the link between superfluidity and Bose–Einstein statistics. Indeed, as a by-product of the nuclear activities started at that time, a sizeable quantity of the isotope ^{3}He became available (it comes from tritium disintegration), so that low-temperature liquid ^{3}He could be produced and its physical properties explored. This showed unambiguously [13] that in the temperature range where liquid ^{4}He becomes superfluid, nothing of this sort occurs in liquid ^{3}He. On the other hand, the only difference between the two isotopes is that a neutron is missing in the ^{3}He nucleus with respect to the ^{4}He nucleus. This means that ^{3}He atoms, being made of an odd number of fermions (three nucleons and two electrons), are themselves fermions while ^{4}He atoms, having an additional neutron, are bosons (as already emphasized). Otherwise, just as for any isotopes, ^{3}He and ^{4}He have the same electronic structure and accordingly the same chemical and interaction properties.

Hence it is natural to ascribe the marked difference in physical properties between the two liquids to the only qualitative difference[6] between them, namely the different statistics they obey. This is a very strong argument in favor of a link between superfluidity in ^{4}He and Bose-Einstein statistics, and hence a possible relation with Bose–Einstein condensation. However, this did not improve the microscopic understanding of superconductivity. Indeed, although superconductivity is very much like electronic superfluidity in metals, one cannot merely argue that electrons behave as ^{4}He atoms in superfluid ^{4}He, since electrons are fermions and cannot undergo anything like Bose–Einstein condensation.

Another very important step was made about the same time, pointing toward an essential role played by lattice vibrations in the occurrence of superconductivity. On one hand, Fröhlich [14] pointed out that an indirect attractive interaction between two electrons could result from the exchange of lattice vibrations. This is completely analogous to the phenomena explored at the same period of time in quantum electrodynamics, where a proper physical description of the interaction between two charges has to include electromagnetic fluctuations, namely the possibility of emission or absorption of (virtual) photons. Such processes give rise, for example, to the Lamb shift in atomic physics. The quanta of the electromagnetic field are photons. Similarly the quanta of the field of lattice displacements are phonons, and the indirect interaction pointed out by Fröhlich comes from the exchange of phonons between the two electrons.

On the other hand, almost simultaneously, experiments [15] with the various isotopes of mercury available at the time showed that the critical temperature had a clear dependence on the isotope mass, with T_c increasing with decreasing mass. This definitively proved that lattice vibrations played a role in the appearance of superconductivity. Indeed, if the mercury nuclei are assumed to be fixed, the electronic properties of the various isotopes are identical, as we have already stressed above in the case of helium. Accordingly, the superconducting properties of these isotopes should also be identical. A dependence on the nucleus mass can only appear because the motion[7] of the nucleus comes into play.

The idea of an attractive interaction between electrons drew attention to the possible existence of a bound state of two electrons [16]. Such a pair of electrons, made of two fermions, would be a boson. This would make possible a Bose–Einstein condensation of these charged bosonic pairs. But it is not easy to obtain a quantitative microscopic description from such ideas, and it is hard to reconcile them with the existence of the electronic Fermi sea in the normal metal. Since a quantitative evaluation of the thermodynamics shows that the energy of the superconducting state is not very different from that of the normal state, the superconducting ground state cannot be very different from the normal-state Fermi sea.

The decisive breakthrough occurred in 1956, when Cooper [17] found that, in the presence of the electronic Fermi sea, an attractive interaction between two electrons leads to a bound state, however weak the attraction between the two electrons. This made

---

[6] Actually, ^{3}He is also markedly lighter than ^{4}He, but it is hard to ascribe a qualitative difference such as the absence of superfluidity in ^{3}He to the mere quantitative mass difference.

[7] Actually there is, in principle, the effect of zero point motion of the nucleus on the electronic properties, but for such a heavy element as mercury, this is an exceedingly small effect.

electronic pairing compatible with the presence of other electrons, obeying Fermi-Dirac statistics. The next year, in 1957, the fundamental paper [18] of BCS theory by Bardeen, Cooper, and Schrieffer[8] extended this idea of electronic pairing to all the electrons of the metal. As we will see, the resulting physical picture is quite strange and very far from a gas of charged bosons, since there is actually a huge overlap between all the electronic pairs.

1.5 Further Reading: Magnetic Field Penetration in the General Case

Here we take over the simple one-dimensional situation considered in Section 1.2 and extend the treatment to the case of a general response function $K(q)$, instead of a constant one. However, handling exactly this situation is not very convenient because the result for the field is asymmetrical, as displayed in Fig. (1.6) b). Actually we are only interested in the field distribution in the superconductor – that is, for $x > 0$. Hence it is more convenient to have a (anti)symmetrical situation where on the $x < 0$ side we also have a superconductor, as displayed in Fig. (1.7)a), with a field distribution $\mathbf{B}(-x) = -\mathbf{B}(x)$ antisymmetrical of the one for $x > 0$, as seen in Fig. (1.7)b). This can be produced by an external current sheet at $x = 0$. Hence we consider a space filled by a superconductor in the presence of uniform current sheet $\mathbf{j}_e(\mathbf{r}) = \mathbf{J}_e\delta(x)$, where for example $\mathbf{J}_e$ is along the y-axis and $\delta(x)$ is the Dirac function.

In the absence of a superconductor, the current sheet creates a uniform magnetic field, with opposite values for $x > 0$ and $x < 0$. For the present geometry with $\mathbf{B}(x) = \text{curl}\,\mathbf{A}(x)$ along z, and $\mathbf{A}(x)$ together with $\mathbf{j}(x)$ along y, the Maxwell–Ampère equation

$$\text{curl}\,\mathbf{B}(\mathbf{r}) = \mu_0\,\mathbf{j}(\mathbf{r}) \tag{1.42}$$

gives

$$-\frac{d^2}{dx^2}A_y(x) = \mu_0\,j_y(x) \tag{1.43}$$

where the total current $\mathbf{j}(x)$ contains both the current sheet and the supercurrent $\mathbf{j}(x) = \mathbf{j}_e(x) + \mathbf{j}_s(x)$. Going to the Fourier transform with respect to the variable x, Eq. (1.43) gives for the Fourier transform $A(q)$ of $A_y(x)$

$$q^2A(q) = \mu_0\,(J_e + j_s(q)) \tag{1.44}$$

where the Fourier transform $j_e(q) = J_e$ of $j_{e,y}(x) = J_e\delta(x)$ is independent of q. Now the supercurrent is given by Eq. (1.35), $j_s(q) = -K(q)A(q)$, in terms of the vector potential, and Eq. (1.44) leads to the explicit solution:

$$A(q) = \mu_0\frac{J_e}{q^2 + \mu_0\,K(q)} \tag{1.45}$$

---

[8] Bardeen, Cooper, and Schrieffer received the 1972 Nobel Prize for their work. This reward came fairly late because Bardeen had already received the Nobel Prize in Physics in 1956 for his work on semiconductors, and it was against the principles of the Nobel Committee to award the same prize to the same recipient twice in a given field.

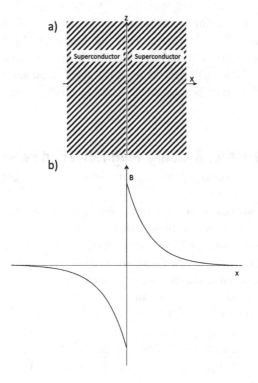

Fig. 1.7 (a) Sketch of a simple symmetrical one-dimensional situation: superconductor for $x > 0$ and for $x < 0$.
(b) Antisymmetrical behavior of the magnetic field in the superconductor.

Performing the inverse Fourier transform and going back to the magnetic field yields

$$B_z(x) = \frac{\mu_0 J_e}{2\pi} \int_{-\infty}^{\infty} dq \, \frac{iq}{q^2 + \mu_0 K(q)} e^{iqx} = -\frac{\mu_0 J_e}{2\pi} \int_{-\infty}^{\infty} dq \, \frac{q \sin(qx)}{q^2 + \mu_0 K(q)} \qquad (1.46)$$

where the imaginary part of $B_z(x)$ is zero because it is the integral of an odd function of q.

In the absence of superconductor $K(q) = 0$, and since $\int_{-\infty}^{\infty} du \sin(u)/u = \pi$, Eq. (1.46) gives $B_z(x) = -(\mu_0 J_e/2) \, \text{sgn}(x)$, which takes indeed constant opposite values for $x > 0$ and $x < 0$.

For a London superconductor where $K(q) = K_0 = n_s e^2/m = 1/(\mu_0 \lambda_L^2)$ is constant, the result is conveniently obtained from the first expression in Eq. (1.46) by residue integration, as shown in Fig. (1.8). For $x > 0$, the integration contour from $-\infty$ to $+\infty$ can be closed by the half-circle at infinity in the upper complex plane of the q variable, which brings no contribution since e^{iqx} is zero, for $x > 0$, when $\text{Im} \, q$ is positive and goes to $+\infty$. The only singularity of the integrand in the upper complex plane is a pole at $q = i\sqrt{\mu_0 K_0} = i/\lambda_L \equiv iq_0$; the corresponding residue is $ie^{-q_0 x}/2$. The integral is equal to $2i\pi$ times this residue, which gives for the field $B_z(x) = -(\mu_0 J_e/2)e^{-x/\lambda_L}$. This is the expected result identical to Eq. (1.28).

In the general case, we may proceed in the same way, adding the half-circle at infinity in the upper complex plane to the integration contour and then trying to deform continuously

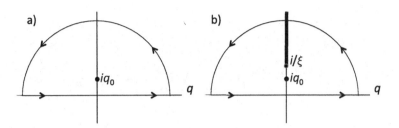

Fig. 1.8 (a) Sketch of the contour for the q integration, with a simple pole in the upper complex plane, corresponding to the case where $K(q)$ is a constant. (b) Sketch of the contour for the q integration, with a simple pole and a cut in the upper complex plane, corresponding to the case where $K(q)$ is given by Eq. (1.37) (the figure assumes $\lambda_L > \xi$).

the resulting closed contour. However, we no longer know the singularities of the integrand in the upper complex plane. In the case where $K(q = 0) \neq 0$, the integrand has a regular expansion in powers of q at $q = 0$. This is in contrast with the case $K(q = 0) = 0$, where there is a singularity at $q = 0$. Hence (barring some physically unexpected singularity coming from $K(q)$ itself) the integrand is regular on the real q-axis. So we can shift, parallel to itself, this part of the integration contour toward the upper complex plane, until it meets the singularity nearest to the real axis. If iq_0 is the imaginary part of this singular point, we can make the change of variable $q = iq_0 + q'$. In this way, a factor $e^{-q_0 x}$ comes out of the integral in Eq. (1.46) and the field decreases at least as fast as $e^{-q_0 x}$. This means that there is a Meissner effect. From this kind of procedure, we see that when x gets very large, the decrease of the magnetic is fully controlled by the singularity that is nearest from the real q-axis, with the contributions from singularities located farther in the complex being exponentially smaller.

Clearly the actual calculation of the field in the superconductor is more complicated when $K(q)$ is not constant, and generally it has to be done numerically. If we take the specific case of Eq. (1.37) (with $\xi = \xi_0$ to be definite), there is, as singularities in the upper complex plane, a branch cut starting from $q = i/\xi$ due to the arctan($q\xi$). Additionally there is a pole located at iq_0 with $q_0^2 = \mu_0 K(iq_0)$. When $\lambda_L \gg \xi$, $K(q)$ is small and the nearest singularity from the real axis is the pole at iq_0. Since $q_0\xi \ll 1$, $K(iq_0)$ can be approximated by its London limit $K(0)$. Hence the pole is merely given by $q_0 = 1/\lambda_L$. We recover, as expected, a simple exponential decay with λ_L as the penetration depth. On the other hand, when $\lambda_L \ll \xi$, the dominant singularity nearest to the real axis is the branch point at i/ξ. The decrease of the field is now more complicated, but it has a dominant $e^{-x/\xi}$ dependence, corresponding to an actual penetration depth ξ. Between these two limits, the actual penetration depth will go from λ_L to ξ. Hence we see that generally the actual penetration depth is a complicated function of λ_L and ξ.

The BCS Theory

2.1 The Cooper Problem

The BCS theory makes an essential use of the physical conclusion coming out of the solution of the Cooper problem. Hence we start our study of BCS theory with this steppingstone, all the more since it is technically a fairly simple matter.

In 1956, in a very short paper [17], Cooper found a somewhat surprising result: in the presence of the electronic Fermi sea, two electrons subject to an attractive interaction form a bound state, however weak the attraction between the two electrons.

There are from the outset two important comments to be made about this statement. First, it is not quite natural physically to think of two electrons attracting each other, since there is a strong Coulomb repulsive contribution coming from their electronic charge. Later on, we will be back at length to understand how the total effective interaction between two electrons may be attractive. For the moment, we may just think of these electrons as two uncharged fermions, although in his paper Cooper was clearly considering electrons in a metal and superconductivity.

Then it is natural that if the attraction is strong enough, it leads to a bound state. However, in the standard situation of two isolated fermions in a three-dimensional space, it is well known that this bound state disappears if the attraction is too weak. The interest in this bound state is that it gives rise in the one-electron energy spectrum to a gap between the ground state and the first excited state, which could be associated with superconductivity, as we have seen in Section 1.2 on London theory. But at the time of Cooper's paper, it was wondered if a possible inter-electronic attraction could be strong enough to lead to a bound state. The Cooper result showed that this reservation was injustified and that, due to the presence of the Fermi sea, any weak attraction could produce a gap and possibly lead to superconductivity.

Turning to the practical matter of handling the Cooper problem, we will take at first the simplest situation for the electrons. Later on we will remove some simplifications. Also we restrict ourselves in this chapter to the zero-temperature case and look for the ground state of the electronic system. First of all, we omit any band structure effect in the metal, and we assume that the non-interacting electronic eigenstates are simple plane waves. The kinetic energy of an electron with wave vector $\mathbf{k}$ is $\epsilon_{\mathbf{k}} = \hbar^2 \mathbf{k}^2 / 2m$, where m is the electronic effective mass (not necessarily equal to the free electron mass).

The Fermi sea is made of all the lower energy single-electron eigenstates occupied by the conduction electrons of the metal. In the present case, it corresponds to all the states with wavevector $k = |\mathbf{k}|$ smaller than the Fermi wavevector k_F. The kinetic energy $E_F = \hbar^2 k_F^2 / 2m$ corresponding to k_F is the Fermi energy. It is the chemical potential μ of the non-interacting electrons, since if we want to add an electron to the electronic system, we must put it at the Fermi level E_F to pay the smallest possible energy because, all the states with lower energy being occupied, we cannot by Pauli exclusion principle put our additional electron in an already occupied state.

In the Cooper problem, the only role of the electronic Fermi sea is to forbid, by Pauli exclusion principle, all the states with $k < k_F$ to the two interacting electrons. Naturally this is physically inconsistent; since all the electrons in a metal are indistinguishable, there is no way to distinguish two specific electrons from all the other ones and all electrons should be treated on an equal footing. This inconsistency will be repaired in the BCS theory, where all the electrons will be handled as indistinguishable.

Once the role of the Fermi sea is settled in this way, the Cooper problem is just a two-body problem, which is a fairly standard problem in quantum mechanics. First of all, we use for the two-body wave function a momentum representation, rather than a direct space representation, since it is quite suited to write the exclusion principle due to the Fermi sea. This is equivalent to perform a Fourier expansion of the wave function $\psi(\mathbf{r}_1, \mathbf{r}_2)$ of the two electrons, with respective position $\mathbf{r}_1$ and $\mathbf{r}_2$ in real space

$$\psi(\mathbf{r}_1, \mathbf{r}_2) = \sum_{\mathbf{k}_1, \mathbf{k}_2} b_{\mathbf{k}_1, \mathbf{k}_2} \, e^{i\mathbf{k}_1 \cdot \mathbf{r}_1} e^{i\mathbf{k}_2 \cdot \mathbf{r}_2} \tag{2.1}$$

In Eq. (2.1) we assume that the metallic sample has the shape of a cube with volume $\mathcal{V}$, and we take the well-known Born-von Karman periodic boundary conditions to obtain the allowed wavevectors $\mathbf{k}$. It is often convenient to keep discrete summations over the wavevectors allowed by these boundary conditions, and go only when this is necessary to the very large volume limit by the standard substitution $\sum_{\mathbf{k}} \rightarrow \mathcal{V}/(2\pi)^3 \int d\mathbf{k}$. Actually, for simplicity, we will take everywhere the sample volume equal to unity $\mathcal{V} = 1$. Pauli exclusion by the Fermi sea implies $k_1 \geq k_F$, $k_2 \geq k_F$ in Eq. (2.1).

In handling the standard two-body problem, it is well known that it is convenient to separate the center of mass problem from the relative motion problem, because these two problems decouple. Actually, the center of mass problem is solved by Galilean invariance, which no longer exists in the Cooper problem because of the presence of the Fermi sea. Nevertheless it is still useful to introduce the wavevectors corresponding to the center of mass motion and to the relative motion respectively, namely $\mathbf{q} = \mathbf{k}_1 + \mathbf{k}_2$ and $\mathbf{k} = (\mathbf{k}_1 - \mathbf{k}_2)/2$. With these wavevectors, Eq. (2.1) reads

$$\psi(\mathbf{r}_1, \mathbf{r}_2) = \sum_{\mathbf{k}, \mathbf{q}} a_{\mathbf{k}, \mathbf{q}} \, e^{i\mathbf{k} \cdot (\mathbf{r}_1 - \mathbf{r}_2)} e^{i\mathbf{q} \cdot \frac{\mathbf{r}_1 + \mathbf{r}_2}{2}} \tag{2.2}$$

where $a_{\mathbf{k}, \mathbf{q}}$ is directly related to $b_{\mathbf{k}_1, \mathbf{k}_2}$. In this way, the relative position $\mathbf{r}_1 - \mathbf{r}_2$ of the two electrons and the position $(\mathbf{r}_1 + \mathbf{r}_2)/2$ of their center of mass appear explicitly.

We will not tackle the Cooper problem at this level of generality. The main reason is that we are interested only in the ground state energy of the two electrons, which is their lowest possible energy. In the absence of the Fermi sea, the center of mass momentum

q is a conserved quantity. A moving center of mass would introduce in the total energy
an additional kinetic energy $\hbar^2 q^2/4m$, linked to the total mass $2m$ of the two electrons.
Hence the lowest energy is obtained when the center of mass of the two electrons is not
moving and the corresponding momentum is zero **q** $= \mathbf{0}$, which is obvious intuitively.
In the presence of the Fermi sea, it sounds physically reasonable to assume that, in an
analogous way, giving some momentum to the center of mass will necessarily increase the
energy, and that again the lowest energy is obtained for **q** $= \mathbf{0}$. When the Cooper problem
is solved for **q** $\neq \mathbf{0}$, one finds indeed that this intuitive idea is correct, although the details
on the energy increase due to the center of mass motion[1] are different from the case in the
absence of the Fermi sea. Accordingly, we will restrict ourselves to the case **q** $= \mathbf{0}$, where
the center of mass of the electronic pair is not moving.

The wave function Eq. (2.2) becomes now

$$\psi(\mathbf{r}_1, \mathbf{r}_2) = \sum_{\mathbf{k}} a_{\mathbf{k}}\, e^{i\mathbf{k}\cdot(\mathbf{r}_1 - \mathbf{r}_2)} \tag{2.3}$$

where $a_{\mathbf{k}}$ is for $a_{\mathbf{k},\mathbf{q}=0}$ and $k \geq k_F$ in the summation, since all the states with $k < k_F$ are
blocked by the Fermi sea (or equivalently $a_{\mathbf{k}} = 0$ for $k < k_F$). This wave function depends
only on the relative position of the two electrons $\mathbf{r} = \mathbf{r}_1 - \mathbf{r}_2$ and is actually the wave
function $\psi(\mathbf{r})$ for their relative motion. From Eq. (2.3), $a_{\mathbf{k}}$ is just the Fourier transform of
this wave function.

Let us now write the Schrödinger equation $H|\psi\rangle = E|\psi\rangle$ for this wave function in
momentum representation. It reads

$$\langle \mathbf{k}|H|\psi\rangle = \sum_{\mathbf{k}'} \langle \mathbf{k}|H|\mathbf{k}'\rangle\langle \mathbf{k}'|\psi\rangle = E\langle \mathbf{k}|\psi\rangle \tag{2.4}$$

where $\langle \mathbf{k}|\psi\rangle$ is the wave function in momentum representation, that is, the Fourier trans-
form of $\psi(\mathbf{r})$, so $\langle \mathbf{k}|\psi\rangle = a_{\mathbf{k}}$ from Eq. (2.3). The Hamiltonian of the two electrons
$H = H_c + H_{int}$ contains the kinetic energy part H_c, which is diagonal in momentum rep-
resentation. For a given wavevector **k**, the eigenvalue is the kinetic energy for this relative
motion $\hbar^2 k^2/2m_r$, where we have to use the reduced mass $1/m_r = 1/m + 1/m = 2/m$ for
this problem. Equivalently this is the sum of the kinetic energy of the two electrons with
momentum **k** and $-\mathbf{k}$, that is, $2\epsilon_k$ where $\epsilon_k = \hbar^2 k^2/2m$. Hence the matrix elements of the
kinetic energy are

$$\langle \mathbf{k}|H_c|\mathbf{k}'\rangle = 2\,\epsilon_k \langle \mathbf{k}|\mathbf{k}'\rangle = 2\,\epsilon_k\,\delta_{\mathbf{k},\mathbf{k}'} \tag{2.5}$$

where $\delta_{\mathbf{k},\mathbf{k}'}$ is the Kronecker delta.

Regarding the interaction energy term H_{int} in H, in the spirit of taking first the simplest
situation, we will take a contact interaction. Later on, we will consider the case of more
realistic interactions. Coherently with our free electron assumption for the non-interacting
electronic eigenstates, the interaction energy $U(\mathbf{r}_1 - \mathbf{r}_2)$ of the two electrons depends only
on the relative position $\mathbf{r} = \mathbf{r}_1 - \mathbf{r}_2$ of the two electrons and is rotationally invariant. We
take for $U(\mathbf{r})$ the contact interaction form

$$U(\mathbf{r}) = -V\delta(\mathbf{r}) \tag{2.6}$$

---

[1] This problem is treated explicitly in Section 8.7.

where $\delta(\mathbf{r})$ is the Dirac function and V is a positive constant since we have in mind an attractive interaction. Hence the matrix elements of the interaction energy are

$$\langle \mathbf{k}|H_{int}|\mathbf{k}'\rangle = \int d\mathbf{r}\, e^{-i\mathbf{k}\cdot\mathbf{r}} U(\mathbf{r}) e^{i\mathbf{k}'\cdot\mathbf{r}} = -V \tag{2.7}$$

Carrying Eq. (2.5) and Eq. (2.7) into Eq. (2.4) yields

$$2\,\epsilon_k a_{\mathbf{k}} - V\sum_{\mathbf{k}'} a_{\mathbf{k}'} = E a_{\mathbf{k}} \tag{2.8}$$

This has the form of an integral equation rather than a differential equation because we have chosen to work in momentum representation. However, as a result of taking a contact interaction, this integral equation has an explicit solution for $a_{\mathbf{k}}$ from Eq. (2.8) since $\sum_{\mathbf{k}'} a_{\mathbf{k}'}$ is a constant independent of $\mathbf{k}$. We will now proceed to write this solution more explicitly.

Since the only explicit $\mathbf{k}$ dependence in Eq. (2.8) comes through $\epsilon_{\mathbf{k}}$, which depends only on $k = |\mathbf{k}|$ and not on the direction of $\mathbf{k}$, $a_{\mathbf{k}}$ depends also on $k = |\mathbf{k}|$ and not on its direction, so we write $a_{\mathbf{k}} = a(k)$ to make this explicit. Anyway, it is physically expected that the ground state wave function of the two electrons we are looking for has this isotropy property. In the summation $\sum_{\mathbf{k}'} \to 1/(2\pi)^3 \int d\mathbf{k}'$, we can integrate over the direction of $\mathbf{k}'$, which gives a factor 4π, and we are left with the integration over k'. We may, instead of k', take $\epsilon_{k'}$ as a variable. This leads to $\sum_{\mathbf{k}'} \to 4\pi/(2\pi)^3 \int k'^2\, dk' = \int n(\epsilon_{k'})d\epsilon_{k'}$, where the (single spin) density of states is given by $n(\epsilon_k) = (1/4\pi^2)(2m/\hbar^2)^{3/2}\, \epsilon_k^{1/2} = mk/(2\pi^2\hbar^2)$.

Finally we rewrite Eq. (2.8) with more appropriate variables. For zero interaction, the lowest energy of the two electrons is $2E_F$ since the Fermi sea is blocking all the states with lower kinetic energy. Hence, with an attractive interaction which is physically expected to be fairly weak, we should find an energy not so far from $2E_F$. So we set $E = 2E_F + \epsilon$. Similarly we set for the kinetic energy $\epsilon_k = E_F + \xi_k$, so ξ_k is the kinetic energy measured from the Fermi level E_F. We introduce the same change of variable in the density of states by setting $n(\epsilon_k) = N(\xi_k)$. In this way, Eq. (2.8) becomes

$$(2\,\xi_k - \epsilon)\, a(k) = V \int d\xi_{k'} N(\xi_{k'})\, a(k') \tag{2.9}$$

We have not yet specified the boundaries for the $\xi_{k'}$ integration. The lower bound is clear. Since all the states $k < k_F$, that is $\xi_k < 0$, are blocked by the Fermi sea the lower bound is $\xi_{k'} = 0$. On the other hand, there is no upper limitation on the wavevector, and the natural upper bound is $\xi_{k'} = \infty$. However, if we try to do this, we encounter a problem. Since the right-hand side of Eq. (2.9) is a constant, this implies that for large k (or ξ_k), we have $a(k) \sim 1/k^2 \sim 1/\xi_k$. When we plug this behavior on the right-hand side, taking into account that $N(\xi_k') \sim k' \sim \sqrt{\xi_k'}$, we find that the integrand behaves as $1/\sqrt{\xi_k'}$ in this limit so that the integral diverges at the boundary $\xi_{k'} = \infty$.

This problem is actually a known pitfall of the contact interaction we have taken. To avoid it we should take, instead of a zero range interaction, an interaction with a very short range r_c. Instead of getting a constant in the Fourier transform Eq. (2.7), we would get a result tending to zero beyond a very large wavevector $k > k_c = 1/r_c$, as it can be checked on simple examples. This would effectively remove the divergence we have found in the large $\xi_{k'}$ limit by having an integrand going to zero, for $\xi_{k'} \to \infty$, much more rapidly than $1/\sqrt{\xi'_k}$. In practice, this means that the divergent behavior of the integrand stops for $k' \sim k_c$. The simplest way to take this into account is to put ξ_{k_c} as an upper boundary in the $\xi_{k'}$ integration, instead of ∞, that is to put a cut-off on the divergent integral. This will provide the dominant divergent behavior in k_c of this integral.

Actually, in the original Cooper problem, there is a physical reason that provides a natural cut-off. The attractive interaction between electrons is an indirect one through lattice vibrations. If the electrons are too energetic, they are moving too rapidly and the coupling to the lattice is very inefficient, because the metallic ions do not have enough time to react to the presence of the electron. In practice, there is no effective interaction in this case. In order to have an efficient coupling, the electrons have to be somehow tuned to the frequency of lattice vibrations. This occurs when the electron energy (divided by $\hbar$) is of the order, or less than a characteristic frequency of lattice vibrations, which we call ω_D. This frequency is often called the Debye frequency, but although the order of magnitude is similar, there is no reason to take it equal to the standard Debye frequency of solid-state physics. Naturally at this stage our discussion is quite rough, but at the level of the Cooper problem there is no point in using a more accurate description. We will explore this problem in more detail when we will address the BCS theory itself. An accurate description and treatment of the indirect interaction will be given in Chapter 9, but unfortunately it will be markedly more complex than our present handling. At the level of our Cooper problem we are thus led, from the above discussion, to use an energy cut-off $\hbar\omega_D$ because there is no effective attractive interaction for electrons with a higher energy.

In this way, Eq. (2.9) becomes explicitly

$$(2\xi_k - \epsilon)\,a(k) = V \int_0^{\hbar\omega_D} d\xi_{k'} N(\xi_{k'})\, a(k') \tag{2.10}$$

This can be simplified by noticing that, in standard situations, $\hbar\omega_D$ is a very small energy scale compared to the Fermi energy. Indeed a typical lattice vibration frequency is $\sim 10^{14}$ Hz, whereas the frequency $E_F/\hbar$ corresponding to a typical Fermi energy of a few eV is $\sim 10^{16}$ Hz. The ratio 10^{-2} between these two frequencies appears in simple solid-state models as $\sqrt{m_0/M}$, the square root of the ratio between the (bare) electron mass m_0 and the ionic mass M. This ratio is indeed typically of order 10^{-4}. As a result $N(\xi_{k'})$ has a very weak variation when, in Eq. (2.10), $\xi_{k'}$ goes from 0 to $\hbar\omega_D$, since $N(0) = n(E_F)$ and $N(\hbar\omega_D) = n(E_F + \hbar\omega_D)$ with $n(\epsilon_k) = mk/(2\pi^2\hbar^2)$. So in the integral we may replace $N(\xi_{k'})$ with the constant $N(0)$ and Eq. (2.10) becomes

$$(2\xi_k - \epsilon)\,a(k) = N_0 V \int_0^{\hbar\omega_D} d\xi_{k'}\, a(k') \tag{2.11}$$

with the simpler notation $N_0 = N(0)$.

It is now quite simple to solve for $a(k)$ and the binding energy of the two electrons. We make explicit that we are looking for a bound state of the two electrons, so $\epsilon < 0$ and $\epsilon = -|\epsilon|$. Setting $C = \int_0^{\hbar\omega_D} d\xi_{k'}\, a(k')$, we obtain

$$a(k) = \frac{N_0 V C}{2\,\xi_k + |\epsilon|} \tag{2.12}$$

Carrying this result into the definition of C, we find

$$C = \int_0^{\hbar\omega_D} d\xi_{k'}\, a(k') = N_0 V C \int_0^{\hbar\omega_D} \frac{d\xi_{k'}}{2\,\xi_{k'} + |\epsilon|} \tag{2.13}$$

so that

$$\frac{1}{N_0 V} = \int_0^{\hbar\omega_D} \frac{d\xi_{k'}}{2\,\xi_{k'} + |\epsilon|} = \frac{1}{2} \ln \frac{2\hbar\omega_D + |\epsilon|}{|\epsilon|} \tag{2.14}$$

This is easily solved as $|\epsilon| = 2\hbar\omega_D/[\exp(2/N_0 V) - 1]$.

However, this has to be even further simplified if we want to keep a physically consistent picture, within our attractive electronic interaction via lattice vibrations. We have replaced this indirect interaction with an instantaneous direct interaction $U(\mathbf{r})$, whereas the interaction via lattice vibrations is a more complex retarded interaction. As we have remarked, if the electrons are too energetic (compared to $\hbar\omega_D$) and too rapid, the interaction is ineffective. But the binding energy $|\epsilon|$ is also a measure of how fast the electrons move, how energetic they are in their relative motion in the bound state. If $|\epsilon|$ is of order of $\hbar\omega_D$, this means that the electrons are already energetic enough for our description of the interaction to be inaccurate, the retarded nature of the interaction should be taken into account for a proper description of the electronic motion. To be on the safe side and have a consistent framework, we have to be in a regime where the electronic motion is slow compared to the lattice vibrations – that is, we have to assume that the binding energy is small compared to $\hbar\omega_D$. This is the so-called "weak coupling" regime $|\epsilon| \ll \hbar\omega_D$, which occurs when $N_0 V \ll 1$. In this case, Eq. (2.14) gives merely

$$|\epsilon| = 2\hbar\omega_D\, e^{-\frac{2}{N_0 V}} \tag{2.15}$$

There are several remarks to be made on this result. The first one is the remarkable dependence of the binding energy on the strength of the interaction. This is typically a $e^{-1/V}$ dependence, which is quite noticeable because it is not possible to write for this quantity an expansion in powers of V (the dependence on V is non-analytic at $V = 0$). This implies that it is not possible to obtain this result by a perturbative calculation since such an approach produces precisely a result which is an expansion in powers of the interaction strength V. This is a relevant point since the weakness of the coupling between electrons and lattice vibrations leads naturally to the idea that it is worth trying a perturbative approach for a theory of superconductivity, and this has indeed been done. By contrast Eq. (2.15) results from a non-perturbative approach.

Next, it is remarkable that, as we indicated at the beginning of this section, there is always a bound state, whatever the weakness of the attractive interaction between the two electrons. Naturally, when V is small, the binding energy Eq. (2.15) gets extremely small, yet it is not vanishing. To see this result, it is not necessary to find the explicit expression

Eq. (2.15) for the binding energy. This is immediately seen at the level of Eq. (2.14), since if one sets $\epsilon = 0$, the integral to be performed diverges logarithmically for $\xi_{\mathbf{k}'} = 0$. This implies that since the left-hand side $1/N_0 V$ is large but not infinite, however small V may be, there is always some small value of ϵ that will make the right-hand side large but not infinite, so the equality is satisfied.

This in turn points toward the physical origin of this result. It is due to the nonzero electronic density of states N_0 at the Fermi surface, due to the existence of the Fermi sea. If there was no Fermi sea, we would have a zero Fermi energy $E_F = 0$, so $\xi_{\mathbf{k}} = \epsilon_{\mathbf{k}}$, and instead of N_0 we should use the density of states around zero energy, which behaves as $N(\xi_k) = N(\epsilon_k) \sim \epsilon_k^{1/2}$. In this case, the integral in Eq. (2.14) would behave for $\epsilon = 0$ as $\int_0 d\xi\, \xi^{1/2}/\xi = \int_0 d\xi\, \xi^{-1/2}$, which is convergent for $\xi = 0$, so the integral cannot become as large as we wish by an appropriate choice of ϵ. Hence if the attractive interaction V is weak enough, there is no way to satisfy the equation $1/V = \int_0^{\hbar\omega_D} d\xi\, N(\xi)/(2\xi + |\epsilon|)$ corresponding to this case and there is no bound state. This is a well-known situation with quantum mechanics in three dimensions. In conclusion, the presence of the Fermi sea produces an effective switch from a $D = 3$ to a $D = 2$ physical situation.

It is worthwhile to recall the simple qualitative argument that leads to this result on the existence of a bound state, depending on space dimension D. If we have an attractive interaction potential (think of a square well potential) of typical strength $-U_0$ (with $U_0 > 0$) and range R, and try a (quasi)-constant wave function ψ as a solution, with typical extension L, we have $\psi^2 L^D \sim 1$ from the normalization of this wave function, where D is the space dimension. On one hand, the typical momentum with this wave function is (this Heisenberg principle) $p \sim \hbar/L$. Hence the typical kinetic energy is $p^2/2m \sim \hbar^2/(mL^2)$. On the other hand, the interaction energy is of order $-U_0\psi^2 R^D$, that is $-U_0$ if $L \sim R$, and $-U_0(R/L)^D$ for $L > R$. (It is not energetically favorable to take $L < R$ since there is no gain in potential energy and the kinetic energy gets higher.) This gives a total energy that is typically $E = \hbar^2/(mL^2) - U_0(R/L)^D$.

For $L \simeq R$ we have a negative energy, and accordingly a bound state, if $U_0 > \hbar^2/(mR^2)$, that is when the attractive potential is strong enough. If this condition is not satisfied, for $D = 3$, there is nothing to gain by increasing L since the interaction energy decreases as $1/L^3$, that is more rapidly than the kinetic energy, which goes as $1/L^2$. However, for $D = 1$ the opposite is true. By going to large enough L, the kinetic energy becomes negligible compared to the interaction energy, which goes as $1/L$. Hence for $D = 1$, there is always a bound state obtained by taking a wave function with a large enough extension L. On the other hand, $D = 2$ appears as a borderline case. Both kinetic and interaction energy vary as $1/L^2$, and the above argument is not precise enough to settle the matter. Since the density of states varies as $N(\epsilon) \sim \epsilon^{D/2-1}$ in D dimensions, this argument is consistent with the integral $\int_0 d\xi\, \xi^{(D/2)-1}/\xi$, which has to be calculated in our above approach to the Cooper problem. It is indeed divergent for $D = 1$, so there is always a bound state for an attractive potential for $D = 1$.

Finally it is worthwhile to stress that we have made a proper treatment of the Schrödinger equation Eq. (2.8) for the relative motion. This is not immediately obvious since we have only found a single bound state. This is due to the fact that we looked immediately for this bound state which has $\epsilon < 0$. However, if we do not go from the discrete summation to the

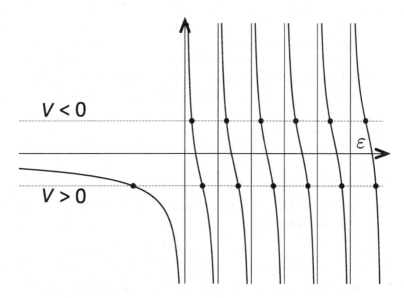

Fig. 2.1 Full line: schematical representation of the left-hand side of Eq. (2.16) as a function of ϵ. The horizontal dashed lines correspond to $-1/V$ with either $V > 0$ (attractive interaction) or $V < 0$ (repulsive interaction). The solutions of Eq. (2.16) are indicated by the black dots.

integration over $\mathbf{k}$, but proceed from Eq. (2.8) directly as in Eq. (2.12) and Eq. (2.13), we obtain for the energy, by following the same procedure,

$$\sum_{\mathbf{k}} \frac{1}{\epsilon - 2\xi_{\mathbf{k}}} = -\frac{1}{V} \tag{2.16}$$

which is analogous to Eq. (2.14). The left-hand side is displayed schematically in Fig. (2.1) as a function of ϵ. It diverges whenever $\epsilon = 2\xi_{\mathbf{k}}$, where $\mathbf{k}$ is any wavevector allowed by the Born-von Karman boundary conditions. The corresponding asymptotes are indicated by the vertical lines. An eigenstate is found when this left-hand side is equal to $-1/V$. The solutions are indicated by black dots in the figure. For an attractive interaction, which corresponds to $V > 0$ with our notations, we find indeed a solution $\epsilon < 0$, which is our bound state, but also many solutions with $\epsilon > 0$, which are the free electrons solutions without interaction $2\xi_{\mathbf{k}}$, slightly modified by the interaction potential (these are the scattering states). In the case of a repulsive interaction $V < 0$, these scattering states are naturally still present, but there is no bound state.

Let us finally come back on our simplifying assumption regarding the interaction, which we have taken as a contact interaction. We consider now a general interaction between the two electrons but still assume coherently with the electronic dispersion relation that it is isotropic; hence it is some function $U(|\mathbf{r}|)$ of the relative position $\mathbf{r}$ between the two electrons. This rotational invariance allows us to break the Cooper problem into subproblems corresponding to specific angular momenta. Indeed, in this case, the matrix element Eq. (2.7) of the interaction energy

$$\langle \mathbf{k}|H_{int}|\mathbf{k}'\rangle = \int d\mathbf{r}\, e^{-i\mathbf{k}\cdot\mathbf{r}} U(\mathbf{r}) e^{i\mathbf{k}'\cdot\mathbf{r}} = \int d\mathbf{r}\, U(|\mathbf{r}|) e^{-i(\mathbf{k}-\mathbf{k}')\cdot\mathbf{r}} \tag{2.17}$$

depends only on $|\mathbf{k} - \mathbf{k}'|$ since there is no preferred direction (the angular integration in Eq. (2.17) is easily performed). Hence the result depends only on $k \equiv |\mathbf{k}|$, $k' \equiv |\mathbf{k}'|$ and on the angle θ between $\mathbf{k}$ and $\mathbf{k}'$. With respect to this angular dependence, one can perform an expansion in Legendre polynomials $P_\ell(\cos\theta)$, leading through the relation $P_\ell(\cos\theta) = (4\pi/2\ell + 1) \sum_m Y_{\ell,m}(\hat{\mathbf{k}})Y^*_{\ell,m}(\hat{\mathbf{k}}')$ to an expansion of the result in spherical harmonics $Y_{\ell,m}(\hat{\mathbf{k}})$

$$\langle \mathbf{k}|H_{int}|\mathbf{k}'\rangle = -4\pi \sum_{\ell,m} V_\ell(k,k')Y_{\ell,m}(\hat{\mathbf{k}})Y^*_{\ell,m}(\hat{\mathbf{k}}') \tag{2.18}$$

where $\hat{\mathbf{k}}$ is the unit vector in the direction of $\mathbf{k}$. We have introduced a minus sign in the definition of $V_\ell(k,k')$ in analogy with Eq. (2.6). Similarly we expand the wave function $a_\mathbf{k}$ in sperical harmonics

$$a_\mathbf{k} = \sum_{\ell,m} \psi_{\ell,m}(k)Y_{\ell,m}(\hat{\mathbf{k}}) \tag{2.19}$$

When we carry these expansions into the Schrödinger equation Eq. (2.4), we obtain

$$(2\epsilon_k - E) \sum_{\ell,m} \psi_{\ell,m}(k)Y_{\ell,m}(\hat{\mathbf{k}}) \tag{2.20}$$

$$= 4\pi \sum_{\mathbf{k}'} \sum_{\ell,m,\ell',m'} V_\ell(k,k')\psi_{\ell',m'}(k')Y_{\ell,m}(\hat{\mathbf{k}})Y^*_{\ell,m}(\hat{\mathbf{k}}')Y_{\ell',m'}(\hat{\mathbf{k}}')$$

$$= \frac{1}{2\pi^2} \sum_{\ell,m} \int_0^\infty dk'\ k'^2 V_\ell(k,k')\psi_{\ell,m}(k')Y_{\ell,m}(\hat{\mathbf{k}})$$

where the last step results from $\sum_{\mathbf{k}'} \to 1/(2\pi)^3 \int d\mathbf{k}' = 1/(2\pi)^3 \int_0^\infty dk'\ k'^2 \int d\Omega_{\mathbf{k}'}$ and the orthogonality of spherical harmonics $\int d\Omega_{\mathbf{k}} Y^*_{\ell,m}(\hat{\mathbf{k}})Y_{\ell',m'}(\hat{\mathbf{k}}) = \delta_{\ell,\ell'}\delta_{m,m'}$. Comparing the coefficients of $Y_{\ell,m}(\hat{\mathbf{k}})$ in this equation, we obtain independent equations satisfied by the $\psi_{\ell,m}(k)$. Actually, they do not depend on m as expected from rotational invariance, so we do not write anymore the index m in $\psi_{\ell,m}(k)$. Moreover, we make the same change of variables $E = 2E_F + \epsilon$ and $\epsilon_k = E_F + \xi_k$ as we have done for the contact interaction, and introduce similarly the density of states $N(\xi_k)$. In this way, we find

$$(2\xi_k - \epsilon)\psi_\ell(k) = \int d\xi_{k'} N(\xi_{k'})\, V_\ell(k,k')\psi_\ell(k') \tag{2.21}$$

Now we can follow closely our discussion for the contact interaction. First, the presence of the Fermi sea forbids to use in the wave function all the states with $k < k_F$, which means that we have to take $\psi_\ell(k) = 0$ for $k < k_F$. This implies that the lower boundary for the integral in Eq. (2.21) is $\xi_{k'} = 0$. Then we argue that the attractive interaction will not operate for electrons that are too energetic. Hence the electronic wavevectors, k and k' should not be much larger than k_F. On the other hand, the effective interaction is of microscopic origin, its typical length scale is an atomic size, and its corresponding typical wavevectors are of order of the Fermi momentum k_F in a standard metal. There is no reason for $V_\ell(k,k')$ to vary much when k and k' stay in the vicinity of k_F. Hence taking $V_\ell(k,k') = V_\ell(k_F,k_F) \equiv V_\ell$ is quite a good approximation. Similarly we may replace

$N(\xi_{k'})$ by $N(0) \equiv N_0$ as we have done for the contact interaction. Finally we can take into account the fact that $\xi_{k'}$ should stay small to have an efficient attractive interaction by putting[2] to the integral in Eq. (2.21) an upper cut-off, which we call again $\hbar\omega_D$. We end up with an equation for $\psi_\ell(k')$

$$(2\,\xi_k - \epsilon)\,\psi_\ell(k) = N_0 V_\ell \int_0^{\hbar\omega_D} d\xi_{k'}\,\psi_\ell(k') \qquad (2.22)$$

which is identical to Eq. (2.11) except that V has been replaced by V_ℓ.

Hence for each angular momentum ℓ for which the corresponding V_ℓ is positive (attractive interaction), we find a bound state. The actual ground state is found for the angular momentum having the largest V_ℓ. It is quite frequent that it is $\ell = 0$ that has the largest V_ℓ because there is no centrifugal energy to pay in this case. However, the actual answer depends naturally on the specific properties of the effective interaction $U(\mathbf{r})$. Finally, let us remark that a similar analysis could be performed if $U(\mathbf{r})$ does not have a rotational invariance, due to restrictions by the crystal. The problem would split into separate problems corresponding to the various wave functions with symmetries compatible with the crystal symmetry.

The net conclusion from the Cooper problem is that in the presence of an attractive interaction, the standard Fermi sea is unstable since the energy is lowered by the formation of a Cooper pair. Therefore the Fermi sea cannot correspond to the ground state. Naturally if the energy is lowered by forming a Cooper pair, the formation of another Cooper pair will lower it further, and so on. The first purpose of the BCS theory is to find the resulting ground state by restoring an equivalent role to all the electrons. However, before adressing this problem, we first need to discuss in detail what is actually the effective interaction between electrons in a real metal. We do this in the next section.

2.2 Effective Interaction

If we want to have a description of a superconductor that is not a simple naive modeling but something closer to the physical reality, then we have to go into the details of the electronic interactions in a metal. This is a very complicated matter since one has to deal with the many-body problem of a system of particles with strong interactions. We will naturally not go here into the details of this matter, but we want only to sketch the essential ideas that make it possible to have a reasonable physical description of what is going on.

The first problem one meets in dealing with electronic interactions in a metal is that the by far dominant interaction is the Coulomb interaction, both between electrons, and between electron and ions. This interaction is both strong and long range, and the physics that may result from such a complicated situation is quite unclear at first. Following the early free electron model of Sommerfeld and its success in describing some properties

---

[2] A more sophisticated and mathematically cleaner possibility is to introduce, as a model, a separable potential $V_\ell(k, k') = V_\ell w(k)w(k')$, where $w(k)$ goes smoothly to zero when $\xi(k)$ gets larger than $\hbar\omega_D$. This is more agreeable, as it allows one to neatly solve the Schrödinger equation, Eq. (2.21). However, in the end, it is not more informative than the sharp cut-off.

of metals, the standard approach is to ignore entirely electron–electron interactions and retain only the periodic potential felt by an electron due to the regular arrangement of ions. This leads to the standard band theory where the eigenstates of non-interacting electrons are described by Bloch's wave functions. But this leaves open the effect ot the Coulomb interaction between electrons.

2.2.1 Screening

A first step in handling the Coulomb interaction is to realize that it gives rise to the screening[3] effect. For example, a fixed positive charge, in the presence of mobile negatively charged electrons, will be surrounded by electrons due to the attractive Coulomb interaction between these opposite charges. The negative charge of the whole cloud of surrounding electrons compensates exactly the positive charge so that, for an additional electron located far away from the positive charge, the total charge seen by this additional electron is zero, and this electron is not attracted by the resulting neutral object. In this way, the positive charge is fully screened by the mobile electrons. Actually, the formation of an ion from the nucleus can be seen as a partial screening of the nuclear charge by electrons, which are very tightly bound to the nucleus. The resulting charged ion may itself be partially screened in a metal by mobile surrounding electrons.

This screening phenomenon occurs also between electrons. Here, because of Coulomb repulsion, an electron tends to be away from another electron. But pushing away negative charges is equivalent to attracting positive charges. (Physically in a metal the positive charges will be due to the positively charged ions.) Hence a given electron is screened by all the other electrons, which tend to avoid it. This implies naturally that the collective motion of the set of all electrons is highly correlated and complex. But, as for the above example of the fixed positive charge, the net result is that the Coulomb field created by an electron is screened by the other electrons. It has no longer the long-range behavior of the bare Coulomb interaction, but it extends only over a typical distance, known as the screening length, which is typically the size of the electron cloud we had in the case of the fixed charge. In standard metals, this screening length happens to be of the order of an interatomic distance. Quantitatively this complex situation for the electron gas in a metal can nevertheless be described by appropriate theoretical schemes, with various possibilities of refinements, all being nevertheless approximate. The simplest form is the random phase approximation we have already mentioned in the preceding chapter. To summarize, screening results with the bare long-range Coulomb potential replaced by a short range interaction; the strength of this interaction is essentially the same as the one of the bare Coulomb potential.

2.2.2 Fermi Liquid Theory

Although screening simplifies the physical picture by replacing a long-range interaction with a short-ranged one, we are still left to handle a system of strongly interacting fermions,

---

[3] Screening is considered more quantitatively in Section 3.7.1.

and the resulting physical behavior is quite unclear. The essential step leading to a physical understanding of this system is the so-called Fermi liquid theory, first developped by Landau to describe the low-temperature properties of liquid ^{3}He, a fermionic system with very strong interactions due to the hard core repulsion between helium atoms. This theory establishes a one-to-one correspondence between the eigenstates of the non-interacting fermionic system and the (quasi) eigenstates of the interacting system.

This is done by starting with the $T = 0$ Fermi sea of the non-interacting fermions and adding a fermion in an eigenstate with wavevector **k** above the Fermi surface. Then one considers what happens when the interaction between fermions is adiabatically switched on. Due to its interactions with the Fermi sea, the **k** fermion will be able to create particle-hole pairs, so this **k** fermion will become "dressed" with a cloud of such virtual particle-hole pairs. This is analogous to what happens in quantum electrodynamics, where a free electron gets dressed with a virtual electron-positron cloud, due to the coupling with the electromagnetic field. This is known as vacuum polarization. Similarly, one can say that the **k** fermion polarizes the Fermi sea. Hence, due to the interactions, the simple **k** fermion becomes a quite complex object.

Nevertheless, since interactions conserve momentum, switching on interactions does not change momentum, so this complex object has still momentum **k**. Correspondingly, it has also an energy $\epsilon(\mathbf{k})$, which is the difference between the energy of the object together with the Fermi sea in presence of interactions, and the energy of the Fermi sea alone, but also modified by the interactions. A very important point is that this complex object is essentially stable when **k** is close to the Fermi surface because it can be seen that the phase space for its possible disintegration processes gets very restricted, due to the Pauli principle, which forbids all the states of the Fermi sea to be used for final states of these processes. This phase space decreases as **k** gets closer to the Fermi surface, so the complex object has a very long lifetime. When **k** is taken at the Fermi surface, this lifetime becomes infinite, so the complex object is an eigenstate, which justifies the introduction of its energy $\epsilon(\mathbf{k})$. The same remains approximately true when **k** is in the vicinity of the Fermi surface.

As a result of this analysis, we come to the conclusion that in the presence of interactions, the free **k** fermion becomes a complex object that has nevertheless a well-defined momentum **k** and well-defined dispersion relation $\epsilon(\mathbf{k})$. Hence the situation is still quite similar to the one we have when we deal with a standard free particle. Accordingly, the usual name for this complex object is "quasi-particle." In the case of electrons, we should consider quasi-electrons. Naturally the above argument can be extended to the case where, instead of starting from a free fermion with wave vector **k** above the Fermi surface, we remove a fermion with wave vector below the Fermi surface and therefore we start with a hole. Dressing of this hole due to interactions will lead to a quasi-hole. It is also possible to start from a few particles and holes. If there are not too many of them, there will be no interference in the various dressings, and we end up with the corresponding quasi-particles and quasi-holes. This is in particular the situation we have to deal with at low temperature, where there are few thermally produced excitations. On the other hand, the concept of quasi-particle loses its meaning at high energy, far above the Fermi surface, because there is no longer any sizeable restriction to the disintegration processes and the lifetime of any complex object is extremely short, so it does not have a well-defined energy. Hence the quasi-particle picture becomes invalid at higher temperature, when the thermal energy $k_B T$ becomes an appreciable fraction of the Fermi energy. Fortunately, in dealing with standard superconductivity, we will be in this low-temperature regime where we can safely use the quasi-particle picture.

In summary it seems that, as far as we are concerned, the effect of the interactions boils down to something very simple since we have merely to replace electrons with quasi-electrons. Actually, this is in general more complicated and Landau has shown that, for a proper description of the physical properties of the fermionic system, one should take into account residual interactions between quasi-particles. Physically these can be understood as the result of the overlap between the clouds corresponding to each quasi-particle. Fortunately, the resulting theoretical description is not complicated since it is formally identical to a mean-field description. However, it introduces parameters that need to be handled as phenomenological quantities describing the residual effect of interactions. Nevertheless these residual interactions are not necessarily small, and in particular in the case of liquid ^{3}He, where it has been possible to obtain them experimentally in a reliable way, they turn out to be fairly large. They are in particular related to the fact that the compressibility of liquid ^{3}He is quite weak because of the strong hard-core repulsion between ^{3}He atoms. A related effect is that the mass of the quasiparticle is not equal to the mass of the bare particle (the corresponding dispersion relations are not identical). This is clear physically since, for the quasi-particle, one has also to take into account the cloud in addition to the bare particle. For electrons in metals, this comes in addition to the modification of the electron mass due to the band structure.

Fortunately, in the case of the standard metals we will be mainly concerned with, these Fermi liquid effects are not strong. The residual Coulomb interactions turn out to be not that important, so that it is reasonable to omit them in a first step, and this is actually what we will do below in practice. Nevertheless Coulomb interactions are present, and in particular when we will be concerned with the formation of Cooper pairs, as we have done in the preceding section, we have to realize that to have a meaningful physical description, we have to form pairs with quasi-electrons, not bare electrons. This makes it obviously quite complicated to realistically evaluate the attractive interaction responsible for the formation of these pairs. On the other hand, although we should in principle from now on write only about "quasi-electrons," not bare electrons, we will keep the standard habit of writing "electrons," because having "quasi-electrons" written everywhere would make reading quite unpleasant. But we have to keep in mind that we are actually dealing with quasi-electrons.

Finally, regarding the validity of Fermi liquid theory, note that it describes only one possible, coherent scenario about the effect of interactions. It gives the result obtained from the non-interacting fermions situation by switching on interactions adiabatically. This physical situation is indeed the one that occurs in many standard metals, and this provides the justification for the use of band structure calculations. On the other hand, one may very well think that for some strength of the interactions, the fermionic system branches out toward a completely different physical behavior. Indeed, in one-dimensional systems where in some cases the effect of strong interactions can be fully taken into account in the theory, there are known examples where Fermi liquid theory fails and elementary excitations are not

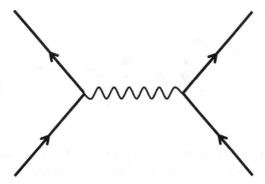

Graphic description of the indirect electron–electron interaction. An electron (full line) emits a phonon (wavy line), which is absorbed by another electron (full line).

Landau's quasiparticles. Actually, superconductivity itself can be considered as an example where Fermi liquid theory fails, since it cannot be obtained by switching on adiabatically interactions on the free electrons system. However, as we will see, it is possible to obtain a perfectly consistent picture of a superconducting metal by starting from the Fermi liquid description of the corresponding simple normal metal.

2.2.3 Electron–Phonon Interaction

As we mentioned already at the end of the last chapter, for standard superconductors, the physical origin of the attractive component of the effective interaction responsible for the formation of the Cooper pairs is the interaction of electrons with lattice vibrations. A clear experimental evidence for this role of the lattice in superconductivity is the existence of the isotope effect, where for example in mercury the superconducting critical temperature depends on the specific mercury isotope. Long wavelength-low frequency lattice vibrations correspond physically to standard sound waves while on the opposite high frequency vibrations correspond to optical modes, which are analogous to internal vibrations for a molecule. Just as for electromagnetic waves where the energy carried by a wave is quantized and appears only in discrete quanta, namely photons, sound waves, and more generally lattice vibrations are quantized, and the quanta are called phonons. Although there are naturally differences in details, there is a strong analogy between phonons and photons.

The interaction of an electron with lattice vibration occurs for example because the negatively charged electron may attract a positively charged ion through Coulomb interaction. This ion motion starts a lattice vibration; hence this process implies an electron–phonon interaction. This vibration may be felt in the same way by another electron; hence as a result, this second electron has indirectly felt the first electron. One can represent this process as in Fig. (2.2) by showing an electron emitting a phonon that is absorbed by another electron. A more naive but telling picture of this whole process is shown in Fig. (2.3). One sees that a negatively charged electron passing by attracts positively charged ions toward its trajectory. This results locally in an excess positive charge

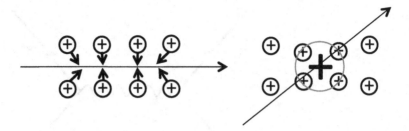

Fig. 2.3 Pictorial explanation of the indirect electron–electron attractive interaction in a metal. Left part: a negatively charged electron (horizontal line) moving through the lattice of positively charged ions attracts these ions toward its trajectory. Right part: the displaced ions create an excess positive charge (large cross), which is attracting another electron (transverse line) moving through the crystal.

which may attract another negatively charged electron. As a result, the second electron has been effectively attracted by the first one.

This second picture has the advantage of making clear a very important feature of this indirect electron–electron attraction: it is a retarded interaction. If the second electron passes by immediately after the first electron, the metallic ions do not have enough time to react and move, so there is no ionic charge build-up and no indirect interaction. In order to react appreciably, the ions need a time that is of order of the inverse of a typical vibration frequency ω_D, as already introduced in the preceding section about the Cooper problem. As a result of this time delay, the two interacting electrons are never located at the same place and there is some distance between them. This produces a reduction of the direct Coulomb repulsion, which is naturally weaker if the two electrons are far apart. Hence this retardation effect allows one to understand how the indirect electron–electron interaction, through phonon exchange, is able to overwhelm the direct Coulomb interaction so that the net result is an electron–electron attractive interaction.

However, this retardation effect, while so interesting physically, introduces a major practical difficulty. When we have addressed the Cooper problem, we have considered an instantaneous attractive interaction. In the process of generalizing the Cooper pair formation to the whole Fermi sea, we clearly want to stay at the same level of complexity for the interaction since dealing with a retarded interaction is a quite complicated problem. This means we would like to be able to consider that the time delay due to the retardation effect is negligible for our purpose. In other words, we would like to treat the ion motion as fast enough to make the interaction quasi-instantaneous. This occurs if the typical ion oscillation frequency ω_D is much larger than any frequency coming in our problem, or equivalently if $\hbar\omega_D$ is large compared to any typical energy ϵ we have to consider, such as the Cooper binding energy. We may also say that the time delay $1/\omega_D$ due to the ion motion should be negligible. If we consider the phase factor $e^{-i\epsilon t/\hbar}$ coming in a wave function linked for example to the Cooper pair, shifting t by $1/\omega_D$ should have a negligible effect, which implies $\epsilon/\hbar\omega_D \ll 1$, which leads again to the condition $\epsilon \ll \hbar\omega_D$. This regime of very small energies involved in superconductivity is called in this context the "weak coupling" regime, and we have already made use of it in our handling of the Cooper problem to simplify coherently the result. This is the simple regime we want to investigate, and fortunately a fair deal of standard superconductors satisfy this weak coupling condition. However, this is not true for some of the earlier superconductors discovered, such as mercury and lead. They belong to the "strong coupling" regime, and a much more complex treatment considered in Chapter 9 is necessary in order to account for their detailed physical properties.

2.2.4 BCS Reduced Hamiltonian

We focus now on the attractive part of the electron–electron interaction that is described, in the weak coupling regime, by an instantaneous interaction. Assuming momentum conservation due to translational invariance, its most general form is

$$H_{int} = \frac{1}{2} \sum_{\mathbf{k},\mathbf{k}',\mathbf{q}} V_{\mathbf{k},\mathbf{k}',\mathbf{q}} c^{\dagger}_{\mathbf{k}'+\mathbf{q}/2} c^{\dagger}_{-\mathbf{k}'+\mathbf{q}/2} c_{-\mathbf{k}+\mathbf{q}/2} c_{\mathbf{k}+\mathbf{q}/2} \tag{2.23}$$

corresponding to the process shown in Fig. (2.4). This expression has been written in second quantization.[4] The operators $c^{\dagger}_{\mathbf{k}}$ and $c_{\mathbf{k}}$ respectively create and annihilate an electron in a plane wave state of wavevector $\mathbf{k}$.

We simplify this expression by making use of the various physical ingredients we have learned from our study of the Cooper problem. To be specific, we make for the electronic dispersion relation $\epsilon_{\mathbf{k}}$ the same simplifying assumption that we have made for the Cooper problem, essentially taking an isotropic dispersion relation. We will take up this matter later on at the end of Chapter 3 and discuss what may happen when the simplifying assumptions are relaxed.

First, in handling the Cooper problem, we have come to the conclusion that the lowest energy is obtained by having a Cooper pair with total momentum $\mathbf{q} = \mathbf{0}$. Here we are similarly interested in the ground state of the electronic system in the presence of the attractive interaction. This will be obtained by forming many Cooper pairs. From our experience with a single pair, it is reasonable to conclude that the lowest energy state is obtained by having all the Cooper pairs with zero total momentum. Hence to generalize the Cooper problem and evaluate the energy, the relevant terms in the interaction are those corresponding to pairs of electrons with zero total momentum. This leads us to retain only the terms with $\mathbf{q} = \mathbf{0}$ in the expression Eq. (2.23) of H_{int}. Later on, we will be back to this question and consider Cooper pairs with nonzero momentum, but coherently this will describe an excited state of the electronic system.

Then, with the electron–phonon interaction as the physical origin of the attractive interaction, this interaction disappears when the electrons are too energetic, as previously noted.

---

[4] It is actually better to have some familiarity with this technique, since it is necessary to handle conveniently systems with a large number of particles. Hence it is a universal language in the corresponding scientific literature. Nevertheless we provide a short introduction in the Appendix. We recall that in second quantization, one makes use of creation $c^{\dagger}_{\mathbf{k}}$ and annihilation $c_{\mathbf{k}}$ operators, which are Hermitian conjugates. For fermions, these operators satisfy anticommutation relations, which are $\{c_{\mathbf{k}}, c_{\mathbf{k}'}\} = 0$, $\{c^{\dagger}_{\mathbf{k}}, c^{\dagger}_{\mathbf{k}'}\} = 0$, and $\{c_{\mathbf{k}}, c^{\dagger}_{\mathbf{k}'}\} = \delta_{\mathbf{k},\mathbf{k}'}$, where $\{A, B\} \equiv AB + BA$ is the anticommutator of the two operators, A and B. These anticommutation relations ensure proper antisymmetrization properties. For bosons the same relations hold, provided the anticommutators are replaced by commutators. They ensure the corresponding symmetry properties.

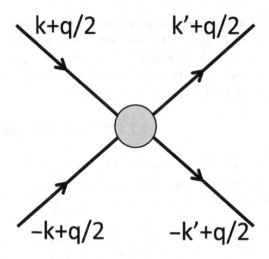

Fig. 2.4 Graphic depiction of an instantaneous electron–electron scattering process, with two entering electrons with wavevectors $\mathbf{k} + \mathbf{q}/2$ and $-\mathbf{k} + \mathbf{q}/2$, and two outgoing electrons with wavevectors $\mathbf{k}' + \mathbf{q}/2$ and $-\mathbf{k}' + \mathbf{q}/2$.

Indeed, if an electron goes too fast through the lattice, ions do not have enough time to react to the presence of the electron which is already gone before they have had a sizeable move. On the other hand, if the electron is slow enough, ions have enough time to move due to the presence of the electron and build for example a positive charge which will be felt by another electron.

This request for a slow-enough electron is modeled in BCS theory in the same way as we have done for the Cooper problem, that is by putting a sharp cut-off on the electron energy. One must realize that the electronic energy we are considering has to be measured with respect to the Fermi energy. Indeed in a metal the full Fermi sea alone is very much like the vacuum in particle physics: nothing is happening. The equivalent of adding a particle to vacuum in particle physics is to add an electron above the Fermi surface, and its energy has to be measured compared to its lowest possible value, that is when the electron is added right at the Fermi surface where its kinetic energy is E_F. Hence the measure of the electronic energy is its kinetic energy $\epsilon_{\mathbf{k}}$ compared to the Fermi energy E_F, that is $\xi_{\mathbf{k}} = \epsilon_{\mathbf{k}} - E_F$. This is indeed what comes out when detailed calculations of the effective interaction are performed. But for the moment, we follow the simple BCS modeling by putting a sharp cut-off on $\xi_{\mathbf{k}}$. So we request $\xi_{\mathbf{k}} < \hbar\omega_D$; otherwise, the attractive interaction in Eq. (2.23) is zero.

Finally we want to treat all the electrons on the same footing. Hence we do not consider only electrons above the Fermi surface with $\xi_{\mathbf{k}} > 0$, but also electrons below the Fermi surface with $\xi_{\mathbf{k}} < 0$. We also request that these electrons not be too energetic, which implies $|\xi_{\mathbf{k}}| < \hbar\omega_D$. Another way to see this is, for electrons below the Fermi surface, to think in terms of holes instead of electrons. Creating a hole is done by removing an electron below the Fermi surface in the Fermi sea. This creates an excited state with energy opposite to the energy of the removed electron, that is $-\xi_{\mathbf{k}} > 0$. Instead of electrons, we may very well think about all our problems in terms of holes, which are very much like anti-electrons. We could for example consider a Cooper problem with two holes instead of two electrons. With this picture we naturally request that the holes are not too energetic in order to have the attractive interaction effective, which leads to the condition $0 < -\xi_\mathbf{k} < \hbar\omega_D$, equivalent to $|\xi_\mathbf{k}| < \hbar\omega_D$. Finally this condition holds for all $\mathbf{k}$'s, and in particular for $\mathbf{k}'$ in Eq. (2.23). In summary, the BCS model requires in Eq. (2.23) $|\xi_\mathbf{k}| < \hbar\omega_D$ and $|\xi_{\mathbf{k}'}| < \hbar\omega_D$ to have a nonzero attractive interaction.

We are now left to consider the amplitude $V_{\mathbf{k},\mathbf{k}',\mathbf{q}=0}$ of the interaction for $\mathbf{k}$ and $\mathbf{k}'$ close to the Fermi surface. However, in principle, this amplitude is obtained by a microscopic calculation where the Fermi surface plays no role, since basically this surface and the Fermi energy are merely related to the total number of electrons in the metal. Hence $V_{\mathbf{k},\mathbf{k}',\mathbf{q}=0}$ is in general fully regular in the vicinity of the Fermi surface, and we can with small error replace $V_{\mathbf{k},\mathbf{k}',\mathbf{q}=0}$ by its value for $|\mathbf{k}| = |\mathbf{k}'| = k_F$. This parallels our discussion of the Cooper problem for a general interaction.

Hence we are similarly left with the fact that the only dependence we have to deal with is the dependence of $V_{\mathbf{k},\mathbf{k}',\mathbf{q}=0}$ on the angle between $\mathbf{k}$ and $\mathbf{k}'$. For our isotropic model, we can expand this angular dependence in Legendre polynomials, and we will have to examine which angular momentum ℓ leads to the lowest energy ground state. The standard assumption is, as we have already explained for the Cooper problem, that the lowest energy is obtained for $\ell = 0$. This turns out to be consistent with the experimental situation of the standard low-temperature superconductors. In this case, we can omit the angular dependence of $V_{\mathbf{k},\mathbf{k}',\mathbf{q}=0}$, which is irrelevant for our search of the ground state, and keep only the $\ell = 0$ component. This is equivalent to replace $V_{\mathbf{k},\mathbf{k}',\mathbf{q}=0}$ with a constant, which we call merely $-V$ to make it clear (with $V > 0$) that the interaction is attractive.

We have a last point to consider, namely the spin of the electrons. We have not even mentioned it up to now, to avoid unnecessary complications. This has the implication that the spin is irrelevant to our preceding considerations. This is indeed true for the electronic dispersion relation of simple metals. The indirect attractive interaction through electron–phonon interaction is also spin independent. However, when considering the Cooper pair formation, our preceding conclusion that they form with $\ell = 0$ angular momentum has immediate consequences for the spin of the involved electrons. Indeed, when we deal with the pair wave function, we have to write an expression that is antisymmetric with respect to the full exchange of the variables of the two electrons, since these electrons are indistinguishable fermions. With respect to the exchange of the orbital variables, the wave function is symmetric, since we take it with $\ell = 0$ angular momentum. This implies that with respect to the exchange of the spin variables it must be antisymmetric, to obtain a global antisymmetry with respect to the exchange of all the variables. This means that with respect to their spins, the electrons must be in a singlet state, a triplet state being excluded since it is symmetric with respect to the exchange of the spin variables. Hence the pair wave function is proportional to $| \uparrow\downarrow \rangle - | \downarrow\uparrow \rangle$. In practice, rather than writing this spin part of the pair wave function, it is more convenient for all the manipulations to merely indicate that the electrons have opposite spins. The even symmetry of the orbital parts ensures in practice that we are not retaining a triplet component. Coming back to the effective Hamiltonian, since we have retained only the terms contributing to the formation of the Cooper pairs, we are led to keep in the attractive interaction only terms involving electrons with opposite spins.

This leads us finally to the reduced Hamiltonian considered by Bardeen, Cooper, and Schrieffer

$$H_{BCS} = H_c + H_{int} \tag{2.24}$$

with the kinetic energy term

$$H_c = \sum_{\mathbf{k}} \epsilon_{\mathbf{k}} (c_{\mathbf{k}\uparrow}^{\dagger} c_{\mathbf{k}\uparrow} + c_{\mathbf{k}\downarrow}^{\dagger} c_{\mathbf{k}\downarrow}) \tag{2.25}$$

and the interaction term[5]

$$H_{int} = -V \sum_{\mathbf{k},\mathbf{k}'} c_{\mathbf{k}'\uparrow}^{\dagger} c_{-\mathbf{k}'\downarrow}^{\dagger} c_{-\mathbf{k}\downarrow} c_{\mathbf{k}\uparrow} \qquad\qquad |\xi_{\mathbf{k}}|, |\xi_{\mathbf{k}'}| < \hbar\omega_D \tag{2.26}$$

Note that we could as well write $\sum_{\mathbf{k},\mathbf{k}'} c_{\mathbf{k}'\downarrow}^{\dagger} c_{-\mathbf{k}'\uparrow}^{\dagger} c_{-\mathbf{k}\uparrow} c_{\mathbf{k}\downarrow}$ in the interaction term, since by $c_{-\mathbf{k}\uparrow} c_{\mathbf{k}\downarrow} = -c_{\mathbf{k}\downarrow} c_{-\mathbf{k}\uparrow}$, and the changes $\mathbf{k} \to -\mathbf{k}$, $\mathbf{k}' \to -\mathbf{k}'$, it reduces to the term written in Eq. (2.26). Note also that (with real V) H_{int} is Hermitian. We will often imply, without writing them explicitly, the cut-off conditions in Eq. (2.26).

To summarize, we have reached this reduced form for the Hamiltonian by keeping only, in the interaction, the terms involved in a qualitative change in the ground state of the electronic system. In this process, we have discarded a large number of contributions. The main idea is that these discarded terms lead to only quantitative changes, which can be taken into account later on by perturbative calculations or by generalizing the Fermi liquid approach to the qualitatively new situation. Another way is to make use of the BCS wave function we will obtain below as a variational wave function for the complete Hamiltonian. The ultimate justification for all this process is anyway the agreement with experiments.

2.3 BCS Wave Function

Just as for the Cooper problem, we want to obtain the ground state of H_{BCS}. Since this is not such an easy matter (which is addressed in Section 2.5), we rather introduce here the corresponding wave function on physical grounds. In the Cooper problem, we formed a pair with the two interacting electrons. Let us call $\Phi(\mathbf{r}_1 - \mathbf{r}_2)$ the corresponding pair wave function, where $\mathbf{r}_1$ and $\mathbf{r}_2$ are the respective positions of the two electrons. (This wave function depends only on the relative position $\mathbf{r}_1 - \mathbf{r}_2$ of the electrons since their center of mass does not move.) Now we want to treat all the electrons on the same footing. So we form another pair with two other electrons, another one with two others, and so on until we have formed pairs with all the available electrons. (We assume an even number N of electrons for simplicity, a lonely additional electron being clearly unimportant.)

---

[5] One can see that a triplet contribution $\sum_{\mathbf{k}} c_{\mathbf{k}\uparrow} c_{-\mathbf{k}\uparrow}$ is automatically excluded since from $c_{-\mathbf{k}\uparrow} c_{\mathbf{k}\uparrow} = -c_{\mathbf{k}\uparrow} c_{-\mathbf{k}\uparrow}$, and $\mathbf{k} \to -\mathbf{k}$, one obtains $\sum_{\mathbf{k}} c_{\mathbf{k}\uparrow} c_{-\mathbf{k}\uparrow} = -\sum_{\mathbf{k}} c_{-\mathbf{k}\uparrow} c_{\mathbf{k}\uparrow} = -\sum_{\mathbf{k}} c_{\mathbf{k}\uparrow} c_{-\mathbf{k}\uparrow}$, which implies $\sum_{\mathbf{k}} c_{\mathbf{k}\uparrow} c_{-\mathbf{k}\uparrow} = 0$. Thus such a term is automatically zero.

In the spirit of Bose condensation, we form a Bose condensate with all these pairs, so we put all these electron pairs in the same state Φ. This leads us to write the wave function of the N electrons as

$$\Psi_N(\mathbf{r}_1, \mathbf{r}_2, \cdots, \mathbf{r}_N) = \Phi(\mathbf{r}_1 - \mathbf{r}_2)\Phi(\mathbf{r}_3 - \mathbf{r}_4)\cdots\Phi(\mathbf{r}_{N-1} - \mathbf{r}_N) \tag{2.27}$$

Unfortunately, this wave function is not acceptable because electrons are fermions, and the wave function should be antisymmetric with respect to the exchange of the position variables of any two identical electrons. (For the sake of clarity, we forget the spin variable for the moment, since it plays no role in the following argumentation; it will be restored at the end of the discussion.) Physically this is extremely important since, as we will see, there is a huge overlap between pairs. This would be much less so if we wanted to deal with a Bose condensate of a dilute gas of molecules made of two fermions.

To properly antisymmetrize the wave function, we should for example subtract from Eq. (2.27) the expression obtained by exchanging $\mathbf{r}_1$ and $\mathbf{r}_3$ and so on, for all possible exchanges. This would lead to a huge number of different terms in the wave function, making any actual calculation completely intractable. This essential technical problem is solved by the use of second quantization, which we have already introduced when writing the interaction Hamiltonian Eq. (2.23). For example, in the second quantization, the state corresponding to two electrons forming a pair with wave function $\Phi(\mathbf{r}_1 - \mathbf{r}_2)$ reads

$$\int d\mathbf{r}_1 d\mathbf{r}_2 \Phi(\mathbf{r}_1 - \mathbf{r}_2)\psi^\dagger(\mathbf{r}_1)\psi^\dagger(\mathbf{r}_2)|0\rangle \tag{2.28}$$

The creation operators $\psi^\dagger(\mathbf{r}_1)$ and $\psi^\dagger(\mathbf{r}_2)$ are acting on the vacuum of particles $|0\rangle$, and they create two electrons at $\mathbf{r}_1$ and $\mathbf{r}_2$, with probability amplitude $\Phi(\mathbf{r}_1 - \mathbf{r}_2)$, and one integrates over $\mathbf{r}_1$ and $\mathbf{r}_2$ to account for the fact that the electrons can be anywhere. These creation operators $\psi^\dagger(\mathbf{r})$ anticommute, which ensures that Fermi statistics is satisfied and that the resulting wave functions are properly antisymmetrized. In this way, the state $|\Psi_N\rangle$ corresponding to the $N/2$ pairs of fermions in Eq. (2.27) is written properly as

$$|\Psi\rangle_N = \left[\int d\mathbf{r}_1 d\mathbf{r}_2 \Phi(\mathbf{r}_1 - \mathbf{r}_2)\psi^\dagger(\mathbf{r}_1)\psi^\dagger(\mathbf{r}_2)\right]^{N/2}|0\rangle \tag{2.29}$$

Above we have written the wave function in the natural $\mathbf{r}$ space representation. However, we expect the Fermi sea to play an essential role, as it did in the Cooper problem. Hence going to the momentum space $\mathbf{k}$ representation is likely to be more appropriate. This is done by Fourier transforming everything. This leads us to use the creation operators $c_\mathbf{k}^\dagger$, already introduced in Eq. (2.23). They are related to $\psi^\dagger(\mathbf{r})$ by

$$c_\mathbf{k}^\dagger = \int d\mathbf{r}\, e^{i\mathbf{k}\cdot\mathbf{r}}\psi^\dagger(\mathbf{r}) \qquad \psi^\dagger(\mathbf{r}) = \sum_\mathbf{k} e^{-i\mathbf{k}\cdot\mathbf{r}}c_\mathbf{k}^\dagger \tag{2.30}$$

Conversely, the annihilation operators $\psi(\mathbf{r})$ and $c_\mathbf{k}$, which are the Hermitian conjugate of the corresponding creation operators, are related by

$$c_\mathbf{k} = \int d\mathbf{r}\, e^{-i\mathbf{k}\cdot\mathbf{r}}\psi(\mathbf{r}) \qquad \psi(\mathbf{r}) = \sum_\mathbf{k} e^{i\mathbf{k}\cdot\mathbf{r}}c_\mathbf{k} \tag{2.31}$$

In Eq. (2.28) we may express $\psi^\dagger(\mathbf{r})$ in terms of $c_\mathbf{k}^\dagger$ to obtain the desired expression. However, a faster way to obtain the same result is to notice that when we write the wave function $\Phi(\mathbf{r}_1 - \mathbf{r}_2)$ in terms of its Fourier transform $\Phi_\mathbf{k}$

$$\Phi(\mathbf{r}_1 - \mathbf{r}_2) = \sum_\mathbf{k} \Phi_\mathbf{k} e^{i\mathbf{k}\cdot(\mathbf{r}_1 - \mathbf{r}_2)} = \sum_\mathbf{k} \Phi_\mathbf{k} e^{i\mathbf{k}\cdot\mathbf{r}_1} e^{-i\mathbf{k}\cdot\mathbf{r}_2} \tag{2.32}$$

we express the fact that in order to put the two electrons in the state $|\Phi\rangle$ corresponding to the wave function $\Phi(\mathbf{r}_1 - \mathbf{r}_2)$, we have to put the first electron in the plane wave state $\exp(i\mathbf{k}\cdot\mathbf{r}_1)$ of wavevector $\mathbf{k}$, the second electron in the plane wave state $\exp(-i\mathbf{k}\cdot\mathbf{r}_2)$ of wavevector $-\mathbf{k}$, with a probability amplitude $\Phi_\mathbf{k}$, and sum over all possible $\mathbf{k}$. Now the state obtained by putting an electron in a plane wave state $\mathbf{k}$ and another one in a plane wave state $-\mathbf{k}$ is just written $c_\mathbf{k}^\dagger c_{-\mathbf{k}}^\dagger |0\rangle$ in second quantization. Hence the expression of $|\Phi\rangle$ we are looking for is

$$|\Phi\rangle = \sum_\mathbf{k} \Phi_\mathbf{k} c_\mathbf{k}^\dagger c_{-\mathbf{k}}^\dagger |0\rangle \tag{2.33}$$

This is just the equivalent of Eq. (2.28) written in $\mathbf{k}$ space instead of $\mathbf{r}$ space. Accordingly we may write, just as in Eq. (2.29), the state corresponding to $N/2$ Bose-condensed pairs as

$$|\Psi\rangle_N = \left(\sum_\mathbf{k} \Phi_\mathbf{k} c_\mathbf{k}^\dagger c_{-\mathbf{k}}^\dagger \right)^{N/2} |0\rangle \tag{2.34}$$

Eq. (2.34) is perfectly fine physically. However, it is terribly inconvenient in practice as soon as one tries to perform actual calculations with it. The trouble is that, when we expand the sum to the power $N/2$ in this equation, we will obtain terms like

$$\sum_{\mathbf{k}_1, \mathbf{k}_2, \cdots, \mathbf{k}_{N/2}} c_{\mathbf{k}_1}^\dagger c_{-\mathbf{k}_1}^\dagger c_{\mathbf{k}_2}^\dagger c_{-\mathbf{k}_2}^\dagger \cdots c_{\mathbf{k}_{N/2}}^\dagger c_{-\mathbf{k}_{N/2}}^\dagger \tag{2.35}$$

but all the terms having (at least) two identical wavevectors $\mathbf{k}_i = \mathbf{k}_j$ will be zero because the Pauli principle forbids putting two fermions in the same state.[6] All the N wavevectors have to be different. This produces an effective link between the different wavevectors, which cannot be manipulated independently. This situation is quite analogous to the one met in statistical physics, when one wants to obtain the Fermi–Dirac distribution for non-interacting fermions.

Indeed, when one deals with N non-interacting fermions, which can occupy plane-wave states $\mathbf{k}$ with energy $\epsilon_\mathbf{k}$, the situation looks simple for the calculation of the partition function $Z_N = \sum_n \exp(-E_n/k_B T)$, with k_B being the Boltzmann constant, since the energy E_n of state n is merely the sum of the energies of the N fermions making up this state: $E_n = \sum_{i=1}^N \epsilon_{\mathbf{k}_i}$. It seems that one just has to factorize the exponential and calculate $\sum_\mathbf{k} \exp(-\epsilon_\mathbf{k}/k_B T)$. However, this is not correct because there is a fixed number N of plane waves $\mathbf{k}_i$, and moreover Pauli principle requires all these N wavevectors $\mathbf{k}_i$ to be different.

---

[6] The Pauli principle is automatically satisfied by the use of second quantization: for any wavevectors $\mathbf{k}$ and $\mathbf{k}'$, the corresponding creation operators anticommute $c_\mathbf{k}^\dagger c_{\mathbf{k}'}^\dagger = -c_{\mathbf{k}'}^\dagger c_\mathbf{k}^\dagger$. If we take $\mathbf{k} = \mathbf{k}'$, we obtain $c_\mathbf{k}^\dagger c_\mathbf{k}^\dagger = -c_\mathbf{k}^\dagger c_\mathbf{k}^\dagger$, which implies $c_\mathbf{k}^\dagger c_\mathbf{k}^\dagger = 0$. Hence we cannot create two fermions in the same state with wavevector $\mathbf{k}$.

This produces a link between the wavevectors $\mathbf{k}_i$, and it is not correct to sum over these wavevectors as if they are independent.

The simple way out of this problem is to go to the grand canonical ensemble. Instead of dealing with a system with a fixed number N of fermions, we consider a system where the number of fermions is free to vary, due to the contact with a reservoir of fermions, but the chemical potential μ of these fermions is fixed. The grand partition function suitable for this ensemble is $\Xi = \sum_N Z_N \exp[\mu N/k_B T] = \sum_{n,N} \exp[-(E_n - \mu N)/k_B T]$. Since in the present case the number of fermions is no longer restricted, it is now possible to factorize the exponential and to write $\Xi = \prod_\mathbf{k} \Xi_\mathbf{k}$ as the product over all available wavevectors $\mathbf{k}$ of grand partition functions $\Xi_\mathbf{k}$ restricted to a given plane wave state $\mathbf{k}$. Within this state, we have to enforce the Pauli principle so that the number of fermions $n_\mathbf{k}$ in state $\mathbf{k}$ is either zero or one. This leads explicitly to $\Xi_\mathbf{k} = 1 + \exp[-(\epsilon_\mathbf{k} - \mu)/k_B T]$, the first term corresponding to the case $n_\mathbf{k} = 0$ and the second to $n_\mathbf{k} = 1$. Hence the probability to have zero fermion is $1/\Xi_\mathbf{k}$, and the probability to have one fermion is $\exp[-(\epsilon_\mathbf{k} - \mu)/k_B T]/\Xi_\mathbf{k}$. This gives $\bar{n}_\mathbf{k} = 1/[\exp[(\epsilon_\mathbf{k} - \mu)/k_B T] + 1]$ as the mean value of the fermion number in state $\mathbf{k}$, which is the Fermi–Dirac distribution.

It is possible to handle the problem of non-interacting fermions within the canonical ensemble, but this is mathematically much more complex than the simple above derivation, and actually the procedure turns out to be quite similar to going from the canonical to the grand canonical ensemble. Nevertheless, the canonical ensemble and the grand canonical ensemble describing different physical situations, they lead in full generality to different results. However, in the case of a very large number of particles, and for standard thermodynamic quantities, the two ensembles lead to the same results. This is due to the fact that when one writes $\Xi = \sum_N Z_N \exp[\mu N/k_B T]$, the quantity to be summed $Z_N \exp[\mu N/k_B T]$ very sharply peaks at a value $\bar{N}$, which is the average value of N in the grand canonical ensemble. The very sharp peak corresponds physically to the fact that in the thermodynamic limit of large N, fluctuations around the average are very small. Mathematically this arises because $Z_N \exp[\mu N/k_B T]$ contains the product of a rapidly growing function of N by a rapidly decreasing one.[7] This implies in practice that to dominant order in $\bar{N}$, $\Xi = Z_{\bar{N}} \exp[\mu \bar{N}/k_B T]$, so that the results from the grand canonical ensemble and the canonical ensemble are identical.

In order to eliminate the mathematical troubles arising from working with Eq. (2.34), we proceed to a step similar to go from the canonical ensemble to the grand canonical ensemble by introducing a wave function that is a sum of wave functions $|\Psi\rangle_N$ corresponding to

---

[7] A simple analogous situation is found when one wants to evaluate for large n the integral $\int_0^\infty dx\, x^n e^{-x}$, which is equal to $n!$ for integer n (as found by part integration). In this case, x^n is a rapidly growing function of x, while e^{-x} is rapidly decreasing. This gives rise to a sharp maximum located where the derivative of $\ln(x^n e^{-x}) = n \ln x - x$ is zero, namely at $\bar{x} = n$. The simplest evaluation of the integral is merely to take the value of the integrand at this maximum, which gives $n^n e^{-n}$. A better one is obtained by taking into account the fact that $x^n e^{-x}$ has a Gaussian shape in the vicinity of this maximum. The coefficient in this Gaussian is obtained from the second derivative of $\ln(x^n \exp^{-x})$ at $\bar{x}$, which is $-1/n$, so that in the vicinity of $\bar{x}$ an excellent approximation for the integrand is $x^n e^{-x} \simeq n^n e^{-n} \exp[-(x - \bar{x})^2/2n]$. Performing the Gaussian integration after having extended the integration from $-\infty$ to ∞ (which brings vanishingly small correction) yields finally $n! = \int_0^\infty dx\, x^n e^{-x} = n^n e^{-n}\sqrt{2\pi n}$, which is the well-known Stirling formula. Amazingly it gives for $n = 1$ the result 0.92, which is quite reasonable since 1 is by no means large.

different number N of fermions. This is analogous to going from the partition function Z to the grand partition function Ξ. However, in much the same way, the terms of the sum will be strongly peaked around some mean value $\bar{N}$, and the introduced wave function will essentially be equal to $|\Psi\rangle_{\bar{N}}$. Our purpose is, by relaxing the requirement to have exactly N fermions, to decouple all the wavevectors $\mathbf{k}$ in the same way as one does in statistical physics by going from Z to Ξ.

Our choice for this decoupling is

$$|\Psi\rangle = \sum_{N=0}^{\infty} \frac{1}{(N/2)!} |\Psi\rangle_N \qquad (2.36)$$

where N runs only over even values. The only justification for the coefficient $1/(N/2)!$ is that it is, as we will see, the one appropriate to obtain the decoupling we are looking for.

We notice that the quantity to be summed up in Eq. (2.36) is indeed the product of $|\Psi\rangle_N$, a rapidly growing function of N, by $1/(N/2)!$ that decreases rapidly with N. This is quite analogous to $\sum_{n=0}^{\infty} x^n/n! = e^x$. Fig. (2.5) represents $x^n/n!$ as a function of n, for $x = 400$, displaying explicitly the sharp peak feature mentioned above. Note that in the dominant order, the location of the maximum is found (by taking the logarithm) at $\bar{n} = x = 400$ from the Stirling formula (see footnote 7), and that taking the term $\bar{n}^{\bar{n}}/\bar{n}!$ of the series for this value gives (to dominant order) precisely $e^{\bar{n}} = e^x$. One can perform a better summation of this series, and go to the next order, by integrating over the Gaussian shape displayed in Fig. (2.5); this gives the same result e^x with a better justification. This concrete example provides a feeling for what Eq. (2.36) implies.

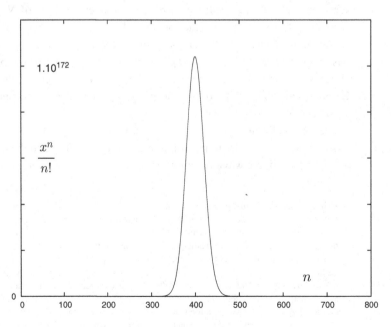

Fig. 2.5   Terms $x^n/n!$, of the series expansion of e^x, as a function of n, for $x = 400$.

Coming back to Eq. (2.36), we insert in it the explicit expression Eq. (2.34) for $|\Psi\rangle_N$. This gives

$$|\Psi\rangle = \sum_{N=0}^{\infty} \frac{1}{(N/2)!} \left(\sum_{\mathbf{k}} \Phi_{\mathbf{k}} c_{\mathbf{k}}^{\dagger} c_{-\mathbf{k}}^{\dagger} \right)^{N/2} |0\rangle \tag{2.37}$$

We see that the summation to be performed is just the series expansion of an exponential, so we obtain

$$|\Psi\rangle = \exp\left(\sum_{\mathbf{k}} \Phi_{\mathbf{k}} c_{\mathbf{k}}^{\dagger} c_{-\mathbf{k}}^{\dagger} \right) |0\rangle \tag{2.38}$$

Now in Eq. (2.38) the various operators in the sum over $\mathbf{k}$ commute:

$$[c_{\mathbf{k}}^{\dagger} c_{-\mathbf{k}}^{\dagger}, c_{\mathbf{k}'}^{\dagger} c_{-\mathbf{k}'}^{\dagger}] = 0 \tag{2.39}$$

This is because all the creation operators anticommute, so when we exchange two of them, there is a sign change. But when we take a pair of them, there are two sign changes, which means there is no sign change at all. This is consistent with the fact that a pair of fermions behaves as a boson. Since these operators commute, they behave as ordinary numbers, and in particular the exponential of a sum is equal to the product of the exponentials corresponding to each term, so we can write Eq. (2.38) as

$$|\Psi\rangle = \prod_{\mathbf{k}} \exp\left(\Phi_{\mathbf{k}} c_{\mathbf{k}}^{\dagger} c_{-\mathbf{k}}^{\dagger} \right) |0\rangle \tag{2.40}$$

We see that we have already obtained at this stage the decoupling between the various wavevectors. This expression can be written in a simpler form by expanding each exponential as

$$\exp\left(\Phi_{\mathbf{k}} c_{\mathbf{k}}^{\dagger} c_{-\mathbf{k}}^{\dagger} \right) = 1 + \Phi_{\mathbf{k}} c_{\mathbf{k}}^{\dagger} c_{-\mathbf{k}}^{\dagger} + \cdots + \frac{1}{n!} \left(\Phi_{\mathbf{k}} c_{\mathbf{k}}^{\dagger} c_{-\mathbf{k}}^{\dagger} \right)^{n} + \cdots \tag{2.41}$$

However, since we deal with pairs of fermions, we are not allowed to create more than one pair in a given pair state $(\mathbf{k}, -\mathbf{k})$. This is the stage where we see clearly that we cannot consider that these pairs of fermions behave as simple elementary bosons; otherwise we would not meet such a restriction. For example, for the creation of two pairs, we have to write $\left(c_{\mathbf{k}}^{\dagger} c_{-\mathbf{k}}^{\dagger} \right)^2 = c_{\mathbf{k}}^{\dagger} c_{-\mathbf{k}}^{\dagger} c_{\mathbf{k}}^{\dagger} c_{-\mathbf{k}}^{\dagger} = -c_{\mathbf{k}}^{\dagger} c_{\mathbf{k}}^{\dagger} c_{-\mathbf{k}}^{\dagger} c_{-\mathbf{k}}^{\dagger} = 0$ since $(c_{\mathbf{k}}^{\dagger})^2 = 0$, expressing the Pauli principle which forbids to create more than one fermion in state $\mathbf{k}$ (see footnote 6). Hence we have $\left(c_{\mathbf{k}}^{\dagger} c_{-\mathbf{k}}^{\dagger} \right)^p = 0$ for $p \geq 2$, and in Eq. (2.41) only the first two terms are nonzero, so that Eq. (2.40) becomes

$$|\Psi\rangle = \prod_{\mathbf{k}} \left(1 + \Phi_{\mathbf{k}} c_{\mathbf{k}}^{\dagger} c_{-\mathbf{k}}^{\dagger} \right) |0\rangle \tag{2.42}$$

Finally it is useful to normalize conveniently the state $|\Psi\rangle$, which we have not considered up to now. Naturally since the various pair states $(\mathbf{k}, -\mathbf{k})$ are decoupled, we do not want to couple them by a global normalization. Accordingly, we normalize within each subspace $(\mathbf{k}, -\mathbf{k})$ by multiplying by an appropriate coefficient $u_{\mathbf{k}}$. Defining

$$v_{\mathbf{k}} = u_{\mathbf{k}} \Phi_{\mathbf{k}} \tag{2.43}$$

we obtain

$$|\Psi\rangle_{BCS} = \prod_{\mathbf{k}} \left(u_{\mathbf{k}} + v_{\mathbf{k}} c_{\mathbf{k}\uparrow}^{\dagger} c_{-\mathbf{k}\downarrow}^{\dagger} \right) |0\rangle \tag{2.44}$$

which is the celebrated BCS wave function. In this final form, we have written explicitly the spin of the electrons in the corresponding creation operators.

We note that we could have written down this expression directly (as Bardeen, Cooper, and Schrieffer did). Indeed, if we require for $|\Psi\rangle$ an expression where the different wavevectors $\mathbf{k}$ are decoupled, it must be the product (acting on vacuum) over all wavevectors $\mathbf{k}$ of operators acting only on the subspace $(\mathbf{k}_{\uparrow}, -\mathbf{k}_{\downarrow})$. Since we want to have the electrons coming in pairs, to obtain the lowest energy, we can either have a pair with probability amplitude $v_{\mathbf{k}}$ or no pair at all with probability amplitude $u_{\mathbf{k}}$. This is just what is written in Eq. (2.44), and we see that this expression is physically quite reasonable.

Nevertheless there is a price to pay for this convenient form, which is the fact that $|\Psi\rangle_{BCS}$ does not correspond to a specific number of electrons; it is rather a superposition of states with the number of electrons taking all the possible (even) values from 0 to ∞. We may always extract from $|\Psi\rangle_{BCS}$ the component corresponding to a specific number of electrons, which leads back to Eq. (2.34), but we lose the convenience of the decoupling. It is clear that Eq. (2.44) cannot be strictly speaking an eigenstate of H_{BCS}, since this would imply that the eigenvalues for different N are the same, whatever the number N of electrons, which obviously cannot be true (as can be easily checked on simple examples). Hence, although we have detailed above which sense should be attributed to Eq. (2.44), it is not surprising that the validity of the BCS wave function has been questioned. A natural reaction is to consider that the physical situation for an isolated superconductor should be described by the canonical ensemble since the number of electrons is fixed, not by the grand canonical ensemble. We will be back to this point in Section 2.5. But we will first investigate the physics linked to the BCS wave function.

Let us write explicitly the normalization condition, which we have not done yet. From the probability amplitudes $u_{\mathbf{k}}$ and $v_{\mathbf{k}}$ to have zero pair or one pair, we obtain that the corresponding probabilities are $|u_{\mathbf{k}}|^2$ and $|v_{\mathbf{k}}|^2$. Since the sum of all the probabilities must be equal to 1, we conclude that the normalization implies

$$|u_{\mathbf{k}}|^2 + |v_{\mathbf{k}}|^2 = 1 \tag{2.45}$$

We can also obtain this result algebraically by requiring explicitly that $\langle\Psi|\Psi\rangle = 1$ (from now on we omit the subscript "BCS" for short). From Eq. (2.44) we have to write

$$\langle 0| \prod_{\mathbf{k}'} (u_{\mathbf{k}'}^{*} + v_{\mathbf{k}'}^{*} c_{-\mathbf{k}'} c_{\mathbf{k}'}) \prod_{\mathbf{k}} (u_{\mathbf{k}} + v_{\mathbf{k}} c_{\mathbf{k}}^{\dagger} c_{-\mathbf{k}}^{\dagger})|0\rangle = 1 \tag{2.46}$$

where again the spins are understood. Since all the operators corresponding to different values of $\mathbf{k}$ commute, we can gather the terms containing a specific value $\mathbf{k}_0$ of the wavevector, so we are left to deal with

$$\langle 0 | (u_{\mathbf{k}_0}^* + v_{\mathbf{k}_0}^* c_{-\mathbf{k}_0} c_{\mathbf{k}_0}) (u_{\mathbf{k}_0} + v_{\mathbf{k}_0} c_{\mathbf{k}_0}^\dagger c_{-\mathbf{k}_0}^\dagger) | 0 \rangle \tag{2.47}$$

where we have not written all the other factors with other wavevectors, which are irrelevant for $\mathbf{k}_0$. Now with the annihilation operator $c_{\mathbf{k}_0}$, we have $c_{\mathbf{k}_0} | 0 \rangle = 0$ since there is no fermion in vacuum $| 0 \rangle$ and we cannot annihilate a fermion that does not exist. As a result, the term with the factor $v_{\mathbf{k}_0}^* u_{\mathbf{k}_0}$ in the expansion of expression (2.47) is zero. Similarly we have also $\langle 0 | c_{\mathbf{k}_0}^\dagger = 0$ for the corresponding bra, which makes the $u_{\mathbf{k}_0}^* v_{\mathbf{k}_0}$ disappear in the expansion. Finally we have $\langle 0 | c_{-\mathbf{k}_0} c_{\mathbf{k}_0} c_{\mathbf{k}_0}^\dagger c_{-\mathbf{k}_0}^\dagger | 0 \rangle = 1$ since we create from vacuum and then annihilate the two electrons of the pair in proper order to go back to vacuum. Since we have $\langle 0 | 0 \rangle = 1$, expression (2.47) reduces to $|u_{\mathbf{k}_0}|^2 + |v_{\mathbf{k}_0}|^2$. Repeating this calculation for all wavevectors, we can rewrite Eq. (2.46) as

$$\prod_{\mathbf{k}} (|u_{\mathbf{k}}|^2 + |v_{\mathbf{k}}|^2) = 1 \tag{2.48}$$

Since we want to normalize in each subspace $(\mathbf{k}, -\mathbf{k})$ independently, this leads us indeed to the normalization condition Eq. (2.45).

Let us finally check explicitly on the BCS wave function Eq. (2.44) that the terms are very sharply peaked around the mean value $\bar{N}$, as we have already argued. We first calculate the mean value $\bar{N} = \langle \Psi | \hat{N} | \Psi \rangle \equiv \langle \hat{N} \rangle$ of the electron number. In second quantization, the operator corresponding to the number of electrons in state $\mathbf{k}$ is[8] $\hat{n}_{\mathbf{k}} = c_{\mathbf{k}}^\dagger c_{\mathbf{k}}$ and the operator corresponding to the total number of electrons is $\hat{N} = \sum_{\mathbf{k}} \hat{n}_{\mathbf{k}} = \sum_{\mathbf{k}} (\hat{n}_{\mathbf{k}\uparrow} + \hat{n}_{-\mathbf{k}\downarrow}) = \sum_{\mathbf{k}} (c_{\mathbf{k}\uparrow}^\dagger c_{\mathbf{k}\uparrow} + c_{-\mathbf{k}\downarrow}^\dagger c_{-\mathbf{k}\downarrow})$. For the BCS wave function, we have for the average number of electrons $\langle \hat{n}_{\mathbf{k}_0\uparrow} \rangle \equiv \langle \Psi | \hat{n}_{\mathbf{k}_0\uparrow} | \Psi \rangle$ in state $\mathbf{k}_0\uparrow$

$$\langle \Psi | \hat{n}_{\mathbf{k}_0\uparrow} | \Psi \rangle = \langle 0 | \prod_{\mathbf{k}'} (u_{\mathbf{k}'}^* + v_{\mathbf{k}'}^* c_{-\mathbf{k}'\downarrow} c_{\mathbf{k}'\uparrow}) \, \hat{n}_{\mathbf{k}_0\uparrow} \prod_{\mathbf{k}} (u_{\mathbf{k}} + v_{\mathbf{k}} c_{\mathbf{k}\uparrow}^\dagger c_{-\mathbf{k}\downarrow}^\dagger) | 0 \rangle \tag{2.49}$$

All the operators in Eq. (2.49) involving wavevectors different from $\mathbf{k}_0$ commute with $\hat{n}_{\mathbf{k}_0\uparrow}$, and we can gather them as we have done in expression Eq. (2.47). But with the normalization Eq. (2.45), this expression is just equal to unity. Hence all the wavevectors different from $\mathbf{k}_0$ give factors equal to 1, and we may completely forget them. In this example, we see the extreme convenience of the factorized form of the BCS wave function that decouples the different wavevectors. The same simplification will appear again and again in all the calculations done in the following. We are just left with the contributions containing $\mathbf{k}_0$

$$\langle \Psi | \hat{n}_{\mathbf{k}_0\uparrow} | \Psi \rangle = \langle 0 | (u_{\mathbf{k}_0}^* + v_{\mathbf{k}_0}^* c_{-\mathbf{k}_0\downarrow} c_{\mathbf{k}_0\uparrow}) \, c_{\mathbf{k}_0\uparrow}^\dagger c_{\mathbf{k}_0\uparrow} (u_{\mathbf{k}_0} + v_{\mathbf{k}_0} c_{\mathbf{k}_0\uparrow}^\dagger c_{-\mathbf{k}_0\downarrow}^\dagger) | 0 \rangle \tag{2.50}$$

With $c_{\mathbf{k}_0\uparrow} | 0 \rangle = 0$ and $c_{\mathbf{k}_0\uparrow} c_{\mathbf{k}_0\uparrow}^\dagger c_{-\mathbf{k}_0\downarrow}^\dagger | 0 \rangle = c_{-\mathbf{k}_0\downarrow}^\dagger c_{\mathbf{k}_0\uparrow} c_{\mathbf{k}_0\uparrow}^\dagger | 0 \rangle = c_{-\mathbf{k}_0\downarrow}^\dagger | 0 \rangle$, together with $\langle 0 | c_{\mathbf{k}_0\uparrow}^\dagger = 0$, we find

$$\langle \Psi | \hat{n}_{\mathbf{k}_0\uparrow} | \Psi \rangle = |v_{\mathbf{k}_0}|^2 \langle 0 | c_{-\mathbf{k}_0\downarrow} c_{\mathbf{k}_0\uparrow} c_{\mathbf{k}_0\uparrow}^\dagger c_{-\mathbf{k}_0\downarrow}^\dagger | 0 \rangle = |v_{\mathbf{k}_0}|^2 \tag{2.51}$$

---

[8] Indeed, if $|\mathbf{k}\rangle = c_{\mathbf{k}}^\dagger | 0 \rangle$ is the state with one electron with wavevector $\mathbf{k}$, $\hat{n}_{\mathbf{k}} | \mathbf{k} \rangle = c_{\mathbf{k}}^\dagger c_{\mathbf{k}} | \mathbf{k} \rangle = c_{\mathbf{k}}^\dagger c_{\mathbf{k}} c_{\mathbf{k}}^\dagger | 0 \rangle = c_{\mathbf{k}}^\dagger (1 - c_{\mathbf{k}}^\dagger c_{\mathbf{k}}) | 0 \rangle = c_{\mathbf{k}}^\dagger | 0 \rangle = | \mathbf{k} \rangle$ from the anticommutation relation $c_{\mathbf{k}}^\dagger c_{\mathbf{k}} + c_{\mathbf{k}} c_{\mathbf{k}}^\dagger = 1$, and $c_{\mathbf{k}} | 0 \rangle = 0$. Hence $| \mathbf{k} \rangle$ is eigenvector of $\hat{n}_{\mathbf{k}}$ with eigenvalue 1, which is the number of electrons. On the other hand, $\hat{n}_{\mathbf{k}} | 0 \rangle = c_{\mathbf{k}}^\dagger c_{\mathbf{k}} | 0 \rangle = 0$ and the eigenvalue is zero when there is no electron.

One finds naturally the same result for $\langle \hat{n}_{-\mathbf{k}_0 \downarrow} \rangle$, so that the average number of electrons in subspace $(\mathbf{k}_{0\uparrow}, -\mathbf{k}_{0\downarrow})$ is $\langle \hat{n}_{\mathbf{k}_0} \rangle = \langle \hat{n}_{\mathbf{k}_0 \uparrow} \rangle + \langle \hat{n}_{-\mathbf{k}_0 \downarrow} \rangle = 2|v_{\mathbf{k}_0}|^2$. But this result is fairly obvious from the start: in subspace $(\mathbf{k}_{0\uparrow}, -\mathbf{k}_{0\downarrow})$ we have a probability $|v_{\mathbf{k}_0}|^2$ to have a pair of two electrons (and probability $|u_{\mathbf{k}_0}|^2$ to have no pair); hence the average number of electrons is $2|v_{\mathbf{k}_0}|^2$. Finally we have for the mean value of N,

$$\bar{N} = \langle \Psi | \hat{N} | \Psi \rangle = \sum_{\mathbf{k}} \langle \hat{n}_{\mathbf{k}} \rangle = \sum_{\mathbf{k}} 2|v_{\mathbf{k}}|^2 = \frac{\mathcal{V}}{(2\pi)^3} \int d\mathbf{k} \, 2|v_{\mathbf{k}}|^2 \tag{2.52}$$

which is as expected proportional to the sample volume $\mathcal{V}$, whatever $v_{\mathbf{k}}$ may be (which will be found in the next paragraph).

Now we want to investigate the departure $\delta N = N - \bar{N}$ of the electron number N from its average value $\bar{N}$, that is the fluctuations of the electron number in the BCS wave function. Since the average of the operator $\delta \hat{N} = \hat{N} - \bar{N}$ corresponding to this departure is zero, we consider its square and calculate the average value of the corresponding operator $(\delta \hat{N})^2$. This can also be written $\langle (\delta \hat{N})^2 \rangle = \langle (\hat{N} - \bar{N})^2 \rangle = \langle \hat{N}^2 - 2\hat{N}\bar{N} + \bar{N}^2 \rangle = \langle \hat{N}^2 \rangle - \bar{N}^2$, that is

$$\langle (\delta \hat{N})^2 \rangle = \langle \Big(\sum_{\mathbf{k}} \hat{n}_{\mathbf{k}} \Big)^2 \rangle - \Big(\langle \sum_{\mathbf{k}} \hat{n}_{\mathbf{k}} \rangle \Big)^2 = \sum_{\mathbf{k},\mathbf{k}'} \langle \hat{n}_{\mathbf{k}} \hat{n}_{\mathbf{k}'} \rangle - \sum_{\mathbf{k},\mathbf{k}'} \langle \hat{n}_{\mathbf{k}} \rangle \langle \hat{n}_{\mathbf{k}'} \rangle \tag{2.53}$$

When $\mathbf{k} \neq \mathbf{k}'$, we can proceed with the calculation of $\langle \hat{n}_{\mathbf{k}} \hat{n}_{\mathbf{k}'} \rangle$, as we did in Eq. (2.50). However, since the different wavevectors are decoupled in the BCS wave function, there is no correlation between the calculation for $\mathbf{k}$ and for $\mathbf{k}'$; each one goes as it did in Eq. (2.50). Hence (taking into account the up and down spin contributions) the result is merely

$$\langle \hat{n}_{\mathbf{k}} \hat{n}_{\mathbf{k}'} \rangle = 4|v_{\mathbf{k}}|^2 |v_{\mathbf{k}'}|^2 = \langle \hat{n}_{\mathbf{k}} \rangle \langle \hat{n}_{\mathbf{k}'} \rangle \tag{2.54}$$

and in Eq. (2.53) this term cancels exactly with the corresponding contribution from the last term. Consequently, only the terms with $\mathbf{k} = \mathbf{k}'$ survive in Eq. (2.53). We have to evaluate $\langle (\hat{n}_{\mathbf{k}})^2 \rangle$, which can be done by a direct calculation or more easily with the same argument as presented above for Eq. (2.52): in subspace $(\mathbf{k}, -\mathbf{k})$ there is a probability $|v_{\mathbf{k}}|^2$ to have a pair of electrons for which $(n_{\mathbf{k}})^2$ is equal to $2^2 = 4$. The case of zero pair gives naturally a contribution to $(n_{\mathbf{k}})^2$ equal to zero. Hence $\langle (\hat{n}_{\mathbf{k}})^2 \rangle = 4|v_{\mathbf{k}}|^2$. On the other hand, we have found $\langle \hat{n}_{\mathbf{k}} \rangle = 2|v_{\mathbf{k}}|^2$, which gives for Eq. (2.53) the result

$$\langle (\delta \hat{N})^2 \rangle = \sum_{\mathbf{k}} \Big[4|v_{\mathbf{k}}|^2 - (2|v_{\mathbf{k}}|^2)^2 \Big] = \sum_{\mathbf{k}} 4|u_{\mathbf{k}}|^2 |v_{\mathbf{k}}|^2 = \frac{\mathcal{V}}{(2\pi)^3} \int d\mathbf{k} \, 4|u_{\mathbf{k}}|^2 |v_{\mathbf{k}}|^2 \tag{2.55}$$

where we have used the normalization condition Eq. (2.45). The result is again proportional to the sample volume $\mathcal{V}$. Now if we take the ratio of the root mean square of the electron number fluctuation $\langle (\delta \hat{N})^2 \rangle^{1/2}$ to the average electron number $\bar{N}$, we obtain that it behaves as $1/\sqrt{\mathcal{V}}$. Or, since from Eq. (2.52) $\mathcal{V}$ and $\bar{N}$ are proportional, we find that compared to $\bar{N}$, the typical departure of the electron number from its average in the BCS wave function is of order $1/\sqrt{\bar{N}}$. This shows that indeed the BCS wave function is very sharply peaked around $\bar{N}$ in the limit of large $\bar{N}$.

Nevertheless what we have done above is to obtain the BCS wave function on physical grounds. Although it sounds quite reasonable that this wave function corresponds to the ground state of H_{BCS}, we have not demonstrated it. We will come back at the end of this chapter to the complicated problem of properly finding the ground state of H_{BCS} in the canonical ensemble. However, in the next paragraph, we will conclude our elaboration of the BCS wave function by finding the explicit expressions of $u_{\mathbf{k}}$ and $v_{\mathbf{k}}$, which we have left undetermined up to now. We will do it, following Bardeen, Cooper, and Schrieffer, by considering $|\Psi\rangle_{BCS}$ as a variational expression for the wave function, and find $u_{\mathbf{k}}$ and $v_{\mathbf{k}}$ by minimizing the energy corresponding to the reduced Hamiltonian H_{BCS} given by Eq. (2.24).

2.4 Determination of $u_{\mathbf{k}}$ and $v_{\mathbf{k}}$. The Gap Equation

2.4.1 Determination of $u_{\mathbf{k}}$ and $v_{\mathbf{k}}$

First of all, we do not exactly minimize the energy $E = \langle H_c \rangle + \langle H_{int} \rangle$ because the BCS wave function does not correspond to a fixed number N of electrons. If we were to minimize just this energy, the minimum energy would be obtained by putting zero electrons, since for each electron we put in the system, the dominant energy cost is the kinetic energy, which is positive. Actually, we also have to take into account the energy of the reservoir of electrons, and we want to minimize the total energy of the system and the reservoir. Just as in statistical physics (we are merely in the zero-temperature limit of statistical physics) this means that instead of E, we have to minimize $E - \mu N$, where μ is the chemical potential imposed by the electron reservoir. The term $-\mu N$ represents the energy variation of the reservoir when the electron number changes.

Since $N = \langle \hat{N} \rangle = \sum_{\mathbf{k}} \langle \hat{n}_{\mathbf{k}} \rangle$, whereas $\langle H_c \rangle = \sum_{\mathbf{k}} \epsilon_{\mathbf{k}} \langle \hat{n}_{\mathbf{k}} \rangle$, the kinetic energy $E_c = \langle H_c \rangle$ and $-\mu N$ combine conveniently into $E_c - \mu N = \sum_{\mathbf{k}} (\epsilon_{\mathbf{k}} - \mu)\langle \hat{n}_{\mathbf{k}} \rangle$. We define

$$\xi_{\mathbf{k}} = \epsilon_{\mathbf{k}} - \mu \tag{2.56}$$

which is just the kinetic energy of the electron measured from the chemical potential μ. Hence the only modification brought by the presence of the reservoir is the replacement of the kinetic energy $\epsilon_{\mathbf{k}}$ by $\xi_{\mathbf{k}}$. Since we have already found in the preceding paragraph that $\langle \hat{n}_{\mathbf{k}} \rangle = 2|v_{\mathbf{k}}|^2$, we obtain

$$E_c - \mu N = 2 \sum_{\mathbf{k}} \xi_{\mathbf{k}} |v_{\mathbf{k}}|^2 \tag{2.57}$$

Let us now consider the average of the interaction energy in the BCS wave function

$$E_{int} = -V \sum_{\mathbf{k},\mathbf{k}'} \langle c_{\mathbf{k}'}^{\dagger} c_{-\mathbf{k}'}^{\dagger} c_{-\mathbf{k}} c_{\mathbf{k}} \rangle \tag{2.58}$$

where again we have not written explicitly the spin variables. We may consider only the terms $\mathbf{k}' \neq \mathbf{k}$ in Eq. (2.58), since it is easily seen that the terms $\mathbf{k} = \mathbf{k}'$ give an unimportant shift in the chemical potential proportional to V, which becomes vanishing small in the limit of large volume $\mathcal{V}$. (We might as well have decided from the start that this term should not be present in the interaction term in Eq. (2.24).) Then for the calculation of the average in Eq. (2.58), the situation is quite similar to the one we had in the preceding paragraph for the calculation of $\langle \hat{n}_{\mathbf{k}} \hat{n}_{\mathbf{k}'} \rangle$ in Eq. (2.54): since the different wavevectors are decoupled in the BCS wave function, there is no correlation between the calculation for $\mathbf{k}$ and for $\mathbf{k}'$; each one can be performed independently. Hence

$$\langle c_{\mathbf{k}'}^{\dagger} c_{-\mathbf{k}'}^{\dagger} c_{-\mathbf{k}} c_{\mathbf{k}} \rangle = \langle c_{\mathbf{k}'}^{\dagger} c_{-\mathbf{k}'}^{\dagger} \rangle \langle c_{-\mathbf{k}} c_{\mathbf{k}} \rangle \tag{2.59}$$

If we had a wave function $|\Psi\rangle$ with a fixed number of electrons, we would have $\langle c_{-\mathbf{k}} c_{\mathbf{k}} \rangle \equiv \langle \Psi | c_{-\mathbf{k}} c_{\mathbf{k}} | \Psi \rangle = 0$ since after the annihilation of two electrons by the action of $c_{-\mathbf{k}} c_{\mathbf{k}}$ on $|\Psi\rangle$, we do not have the same number of electrons as in $|\Psi\rangle$ and the resulting state $c_{-\mathbf{k}} c_{\mathbf{k}} |\Psi\rangle$ is necessarily orthogonal to $|\Psi\rangle$. However, this is not true here since the BCS wave function does not have a fixed number of electrons. For this reason, averages such as $\langle c_{-\mathbf{k}} c_{\mathbf{k}} \rangle$ and $\langle c_{\mathbf{k}}^{\dagger} c_{-\mathbf{k}}^{\dagger} \rangle$ are called "anomalous" averages. Indeed, proceeding as in Eq. (2.50), we have

$$\langle \Psi | c_{-\mathbf{k}} c_{\mathbf{k}} | \Psi \rangle = \langle 0 | (u_{\mathbf{k}}^* + v_{\mathbf{k}}^* c_{-\mathbf{k}} c_{\mathbf{k}}) \, c_{-\mathbf{k}} c_{\mathbf{k}} (u_{\mathbf{k}} + v_{\mathbf{k}} c_{\mathbf{k}}^{\dagger} c_{-\mathbf{k}}^{\dagger}) | 0 \rangle \tag{2.60}$$

When we expand this expression, the only term which is nonzero is the $u_{\mathbf{k}}^* v_{\mathbf{k}}$ term since it multiplies $\langle 0 | c_{-\mathbf{k}} c_{\mathbf{k}} c_{\mathbf{k}}^{\dagger} c_{-\mathbf{k}}^{\dagger} | 0 \rangle = 1$, which corresponds to creating a pair in vacuum by $c_{\mathbf{k}}^{\dagger} c_{-\mathbf{k}}^{\dagger}$ and then annihilating it by $c_{-\mathbf{k}} c_{\mathbf{k}}$ and going back to vacuum. All the other terms are zero because they involve the annihilation of a pair in vacuum, which is not possible. Hence we find

$$\langle c_{-\mathbf{k}} c_{\mathbf{k}} \rangle = u_{\mathbf{k}}^* v_{\mathbf{k}} \tag{2.61}$$

This result is physically easy to understand: the amplitude $\langle c_{-\mathbf{k}} c_{\mathbf{k}} \rangle$ for the destruction of an electronic pair is proportional to the probability amplitude $v_{\mathbf{k}}$ that it is present in the initial state, and to the probability amplitude $u_{\mathbf{k}}^*$ that it is absent in the final state (since it has been destroyed). Similarly (or just by taking the complex conjugate of Eq. (2.61)), we have

$$\langle c_{\mathbf{k}}^{\dagger} c_{-\mathbf{k}}^{\dagger} \rangle = u_{\mathbf{k}} v_{\mathbf{k}}^* \tag{2.62}$$

Collecting the results Eq. (2.57), Eq. (2.59), Eq. (2.61), and Eq. (2.62), we obtain

$$E - \mu N = 2 \sum_{\mathbf{k}} \xi_{\mathbf{k}} |v_{\mathbf{k}}|^2 - V \sum_{\mathbf{k}, \mathbf{k}'} u_{\mathbf{k}'} v_{\mathbf{k}'}^* u_{\mathbf{k}}^* v_{\mathbf{k}} \tag{2.63}$$

in which $u_{\mathbf{k}}$ and $v_{\mathbf{k}}$ are linked by the normalization condition Eq. (2.45). Up to now we have made no further restriction on $u_{\mathbf{k}}$. However, in order to simplify the calculation that follows, we may clearly assume, without loss of generality, that the appropriate coefficient $u_{\mathbf{k}}$ we introduce to ensure normalization is real positive, since any possible phase factor in $u_{\mathbf{k}}$ disappears in the condition Eq. (2.45). Hence the only variables we have at hand to minimize Eq. (2.63) is the set $\{v_{\mathbf{k}}\}$, with $u_{\mathbf{k}} = (1 - |v_{\mathbf{k}}|^2)^{1/2}$. Since $v_{\mathbf{k}}$ is a priori a complex number, it is natural to take as independent variables its real $\mathrm{Re}\, v_{\mathbf{k}} = (1/2)[v_{\mathbf{k}} + v_{\mathbf{k}}^*]$ and imaginary $\mathrm{Im}\, v_{\mathbf{k}} = (1/2i)[v_{\mathbf{k}} - v_{\mathbf{k}}^*]$ parts. However, it is much more convenient to take the independent linear combinations $\mathrm{Re}\, v_{\mathbf{k}} + i\, \mathrm{Im}\, v_{\mathbf{k}} = v_{\mathbf{k}}$ and $\mathrm{Re}\, v_{\mathbf{k}} - i\, \mathrm{Im}\, v_{\mathbf{k}} = v_{\mathbf{k}}^*$ as independent variables. Accordingly, to minimize $E - \mu N$, we will write that its first derivative with respect to v_k and v_k^* is zero. Or equivalently we will write the first variation of $E - \mu N$ and write that the coefficients of the variations δv_k and δv_k^* are zero.

Differentiating Eq. (2.63), we have

$$\delta(E - \mu N) = 2 \sum_k \xi_k (v_k \delta v_k^* + v_k^* \delta v_k) - V \sum_{k,k'} u_{k'} v_{k'}^* \delta(u_k v_k) + u_k v_k \delta(u_{k'} v_{k'}^*) \qquad (2.64)$$

where, from the normalization condition $u_k^2 + |v_k|^2 = 1$, the variation δu_k is given by

$$\delta u_k = -\frac{1}{2u_k}(v_k \delta v_k^* + v_k^* \delta v_k) \qquad (2.65)$$

We see appearing in Eq. (2.64) the quantity

$$\Delta = V \sum_k^{\omega_D} u_k v_k \qquad (2.66)$$

which we call the gap, leaving it to the next chapter to justify this name. We recall that in Eq. (2.66), just as in Eq. (2.64), the summation is restricted to the range of energy $|\xi_k| < \hbar\omega_D$ where the attractive interaction is assumed to be effective. This is explicitly indicated by the superscript ω_D above the sum symbol.

It is always possible to manage to have Δ real and positive. Indeed, if when inserting into Eq. (2.66) the values of u_k and v_k at hand, we find a complex result $\Delta = |\Delta| e^{i\varphi}$, then we may just change the pair wave function Φ_k by multiplying it by the fixed phase factor $e^{-i\varphi}$. This is always possible since, in quantum mechanics, we may multiply a wave function $\Phi(\mathbf{r})$ (and accordingly its Fourier transform Φ_k) by a phase factor (independent of $\mathbf{r}$ or $\mathbf{k}$) without changing the physics. (This is a particular case of gauge invariance.) Since u_k is real, this multiplication results in having all the $v_k = u_k \Phi_k$ also multiplied by the same phase factor $e^{-i\varphi}$. Hence we have $V \sum_k u_k v_k \rightarrow e^{-i\varphi} V \sum_k u_k v_k$, that is $\Delta = |\Delta| e^{i\varphi} \rightarrow e^{-i\varphi} |\Delta| e^{i\varphi} = |\Delta|$, which is real positive.

Introducing this real gap, Eq. (2.64) becomes

$$\delta(E - \mu N) = 2 \sum_k \xi_k (v_k \delta v_k^* + v_k^* \delta v_k) - \Delta \sum_k \delta(u_k v_k) + \delta(u_k v_k^*) \qquad (2.67)$$

Writing $\delta(u_k v_k) = u_k \delta v_k + v_k \delta u_k$, and expressing δu_k through Eq. (2.65), we can explicitly obtain the factors of δv_k and δv_k^* in Eq. (2.67), which are the first derivatives of $E - \mu N$ with respect to v_k and v_k^*. Writing that the factor of δv_k^* is zero gives

$$2\xi_k v_k - \Delta \left(u_k - v_k \frac{v_k + v_k^*}{2u_k} \right) = 0 \qquad (2.68)$$

and writing that the factor of δv_k is zero gives the complex conjugate of this equation. Now, since $v_k + v_k^*$ is real, Eq. (2.68) implies that v_k itself is real[9] since all the terms in this equation are real. This is naturally linked to our choice of a real Δ. This allows one to rewrite Eq. (2.68) as

$$u_k^2 - v_k^2 = \frac{2\xi_k u_k v_k}{\Delta} \qquad (2.69)$$

[9] Explicitly Eq. (2.68) gives $v_k = 2\Delta u_k^2/(4\xi_k u_k + \Delta(v_k + v_k^*))$.

Squaring this equation and subtracting it from the normalization condition squared, we find

$$(u_{\mathbf{k}}^2 + v_{\mathbf{k}}^2)^2 - (u_{\mathbf{k}}^2 - v_{\mathbf{k}}^2)^2 = 4u_{\mathbf{k}}^2 v_{\mathbf{k}}^2 = 1 - \frac{4\xi_{\mathbf{k}}^2 u_{\mathbf{k}}^2 v_{\mathbf{k}}^2}{\Delta^2} \tag{2.70}$$

Introducing the notation

$$E_{\mathbf{k}} = \sqrt{\xi_{\mathbf{k}}^2 + \Delta^2} \tag{2.71}$$

whose physical meaning will appear in the next chapter, this can be rewritten as

$$u_{\mathbf{k}}^2 v_{\mathbf{k}}^2 = \frac{\Delta^2}{4E_{\mathbf{k}}^2} \tag{2.72}$$

which implies $u_{\mathbf{k}} v_{\mathbf{k}} = \pm \Delta/2E_{\mathbf{k}}$. Carrying this result into Eq. (2.69) gives us an explicit result for $u_{\mathbf{k}}^2 - v_{\mathbf{k}}^2$. Since we know $u_{\mathbf{k}}^2 + v_{\mathbf{k}}^2 = 1$ from normalization, this gives the explicit expression for $u_{\mathbf{k}}$ (which is positive)

$$u_{\mathbf{k}} = \sqrt{\frac{1}{2}\left(1 \pm \frac{\xi_{\mathbf{k}}}{E_{\mathbf{k}}}\right)} \tag{2.73}$$

and, from the expression for $u_{\mathbf{k}} v_{\mathbf{k}}$, the corresponding result for $v_{\mathbf{k}}$

$$v_{\mathbf{k}} = \pm \sqrt{\frac{1}{2}\left(1 \mp \frac{\xi_{\mathbf{k}}}{E_{\mathbf{k}}}\right)} \tag{2.74}$$

The proper sign in Eqs. (2.73) and (2.74) is found on physical grounds. If we take the case of $\mathbf{k}$ states with very large kinetic energy $\xi_{\mathbf{k}} \gg \Delta$, so that $E_{\mathbf{k}} \simeq \xi_{\mathbf{k}}$, we find that the probability $v_{\mathbf{k}}^2$ to have a pair is zero with the upper sign while it is 1 with the lower sign. Clearly the kinetic energy, and consequently the total energy, is extremely large in this last case, and it cannot correspond to the ground state. Hence the proper physical choice is the upper sign. This does not mean that there is anything wrong with the lower sign. Rather the conditions we have written imply that $E - \mu N$ is extremal but not necessarily minimal. They can also lead to a maximal energy, and this is the physical meaning of the solution with the lower sign. In conclusion, we obtain for the coefficients coming in the expression Eq. (2.44) of the BCS wave function the results

$$u_{\mathbf{k}} = \sqrt{\frac{1}{2}\left(1 + \frac{\xi_{\mathbf{k}}}{E_{\mathbf{k}}}\right)} \qquad v_{\mathbf{k}} = \sqrt{\frac{1}{2}\left(1 - \frac{\xi_{\mathbf{k}}}{E_{\mathbf{k}}}\right)} \tag{2.75}$$

Let us now consider the physical implications of Eq. (2.75) for the electronic distribution. We have already found that for a specific spin, say $\uparrow$ (the result is the same for $\downarrow$ spins), the average number of electrons is given by $n_{\mathbf{k}\uparrow} = \langle \hat{n}_{\mathbf{k}\uparrow} \rangle = v_{\mathbf{k}}^2$, that is from Eq. (2.75):

$$n_{\mathbf{k}\uparrow} = n_{\mathbf{k}\downarrow} = \frac{1}{2}\left(1 - \frac{\xi_{\mathbf{k}}}{\sqrt{\xi_{\mathbf{k}}^2 + \Delta^2}}\right) \tag{2.76}$$

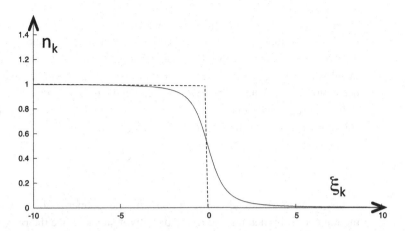

Fig. 2.6   Average number of electrons $n_\mathbf{k}$ in state $\mathbf{k}$ from Eq. (2.76), as a function of its kinetic energy $\xi_\mathbf{k} = \epsilon_\mathbf{k} - \mu$ measured from the chemical potential. The unit on the x-axis is the gap Δ. The dashed line shows the corresponding steplike Fermi–Dirac distribution for non-interacting electrons.

This distribution depends on the single energy scale Δ and is represented in Fig. (2.6). We see that it goes smoothly, over a typical energy of order Δ, from $n_{\mathbf{k}\uparrow} = 1$ at energy much below the chemical potential to $n_{\mathbf{k}\uparrow} = 0$ at energy much above μ, in contrast with the Fermi–Dirac distribution, valid for non-interacting electrons, which goes discontinuously from 1 to 0 at the chemical potential. Naturally these limiting values much above or below μ are physically expected since any departure would entail a too high kinetic energy cost. This modification with respect to the Fermi–Dirac distribution means that some electrons have been promoted from states with energy below the chemical potential to states with energy above it. Naturally there is a positive kinetic energy cost due to this transfer, but it is necessary to take advantage of the attractive interaction by forming Cooper pairs, and reduce in this way the total energy, as we will see explicitly below. This is not possible with the Fermi–Dirac distribution, which can actually be considered as a particular case of Eq. (2.44), with $v_\mathbf{k} = 1$, $u_\mathbf{k} = 0$ for $\xi_\mathbf{k} < 0$, and $v_\mathbf{k} = 0$, $u_\mathbf{k} = 1$ for $\xi_\mathbf{k} > 0$, and which is also the limiting case of Eq. (2.75) for $\Delta \to 0$. Indeed, in this case, one has $u_\mathbf{k} v_\mathbf{k} = 0$ for any $\mathbf{k}$, so that the interaction energy $E_{int} = 0$ from Eqs. (2.58), (2.59), and (2.61).

2.4.2 The Gap Equation

We have not yet quite finished with finding explicitly the coefficients $u_\mathbf{k}$ and $v_\mathbf{k}$ entering the BCS wave function. Indeed Eq. (2.75) depends on the gap Δ, which is defined by Eq. (2.66), but $u_\mathbf{k}$ and $v_\mathbf{k}$ which appear in the r.h.s. of this equation depend also on Δ. Hence Eq. (2.66) is actually an implicit equation for Δ, which is called the gap equation. It can be solved easily. Carrying into Eq. (2.66) the results Eq. (2.75) for $u_\mathbf{k}$ and $v_\mathbf{k}$, or rather the expression $u_\mathbf{k} v_\mathbf{k} = \Delta/2E_\mathbf{k}$ found above in their derivation, we obtain

$$\Delta = V \sum_\mathbf{k}^{\omega_D} \frac{\Delta}{2E_\mathbf{k}} \tag{2.77}$$

We may simplify both sides of the equation by Δ, but this assumes that Δ is different from zero. Actually, $\Delta = 0$ is a perfectly acceptable solution, but it leads as we have just seen above to the Fermi–Dirac distribution. This describes physically the state of the metal where the attractive interaction is ineffective, and nothing like superconductivity is occurring. This is what we have already called the normal state of the metal. Since we want to find a state corresponding to superconductivity, we discard this solution and look for a solution with $\Delta \neq 0$. Hence Eq. (2.77) becomes

$$1 = V \sum_{\mathbf{k}}^{\omega_D} \frac{1}{2E_{\mathbf{k}}} \tag{2.78}$$

Just as we have already done in handling the Cooper problem, we transform the summation over $\mathbf{k}$ by making use of the density of states. Since the range for summation over energy is restricted in the BCS model to be the small domain $|\xi_{\mathbf{k}}| < \hbar\omega_D$, the density of states is essentially constant over this range, and we may replace $N(\xi_{k'})$ with its value at the chemical potential $N(0) \equiv N_0$. Replacing $E_{\mathbf{k}}$ with its explicit expression Eq. (2.71), we obtain

$$1 = N_0 V \int_{-\hbar\omega_D}^{\hbar\omega_D} d\xi_{\mathbf{k}} \, \frac{1}{2\sqrt{\xi_{\mathbf{k}}^2 + \Delta^2}} \tag{2.79}$$

The integration is easily performed and leads to

$$1 = N_0 V \, \text{arcsinh}\left(\frac{\hbar\omega_D}{\Delta}\right) \tag{2.80}$$

which leads immediately to the explicit expression $\Delta = \hbar\omega_D / \sinh(1/N_0 V)$. Again, just as in the Cooper problem, to have a consistent framework where taking an instantaneous interaction between electrons is physically justified, we assume that we are in the weak coupling regime, where all energies characterizing the superconducting state, in the present case Δ, are small compared to the typical phonon energy $\hbar\omega_D$. This leads to the simpler expression

$$\Delta = 2\hbar\omega_D \exp\left(-\frac{1}{N_0 V}\right) \tag{2.81}$$

This result is remarkably similar to the expression of the binding energy Eq. (2.15) in the Cooper problem. From this formula, the weak coupling condition $\Delta \ll \hbar\omega_D$ implies $N_0 V \ll 1$.

The fact that Δ is very small implies that the chemical potential μ in the superconducting state is essentially unchanged with respect to its value in the normal state. Indeed the value of μ is obtained by writing that the number of electrons in the superconducting state is the same as in the normal state. However, in Fig. (2.6), the number of electrons that are promoted from below the chemical potential (this is the area, with $\xi_{\mathbf{k}} < 0$, between the full line and the horizontal dashed line) is

$$n_< = \sum_{\mathbf{k},\xi_\mathbf{k}<0} \left(1 - \frac{1}{2}(1 - \frac{\xi_\mathbf{k}}{\sqrt{\xi_\mathbf{k}^2 + \Delta^2}})\right) = N_0 \int_{-\infty}^0 d\xi \, \frac{1}{2}\left(1 + \frac{\xi}{\sqrt{\xi^2 + \Delta^2}}\right) \quad (2.82)$$

$$= N_0 \int_0^\infty d\xi \, \frac{1}{2}\left(1 - \frac{\xi}{\sqrt{\xi^2 + \Delta^2}}\right)$$

where we have approximated the density of states with the constant N_0 because the integral is rapidly convergent over a ξ interval of a few Δ. For the same reason, we have then extended the lower boundary to $-\infty$. The last step is obtained by the change of variables $\xi \to -\xi$. Similarly we have for the number of electrons above the chemical potential (this is the area, with $\xi_\mathbf{k} > 0$, between the x-axis and the full line)

$$n_> = \sum_{\mathbf{k},\xi_\mathbf{k}>0} \frac{1}{2}(1 - \frac{\xi_\mathbf{k}}{\sqrt{\xi_\mathbf{k}^2 + \Delta^2}}) = N_0 \int_0^\infty d\xi \, \frac{1}{2}\left(1 - \frac{\xi}{\sqrt{\xi^2 + \Delta^2}}\right) \quad (2.83)$$

Hence we have $n_< = n_>$. This implies that the number of electrons in the superconducting state is merely given by $2\sum_{\xi_\mathbf{k}<0} 1$ (the factor 2 accounts for the two possible spins). On the other hand, the number of electrons in the normal state is given by $2\sum_{\epsilon_\mathbf{k}<E_F} 1$. As the number of electrons in the superconducting state and in the normal state are equal, this shows that $\mu = E_F$ since $\xi_\mathbf{k} = \epsilon_\mathbf{k} - \mu$. Note that this result has been obtained because we have been able to approximate the density of states by a constant in Eqs. (2.82) and (2.83), using the fact that Δ is very small compared to μ. This would no longer hold for a larger Δ, and the chemical potential would shift upon going from the normal to the superconducting state, in much the same way as it does in the normal metal when the temperature is increased. Since we have found $\mu = E_F$, $\xi_\mathbf{k} = 0$ is equivalent to $k = k_F$, which we will use from now on.

2.4.3 Condensation Energy

We can now calculate the condensation energy, that is the lowering in energy upon going from the normal state to the superconducting state, given by the BCS wave function. The energy E_S in the superconducting state is given by Eq. (2.63) with $u_\mathbf{k}$ and $v_\mathbf{k}$ obtained from Eq. (2.75). On the other hand, we have noticed that the normal state is obtained from the BCS wave function by letting $\Delta \to 0$, so it is convenient to use for the normal state energy E_N the same formulae with $\Delta = 0$. Hence, since $\mu = E_F$, we have

$$E_S - E_N = (E_S - \mu N) - (E_N - E_F N) \quad (2.84)$$

$$= 2\sum_\mathbf{k} \xi_\mathbf{k} v_\mathbf{k}^2 - V\sum_{\mathbf{k},\mathbf{k}'} u_{\mathbf{k}'} v_{\mathbf{k}'} u_\mathbf{k} v_\mathbf{k} - 2\sum_{k<k_F} \xi_\mathbf{k}$$

where we have used the fact that $u_\mathbf{k}$ and $v_\mathbf{k}$ are real. Gathering the contributions for $k < k_F$ and $k > k_F$, and making use of the definition $\Delta = V\sum_{\mathbf{k}'} u_{\mathbf{k}'} v_{\mathbf{k}'}$ leads to

$$E_S - E_N = \sum_{k<k_F}\left[2\xi_\mathbf{k}(v_\mathbf{k}^2 - 1) - \Delta u_\mathbf{k} v_\mathbf{k}\right] + \sum_{k>k_F}\left[2\xi_\mathbf{k} v_\mathbf{k}^2 - \Delta u_\mathbf{k} v_\mathbf{k}\right] \quad (2.85)$$

Note that in Eq. (2.85) each of the separate contributions $\sum_{\mathbf{k}} u_{\mathbf{k}} v_{\mathbf{k}}$, $\sum_{k>k_F} 2\xi_{\mathbf{k}} v_{\mathbf{k}}^2$, and $\sum_{k<k_F} 2\xi_{\mathbf{k}}(v_{\mathbf{k}}^2 - 1)$ behaves as $\sum_{\mathbf{k}} 1/\xi_{\mathbf{k}}$ for large $\xi_{\mathbf{k}}$, which leads to a logarithmically divergent behavior. This is only avoided if we take into account the cut-off at $\hbar\omega_D$. However, these divergent behaviors cancel between the various terms if they are grouped as done in Eq. (2.85), as we see now.

Since $\xi_{\mathbf{k}}$ is anyhow bounded by $\hbar\omega_D$, we may as before replace the density of states with the constant N_0. Writing the explicit expressions for $u_{\mathbf{k}} v_{\mathbf{k}}$ and $v_{\mathbf{k}}^2$, we obtain

$$E_S - E_N = N_0 \int_{-\infty}^{0} d\xi \left[-\xi(1 + \frac{\xi}{E}) - \frac{\Delta^2}{2E} \right] + N_0 \int_{0}^{\infty} d\xi \left[\xi(1 - \frac{\xi}{E}) - \frac{\Delta^2}{2E} \right] \quad (2.86)$$

where $E = \sqrt{\xi^2 + \Delta^2}$, and we have again extended the boundaries to $\pm\infty$ since the integrals are convergent. The change of variables $\xi \rightarrow -\xi$ in the first term gives a result identical to the second term. The further change $\xi = x\Delta$ leads to

$$E_S - E_N = 2N_0\Delta^2 \int_{0}^{\infty} dx \left[x(1 - \frac{x}{\sqrt{x^2+1}}) - \frac{1}{2\sqrt{x^2+1}} \right] = N_0\Delta^2 [x^2 - x\sqrt{x^2+1}] \Big|_{0}^{\infty} \quad (2.87)$$

With $x^2 - x\sqrt{x^2+1} = -x/(x + \sqrt{x^2+1})$, we obtain finally

$$E_S - E_N = -\frac{1}{2} N_0 \Delta^2 \quad (2.88)$$

This result is physically easy to understand qualitatively. Each electron in the vicinity of the Fermi surface gets an energy lowering of order Δ, which is the single new characteristic energy introduced by the BCS wave function. Naturally this lowering is the net result of an increase of kinetic energy and a decrease due to the attractive interaction, as seen explicitly in Eq. (2.85) and Eq. (2.86). The electrons that are mostly involved are those in the vicinity of the Fermi surface with their $|\xi_{\mathbf{k}}|$ of order Δ. Hence, by definition of the density of states N_0, their number is of order $N_0\Delta$. Hence the total energy lowering is of order $\Delta(N_0\Delta) = N_0\Delta^2$, which is within a numerical factor the result for Eq. (2.88) we have just found.

It is also of interest to compare this condensation energy to the total energy of the electron gas, which is within a multiplicative factor the typical energy E_F multiplied by the electron number N. Since for dimensional reasons the density of states is of order the electron number N divided by the typical energy E_F, we find that the ratio between the condensation energy and the total energy of the Fermi sea is of order $|E_S - E_N|/NE_F \sim (N/E_F)\Delta^2/NE_F = (\Delta/E_F)^2$. We can evaluate Δ from Eq. (2.81), but we will also see in the next chapter that it is of order of the critical temperature of the superconductor, which is typically 10 K for a standard low-temperature superconductor. On the other hand, a typical Fermi energy converted on the temperature scale is rather of order of 10^4 K. Hence the condensation energy is typically of order 10^{-6} times the total energy of the electron gas. This is remarkably small, and it is fascinating that such a small change in energy is associated with such a huge difference in the physical properties between a superconductor and a normal metal.

Finally let us also remark that, just as for the Cooper problem, this condensation energy cannot be found by a perturbative calculation because of the non-analytic dependence $\sim \exp(-1/N_0 V)$ of Δ on V in Eq. (2.81).

2.4.4 Pair Wave Function

We now have all the information we need to obtain the pair wave function $\Phi(\mathbf{r})$ we started from in Eqs. (2.27) and (2.28). Unfortunately, the explicit result does not have a simple form, but in practice this is not important. First of all, the actual calculations performed within BCS theory involve the simple $u_{\mathbf{k}}$ and $v_{\mathbf{k}}$ given by Eq. (2.75), rather than $\Phi(\mathbf{r})$. Moreover, the essential physical feature of $\Phi(\mathbf{r})$ can be readily obtained by a very simple calculation.

Indeed $\Phi(\mathbf{r})$ is found by calculating the Fourier transform of $\Phi_{\mathbf{k}} = v_{\mathbf{k}}/u_{\mathbf{k}}$, from Eq. (2.43). The essential feature in this expression is the appearance, with respect to the normal state, of a new energy scale, namely Δ. Corresponding to this energy scale, there is a new wavevector scale δk which is appearing if we indicate on the x-axis in Fig. (2.6) the wavevector instead of the corresponding kinetic energy $\xi_{\mathbf{k}} = \epsilon_{\mathbf{k}} - E_F$. If, starting from E_F, we increase this energy by Δ, the corresponding increase δk in wavevector is given by $\Delta = \delta \xi_{\mathbf{k}} = \epsilon_{\mathbf{k_F}+\delta\mathbf{k}} - \epsilon_{\mathbf{k_F}} \simeq \hbar^2 k_F \delta k/m = \hbar v_F \delta k$, where $v_F = \hbar k_F/m$ is the Fermi velocity. This gives $\delta k = \Delta/\hbar v_F$. When we calculate the Fourier transform of $\Phi_{\mathbf{k}}$, a new length scale $1/\delta k = \hbar v_F/\Delta$ will appear corresponding to the wavevector scale δk. This new length scale is clearly the typical extension of the pair wave function, since we expect $\Phi(\mathbf{r})$ to have a limited range in space because it corresponds to a bound state. We may naturally guess that this microscopic scale is directly related to the coherence length which we have seen appearing in the Pippard theory in Chapter 1. This will be proved explicitly in Chapter 4 when we will deal with the response function within BCS theory. The standard precise definition of the coherence length ξ_0 coming in BCS theory is

$$\xi_0 = \frac{\hbar v_F}{\pi \Delta} \tag{2.89}$$

We can compare this coherence length to the typical distance d between two electrons in the metal. If the number of electrons per unit volume is n, the typical inter-electronic distance is $d \sim 1/n^{1/3}$. Since the total electronic density is related to the Fermi wavevector k_F by $n = k_F^3/3\pi^2$, this leads merely to $d \sim 1/k_F$. Hence

$$\frac{\xi_0}{d} \sim \frac{\hbar v_F k_F}{\pi \Delta} = \frac{2}{\pi} \frac{E_F}{\Delta} \tag{2.90}$$

We have already evaluated this ratio E_F/Δ in the preceding subsection and found it to be of order 10^3 for a standard low-temperature superconductor. This means that ξ_0 is very large compared to d and is typically of order 100 nm. Physically this implies that there is a huge overlap between all the Cooper pairs making up the superconductor. This is a very strange situation, and it is not easy to get used to this physical picture. On the other hand, we recall that the overall wave function is fully antisymmetrized since electrons are indistinguishable particles, so we do not have to face a situation where a definite spin-up electron is paired with a definite spin-down electron, which is too naive a picture.

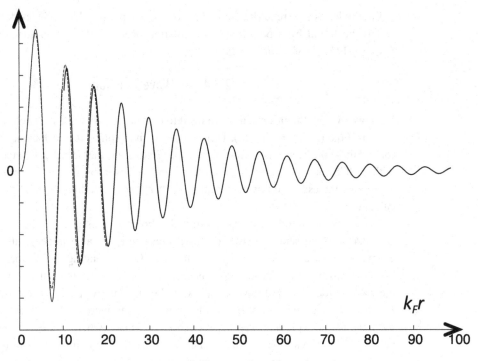

Fig. 2.7 Pair wave function $\Phi(r)$, in arbitrary units, as a function of the inter-electronic distance r, for $\Delta/E_F = 0.1$. The x-axis gives $\rho = k_F r$. For better readability, this is actually $r^3 \Phi(r)$, which is plotted. The thick full line is the asymptotic result for large ρ. The thin full line is the normal state result, which is valid for small ρ. The dashed line is the exact numerical calculation.

Let us now see more precisely how the above feature arises from the calculation of the pair wave function. From Eq. (2.75) we have

$$\Phi_{\mathbf{k}} = \frac{v_{\mathbf{k}}}{u_{\mathbf{k}}} = \left[\frac{1 - \frac{\xi_{\mathbf{k}}}{E_{\mathbf{k}}}}{1 + \frac{\xi_{\mathbf{k}}}{E_{\mathbf{k}}}}\right]^{1/2} = \left[\frac{(1 - \frac{\xi_{\mathbf{k}}}{E_{\mathbf{k}}})^2}{1 - (\frac{\xi_{\mathbf{k}}}{E_{\mathbf{k}}})^2}\right]^{1/2} = \frac{E_{\mathbf{k}} - \xi_{\mathbf{k}}}{\Delta} \tag{2.91}$$

and the pair wave function is obtained by

$$\Phi(\mathbf{r}) = \frac{1}{(2\pi)^3} \int d\mathbf{k} \, e^{i\mathbf{k}\cdot\mathbf{r}} \Phi_{\mathbf{k}} \tag{2.92}$$

Since the s-wave pair wave function $\Phi_{\mathbf{k}}$ depends only on $|\mathbf{k}|$, we can integrate the exponential in Eq. (2.92) over the angular variables of $\mathbf{k}$, that is its polar θ and azimuthal φ angles with respect to $\mathbf{r}$

$$\int_0^{2\pi} d\varphi \int_0^{\pi} \sin\theta \, d\theta \, e^{ikr\cos\theta} = 2\pi \left[-\frac{e^{ikr\cos\theta}}{ikr}\right]_0^{\pi} = 4\pi \frac{\sin(kr)}{kr} \tag{2.93}$$

and we are left with

$$\Phi(\mathbf{r}) = \frac{1}{2\pi^2 r} \int_0^{\infty} dk \, k \sin(kr) \, \Phi_{\mathbf{k}} = \frac{1}{4\pi^2 r} \, \mathrm{Im} \int_{-\infty}^{\infty} dk \, k \, e^{ikr} \, \Phi_{\mathbf{k}} \tag{2.94}$$

where in the last step we have used the fact that $k \sin(kr)\, \Phi_{\mathbf{k}} = k \operatorname{Im} e^{ikr}\, \Phi_{\mathbf{k}}$ is an even function of k.

A convenient way to calculate the integral in Eq. (2.94) is to extend the domain of variation of the variable k to the complex plane (see Fig. (2.9)). A first step is then to close the integration contour, which goes from $-\infty$ to $+\infty$, by adding the half-circle at infinity in the upper complex plane. The contribution to the integral from this half-circle is zero since e^{ikr} is zero for $\operatorname{Im} k$ going to $+\infty$. Then one deforms this closed integration contour continuously in the domain where the integrand $k\, e^{ikr}\, \Phi_{\mathbf{k}}$ is analytic. The result for the integral comes from the singularities of this integrand in the upper complex plane. Here the situation is quite simple since the only singularities are the two branch points coming from the square root function $E_{\mathbf{k}} = \sqrt{\xi_{\mathbf{k}}^2 + \Delta^2}$ (the $\xi_{\mathbf{k}}/\Delta$ term in Eq. (2.91) has no singularity, and so it disappears from this calculation). These two branch points in the upper complex plane, satisfying $E_{\mathbf{k}} = 0$, are located at $k_0 = \sqrt{k_F^2 + 2im\Delta/\hbar^2}$ and at $-k_0^*$. Hence the contour can be deformed into two contours encircling the cuts starting from the branch points and going toward $\operatorname{Im} k \to \infty$. Actually, the two branch points lead to identical contributions in the integral, so let us just consider k_0. Since e^{ikr} decreases exponentially to zero when $\operatorname{Im} k$ goes to $+\infty$, the dominant contribution comes from the vicinity of the branch point. Hence the integral behaves essentially as $e^{ik_0 r}$. Rewriting $k_0 = k_F \sqrt{1 + 2im\Delta/(\hbar k_F)^2} \simeq k_F(1 + i\Delta/2E_F) = k_F + i/(\pi \xi_0)$, we see that the dominant behavior of the integral is $e^{ik_F r} e^{-r/(\pi \xi_0)}$, leading to a result proportional to $\sin(k_F r)\, e^{-r/(\pi \xi_0)}$ for the pair wave function. The first factor $e^{ik_F r}$ corresponds to the fact that the Cooper pair is physically made from a superposition of plane waves states in the vicinity of the Fermi surface, with wave vectors around k_F. The exponential decay coming from the second factor $e^{-r/(\pi \xi_0)}$ is just the one we were looking for. It corresponds to the fact that the Cooper pair has a finite spatial extension of order of the coherence length ξ_0.

The calculation of $\Phi(\mathbf{r})$ can be completed precisely along the lines we have just described. We leave these mathematical details and the specific result to the "Further Reading" section at the end of this chapter. The resulting $r^3\, \Phi(r)$ is displayed in Fig. (2.7) for $\Delta/E_F = 0.1$, which is not such a realistic value but is chosen for figure clarity.

2.5 Further Analysis of the Ground State

2.5.1 The Mattis and Lieb Analysis

In this section, following Mattis and Lieb [20], we handle in a simplified situation the problem of the ground state of the BCS Hamiltonian, Eq. (2.24) in the canonical ensemble and show that for a large number of electrons one finds results in agreement with those obtained from the BCS wave function.

We first rewrite the Hamiltonian, Eq. (2.24) in terms of the operators $b_{\mathbf{k}}^{\dagger}$ and $b_{\mathbf{k}}$ creating and annihilating a pair, defined by $b_{\mathbf{k}}^{\dagger} = c_{\mathbf{k}\uparrow}^{\dagger} c_{-\mathbf{k}\downarrow}^{\dagger}$ and $b_{\mathbf{k}} = c_{-\mathbf{k}\downarrow} c_{\mathbf{k}\uparrow}$. This can be done only because the ground state we are looking for is clearly made of pair states (to take the best advantage of the attractive interaction) and there is no occurrence of singly occupied states where, for example, there is an electron in $\mathbf{k} \uparrow$ but no electron in $\mathbf{k}_{\downarrow}$. In this case, when a pair $(\mathbf{k} \uparrow, -\mathbf{k} \downarrow)$ is present, there is a kinetic energy $2\epsilon_{\mathbf{k}}$ corresponding to the presence of the two electrons. Accordingly, to obtain the kinetic energy, we just need to count the number of pairs, which is done by the operator $b_{\mathbf{k}}^{\dagger} b_{\mathbf{k}}$, and in this restricted situation the Hamiltonian, Eq. (2.24), can also be written as

$$H = \sum_{\mathbf{k}} 2\epsilon_{\mathbf{k}} b_{\mathbf{k}}^{\dagger} b_{\mathbf{k}} - V \sum_{\mathbf{k} \neq \mathbf{k}'} b_{\mathbf{k}'}^{\dagger} b_{\mathbf{k}} \qquad (2.95)$$

For simplicity, we assume that all the $\mathbf{k}$ states in all the sums correspond to states for which the attractive interaction is effective (i.e., they are within the $[-\omega_D, \omega_D]$ range of the BCS model). We have also taken for convenience $\mathbf{k} \neq \mathbf{k}'$ in the interaction term, since the $\mathbf{k} = \mathbf{k}'$ produces only a shift in the kinetic energy, negligible in the limit of large number of states.

It is interesting to note that the operators $b_{\mathbf{k}}$ are very similar to bosonic operators since they commute when $\mathbf{k} \neq \mathbf{k}'$: we have $[b_{\mathbf{k}}, b_{\mathbf{k}'}] = 0$ and $[b_{\mathbf{k}}, b_{\mathbf{k}'}^{\dagger}] = 0$. But they are not bosonic operators since they do not satisfy the standard bosonic commutation relation when the wavevectors are equal. This is because the underlying electrons making up the pairs are fermions and obey the Pauli principle. In particular, in contrast with bosons, it is not allowed to put two pairs in the same state $\mathbf{k}$, that is $(b_{\mathbf{k}}^{\dagger})^2 = 0$, which results directly from the corresponding relation for the fermions in the pair $(c_{\mathbf{k}}^{\dagger})^2 = 0$.

We will consider the situation where the number of possible pair states $\mathbf{k}$ is $2n$, and the number of electron pairs to be put in these states is n. The simplest case is the one where the $\mathbf{k}$ states have all the same kinetic energy, which can be taken as equal to zero for simplicity $\epsilon_{\mathbf{k}} = 0$. Since all the $\mathbf{k}$ states are on equal footing, with no energetic reason to choose one rather than another, it is physically clear (and it can be proved rigorously [20]) that the ground state is made up of the sum of all the possible terms obtained by chosing n pair states to be occupied among the $2n$ available (there are $\binom{2n}{n} = (2n)! / (n!)^2$ such terms), all the terms having the same coefficient. We note that this is just what we find from the BCS wave function if we extract from Eq. (2.44) the component corresponding to n pairs, since with $\xi_{\mathbf{k}} = 0$ (taking $\mu = 0$) we have from Eq. (2.75) $u_{\mathbf{k}} = v_{\mathbf{k}} = 1/\sqrt{2}$ so that all possible terms have the same coefficient.

If the interaction term in Eq. (2.95) acts on one term in the sum corresponding to this ground state, each of the n occupied pair state $\mathbf{k}$ can be scattered into one of the n empty pair state $\mathbf{k}'$. Hence, acting on one term of the ground state, the interaction produces n^2 terms. Repeating this reasoning for all the other terms of the ground state, we end up with the conclusion that the interaction acting on the ground state yields the ground state multiplied by $-Vn^2$. So the energy of the ground state is $-Vn^2$ in this simple case.

Let us now go to the "one step model," which is the simplest complication we can think of with respect to the above $\epsilon_{\mathbf{k}} = 0$ elementary case. It is actually much more difficult to solve, but it is quite interesting since it contains the essential ingredients of the solution to the full problem. In this model, n of the $\mathbf{k}$ pair states corresponds to kinetic energy 0, and the other n states correspond to kinetic energy ϵ. Compared to the above $\epsilon_{\mathbf{k}} = 0$ model, it is clear that the pair states belonging to the same part of the step (ϵ or 0) should be on equal footing, which implies again that all the terms obtained from each other by shuffling pair states within the same part of the step have the same coefficient. On the other hand, the coefficient of the various terms will depend on how many pairs are in the upper step ϵ. Let us call this number p and the corresponding coefficient in the (unnormalized) ground state f_p. To be concrete, consider explicitly the example $n = 2$, where we have two states (states 1 and 2) with kinetic energy 0, two states (states 3 and 4) with kinetic energy ϵ and we are looking for the ground state with two pairs. Its general form is

$$|\Psi\rangle_0 = \left[f_0 b_1^\dagger b_2^\dagger + f_1 (b_1^\dagger + b_2^\dagger)(b_3^\dagger + b_4^\dagger) + f_2 b_3^\dagger b_4^\dagger \right] |0\rangle \qquad (2.96)$$

where the $b_i^\dagger$ create pairs in the various states i. Writing the Schrödinger equation $H|\Psi\rangle_0 = E|\Psi\rangle_0$ with H given by Eq. (2.95), we find the three following linear equations for the three coefficients $f_0, f_1,$ and f_2:

$$\begin{cases} -E f_0 = 4V f_1 \\ (2\epsilon - E) f_1 = V (f_0 + 2f_1 + f_2) \\ (4\epsilon - E) f_2 = 4V f_1 \end{cases} \qquad (2.97)$$

We check that for $\epsilon = 0$ we recover the solution from the preceding paragraph with $f_0 = f_1 = f_2$ and $E = -4V$.

When we go to the general case with n pairs, n states with kinetic energy 0, and n states with kinetic energy ϵ, we find that this set of equations is generalized into

$$\left[2\epsilon p - E - 2Vp(n - p) \right] f_p = V(n - p)^2 f_{p+1} + Vp^2 f_{p-1} \qquad (2.98)$$

with $p = 0, 1, \cdots, n$. This is obtained from the coefficient in the Schrödinger equation of the terms with p pairs in the upper step ϵ and $n - p$ pairs in the lower step 0. The kinetic energy is clearly $2\epsilon p$ for these terms. Such a specific term can also arise from the action of the interaction in Eq. (2.95), which scatters a pair within the upper step. Since there are p pairs that can be scatters and $n - p$ empty states where the scattered pair can end up, the action of the interaction on a single term gives rise to $p(n - p)$ terms. However, such a scattering can also occur in the lower step, giving rise to $(n - p)p$ terms by the same reasoning. Hence the factor $2p(n - p)$ for f_p on the left hand side of Eq. (2.98).

Moreover, a specific term with p pairs can arise from a term with $p + 1$ pairs in the upper step (hence the factor f_{p+1}), where a pair has been scattered by the interaction from the upper step to the lower step. There are $(n - p)$ choices for the pair in the lower step that could originate from such a scatttering, and there are $(n - p)$ choices for the empty states in the upper step (in the final situation) from which this pair could originate. Hence the term $V(n - p)^2 f_{p+1}$ on the right hand side of Eq. (2.98). Finally, the specific term with p pairs can also arise from a term with $p - 1$ pairs in the upper step (hence the factor f_{p-1}), where a pair has been scattered by the interaction from the lower step to the upper step. There are p choices for the pair in the upper step that could originate from such a scatttering, and there are p choices for the empty states in the lower step (in the final situation) from which this pair could originate. Hence the term $Vp^2 f_{p-1}$ on the right hand side of Eq. (2.98).

This reasoning fails when $p = 0$ or $p = n$, where the upper step is either empty or full, but this is easily taken care of by deciding to take the formal convention $f_{-1} = f_{n+1} = 0$.

For these two limiting cases, one obtains explicitly

$$\begin{cases} -Ef_0 = Vn^2f_1 \\ (2\epsilon n - E)f_n = Vn^2f_{n-1} \end{cases} \tag{2.99}$$

Naturally we can check that, for $n = 2$, Eq. (2.98) reduces to Eq. (2.97). One also checks that, for $\epsilon = 0$, Eq. (2.98) admits the solution $f_0 = f_1 = \cdots = f_n$ with $E = -Vn^2$.

Eq. (2.98) is a set of homogeneous linear equations for f_p, which generally has no solution. In principle, the condition for the compatibility between these equations is obtained by writing that the corresponding determinant is zero, which provides an equation to be solved for the energy E. Naturally in the limit of very large n, this method is not a practical one. One can also consider Eq. (2.98) as a difference equation, with the two limiting cases Eq. (2.99) serving as boundary conditions. For example, starting with $f_0 = 1$, the first equation in Eq. (2.99) gives f_1; then the $p = 1$ equation in Eq. (2.98) gives f_2; and so on. One obtains finally E by requiring that the second equation in Eq. (2.99) is satisfied.

It is convenient to introduce $x = p/n$, which goes from 0 to 1 and which is, in this large n limit, a quasi-continuous variable. Since it is typical for f_p in these kinds of equations to vary quite rapidly, it is also convenient to introduce the ratio between successive values $P(x) \equiv f_p/f_{p-1}$. Dividing Eq. (2.98) by nf_p we find:

$$[2\epsilon x - W - 2x(1 - x)] = (1 - x)^2 P(x + \frac{1}{n}) + \frac{x^2}{P(x)} \tag{2.100}$$

where we have set $W = E/n$ and we have taken for simplicity $Vn = 1$ as an energy unit. We assume a regular behavior for $P(x)$. In the large n limit, we thus have $P(x+1/n) \simeq P(x)$, so Eq. (2.100) reduces to the simple quadratic equation for $P(x)$

$$(1 - x)^2 P^2(x) - [2\epsilon x - 2x(1 - x) - W]P(x) + x^2 = 0 \tag{2.101}$$

With these notations, the boundary conditions Eq. (2.99) become $P(0) = -W$ and $1/P(1) = 2\epsilon - W$.

It is of interest to consider again the situation $\epsilon = 0$ in terms of this equation. Since in this case all the f_p are equal, we have $P(x) = 1$. Indeed $P(x) = 1$ is solution of Eq. (2.101) provided $W = -1$, which corresponds to the result $E = -Vn^2$ we have found, taking $nV = 1$ into account. With this value, the boundary conditions $P(0) = -W = 1$ and $P(1) = -1/W = 1$ are correctly satisfied. However, the quadratic equation Eq. (2.101) has another solution, which is merely $P(x) = x^2/(1-x)^2$ since the product of the two roots has precisely this value. However, it does not satisfy the boundary conditions and accordingly it is not acceptable physically. The two solutions cross for $x = 1/2$, for which Eq. (2.101) has the double root $P(x) = 1$. This situation is displayed in Fig. (2.8)a).

Coming back to the general case, it is easy to find the behavior of the two solutions in the vicinity of $x = 0$ and $x = 1$. For $x \to 0$, Eq. (2.101) reduces to $P^2(x) + WP(x) + x^2 = 0$, which has the two solutions $P(x) \simeq -W$ and $P(x) \simeq -x^2/W$. Only the first one satisfies the boundary condition $P(0) = -W$. For $x \to 1$, it is simpler to write the equation for $1/P(x)$, which reads $1/P^2(x) + (W - 2\epsilon)/P(x) + (1 - x)^2 = 0$. It has the two solutions

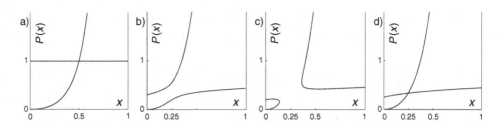

Fig. 2.8 Solution $P(x)$ of Eq. (2.101) as a function of x. (a) For $\epsilon = 0$ and $W = -1$. (b) For $\epsilon = 1$ and $W = -0.3$. (c) For $\epsilon = 1$ and $W = -0.2$. (d) For $\epsilon = 1$ and $W = -0.25$.

$1/P(x) \simeq (2\epsilon - W)$ and $1/P(x) \simeq (1 - x)^2/(2\epsilon - W)$. Again only the first one satisfies the boundary condition $1/P(1) = 2\epsilon - W$.

However, in the general case, the good solution that behaves as $P(x) \simeq -W$ for $x \to 0$ does not go continuously to the good solution, that behaves as $P(x) \simeq 1/(2\epsilon - W)$ for $x \to 1$. This is easily seen since (taking into account that the total energy W is negative), the good solution is above the other one for $x \to 0$ since $-W \gg -x^2/W$, while it is below the other one for $x \to 1$ since $1/(2\epsilon - W) \ll (2\epsilon - W)/(1 - x)^2$. In order to have the good solution for $x = 0$ going to the good solution for $x = 1$, the two solutions should cross somewhere for $0 < x < 1$, as they do for $\epsilon = 0$.

However, this is not the generic behavior of the two solutions of Eq. (2.101). In general, either the two solutions anticross, as shown in Fig. (2.8)b (which implies that the good solution goes to the bad one), or they become complex numbers, so that there is no physical solution as shown in Fig. (2.8)c. These two cases are not acceptable physically. The only acceptable situation is that the two solutions do cross, as they do for $\epsilon = 0$, as we have seen, which is a quite specific case and will lead to a condition providing the energy. Indeed a crossing implies that there is for some value of x for which the discriminant Δ of Eq. (2.101) is zero and in the vicinity of which it stays positive, in order to avoid complex solutions. From Eq. (2.101) we have

$$\Delta = [2\epsilon x - 2x(1 - x) - W]^2 - 4x^2(1 - x)^2 \tag{2.102}$$

which (taking into account $\epsilon > 0$ and $W < 0$) is positive or zero provided

$$2\epsilon x - W \geq 4x(1 - x) \tag{2.103}$$

This means that the parabola $4x^2 + (2\epsilon - 4)x - W$ should take positive values, except at its minimum where it goes to 0. The minimum is located at

$$x_m = \frac{1}{2}(1 - \frac{\epsilon}{2}) \tag{2.104}$$

and the minimum of the parabola is zero provided

$$W = -(1 - \frac{\epsilon}{2})^2 \tag{2.105}$$

Note that, since in this case Eq. (2.101) has double root, its value is merely given by the square root of their product $x^2/(1-x)^2$, so that

$$P(x_m) = \frac{x_m}{1-x_m} \tag{2.106}$$

This situation is displayed in Fig. (2.8)d) for $\epsilon = 1$ and $W = -1/4$. Going back to our earlier notations this gives the energy

$$E = -Vn^2(1 - \frac{\epsilon}{2nV})^2 \tag{2.107}$$

Since $x \geq 0$, the above solution does not work for $\epsilon > 2$ because it gives $x_m < 0$. For $\epsilon = 2$ one finds already that $x_m = 0$ and $W = 0$, which means physically that the step ϵ gets "too high." It is best energetically to put all the pairs in the lower step 0, and there is no energy gain from the interaction so that $W = 0$. For $\epsilon > 2$ one finds again $W = 0$, which is physically reasonable since the ϵ step is even higher.

Finally let us consider the pair distribution in the ground state wave function resulting from our solution. The probability to have p pairs in the upper step and $n - p$ pairs in the lower step is proportional to f_p^2 from our definition of f_p. However, we have also to take into account the number of specific states corresponding to a given value of p. There are $\binom{n}{p} = n!/[p!\,(n-p)!\,]$ ways to choose the p occupied pair states among the n pair states in the upper step. Similarly there are there are $\binom{n}{n-p} = n!/[(n-p)!\,p!\,]$ ways to choose the $n - p$ occupied pair states among the n pair states in the lower step. Hence, if we want for example to normalize the ground state wave function, the weight corresponding to p pairs in the upper step is proportional to $w(p) = (n!/[p!\,(n-p)!\,])^2 f_p^2$.

Taking first the case $\epsilon = 0$, where all the f_p are equal to 1, we have a distribution proportional to $(n!/[p!\,(n-p)!\,])^2$. It is well known that, for large n, such a distribution is very strongly peaked around $p = n/2$. Specifically we can use in this case the Stirling formula $n! = n^n e^{-n}\sqrt{2\pi n}$ (see footnote 7 in Section 2.3) to evaluate the various factorials. This leads to

$$\ln \frac{n!}{p!\,(n-p)!} = -nx \ln x - n(1-x)\ln(1-x) - \frac{1}{2}\ln[2\pi nx(1-x)] \tag{2.108}$$

where we have used $p = nx$. The first derivative with respect to x of the right-hand side of Eq. (2.108) is $-n\ln[x/(1-x)] - (1/2)[1/x - 1/(1-x)]$, which is zero when $x = x_m = 1/2$, as expected. The second derivative with respect to x evaluated at $x = 1/2$ is equal to $-4n$, for large n. This leads in the vicinity of $x = 1/2$ to the approximate expression

$$\ln \frac{n!}{p!\,(n-p)!} = n\ln 2 - \frac{1}{2}\ln\frac{\pi n}{2} - 2n(x - \frac{1}{2})^2 \tag{2.109}$$

which gives

$$\left[\frac{n!}{p!\,(n-p)!}\right]^2 = \frac{2^{2n+1}}{\pi n} e^{-4n(x-\frac{1}{2})^2} \tag{2.110}$$

This distribution has the shape of a Gaussian with a very sharp and narrow peak around its maximum at $x = 1/2$, its width of order $1/\sqrt{n}$ going to zero in the limit of large n.

Coming back to the general case, the weight corresponding to $p_m = nx_m$, with x_m given by Eq. (2.104), is proportional to $w(p_m) = (n! / [p_m! (n - p_m)!])^2 f_{p_m}^2$. If we compare $w(p_m - 1)$ to $w(p_m)$, we find

$$\frac{w(p_m - 1)}{w(p_m)} = \left[\frac{n!}{(p_m - 1)! (n - (p_m - 1))!} \frac{p_m! (n - p_m)! f_{p_m - 1}}{n!} \frac{f_{p_m - 1}}{f_{p_m}} \right]^2 \tag{2.111}$$

$$= \left[\frac{p_m}{n - p_m + 1} \frac{1}{P(x_m)} \right]^2 = 1$$

from the definition $P(x) \equiv f_p / f_{p-1}$. The last equality in Eq. (2.111) results from Eq. (2.106) and $p_m / (n - p_m + 1) \simeq x_m / (1 - x_m)$ for large n and p_m. Naturally we find in the same way $w(p_m + 1) / w(p_m) = 1$. These results imply that, in the vicinity of $p_m = nx_m$, the distribution $w(p)$ is flat, which means that $w(p)$ is at its maximum for $p = p_m$. This is the same situation as the one we have just found above for $\epsilon = 0$ at $x_m = 1/2$ in Eq. (2.110). Mostly because we have in $w(p)$ the same factor $[\binom{n}{p}]^2$ as in the case $\epsilon = 0$, we would similarly find that, around this maximum, the distribution has a Gaussian shape with a very narrow width. The qualitative situation is the same as for $\epsilon = 0$; the location of the maximum has just been shifted to its actual value x_m. In the limit $n \to \infty$, such a very sharp peak in the distribution means that the number of pairs in the upper step takes a well-defined value equal to p_m. This is in agreement with the expectation from the BCS wave function.

To complete the comparison, we have to check that this BCS wave function leads to the same number of pairs in the upper step, and to the same energy, as the exact results we have found above. Since it is easier to take a $\mu = 0$ chemical potential, we shift the kinetic energies of the two steps by $-\epsilon/2$ to have a symmetrical situation with the lower step at $-\epsilon/2$ and the upper one at $\epsilon/2$. This merely lowers the total energy by $-n\epsilon$. Since the n pair states of the lower step are equivalent, they have obviously the same coefficients u_0 and v_0 in the BCS wave function Eq. (2.44). Similarly the n pair states of the upper step have the same coefficients u_1 and v_1. The product over $\mathbf{k}$ in Eq. (2.44) becomes a product over the n states of the lower step and over the n states of the upper step. Hence the BCS wave function reads in the present case

$$|\Psi\rangle_{BCS} = \prod_{\mathbf{k}, \epsilon_\mathbf{k} = -\epsilon/2} \left(u_0 + v_0 b_\mathbf{k}^\dagger \right) \prod_{\mathbf{k}, \epsilon_\mathbf{k} = \epsilon/2} \left(u_1 + v_1 b_\mathbf{k}^\dagger \right) |0\rangle \tag{2.112}$$

Actually, because of the symmetrical situation, it is clear that the occupation of a pair state in the lower step with amplitude v_0 implies an equivalent lack of pair in the upper step, so $u_1 = v_0$ (assuming from the start real amplitudes). Similarly we have $u_0 = v_1$. These relations are similar to the relation $u(\xi_\mathbf{k}) = v(-\xi_\mathbf{k})$ in Eq. (2.75), and they could be derived in the same way by minimizing the energy. But it simplifies our calculation to take them into account from the start.

Hence, similarly to Eq. (2.57), the kinetic energy reads

$$E_c = -n\epsilon v_0^2 + n\epsilon v_1^2 = n\epsilon(u_0^2 - v_0^2) \tag{2.113}$$

and the interaction energy, similar to Eq. (2.63), is

$$E_{int} = -n^2 V(u_0 v_0 + u_1 v_1)(u_0 v_0 + u_1 v_1) = -4n^2 V u_0^2 v_0^2 \tag{2.114}$$

In contrast with the condition $\mathbf{k} \neq \mathbf{k}'$ in Eq. (2.95), this includes a term $\mathbf{k} = \mathbf{k}'$ in the interaction, but in the limit of large n, this is a negligible change. This leads to the total energy

$$E = n\epsilon(u_0^2 - v_0^2) - 4n^2 V u_0^2 v_0^2 \tag{2.115}$$

We may minimize this energy with respect to v_0, with the normalization condition $u_0^2 + v_0^2 = 1$ (actually, $u_0^2 - v_0^2$ is a more convenient variable than v_0), but this will merely reproduce the calculation below Eq. (2.63). Hence we rather make direct use of it. From Eq. (2.66) the gap Δ becomes

$$\Delta = Vn(u_0 v_0 + u_1 v_1) = 2Vn u_0 v_0 = 2u_0 v_0 \tag{2.116}$$

where we take into account that $Vn = 1$ is our energy unit. For the lower step, Eq. (2.69) becomes

$$u_0^2 - v_0^2 = -\frac{\epsilon u_0 v_0}{\Delta} = -\frac{\epsilon}{2} \tag{2.117}$$

The corresponding equation for the upper step would just confirm that our statement $u_1 = v_0$ and $u_0 = v_1$ is correct. Combined with the normalization condition, Eq. (2.117) leads to

$$u_0^2 = v_1^2 = \frac{1}{2}(1 - \frac{\epsilon}{2}) \tag{2.118}$$

Since nv_1^2 is, from Eq. (2.112), the number of pairs in the upper step, we see that the average number of pairs per pair state v_1^2 in the upper step agrees perfectly with our exact result Eq. (2.104) for x_m. When this result is carried in Eq. (2.115), taking again into account $Vn = 1$, we find

$$\frac{E}{n} = -\frac{\epsilon^2}{2} - (1 - \frac{\epsilon^2}{4}) = -1 - \frac{\epsilon^2}{4} \tag{2.119}$$

When we shift this energy by an amount of ϵ per pair, to compensate for the shift by $-\epsilon/2$ in the kinetic energies we have made at the beginning, we find

$$W = \frac{E}{n} + \epsilon = -1 - \frac{\epsilon^2}{4} + \epsilon = -(1 - \frac{\epsilon}{2})^2 \tag{2.120}$$

in agreement with Eq. (2.105).

In conclusion, let us mention that Mattis and Lieb [20] have been able to extend their approach to a general dispersion relation, by representing it as a multistep structure for the kinetic energy. The generalization is possible, with equations analogous to Eq. (2.98) and Eq. (2.100), because one can still analyze the effect of the interaction by considering its scattering action between two specific steps of the multisteps structure. In this way, one obtains results that are, in the large number limit, in agreement with those obtained from the BCS wave function.

2.5.2 Richardson Equations

Although the Mattis and Lieb approach has provided us with results on the ground state energy within the canonical ensemble, it is not a full solution of the Hamiltonian Eq. (2.95). Hence it is interesting to show that it is possible to write explicit equations which give not only the ground state energy, together with the corresponding wave functions, but also all the excited states energies. These equations have been found by Richardson [21], with the aim of studying atomic nuclei. Indeed pairing may also occur in nuclei and is an important part of the physical understanding in nuclear physics.

The general idea behind these equations is that although pairs of fermions do not behave exactly as bosons as we indicated in the preceding section, they are nevertheless very similar to bosons. Hence one can obtain Richardson equations by trying to solve for the eigenstates of Eq. (2.95) as if the operators $b_{\mathbf{k}}$ were bosonic operators, but taking care properly of the Pauli principle which forbids to put two pairs in the same state $\mathbf{k}$.

Let us start by solving the problem for a single pair. This is very easy because this problem is the same as the Cooper problem we considered at the beginning of this chapter, without the constraint due to the Fermi sea. For convenience, let us repeat briefly the derivation with our present notations. We look for an eigenstate with wave function $a_{\mathbf{k}}$ in momentum representation, that is a state

$$|\psi\rangle = \sum_{\mathbf{k}} a_{\mathbf{k}} b_{\mathbf{k}}^{\dagger} |0\rangle \tag{2.121}$$

The action of the kinetic energy term

$$H_c = \sum_{\mathbf{k}} 2\epsilon_{\mathbf{k}} b_{\mathbf{k}}^{\dagger} b_{\mathbf{k}} \tag{2.122}$$

is merely

$$H_c |\psi\rangle = \sum_{\mathbf{k},\mathbf{k}'} 2\epsilon_{\mathbf{k}} a_{\mathbf{k}'} b_{\mathbf{k}}^{\dagger} b_{\mathbf{k}} b_{\mathbf{k}'}^{\dagger} |0\rangle = \sum_{\mathbf{k}} 2\epsilon_{\mathbf{k}} a_{\mathbf{k}} b_{\mathbf{k}}^{\dagger} |0\rangle \tag{2.123}$$

because for $\mathbf{k} \neq \mathbf{k}'$, $b_{\mathbf{k}}$ and $b_{\mathbf{k}'}^{\dagger}$ commute, and $b_{\mathbf{k}}|0\rangle = 0$, while for $\mathbf{k} = \mathbf{k}'$, we use $b_{\mathbf{k}} b_{\mathbf{k}}^{\dagger} |0\rangle = |0\rangle$ (creating, then destroying, a pair in vacuum leads back to vacuum). The action of the interaction term

$$H_{int} = -V \sum_{\mathbf{k},\mathbf{k}'} b_{\mathbf{k}'}^{\dagger} b_{\mathbf{k}} \tag{2.124}$$

(here it is more convenient to keep the $\mathbf{k} = \mathbf{k}'$ term in the interaction) gives

$$H_{int}|\psi\rangle = -V \sum_{\mathbf{k},\mathbf{k}',\mathbf{k}''} a_{\mathbf{k}''} b_{\mathbf{k}'}^{\dagger} b_{\mathbf{k}} b_{\mathbf{k}''}^{\dagger} |0\rangle = -V \sum_{\mathbf{k},\mathbf{k}'} a_{\mathbf{k}} b_{\mathbf{k}'}^{\dagger} |0\rangle \tag{2.125}$$

since again the result is zero for $\mathbf{k} \neq \mathbf{k}''$ and $b_{\mathbf{k}} b_{\mathbf{k}}^{\dagger} |0\rangle = |0\rangle$. Gathering these results to write the Schrödinger equation $(H - E)|\psi\rangle = 0$ for $H = H_c + H_{int}$ leads to

$$(H - E)|\psi\rangle = \sum_{\mathbf{k}} \left[(2\epsilon_{\mathbf{k}} - E)a_{\mathbf{k}} - V \sum_{\mathbf{k}'} a_{\mathbf{k}'} \right] b_{\mathbf{k}}^{\dagger} |0\rangle = 0 \tag{2.126}$$

(after exchanging the dummy variables in Eq. (2.125)). This implies

$$a_{\mathbf{k}} = \frac{C}{2\epsilon_{\mathbf{k}} - E} \qquad C = V \sum_{\mathbf{k}'} a_{\mathbf{k}'} = C \sum_{\mathbf{k}'} \frac{V}{2\epsilon_{\mathbf{k}'} - E} \tag{2.127}$$

which gives for the energy E the equation

$$\frac{1}{V} = \sum_{\mathbf{k}} \frac{1}{2\epsilon_{\mathbf{k}} - E} \tag{2.128}$$

we have found in the Cooper problem. Hence, from Eq. (2.121) and Eq. (2.127),

$$B^{\dagger}(E) \equiv \sum_{\mathbf{k}} \frac{b_{\mathbf{k}}^{\dagger}}{2\epsilon_{\mathbf{k}} - E} \tag{2.129}$$

acting on vacuum creates a single pair eigenstate with energy E given by Eq. (2.128).

Now let us see what happens if, instead of creating a single boson with energy E, we try to create two bosons with different energies E_1 and E_2. That is, we want to check if it is possible that $B^{\dagger}(E_1)B^{\dagger}(E_2)|0\rangle$ may be an eigenstate with energy $E = E_1 + E_2$. This leads us to calculate

$$[H - E_1 - E_2]B^{\dagger}(E_1)B^{\dagger}(E_2)|0\rangle = \sum_{\mathbf{k},\mathbf{k}'}[H - E_1 - E_2]\frac{b_{\mathbf{k}}^{\dagger}b_{\mathbf{k}'}^{\dagger}}{(2\epsilon_{\mathbf{k}} - E_1)(2\epsilon_{\mathbf{k}'} - E_2)}|0\rangle \tag{2.130}$$

to see if it is possible to have it equal to zero by some proper choice of E_1 and E_2. Note that in the above expression the term with $\mathbf{k} = \mathbf{k}'$ is zero since Pauli exclusion implies $(b_{\mathbf{k}}^{\dagger})^2 = 0$, so only $\mathbf{k} \neq \mathbf{k}'$ needs to be considered. We may evaluate $H_c b_{\mathbf{k}}^{\dagger}b_{\mathbf{k}'}^{\dagger}|0\rangle$ as above

$$H_c b_{\mathbf{k}}^{\dagger}b_{\mathbf{k}'}^{\dagger}|0\rangle = \sum_{\mathbf{k}''} 2\epsilon_{\mathbf{k}''} b_{\mathbf{k}''}^{\dagger}b_{\mathbf{k}''}b_{\mathbf{k}}^{\dagger}b_{\mathbf{k}'}^{\dagger}|0\rangle = (2\epsilon_{\mathbf{k}} + 2\epsilon_{\mathbf{k}'})b_{\mathbf{k}}^{\dagger}b_{\mathbf{k}'}^{\dagger}|0\rangle \tag{2.131}$$

since we must have either $\mathbf{k}'' = \mathbf{k}$ or $\mathbf{k}'' = \mathbf{k}'$; otherwise the result is zero because $b_{\mathbf{k}''}|0\rangle = 0$.

Setting $\bar{b}^{\dagger} = \sum_{\mathbf{k}'} b_{\mathbf{k}'}^{\dagger}$, the calculation for $H_{int} = -V\bar{b}^{\dagger}\sum_{\mathbf{k}} b_{\mathbf{k}}$ is similar. One finds

$$H_{int} b_{\mathbf{k}}^{\dagger}b_{\mathbf{k}'}^{\dagger}|0\rangle = -V\bar{b}^{\dagger}\sum_{\mathbf{k}''} b_{\mathbf{k}''}b_{\mathbf{k}}^{\dagger}b_{\mathbf{k}'}^{\dagger}|0\rangle = -V\bar{b}^{\dagger}\left[b_{\mathbf{k}}^{\dagger} + b_{\mathbf{k}'}^{\dagger} - 2\delta_{\mathbf{k},\mathbf{k}'}b_{\mathbf{k}}^{\dagger}\right]|0\rangle \tag{2.132}$$

The essential difference with Eq. (2.131) is that we have taken care explicitly of the fact that the result has to be zero when $\mathbf{k} = \mathbf{k}'$, due to Pauli principle, by putting the additional term $2\delta_{\mathbf{k},\mathbf{k}'}b_{\mathbf{k}}^{\dagger}$ on the right hand side. In contrast, Eq. (2.131) is automatically satisfied when $\mathbf{k} = \mathbf{k}'$ since all the terms vanish.

When these results are carried into Eq. (2.130), we obtain

$$[H - E_1 - E_2]B^{\dagger}(E_1)B^{\dagger}(E_2)|0\rangle = \sum_{\mathbf{k},\mathbf{k}'} \frac{(2\epsilon_{\mathbf{k}} - E_1) + (2\epsilon_{\mathbf{k}'} - E_2)}{(2\epsilon_{\mathbf{k}} - E_1)(2\epsilon_{\mathbf{k}'} - E_2)}b_{\mathbf{k}}^{\dagger}b_{\mathbf{k}'}^{\dagger}|0\rangle \tag{2.133}$$

$$-V\bar{b}^{\dagger}\sum_{\mathbf{k},\mathbf{k}'} \frac{b_{\mathbf{k}}^{\dagger} + b_{\mathbf{k}'}^{\dagger}}{(2\epsilon_{\mathbf{k}} - E_1)(2\epsilon_{\mathbf{k}'} - E_2)}|0\rangle + 2V\bar{b}^{\dagger}\sum_{\mathbf{k}} \frac{b_{\mathbf{k}}^{\dagger}}{(2\epsilon_{\mathbf{k}} - E_1)(2\epsilon_{\mathbf{k}} - E_2)}|0\rangle$$

Taking into account the definition Eq. (2.129) of $B^\dagger(E)$, and the definition $\bar{b}^\dagger = \sum_{\mathbf{k}'} b_{\mathbf{k}'}^\dagger$, and writing in the last term

$$\frac{1}{(2\epsilon_\mathbf{k}-E_1)(2\epsilon_\mathbf{k}-E_2)} = \frac{1}{2\epsilon_\mathbf{k}-E_1}\frac{1}{E_1-E_2} + \frac{1}{2\epsilon_\mathbf{k}-E_2}\frac{1}{E_2-E_1} \tag{2.134}$$

one can rewrite Eq. (2.133) as

$$[H-E_1-E_2]B^\dagger(E_1)B^\dagger(E_2)|0\rangle = \bar{b}^\dagger B^\dagger(E_1)|0\rangle\left[1-V\sum_\mathbf{k}\frac{1}{2\epsilon_\mathbf{k}-E_2}+\frac{2V}{E_1-E_2}\right] \tag{2.135}$$

$$+\bar{b}^\dagger B^\dagger(E_2)|0\rangle\left[1-V\sum_\mathbf{k}\frac{1}{2\epsilon_\mathbf{k}-E_1}+\frac{2V}{E_2-E_1}\right]$$

Hence the Schrödinger equation is satisfied, provided the energies E_1 and E_2 verify the two equations

$$\frac{1}{V} = \sum_\mathbf{k}\frac{1}{2\epsilon_\mathbf{k}-E_1}+\frac{2}{E_1-E_2} = \sum_\mathbf{k}\frac{1}{2\epsilon_\mathbf{k}-E_2}+\frac{2}{E_2-E_1} \tag{2.136}$$

It is interesting to note that except for the last term on the right hand side, these equations are the same as Eq. (2.128) for each boson. The last term stems directly from Pauli exclusion, coming directly from the last term in Eq. (2.132). This last term excludes explicitly to take equal values $E_1 = E_2$ for the boson energies.

This derivation can be extended without basic modifications to the case of n bosons. The essential step is to generalize Eq. (2.132). One has to write

$$b_\mathbf{q}\prod_{p=1}^n b_{\mathbf{k}_p}^\dagger|0\rangle = \left(\sum_{p'=1}^n \delta_{\mathbf{q},\mathbf{k}_{p'}}\prod_{p=1,\neq p'}^n b_{\mathbf{k}_p}^\dagger - \sum_{p',p'',p'\neq p''}\delta_{\mathbf{k}_{p'},\mathbf{k}_{p''}}\delta_{\mathbf{q},\mathbf{k}_{p'}}\prod_{p=1,\neq p'}^n b_{\mathbf{k}_p}^\dagger\right)|0\rangle \tag{2.137}$$

One checks that, just as the left-hand side, the right-hand side is zero whenever two wavevectors $\mathbf{k}_{p_1}$ and $\mathbf{k}_{p_2}$ (with $p_1 \neq p_2$) in the set $\mathbf{k}_p$ are equal. (In the last term, one can have $p' = p_1$ and $p'' = p_2$, or $p' = p_2$ and $p'' = p_1$, which gives the factor 2 in the last term of Eq. (2.132)). In this way, one obtains that $\prod_{p=1}^n B^\dagger(E_p)|0\rangle$ is an eigenstate of H with energy $E = \sum_{p=1}^n E_p$, provided that the set of n partial energies E_p satisfies the following set of n equations:

$$\frac{1}{V} = \sum_\mathbf{k}\frac{1}{2\epsilon_\mathbf{k}-E_i}+\sum_{j=1,\neq i}^n\frac{2}{E_i-E_j} \qquad i = 1,\cdots,n \tag{2.138}$$

It is clear that these equations provide all the eigenstates since in the case where there is no interaction $V \to 0$ the solutions are $E_p = 2\epsilon_{\mathbf{k}_p}$, where the $\mathbf{k}_p$ are any set of (different) plane waves, which is physically the expected general result.

Unfortunately, these nonlinear equations are quite hard to solve, not only analytically but also numerically. In particular, the contact with BCS theory in the limit of large number of pairs is quite complicated [22, 23]. Also, in contrast with simple expectations, some E_p may become complex, although the resulting energy E is naturally real. Another annoying feature is that when one goes in Eq. (2.138) from discrete summation over $\mathbf{k}$ to integration, a large number of solutions is lost. Hence these equations appear to be more suited to

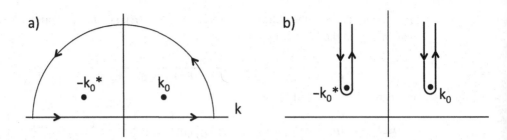

Fig. 2.9 Contour in the complex plane for the calculation of Eq. (2.139). (a) Original contour. (b) Deformed contour encircling the cuts starting from the branch points at k_0 and $-k_0^*$.

the study of small systems such as atomic nuclei, quantum dots, or microscopic superconducting systems than to the description of bulk superconductivity. In this respect, they are complementary to the BCS wave function, since the latter does not give the exact ground state of H when the number of pairs is not large. Nevertheless, even in these cases, the BCS wave function can always be used as a reasonable variational wave function.

2.6 Further Reading: Pair Wave Function, Mathematical Details

Starting from Eq. (2.94) with Eq. (2.91), we want to calculate

$$4\pi^2 r \Delta \Phi(r) = \mathrm{Im} \int_C dk\, k\, e^{ikr} \left[\sqrt{\xi_{\mathbf{k}}^2 + \Delta^2} - \xi_{\mathbf{k}}\right] \qquad (2.139)$$

where contour C, shown in Fig. (2.9)a), encircles the upper half complex plane. It is obtained by adding to the original integration interval $[-\infty, +\infty]$ the half-circle at infinity, which gives a zero contribution to the integral since the integrand is zero on this half-circle because of the factor e^{ikr}. The singularities of the integrand in this upper complex plane are at $k_0 = \sqrt{k_F^2 + 2im\Delta/\hbar^2}$ and $-k_0^*$, where $E_k = \sqrt{\xi_{\mathbf{k}}^2 + \Delta^2} = 0$. The contour C is then deformed to encircle these two branch points, as shown in Fig. (2.9)b). Since the integrand decreases exponentially when $\mathrm{Im}\,k$ goes to large values, the dominant contribution to the integral comes from the vicinity of k_0 and $-k_0^*$.

One can then easily see, by making the change of variable $k' = -k^*$, that the contour integral around $-k_0^*$ is just the complex conjugate of the opposite of the contour integral around k_0 (one has to take properly into account the sense in which one goes on the contours). Since in Eq. (2.139) we are only interested in the imaginary part, this implies that the contributions to $\Phi(r)$ from k_0 and $-k_0^*$ are equal, and we are only left to deal with the k_0 contribution.

It is then convenient to switch variables in order to have the cut along the real positive axis and the branch point at the origin. This is done by the change of variable $k^2 = k_F^2(1 + i\epsilon + iz)$, where we have set $\epsilon = \Delta/E_F$. In the resulting contour integral around the cut, the only difference between the parts above and below the cut is the change of sign in the determination of $\sqrt{z}$. Hence the result is just twice the integral below the cut. Keeping track of the proper determination for this square root in the change of variable, this leads to

$$\frac{2\pi^2}{k_F^3} \epsilon\rho \, \Phi(r) = -\text{Im} \int_0^\infty dz \, \sqrt{z(z+2\epsilon)} \, e^{i\rho(1+i\epsilon+iz)^{1/2}} \tag{2.140}$$

where we have set $\rho = k_F r$, which implies $\epsilon\rho/2 = r/(\pi\xi_0)$.

We are mostly interested in the regime where $r \sim \xi_0$, which implies $\rho \gg 1$. In this case, the integral in Eq. (2.140) converges already exponentially for small values of z. Since we have also $\epsilon \ll 1$, the square root in the exponential can be properly approximated as $(1 + i\epsilon + iz)^{1/2} \simeq 1 + i(\epsilon + z)/2$, which gives

$$\frac{2\pi^2}{k_F^3} \epsilon\rho \, \Phi(r) = -e^{-\epsilon\rho/2} \sin\rho \int_0^\infty dz \, \sqrt{z(z+2\epsilon)} \, e^{-\rho z/2} \tag{2.141}$$

Hence we recover the dominant dependence in $e^{-r/\pi\xi_0} \sin(k_F r)$ for $\Phi(r)$ indicated in the main text. The integral in Eq. (2.141) can actually be expressed [19] in terms of the Bessel function $K_1(x)$ as

$$\int_0^\infty dz \, \sqrt{z(z+2\epsilon)} \, e^{-\rho z/2} = \frac{2\epsilon}{\rho} e^{\epsilon\rho/2} K_1(\frac{\epsilon\rho}{2}) \tag{2.142}$$

where $K_1(x) \simeq 1/x$ for $x \to 0$, and $K_1(x) \simeq \sqrt{\pi/2x} \, e^{-x}$ for $x \to \infty$. This gives in this regime $\rho \gg 1$ the final result

$$\Phi(r) = -\frac{k_F^3}{\pi^2} \frac{\sin\rho}{\rho^2} K_1(\frac{\epsilon\rho}{2}) \tag{2.143}$$

which reduces to

$$\Phi(r) = -\frac{k_F^3}{(\epsilon\pi^3)^{1/2}} \frac{\sin\rho}{\rho^{5/2}} \exp(-\frac{\epsilon\rho}{2}) \tag{2.144}$$

when $\epsilon\rho \gg 1$ (i.e. $r \gg \xi_0$), and to

$$\Phi(r) = -\frac{2k_F^3}{\pi^2\epsilon} \frac{\sin\rho}{\rho^3} \tag{2.145}$$

when $\epsilon\rho \ll 1$ (i.e., $r \ll \xi_0$). This corresponds to the thick line in Fig. (2.7).

The above result is no longer valid when $\rho \lesssim 1$ since in this case the important domain for the integration in Eq. (2.140) is $z \gtrsim 1$. However, in this case since $\epsilon \ll 1$, we can neglect ϵ compared to z on the right hand side of Eq. (2.140). This amounts merely to set $\epsilon = 0$, which corresponds to go to the normal state. Note that it is natural to recover the normal state behavior at short distance r (corresponding to large wavevectors k) since the effect of pairing is rather felt for typical distances of order of the coherence length $r \sim \xi_0$. The result is then easily obtained by setting $\Delta = 0$ on the right hand side of Eq. (2.139) and merely integrating from $-k_F$ to k_F (but one can also integrate with the contour Fig. (2.9)b). One finds

$$\Phi(r) = \frac{2k_F^3}{\pi^2\epsilon} \frac{1}{\rho^5} (3\sin\rho - 3\rho\cos\rho - \rho^2\sin\rho) \tag{2.146}$$

This corresponds to the thin line in Fig. (2.7). In particular one finds for $r \to 0$ the limit $\Phi(0) = 2k_F^3/(15\pi^2\epsilon)$. Note that for large ρ this result matches, as it should, the limiting behavior Eq. (2.145) found above. To find the exact result in the regime $\rho \sim 1$, one has to numerically perform the integration in Eq. (2.140). This corresponds to the dashed line in Fig. (2.7).

3 Thermodynamics of the BCS Theory

3.1 Excited States

In the preceding chapter, we dealt exclusively with the ground state of the BCS Hamiltonian. Physically this is the only relevant state at zero temperature. In the present chapter, we consider the extension of BCS theory to nonzero temperature and the corresponding thermodynamics properties. Naturally this implies that we have first to consider the excited states of the BCS Hamiltonian.

We keep the framework we used for the ground state by considering states without a fixed number of electrons, corresponding to a grand canonical ensemble. This allows us again to decouple what happens in the various subspaces $(\mathbf{k}_\uparrow, -\mathbf{k}_\downarrow)$. This leads to a huge simplification since, within such a subspace, the Fock space of all the possible electronic states, whatever the number of electrons, is just made of four states:

$$|0\rangle \quad , \quad c_{\mathbf{k}\uparrow}^\dagger |0\rangle \quad , \quad c_{-\mathbf{k}\downarrow}^\dagger |0\rangle \quad , \quad c_{\mathbf{k}\uparrow}^\dagger c_{-\mathbf{k}\downarrow}^\dagger |0\rangle \tag{3.1}$$

where $|0\rangle$ is the vacuum for our subspace $(\mathbf{k}_\uparrow, -\mathbf{k}_\downarrow)$. The contribution from this $\mathbf{k}$ subspace to the BCS ground state Eq. (2.44) is just the linear combination

$$|\psi_0\rangle_{\mathbf{k}} \equiv u_{\mathbf{k}} |0\rangle + v_{\mathbf{k}} c_{\mathbf{k}\uparrow}^\dagger c_{-\mathbf{k}\downarrow}^\dagger |0\rangle \tag{3.2}$$

Hence we are left with identifying the three excited eigenstates, which must be orthogonal to this ground state.

It is easy to see that the single electron states $c_{\mathbf{k}\uparrow}^\dagger |0\rangle$ and $c_{-\mathbf{k}\downarrow}^\dagger |0\rangle$ are two of them. Indeed they are clearly orthogonal to the ground state since the electron number are different. Moreover, the interaction term in the Hamiltonian gives a null result when acting on them, since this interaction must act on states having pair of electrons to give a nonzero result. On the other hand, these two states are obviously eigenstates of the kinetic energy term of the Hamiltonian, with eigen-energy $\xi_{\mathbf{k}}$. This is the energy that must be paid to add electron $\mathbf{k} \uparrow$, for example, to vacuum $|0\rangle$.

However, rather than knowing the energy to be paid to create this electron from vacuum, we would rather like to know the energy necessary to go from the ground state to this excited state. This leads us to evaluate the energy cost to go from vacuum $|0\rangle$ to the ground state $|\psi_0\rangle_{\mathbf{k}}$. Since formally the vacuum $|0\rangle$ corresponds to the particular case where we would have $u_{\mathbf{k}} = 1$ and $v_{\mathbf{k}} = 0$ in the expression Eq. (3.2) of $|\psi_0\rangle_{\mathbf{k}}$, we have just to find what is the change in energy when $u_{\mathbf{k}}$ and $v_{\mathbf{k}}$ go from $(u_{\mathbf{k}} = 1, v_{\mathbf{k}} = 0)$ to their actual value, which we call $u_{\mathbf{k}_0}$ and $v_{\mathbf{k}_0}$ for the moment. The result is obtained from Eq. (2.63), taking into account that $u_\mathbf{k}$ and $v_\mathbf{k}$ are real and that the only change occurs for $\mathbf{k} = \mathbf{k}_0$. This gives

$$\delta(E - \mu N) = 2 \sum_\mathbf{k} \xi_\mathbf{k} \delta(v_\mathbf{k}^2) - 2V \sum_{\mathbf{k},\mathbf{k}'} u_{\mathbf{k}'} v_{\mathbf{k}'} \delta(u_\mathbf{k} v_\mathbf{k}) \tag{3.3}$$

$$= 2\xi_{\mathbf{k}_0} v_{\mathbf{k}_0}^2 - 2\Delta u_{\mathbf{k}_0} v_{\mathbf{k}_0} = \xi_{\mathbf{k}_0} - E_{\mathbf{k}_0}$$

where in the last step we have taken into account that $\delta(v_\mathbf{k}^2) = v_{\mathbf{k}_0}^2 \delta_{\mathbf{k},\mathbf{k}_0}$, $\delta(u_\mathbf{k} v_\mathbf{k}) = u_{\mathbf{k}_0} v_{\mathbf{k}_0} \delta_{\mathbf{k},\mathbf{k}_0}$ and $V \sum_{\mathbf{k}'} u_{\mathbf{k}'} v_{\mathbf{k}'} = \Delta$, and the final result is obtained by making use of the explicit expression Eq. (2.75) of $u_{\mathbf{k}_0}$ and $v_{\mathbf{k}_0}$.

Putting the two above results together, we conclude that the energy difference between the ground state $|\psi_0\rangle_\mathbf{k}$ and the excited state $c_{\mathbf{k}\uparrow}^\dagger |0\rangle$ or $c_{-\mathbf{k}\downarrow}^\dagger |0\rangle$ is

$$\xi_\mathbf{k} - (\xi_\mathbf{k} - E_\mathbf{k}) = E_\mathbf{k} \tag{3.4}$$

One may rightfully feel uneasy with the above calculation, since in calculating this excitation energy we compare the energy of states that do not have exactly the same number of particles. Indeed, even if we have taken care to extract the component of the ground state with a definite number N of electrons, by creating $c_{\mathbf{k}\uparrow}^\dagger |0\rangle$ we add definitely one electron, while by creating $|\psi_0\rangle_\mathbf{k}$ we add either zero or two electrons. One can go around this difficulty by enforcing the same number of electrons in the two states we compare. However, this will produce slight differences, of order $1/N$, in the values of $u_\mathbf{k}$ and $v_\mathbf{k}$ corresponding to these two states. Fortunately, since the actual values of $u_\mathbf{k}$ and $v_\mathbf{k}$ minimize the total energy, one can show that these slight differences have an effect on the excitation energy which goes to zero when N gets very large. We will not enter the lengthy details of this argumentation, but they allow us to justify properly our above result.

We are left with identifying the last excited state in the Fock space. This is easily done since all the eigenstates must be orthogonal. Hence this last state is just the linear combination $|\psi_1\rangle_\mathbf{k}$ analogous to Eq. (3.2), but with coefficients chosen in such a way that it is orthogonal to $|\psi_0\rangle_\mathbf{k}$. A properly normalized expression is

$$|\psi_1\rangle_\mathbf{k} = v_\mathbf{k} |0\rangle - u_\mathbf{k} c_{\mathbf{k}\uparrow}^\dagger c_{-\mathbf{k}\downarrow}^\dagger |0\rangle \tag{3.5}$$

We can calculate its energy with respect to vacuum in the same way as above. But it is even easier to notice that $|\psi_1\rangle_\mathbf{k}$ is obtained from $|\psi_0\rangle_\mathbf{k}$ by the exchange $u_{\mathbf{k}_0} \to v_{\mathbf{k}_0}$ and $v_{\mathbf{k}_0} \to -u_{\mathbf{k}_0}$. Hence Eq. (3.3) becomes

$$\delta(E - \mu N) = 2\xi_{\mathbf{k}_0} u_{\mathbf{k}_0}^2 + 2\Delta u_{\mathbf{k}_0} v_{\mathbf{k}_0} = \xi_{\mathbf{k}_0} + E_{\mathbf{k}_0} \tag{3.6}$$

It is worth noticing that in this case, the interaction contribution $2\Delta u_{\mathbf{k}_0} v_{\mathbf{k}_0}$ is positive rather than negative, as we found in Eq. (3.3) when dealing with the ground state and obtaining the contribution $-2\Delta u_{\mathbf{k}_0} v_{\mathbf{k}_0}$. Eq. (3.5) is the wrong choice of coefficients to take advantage of the interaction to lower the energy, and it leads naturally to an excited state. Hence this state may be called the "excited pair" state, to make contrast with the ground state, Eq. (3.2), where a pair is present with the appropriate coefficients. The two states $c_{\mathbf{k}\uparrow}^\dagger |0\rangle$ and $c_{-\mathbf{k}\downarrow}^\dagger |0\rangle$ are also often called "broken pair" states, since they have only lonely electrons.

When we compare the energy Eq. (3.6) necessary to create this excited pair state from vacuum to the corresponding expression for the ground state Eq. (3.3), we find that the

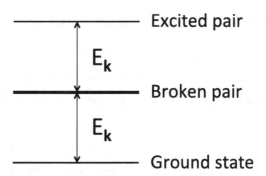

Fig. 3.1 Energy spectrum within the $(\mathbf{k}_\uparrow, -\mathbf{k}_\downarrow)$ subspace. The ground state is described by $|\psi_0\rangle_\mathbf{k}$ Eq. (3.2), while the excited pair state corresponds to $|\psi_1\rangle_\mathbf{k}$ Eq. (3.5). The two broken pair states are $c^\dagger_{\mathbf{k}\uparrow}|0\rangle$ and $c^\dagger_{-\mathbf{k}\downarrow}|0\rangle$.

energy difference between the ground state and the excited pair state is $2E_\mathbf{k}$. Hence we arrive at a quite simple structure for the energy spectrum of the excited states, which is represented in Fig. (3.1).

While it is a natural one, the above presentation of the excited states is nevertheless not the most convenient; nor does it display the full simplicity of the excitation spectrum. We first note that although when operating on vacuum, operators $c^\dagger_{\mathbf{k}\uparrow}$ and $c^\dagger_{-\mathbf{k}\downarrow}$ give the broken pair states, they obviously do not do so upon operating on the ground state. On the other hand, the vacuum does not play any physical role in the energy spectrum, and it is rather highly desirable to find operators $\gamma^\dagger_{\mathbf{k}\uparrow}$ creating the broken pair states while acting on the ground state

$$\gamma^\dagger_{\mathbf{k}\uparrow}|\psi_0\rangle_\mathbf{k} = c^\dagger_{\mathbf{k}\uparrow}|0\rangle \tag{3.7}$$

rather than on vacuum, since they will directly create the excited states from the ground state.

Starting from the ground state, one can obtain a broken pair state $\mathbf{k}\uparrow$ either by merely adding an electron $\mathbf{k}\uparrow$ to the vacuum contribution to the ground state, but also by removing an electron $-\mathbf{k}\downarrow$ from the pair contribution to the ground state. Explicitly

$$c^\dagger_{\mathbf{k}\uparrow}|\psi_0\rangle_\mathbf{k} = c^\dagger_{\mathbf{k}\uparrow}(u_\mathbf{k}|0\rangle + v_\mathbf{k}c^\dagger_{\mathbf{k}\uparrow}c^\dagger_{-\mathbf{k}\downarrow}|0\rangle) = u_\mathbf{k}c^\dagger_{\mathbf{k}\uparrow}|0\rangle \tag{3.8}$$
$$c_{-\mathbf{k}\downarrow}|\psi_0\rangle_\mathbf{k} = c_{-\mathbf{k}\downarrow}(u_\mathbf{k}|0\rangle + v_\mathbf{k}c^\dagger_{\mathbf{k}\uparrow}c^\dagger_{-\mathbf{k}\downarrow}|0\rangle) = -v_\mathbf{k}c^\dagger_{\mathbf{k}\uparrow}|0\rangle$$

Hence any linear combination $\gamma^\dagger_{\mathbf{k}\uparrow} = \lambda c^\dagger_{\mathbf{k}\uparrow} - \mu c_{-\mathbf{k}\downarrow}$ can satisfy our requirement Eq. (3.7).

On the other hand, when we try to destroy an electron in vacuum, where there is no electron, we obtain a null result $c_{\mathbf{k}\uparrow}|0\rangle = 0$. If we want $\gamma^\dagger_{\mathbf{k}\uparrow}$ to behave similarly like a fermionic operator, we would like that $\gamma_{\mathbf{k}\uparrow}|\psi_0\rangle_\mathbf{k} = 0$, for the corresponding annihilation operator $\gamma_{\mathbf{k}\uparrow}$ acting on the ground state, since there is no way to destroy an excited state which is not present. This gives for λ and μ the additional condition

$$\gamma_{\mathbf{k}\uparrow}|\psi_0\rangle_\mathbf{k} = (\lambda^* c_{\mathbf{k}\uparrow} - \mu^* c^\dagger_{-\mathbf{k}\downarrow})(u_\mathbf{k}|0\rangle + v_\mathbf{k}c^\dagger_{\mathbf{k}\uparrow}c^\dagger_{-\mathbf{k}\downarrow}|0\rangle) \tag{3.9}$$
$$= \lambda^* v_\mathbf{k}c^\dagger_{-\mathbf{k}\downarrow}|0\rangle - \mu^* u_\mathbf{k}c^\dagger_{-\mathbf{k}\downarrow}|0\rangle = 0$$

which implies $\lambda^* v_{\mathbf{k}} = \mu^* u_{\mathbf{k}}$. Together with the definition Eq. (3.7), which gives

$$\gamma_{\mathbf{k}\uparrow}^{\dagger} |\psi_0\rangle_{\mathbf{k}} = (\lambda c_{\mathbf{k}\uparrow}^{\dagger} - \mu c_{-\mathbf{k}\downarrow})(u_{\mathbf{k}}|0\rangle + v_{\mathbf{k}} c_{\mathbf{k}\uparrow}^{\dagger} c_{-\mathbf{k}\downarrow}^{\dagger} |0\rangle) \qquad (3.10)$$
$$= \lambda u_{\mathbf{k}} c_{\mathbf{k}\uparrow}^{\dagger} |0\rangle + \mu v_{\mathbf{k}} c_{\mathbf{k}\uparrow}^{\dagger} |0\rangle = c_{\mathbf{k}\uparrow}^{\dagger} |0\rangle$$

and implies $\lambda u_{\mathbf{k}} + \mu v_{\mathbf{k}} = 1$, this leads us to the solution $\lambda = u_{\mathbf{k}}$ and $\mu = v_{\mathbf{k}}$ and the explicit expressions

$$\gamma_{\mathbf{k}\uparrow}^{\dagger} = u_{\mathbf{k}} c_{\mathbf{k}\uparrow}^{\dagger} - v_{\mathbf{k}} c_{-\mathbf{k}\downarrow} \qquad (3.11)$$

$$\gamma_{\mathbf{k}\uparrow} = u_{\mathbf{k}} c_{\mathbf{k}\uparrow} - v_{\mathbf{k}} c_{-\mathbf{k}\downarrow}^{\dagger} \qquad (3.12)$$

Similarly we can exchange the roles of $\mathbf{k}\uparrow$ and $-\mathbf{k}\downarrow$. Since the two operators $c_{\mathbf{k}\uparrow}^{\dagger}$ and $c_{-\mathbf{k}\downarrow}^{\dagger}$ anticommute, exchanging them in Eq. (2.44) is equivalent to change $v_{\mathbf{k}}$ into $-v_{\mathbf{k}}$, and leave $u_{\mathbf{k}}$ unchanged. This leads to the operators

$$\gamma_{-\mathbf{k}\downarrow}^{\dagger} = u_{\mathbf{k}} c_{-\mathbf{k}\downarrow}^{\dagger} + v_{\mathbf{k}} c_{\mathbf{k}\uparrow} \qquad (3.13)$$

$$\gamma_{-\mathbf{k}\downarrow} = u_{\mathbf{k}} c_{-\mathbf{k}\downarrow} + v_{\mathbf{k}} c_{\mathbf{k}\uparrow}^{\dagger} \qquad (3.14)$$

The interpretation of these expressions for these γ operators is quite clear. For example, from the expression Eq. (3.11) of $\gamma_{\mathbf{k}\uparrow}^{\dagger}$, one can create a broken pair excitation $c_{\mathbf{k}\uparrow}^{\dagger}|0\rangle$ either by merely creating an electron $\mathbf{k}\uparrow$ in vacuum, with probability amplitude $u_{\mathbf{k}}$, or by annihilating an electron $-\mathbf{k}\downarrow$ in an electronic pair, with probability amplitude $-v_{\mathbf{k}}$. One could almost write directly these expressions by looking at the structure of the BCS wave function Eq. (2.44). It is worth stressing that the strange nature of these expressions, with their mixture of creation and annihilation operators, is not due to the physical nature of the excited states themselves, which are merely single electrons states. It is due to the nature of the BCS ground state, with its coherent superposition of pair state and vacuum state.

It is possible to write the relation between the γ's and the c's operators by introducing a two-by-two matrix $U_{\mathbf{k}}$. Eqs. (3.12) and (3.13) are equivalent to

$$\begin{pmatrix} \gamma_{\mathbf{k}\uparrow} \\ \gamma_{-\mathbf{k}\downarrow}^{\dagger} \end{pmatrix} = U_{\mathbf{k}} \begin{pmatrix} c_{\mathbf{k}\uparrow} \\ c_{-\mathbf{k}\downarrow}^{\dagger} \end{pmatrix} \qquad U_{\mathbf{k}} \equiv \begin{pmatrix} u_{\mathbf{k}} & -v_{\mathbf{k}} \\ v_{\mathbf{k}} & u_{\mathbf{k}} \end{pmatrix} \qquad (3.15)$$

It is immediately seen that $U_{\mathbf{k}}$ is unitary since $U_{\mathbf{k}} U_{\mathbf{k}}^{\dagger} = U_{\mathbf{k}}^{\dagger} U_{\mathbf{k}} = \mathbb{1}$. More precisely, it is orthogonal since $U_{\mathbf{k}}$ is real. Hence $U_{\mathbf{k}}^{-1} = U_{\mathbf{k}}^{\dagger} = {}^{\mathsf{T}}U_{\mathbf{k}}$, and the inverse of $U_{\mathbf{k}}$ is just its transpose ${}^{\mathsf{T}}U_{\mathbf{k}}$. So the inverse relation between the γ's and the c's reads

$$\begin{pmatrix} c_{\mathbf{k}\uparrow} \\ c_{-\mathbf{k}\downarrow}^{\dagger} \end{pmatrix} = \begin{pmatrix} u_{\mathbf{k}} & v_{\mathbf{k}} \\ -v_{\mathbf{k}} & u_{\mathbf{k}} \end{pmatrix} \begin{pmatrix} \gamma_{\mathbf{k}\uparrow} \\ \gamma_{-\mathbf{k}\downarrow}^{\dagger} \end{pmatrix} \qquad (3.16)$$

This unitary transformation is called the Bogoliubov–Valatin transformation [24, 25] and is quite analogous to the transformation introduced earlier by Bogoliubov in his theory of the weakly interacting condensed Bose gas, which we will consider in Chapter 7.

The unitarity of the transformation from the c's to the γ's implies that the γ's satisfy the same standard anticommutation relation for fermionic operators as the c's, as this is easily checked. This means that the excitations created by the γ's behave as standard fermions, completely analogous to electrons. The major difference is that the energy necessary to create these excitations is $E_\mathbf{k}$, instead of the energy $\xi_\mathbf{k}$ necessary to add an electron in the normal metal.

Let us finally see what is the result obtained if, in the subspace $(\mathbf{k}_\uparrow, -\mathbf{k}_\downarrow)$, we create both excitations corresponding to $\gamma_{\mathbf{k}\uparrow}^\dagger$ and $\gamma_{-\mathbf{k}\downarrow}^\dagger$. We find from Eq. (3.7), Eq. (3.13) and Eq. (3.5)

$$\gamma_{-\mathbf{k}\downarrow}^\dagger \gamma_{\mathbf{k}\uparrow}^\dagger |\psi_0\rangle_\mathbf{k} = \gamma_{-\mathbf{k}\downarrow}^\dagger c_{\mathbf{k}\uparrow}^\dagger |0\rangle = (u_\mathbf{k} c_{-\mathbf{k}\downarrow}^\dagger + v_\mathbf{k} c_{\mathbf{k}\uparrow}) c_{\mathbf{k}\uparrow}^\dagger |0\rangle \tag{3.17}$$

$$= v_\mathbf{k}|0\rangle - u_\mathbf{k} c_{\mathbf{k}\uparrow}^\dagger c_{-\mathbf{k}\downarrow}^\dagger |0\rangle = |\psi_1\rangle_\mathbf{k}$$

This is just the excited pair state, with energy $2E_\mathbf{k}$ with respect to the ground state. Hence by creating two fermions by the action of $\gamma_{\mathbf{k}\uparrow}^\dagger$ and $\gamma_{-\mathbf{k}\downarrow}^\dagger$, with each one corresponding to an excited state with energy $E_\mathbf{k}$, we obtain an excited state with energy $2E_\mathbf{k}$, that is the sum of the energies of the two fermions. This shows that these two fermions are not interacting. In this way, we reach the following very simple picture for the excited states of the BCS Hamiltonian: they are made of non-interacting fermions of energy $E_\mathbf{k}$. This is quite analogous to the situation in a normal state metal with non-interacting electrons.

We can rederive this result in a more efficient and formal way. Writing again the BCS Hamiltonian (with kinetic energy measured from the chemical potential μ)

$$H_{BCS} = \sum_\mathbf{k} \xi_\mathbf{k}(c_{\mathbf{k}\uparrow}^\dagger c_{\mathbf{k}\uparrow} + c_{-\mathbf{k}\downarrow}^\dagger c_{-\mathbf{k}\downarrow}) - V \sum_{\mathbf{k},\mathbf{k}'} c_{\mathbf{k}'\uparrow}^\dagger c_{-\mathbf{k}'\downarrow}^\dagger c_{-\mathbf{k}\downarrow} c_{\mathbf{k}\uparrow} \tag{3.18}$$

we notice that when we consider its action on a specific subspace $(\mathbf{k}_{0\uparrow}, -\mathbf{k}_{0\downarrow})$ for the purpose of calculating the energies of the excited states, in the calculations we have performed above, all the subspaces with $\mathbf{k} \neq \mathbf{k}_0$ were coming in only through the average $\langle c_{-\mathbf{k}\downarrow} c_{\mathbf{k}\uparrow}\rangle = u_\mathbf{k} v_\mathbf{k}$ and $\langle c_{\mathbf{k}'\uparrow}^\dagger c_{-\mathbf{k}'\downarrow}^\dagger\rangle = u_{\mathbf{k}'} v_{\mathbf{k}'}$. Through definition Eq. (2.66), this leads to the appearance of $\Delta = V \sum_\mathbf{k} u_\mathbf{k} v_\mathbf{k}$. We may as well take this into account from the start and replace in Eq. (3.18) the interaction terms relevant for subspace $(\mathbf{k}_{0\uparrow}, -\mathbf{k}_{0\downarrow})$, namely $V \sum_{\mathbf{k}'} c_{\mathbf{k}'\uparrow}^\dagger c_{-\mathbf{k}'\downarrow}^\dagger c_{-\mathbf{k}_0\downarrow} c_{\mathbf{k}_0\uparrow}$ and $V \sum_\mathbf{k} c_{\mathbf{k}_0\uparrow}^\dagger c_{-\mathbf{k}_0\downarrow}^\dagger c_{-\mathbf{k}\downarrow} c_{\mathbf{k}\uparrow}$, by the averages $\Delta c_{-\mathbf{k}_0\downarrow} c_{\mathbf{k}_0\uparrow}$ and $\Delta c_{\mathbf{k}_0\uparrow}^\dagger c_{-\mathbf{k}_0\downarrow}^\dagger$ (the term with $\mathbf{k} = \mathbf{k}_0$ is anyway negligible for large state number). Writing also $c_{-\mathbf{k}\downarrow}^\dagger c_{-\mathbf{k}\downarrow} = -c_{-\mathbf{k}\downarrow} c_{-\mathbf{k}\downarrow}^\dagger + 1$ we obtain, summing over all possible subspaces,

$$h_{BCS} = \sum_\mathbf{k} \xi_\mathbf{k}(c_{\mathbf{k}\uparrow}^\dagger c_{\mathbf{k}\uparrow} - c_{-\mathbf{k}\downarrow} c_{-\mathbf{k}\downarrow}^\dagger) - (\Delta c_{-\mathbf{k}\downarrow} c_{\mathbf{k}\uparrow} + \Delta c_{\mathbf{k}\uparrow}^\dagger c_{-\mathbf{k}\downarrow}^\dagger) \tag{3.19}$$

$$= \sum_\mathbf{k} \begin{pmatrix} c_{\mathbf{k}\uparrow}^\dagger & c_{-\mathbf{k}\downarrow} \end{pmatrix} \mathcal{E}_\mathbf{k} \begin{pmatrix} c_{\mathbf{k}\uparrow} \\ c_{-\mathbf{k}\downarrow}^\dagger \end{pmatrix} \qquad \mathcal{E}_\mathbf{k} \equiv \begin{pmatrix} \xi_\mathbf{k} & -\Delta \\ -\Delta & -\xi_\mathbf{k} \end{pmatrix}$$

Here we have written h_{BCS} instead of H_{BCS} to stress that h_{BCS} is appropriate only to calculate the excited states, not the ground state energy. This is because we have omitted in h_{BCS} all the constant terms. Indeed, in anti-commuting $c_{-\mathbf{k}\downarrow}^\dagger$ and $c_{-\mathbf{k}\downarrow}$, we have not written the term 1 corresponding to the value of the anti-commutator. Moreover, in writing the two terms coming from the interaction, we would have twice the interaction energy if we wanted to calculate the ground state energy, and this should be corrected by an additional constant term.

The eigenvalues of the 2×2 matrix $\mathcal{E}_\mathbf{k}$ in Eq. (3.19) are easily found. Their sum is zero since it is equal to the trace of the matrix. Their product is equal to the determinant of the matrix, that is, $-(\xi_\mathbf{k}^2 + \Delta^2) = -E_\mathbf{k}^2$. Hence these eigenvalues are $\pm E_\mathbf{k}$. On the other hand, substitution of c's in terms of γ's from Eq. (3.16) gives

$$h_{BCS} = \sum_\mathbf{k} \begin{pmatrix} \gamma_{\mathbf{k}\uparrow}^\dagger & \gamma_{-\mathbf{k}\downarrow} \end{pmatrix} \begin{pmatrix} u_\mathbf{k} & -v_\mathbf{k} \\ v_\mathbf{k} & u_\mathbf{k} \end{pmatrix} \mathcal{E}_\mathbf{k} \begin{pmatrix} u_\mathbf{k} & v_\mathbf{k} \\ -v_\mathbf{k} & u_\mathbf{k} \end{pmatrix} \begin{pmatrix} \gamma_{\mathbf{k}\uparrow} \\ \gamma_{-\mathbf{k}\downarrow}^\dagger \end{pmatrix} \tag{3.20}$$

However, the column vector $\begin{pmatrix} u_\mathbf{k} \\ -v_\mathbf{k} \end{pmatrix}$ is eigenvector of $\mathcal{E}_\mathbf{k}$ with eigenvalue $E_\mathbf{k}$, since we have seen in Eq. (2.91) that $(E_\mathbf{k} - \xi_\mathbf{k})u_\mathbf{k} = \Delta v_\mathbf{k}$, and (changing $\xi_\mathbf{k}$ into $-\xi_\mathbf{k}$) $(E_\mathbf{k} + \xi_\mathbf{k})v_\mathbf{k} = \Delta u_\mathbf{k}$. Similarly the orthogonal column vector $\begin{pmatrix} v_\mathbf{k} \\ u_\mathbf{k} \end{pmatrix}$ is eigenvector of $\mathcal{E}_\mathbf{k}$ with eigenvalue $-E_\mathbf{k}$. Hence in Eq. (3.20) $\mathcal{E}_\mathbf{k}$ is post-multiplied by the matrix made of its eigenvector, and pre-multiplied by the transpose of this same matrix. The result is merely a diagonal matrix with the eigenvalues $\pm E_\mathbf{k}$ on the diagonal, as may be easily checked directly:

$$h_{BCS} = \sum_\mathbf{k} \begin{pmatrix} \gamma_{\mathbf{k}\uparrow}^\dagger & \gamma_{-\mathbf{k}\downarrow} \end{pmatrix} \begin{pmatrix} E_\mathbf{k} & 0 \\ 0 & -E_\mathbf{k} \end{pmatrix} \begin{pmatrix} \gamma_{\mathbf{k}\uparrow} \\ \gamma_{-\mathbf{k}\downarrow}^\dagger \end{pmatrix} \tag{3.21}$$

$$= \sum_\mathbf{k} E_\mathbf{k}(\gamma_{\mathbf{k}\uparrow}^\dagger \gamma_{\mathbf{k}\uparrow} + \gamma_{-\mathbf{k}\downarrow}^\dagger \gamma_{-\mathbf{k}\downarrow})$$

where again in the last step we have omitted the constant coming from the anti-commutator $\{\gamma_{-\mathbf{k}\downarrow}, \gamma_{-\mathbf{k}\downarrow}^\dagger\} = 1$. This last expression Eq. (3.21) shows that the excited states of the BCS Hamiltonian are indeed non-interacting fermionic excitations with energy $E_\mathbf{k}$, created by the operators $\gamma_{\mathbf{k}\uparrow}^\dagger$ and $\gamma_{-\mathbf{k}\downarrow}^\dagger$. We see that the Bogoliubov–Valatin transformation shown in Eq. (3.16) is just the one which effectively diagonalizes the BCS Hamiltonian.

These fermionic excitations are very much like free particles, although from the definitions given in Eq. (3.11), they naturally look quite different from free electrons. As a result they are often called "quasi-particles." However, this word presents a vocabulary problem, since we have already used the wording "quasi-particles" with a completely different meaning, when in Section 2.2.2 we discussed Fermi liquid theory. Hence in the following, in order to avoid any confusion, we will rather call these fermionic excitations Bogoliubov quasi-particles, or for short "bogolons."

Let us now come back to the expression of the dispersion relation of these fermionic excitations. We have explicitly

$$E_\mathbf{k} = \sqrt{\xi_\mathbf{k}^2 + \Delta^2} \tag{3.22}$$

It is plotted in Fig. (3.2) as a function of $\xi_\mathbf{k}$. The most striking feature of this expression is the existence of a gap in the excitation spectrum $E_\mathbf{k} \geq \Delta$, as displayed in Fig. (3.2). In order to go from the ground state to any excited state, one has to pay at least an energy Δ. This is in contrast with the situation in a normal metal, where we can find excited states

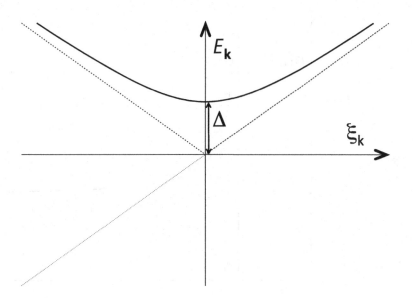

Fig. 3.2 Full line: dispersion relation $E_{\mathbf{k}} = \sqrt{\xi_{\mathbf{k}}^2 + \Delta^2}$ of the elementary excitations of the BCS Hamiltonian as a function of $\xi_{\mathbf{k}}$. The normal state limit, corresponding to $\Delta = 0$, is shown as the dashed line, while the standard electron dispersion relation $\xi_{\mathbf{k}} = \epsilon_{\mathbf{k}} - \mu$ is displayed as the dotted line.

with energies as close to the ground state as we like. As we have already mentioned it was believed, following London, that the existence of a gap was essential to a proper theory of superconductivity. Hence it is rewarding that the BCS theory naturally displays this property.

It is interesting to note that when $\Delta \to 0$, the excitation spectrum displayed in Fig. (3.2) goes continuously to the proper limit $|\xi_{\mathbf{k}}|$ in the normal state, which is shown in Fig. (3.2) as the dashed line. Indeed, for $\xi_{\mathbf{k}} > 0$, $|\xi_{\mathbf{k}}| = \xi_{\mathbf{k}}$ is the energy (measured from the chemical potential) necessary to create an additional electron above the Fermi energy. On the other hand, for $\xi_{\mathbf{k}} < 0$, $|\xi_{\mathbf{k}}| = -\xi_{\mathbf{k}} > 0$ is the energy necessary to create a hole in the Fermi sea, which is the opposite of the energy $\xi_{\mathbf{k}}$ necessary to create an electron with the same wavevector (assuming the state to be empty). This is not the most natural representation in a normal metal, where one rather tends to speak always in terms of the electronic energy $\xi_{\mathbf{k}} = \epsilon_{\mathbf{k}} - \mu$, both above and below the Fermi energy (as shown by the dotted line in Fig. 3.2). But this is the proper formulation when one wants to consider elementary excitations. It is worth noticing that the singular behavior found at $\xi_{\mathbf{k}} = 0$ in the normal state (where the excitation dispersion relation has a jump in its first derivative) disappears in the superconducting state, where the presence of the attractive interaction makes the dispersion relation $E_{\mathbf{k}}$ perfectly smooth.

It is also interesting to remark that for large values of the energy, the dispersion relation in the superconducting state goes asymptotically to the one corresponding to the normal state $E_{\mathbf{k}} \simeq |\xi_{\mathbf{k}}|$. Correspondingly we see that for large positive $\xi_{\mathbf{k}} \gg \Delta$, from Eq. (3.11) we have $\gamma_{\mathbf{k}\uparrow}^{\dagger} \simeq c_{\mathbf{k}\uparrow}^{\dagger}$ since $u_{\mathbf{k}} \to 1$ and $v_{\mathbf{k}} \to 0$ in this limit. Hence, in this case, creating an excitation in the superconductor is essentially identical to create an electron, in agreement

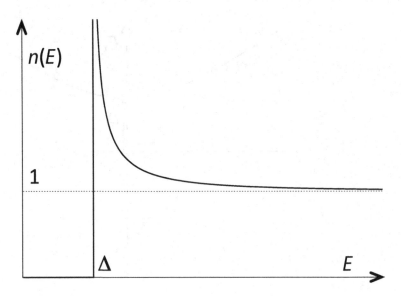

Fig. 3.3 Reduced BCS density of states $n(E) = \mathcal{N}(E)/N_0$ for the excitations as a function of their energy E.

with $E_{\mathbf{k}} \simeq \xi_{\mathbf{k}}$. In the same way, for large negative $\xi_{\mathbf{k}} \ll -\Delta$, we have $\gamma^{\dagger}_{\mathbf{k}\uparrow} \simeq -c_{-\mathbf{k}\downarrow}$ since in this case $u_{\mathbf{k}} \to 0$ and $v_{\mathbf{k}} \to 1$; so creating an excitation is the same as destroying an electron, that is creating a hole, in agreement with $E_{\mathbf{k}} \simeq -\xi_{\mathbf{k}}$. In the general case the operator $\gamma^{\dagger}_{\mathbf{k}\uparrow}$ interpolates between these two limiting situations.

We can now calculate the density of states of excitations $\mathcal{N}(E)$, defined from the number of excitations $dN_{exc} = \mathcal{N}(E)\,dE$ having their energy falling in the interval $[E, E + dE]$. Naturally this number is equal to the number of states having their $\xi_{\mathbf{k}}$ in the corresponding interval of width $d\xi$ obtained from Eq. (3.22), since in the superconducting state we have just a modification of the excitation energy corresponding to a specific $\mathbf{k}$ given by this relation. By differentiating Eq. (3.22), the relation between the two intervals is given by $E\,dE = \xi\,d\xi$. In turn, the number of states having their $\xi_{\mathbf{k}}$ in the corresponding interval of width $d\xi$ is merely given by the density of states of the normal metal $dN_{exc} = N(\xi)\,d\xi \simeq N_0\,d\xi$, as introduced in Eq. (2.9), where we can again approximate this density of states by its value at the Fermi energy, since we always deal with $|\xi| \ll E_F$. Putting these relations together, we find $dN_{exc} = \mathcal{N}(E)\,dE = N_0\,d\xi$; hence

$$\mathcal{N}(E) = N_0\,n(E) \qquad n(E) \equiv \frac{E}{\sqrt{E^2 - \Delta^2}} \qquad (3.23)$$

This reduced density of states $n(E)$ is plotted in Fig. (3.3). For $E < \Delta$, $n(E) = 0$ since due to the gap there are no states with energy less than Δ. For large energy $E \gg \Delta$ we recover as above the normal state result with $n(E) \simeq 1$, implying $\mathcal{N}(E) \simeq N_0$. The divergence of $n(E)$ for $E \to 1_+$ corresponds physically to the fact that as can be seen from Fig. 3.2, there is for small $\xi_{\mathbf{k}}$ an accumulation of states with energy essentially equal to Δ.

Actually, as can be expected physically, there is no real divergence for $E \to 1_+$ since any effect producing a finite lifetime for the excited states will smear out this divergence, as it can be guessed from the Heisenberg time-energy uncertainty relation. No such lifetime effects are included in our description, but they do obviously occur in real systems. This increase in density of excited states may also be understood qualitatively by saying that since a gap has been opened in the energy spectrum, the states with energy less than Δ have been pushed to higher energy by this gap opening; hence the increase in their number for energy higher than Δ. This is coherent with the fact that there is no increase in the total number of excited states; there is just a shift in their energy. This number conservation results directly from the relation $\int dN_{exc} = \int \mathcal{N}(E) \, dE = \int N(\xi) \, d\xi$. In Fig. (3.3), it means merely that the surface below the curve $n(E)$, but above the asymptote $n(E) = 1$ (corresponding to the sum of the additional states) is just equal to Δ, that is the surface of the rectangle between 0 and 1 with $E < \Delta$, which corresponds to the states which have been removed by the gap opening.

This BCS density of states is a fairly striking feature of the BCS theory. Hence it is very nice that it could be observed experimentally quite directly by tunneling experiments. This was done by I. Giaever [26] only a few years after the appearance of BCS theory, and was rewarded by the Nobel Prize. We describe the principle of these experiments in the following section.

3.2 Tunneling

Tunneling is a very specific effect arising in quantum mechanics. For a single particle moving in an external potential, it corresponds to the possibility for this particle to be present in domains where this would be fully excluded according to classical mechanics, namely in domains where its kinetic energy would be negative. In quantum mechanics, in such a situation, the wave function decays typically exponentially. If the thickness of the involved domain is small enough, the exponential decay of the wave function is not too strong, and it may yet allow the particle to cross the classically forbidden regions with an appreciable probability. In quantum mechanics textbooks, this tunneling effect is usually presented for a particle, an electron for example, moving in a vacuum in the presence of an external potential due to an applied electric field.

However, vacuum is by no means necessary, and tunneling may also occur for an electron moving in a solid. For example, if an electron moving in a metal meets an insulating domain, this means that its energy is not high enough to allow it to have a positive kinetic energy in this insulating domain, which would make it possible to propagate in the conduction band of this insulator. (Naturally it all depends on the electron energy: if it is high enough it will be able to propagate in this conduction band.) In this case, just as above in vacuum, the negative kinetic energy implies an exponentially decreasing wave function for the electron. However, if the thickness of the insulator is small enough, the electron will have some probability to go across the insulating region. The tunneling junctions used in solid-state experiments are typically of this kind.

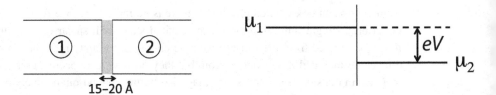

Fig. 3.4 Left: schematic geometry of a tunnel junction (not to scale); the shaded area describes the very thin insulating domain, of typical thickness 15–20 Å, separating the two metallic regions 1 and 2. Right: relative position of the chemical potentials μ_1 and μ_2 in the two metals; their difference is linked to the applied voltage V by $\mu_1 - \mu_2 = eV$ (the figure is drawn assuming the electronic charge $e > 0$).

The junctions used by Giaever are represented schematically in Fig. (3.4). They were made by first forming a thin aluminum film. Then the surface of this aluminum was oxidized by letting it briefly be in presence of air. This produced a very thin layer of insulating aluminum oxide, of typical thickness 15–20 Å. Finally this layer was covered by a deposit of lead. Naturally the insulator thickness is required to be extremely small in order to have the electrons tunnel through it with a significant probability. It is quite fortunate that this simple procedure produced at the time a very efficient tunnel junction, which is in principle extremely difficult to achieve. Indeed, in a standard situation (using other compounds for example), one rather obtains an irregular oxide. Together with the need to have it extremely thin, this results usually in holes or cracks, which short-circuit the expected tunneling current. If one tries to have a thicker insulator to avoid these pitfalls, this results usually in an unobservably small tunneling current. Nowadays it is much easier to produce good tunnel junctions, in particular, with molecular-beam epitaxy machines that allow layer-by-layer deposits. Nevertheless obtaining a good tunnel junction is still a delicate matter, since any defect at the atomic level may be a problem.

The standard procedure in analyzing these tunneling experiments is to disregard the details of the tunneling process, which are in the present case unimportant. One looks only at the net result that as a consequence of this process, an electron may have been transferred from a state with wavevector $\mathbf{k}_1$ in metal 1 to a state with wavevector $\mathbf{k}_2$ in metal 2. One introduces the probability amplitude $\mathcal{T}_{\mathbf{k}_1\mathbf{k}_2}$ for this transfer, together with its complex conjugate for the reverse process. Summing over all wavevectors this leads to the introduction of the tunneling Hamiltonian H_T

$$H_T = \sum_{\mathbf{k}_1\mathbf{k}_2} \mathcal{T}_{\mathbf{k}_1\mathbf{k}_2}(c_{\mathbf{k}_2\uparrow}^{\dagger} c_{\mathbf{k}_1\uparrow} + c_{\mathbf{k}_2\downarrow}^{\dagger} c_{\mathbf{k}_1\downarrow}) + h.c. \qquad (3.24)$$

which is an effective Hamiltonian describing all the processes of interest due to tunneling. We have assumed that the insulator has no magnetic property so the spin of the electron is conserved in the tunneling process. The total Hamiltonian of the junction is

$$H = H_1 + H_2 + H_T \qquad (3.25)$$

where H_1 and H_2 describe, respectively, the isolated metal 1 and 2, and their specific expression is just the BCS Hamiltonian we have studied earlier for an isolated superconductor.

Actually, we will further simplify our analysis by assuming that the matrix element of the tunneling Hamiltonian is a constant independent of $\mathbf{k_1}$ and $\mathbf{k_2}$, that is, $\mathcal{T}_{\mathbf{k_1 k_2}} = \mathcal{T}$. This simplification appears to be a convenience since it is clear that the amplitude $\mathcal{T}_{\mathbf{k_1 k_2}}$ is larger for wavevector $\mathbf{k_1}$ and $\mathbf{k_2}$ perpendicular to the insulating domain, because the length of the path of the electron in the classically forbidden insulating region is shorter in this case. If $\mathbf{k_1}$ and $\mathbf{k_2}$ have sizeable components parallel to the insulating barrier, the corresponding amplitude $\mathcal{T}_{\mathbf{k_1 k_2}}$ is much smaller since it depends typically exponentially on the length of the path in the forbidden region. Nevertheless the simplification we make is not a problem in the case of many superconductors which turn out to be in practice fairly isotropic, so their properties do not depend on the directions of $\mathbf{k_1}$ or $\mathbf{k_2}$. Hence, in this case, if we keep properly the dependence of $\mathcal{T}_{\mathbf{k_1 k_2}}$ on $\mathbf{k_1}$ and $\mathbf{k_2}$, the result would merely contain an average of $|\mathcal{T}_{\mathbf{k_1 k_2}}|^2$ over the wavevectors directions, and it would be otherwise unchanged. On the other hand, in the case of anisotropic superconductors our simple analysis is invalid.

In any case, since tunneling is a process with low probability, we can consider H_T in Eq. (3.25) as small compared to H_1 and H_2 and find its effect by perturbation theory at lowest order. So we can find the probability for an electron to go from metal 1 to metal 2 by applying Fermi's golden rule. For simplicity, we will consider here only the zero temperature case $T = 0$, but there is no basic problem to generalize to nonzero temperature. We note that we may consider the two metals as superconductors, since the case of a normal metal is merely obtained from the superconducting one by setting the gap $\Delta = 0$.

At $T = 0$ the two superconductors are in their BCS ground state. If we consider a specific term $c^\dagger_{\mathbf{k_2}\uparrow} c_{\mathbf{k_1}\uparrow}$ in the tunneling Hamiltonian, by acting on the ground state it creates a broken pair state $c^\dagger_{\mathbf{k_2}\uparrow}|0\rangle_2$ in superconductor 2 with energy $E_{\mathbf{k_2}}$, where $|0\rangle_2$ is the vacuum for superconductor 2 and we have as usual written only the term relevant for the subspace $(\mathbf{k_2}\uparrow, -\mathbf{k_2}\downarrow)$. Similarly this same term of the tunneling Hamiltonian, by annihilating an electron $\mathbf{k_1}$, breaks a pair and creates a broken pair state $c^\dagger_{-\mathbf{k_1}\downarrow}|0\rangle_1$ in superconductor 1 with energy $E_{\mathbf{k_1}}$, where $|0\rangle_1$ is the vacuum for superconductor 1. Hence for the process due to this term, the final state is $|f\rangle = c^\dagger_{-\mathbf{k_1}\downarrow}|0\rangle_1 \oplus c^\dagger_{\mathbf{k_2}\uparrow}|0\rangle_2$, where again only the relevant terms are written. Correspondingly for the initial state made of the BCS ground state in superconductor 1 and 2, we have $|i\rangle = (u_{\mathbf{k_1}} + v_{\mathbf{k_1}} c^\dagger_{\mathbf{k_1}\uparrow} c^\dagger_{-\mathbf{k_1}\downarrow})|0\rangle_1 \oplus (u_{\mathbf{k_2}} + v_{\mathbf{k_2}} c^\dagger_{\mathbf{k_2}\uparrow} c^\dagger_{-\mathbf{k_2}\downarrow})|0\rangle_2$. Conversely, if we consider these specific initial and final states $|i\rangle$ and $|f\rangle$, only the $c^\dagger_{\mathbf{k_2}\uparrow} c_{\mathbf{k_1}\uparrow}$ in H_T has nonzero matrix elements between them. Therefore this leads to

$$\langle f|H_T|i\rangle = \mathcal{T} \,_1\langle 0|c_{-\mathbf{k_1}\downarrow} c_{\mathbf{k_1}\uparrow} (u_{\mathbf{k_1}} + v_{\mathbf{k_1}} c^\dagger_{\mathbf{k_1}\uparrow} c^\dagger_{-\mathbf{k_1}\downarrow})|0\rangle_1 \qquad (3.26)$$
$$\times \,_2\langle 0|c_{\mathbf{k_2}\uparrow} c^\dagger_{\mathbf{k_2}\uparrow} (u_{\mathbf{k_2}} + v_{\mathbf{k_2}} c^\dagger_{\mathbf{k_2}\uparrow} c^\dagger_{-\mathbf{k_2}\downarrow})|0\rangle_2 = \mathcal{T} v_{\mathbf{k_1}} u_{\mathbf{k_2}}$$

As usual, all the other wavevector states participating in the ground states, and which have not been written explicitly for simplicity, give factors equal to 1, due to the normalization. This calculation is quite analogous to the one we have made for the interaction energy of the BCS wave function Eq. (2.58), and the physical interpretation is similar. In order to transfer an electron from $\mathbf{k_1}$ to $\mathbf{k_2}$, we need to have an electron initially in $\mathbf{k_1}$ (probability amplitude $v_{\mathbf{k_1}}$) and have an empty place in $\mathbf{k_2}$ (probability amplitude $u_{\mathbf{k_2}}$) in the final state. Note that we could also have done this calculation by expressing in H_T the c's in terms of the γ's through Eq. (3.16). This is the path we will follow in the next chapter.

In order to obtain the transition rate W_{12} of electrons from superconductor 1 to superconductor 2 through the Fermi golden rule, we have to sum over all wavevectors $\mathbf{k}_1$ and $\mathbf{k}_2$ of the final state, and over spin, which leads to

$$W_{12} = 2 \times \frac{2\pi}{\hbar} \sum_{\mathbf{k}_1 \mathbf{k}_2} |\langle f|H_T|i\rangle|^2 \delta(\mathcal{E}_f - \mathcal{E}_i) \tag{3.27}$$

$$= \frac{4\pi}{\hbar} |T|^2 \sum_{\mathbf{k}_1 \mathbf{k}_2} v_{\mathbf{k}_1}^2 u_{\mathbf{k}_2}^2 \delta(E_{\mathbf{k}_1} + E_{\mathbf{k}_2} - eV)$$

Indeed, in writing $\mathcal{E}_f - \mathcal{E}_i$ in the energy conservation, we have to take into account that as shown in Fig. (3.4), the transfer of an electron from superconductor 1 (with chemical potential μ_1) to superconductor 2 (with chemical potential μ_2) leads to a contribution $\mu_2 - \mu_1 = -eV$ to this energy difference, leading to $\mathcal{E}_f - \mathcal{E}_i = E_{\mathbf{k}_1} + E_{\mathbf{k}_2} - eV$. The physical meaning is clear: the energy $E_{\mathbf{k}_1} + E_{\mathbf{k}_2}$ necessary to go from the ground states to the excited states is provided by the work eV of the electric field in the tunneling junction.

To perform the summation over $\mathbf{k}_1$ and $\mathbf{k}_2$, we may just sum over the energies $\xi_{\mathbf{k}_1}$ and $\xi_{\mathbf{k}_2}$ since these are the only variables appearing. Anticipating that $\xi_{\mathbf{k}_1}$ and $\xi_{\mathbf{k}_2}$ will be restricted to a small range around zero, we may replace as usual the density of states in superconductor 1 and 2 by their values at their Fermi energy, N_{01} and N_{02}, respectively. Substituting for $u_{\mathbf{k}}$ and $v_{\mathbf{k}}$ their explicit expressions Eq. (2.75), we have

$$W_{12} = \frac{4\pi}{\hbar} |T|^2 N_{01} N_{02} \int_{-\infty}^{\infty} d\xi_{\mathbf{k}_1} \int_{-\infty}^{\infty} d\xi_{\mathbf{k}_2} \frac{1}{4}(1 - \frac{\xi_{\mathbf{k}_1}}{E_{\mathbf{k}_1}})(1 + \frac{\xi_{\mathbf{k}_2}}{E_{\mathbf{k}_2}}) \delta(E_{\mathbf{k}_1} + E_{\mathbf{k}_2} - eV) \tag{3.28}$$

where we have extended the integral boundaries to $\pm\infty$ by anticipating again an actual small range of integration around zero. The terms containing $\xi_{\mathbf{k}_1}/E_{\mathbf{k}_1}$ and $\xi_{\mathbf{k}_2}/E_{\mathbf{k}_2}$, which are odd with respect to $\xi_{\mathbf{k}_1}$ and $\xi_{\mathbf{k}_2}$, disappear by parity in the integrations over $\xi_{\mathbf{k}_1}$ and $\xi_{\mathbf{k}_2}$. The remaining term is even with respect to $\xi_{\mathbf{k}_1}$ and $\xi_{\mathbf{k}_2}$, so the integrals from $-\infty$ to 0, and from 0 to ∞, are equal; hence we can also write

$$W_{12} = \frac{4\pi}{\hbar} |T|^2 N_{01} N_{02} \int_0^{\infty} d\xi_{\mathbf{k}_1} \int_0^{\infty} d\xi_{\mathbf{k}_2} \, \delta(E_{\mathbf{k}_1} + E_{\mathbf{k}_2} - eV) \tag{3.29}$$

Since only $E_{\mathbf{k}}$ appears, it is more convenient to take it as variable instead of $\xi_{\mathbf{k}}$. Since $d\xi/dE = E/\xi = n(E)$ is precisely the reduced BCS density of states, we obtain finally

$$W_{12} = \frac{4\pi}{\hbar} |T|^2 N_{01} N_{02} \int_{\Delta_1}^{\infty} dE_1 \, n_1(E_1) \int_{\Delta_2}^{\infty} dE_2 \, n_2(E_2) \, \delta(E_1 + E_2 - eV) \tag{3.30}$$

where Δ_1 and Δ_2 are the respective gaps of superconductor 1 and 2.

Let us now consider several specific cases. We assume that superconductor 1 is in the normal state, so we set $\Delta_1 = 0$, which implies naturally $n_1(E_1) = 1$. We can also integrate the δ function over E_1. Since E_1 is in the range $[0, \infty]$, we have $E_1 \geq 0$, which implies from the δ function $E_2 \leq eV$. This leads to the simple expression

$$W_{12} = \frac{4\pi}{\hbar} |T|^2 N_{01} N_{02} \int_{\Delta_2}^{eV} dE_2 \, n_2(E_2) \tag{3.31}$$

The current I through the tunneling junction is obviously proportional to the transition rate W_{12}, so it is proportional to the integral on the right-hand side of Eq. (3.31). Let us call $1/eR$ the coefficient of proportionality between I and this integral. So by definition of R, we have

$$I = \frac{1}{eR} \int_{\Delta_2}^{eV} dE_2 \, n_2(E_2) \tag{3.32}$$

If we assume superconductor 2 to be also in the normal state, so that $\Delta_2 = 0$ and $n_2(E_2) = 1$, Eq. (3.32) becomes merely $I = V/R$, that is

$$V = RI \tag{3.33}$$

Hence we find that when the two metals are in their normal state, the junction behaves as a standard resistor following Ohm's law, and that R is its resistance, which provides a physical meaning to this constant we have just introduced.

Coming back to Eq. (3.32) in the superconducting case for metal 2, we can take its derivative with respect to voltage V, which gives

$$\frac{dI}{dV} = \frac{1}{R} \, n(eV) \tag{3.34}$$

where in this final result we have dropped the index 2, referring to the superconductor that we considered. Hence the differential conductance of the tunneling junction gives directly the reduced density of states for excitations. This is, after all, a fairly natural result: the tunneling current is physically due to the transfer of fermionic excitations from one superconductor to the other. It is natural that this current is proportional to the available "space" for these excitations, which is quantitatively given by their density of states.

The experimental results are in general in quite good agreement with the prediction of BCS theory, taking into account that various effects, and in particular temperature, contribute to smooth out the sharp features of the theoretical result seen in Fig. (3.3). An extremely attractive aspect of tunneling experiments is that they allow a quite direct measurement of the gap, and the voltage range for these measurements is a quite convenient one since it is in the millivolt range. Associated with this point is the fact that this range is also the typical range for the variables $\xi_{\mathbf{k}}$ and $E_{\mathbf{k}}$ we have used above, as they are constrained by energy conservation coming from the δ function. This range is indeed small as we anticipated.

To obtain the relation $I(V)$ for the junction, it is easy to integrate Eq. (3.34) when one takes the BCS density of states. This leads to

$$I(V) = \frac{1}{eR} \sqrt{(eV)^2 - \Delta^2} \tag{3.35}$$

which is displayed in Fig. (3.5). Its most striking feature is that for $eV < \Delta$, there is no current flowing through the tunneling junction. The physical reason is quite clear. In order to have an electron going from one side to the other, one needs to create an excited state in the superconductor side, corresponding physically to break a Cooper pair. This requires an energy equal at least to the gap, and this energy has to be supplied by the voltage drop of the junction. Hence one has the condition $eV > \Delta$ to have a nonzero current through the

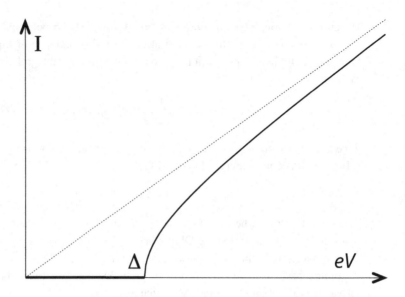

Fig. 3.5 Current I versus voltage V for a tunneling junction between a normal and a superconducting metal, with gap Δ, at zero temperature.

junction. The other noticeable feature of Fig. (3.5) is that for a large voltage, one recovers (as usual) the normal metal properties with its Ohmic behavior.

Let us finally note that Eq. (3.30) can naturally be used when both metal 1 and 2 are superconductors. Since the density of states $n(E)$ of each superconductor can be measured with a normal-superconducting junction, the result of such an experiment can be predicted without any adjustable parameter, which provides a nice coherence check of the theory.

3.3 Free Energy

Since we have the excited states spectrum, we are now able to calculate the free energy. A fairly obvious point can be made from the start, which we have not stressed yet. Temperature will create excited states, which implies that in contrast to the $T = 0$ case, we do not have the ground state in all the $(\mathbf{k}_\uparrow, -\mathbf{k}_\downarrow)$ subspaces. Since the lowering of the energy through the interaction comes from this very ground state, it is clear that the interaction energy, and accordingly the gap Δ, are modified by raising temperature.

In a specific subspace $(\mathbf{k}_\uparrow, -\mathbf{k}_\downarrow)$, the probabilities of having the ground state, the two broken pair states and the excited pair state occupied are proportional to their respective Boltzmann factor, namely 1, $e^{-\beta E_\mathbf{k}}$, $e^{-\beta E_\mathbf{k}}$, and $e^{-2\beta E_\mathbf{k}}$, where $\beta = 1/k_B T$. To have the probabilities themselves, we have to normalize properly by dividing these factors by the appropriate partition function for the $(\mathbf{k}_\uparrow, -\mathbf{k}_\downarrow)$ subspace, namely $Z = 1 + 2e^{-\beta E_\mathbf{k}} + e^{-2\beta E_\mathbf{k}} = (1 + e^{-\beta E_\mathbf{k}})^2$. In Section 3.1, we have already found that the kinetic energy contribution (measured from chemical potential) to the ground state, the broken pair state, and the excited pair state are, respectively, $2\xi_\mathbf{k} v_\mathbf{k}^2$, $\xi_\mathbf{k}$, and $2\xi_\mathbf{k} u_\mathbf{k}^2$. Multiplying by their respective probability, and summing over all subspaces, we obtain for the average of the kinetic energy

$$E_c = \sum_{\mathbf{k}} \frac{\xi_{\mathbf{k}}}{(1 + e^{-\beta E_{\mathbf{k}}})^2} \left[2v_{\mathbf{k}}^2 + 2e^{-\beta E_{\mathbf{k}}} + 2u_{\mathbf{k}}^2 e^{-2\beta E_{\mathbf{k}}} \right] \tag{3.36}$$

$$= \sum_{\mathbf{k}} \frac{2\xi_{\mathbf{k}}}{(1 + e^{-\beta E_{\mathbf{k}}})^2} \left[v_{\mathbf{k}}^2 \left(1 - e^{-2\beta E_{\mathbf{k}}} \right) + \left(e^{-\beta E_{\mathbf{k}}} + e^{-2\beta E_{\mathbf{k}}} \right) \right]$$

$$= \sum_{\mathbf{k}} 2\xi_{\mathbf{k}} \left[v_{\mathbf{k}}^2 \left(1 - \frac{2e^{-\beta E_{\mathbf{k}}}}{1 + e^{-\beta E_{\mathbf{k}}}} \right) + \frac{e^{-\beta E_{\mathbf{k}}}}{1 + e^{-\beta E_{\mathbf{k}}}} \right] = \sum_{\mathbf{k}} 2\xi_{\mathbf{k}} \left[v_{\mathbf{k}}^2 (1 - 2f_k) + f_k \right]$$

where in the second step we have used the normalization condition $u_{\mathbf{k}}^2 = 1 - v_{\mathbf{k}}^2$, and in the last one we have introduced the Fermi distribution for the fermionic excitations

$$f_k = \frac{1}{e^{\beta E_{\mathbf{k}}} + 1} \tag{3.37}$$

Note that this distribution is zero for zero temperature $T = 0$, since in this case β goes to infinity while $E_{\mathbf{k}} \geq \Delta$ is positive: there are no fermionic excitations at $T = 0$.

The last expression may readily be interpreted in terms of the fermionic excitations of energy $E_{\mathbf{k}}$ created by the γ's. Being non-interacting fermions, their average number is given by the Fermi distribution Eq. (3.37). Their kinetic energy is $\xi_{\mathbf{k}}$, the kinetic energy of a broken pair. And the factor 2 comes because we have two kinds $\mathbf{k} \uparrow$ and $-\mathbf{k} \downarrow$ of excitations, differing by their spin. This gives directly the last term contribution $2\xi_{\mathbf{k}} f_k$. The first term is also simple to understand. This is the ground state contribution $2\xi_{\mathbf{k}} v_{\mathbf{k}}^2$ we had found in the preceding chapter, multiplied by the probability $1 - 2f_k$ that the ground state is still occupied. This last probability is indeed 1, minus the probability f_k that we have rather fermionic excitations, multiplied by 2 since we have the two kinds of excitations. Hence one could write directly this result, but it is nevertheless interesting to see in detail, as we have done, how it comes from the description in terms of broken and excited pairs.

We can now deal in a similar way with the interaction energy. As we have seen the broken pairs do not contribute to the interaction energy, so we have only to consider the ground state and the excited pair state. At $T = 0$, we have found in Eq. (2.63)

$$E_{int} = -V \sum_{\mathbf{k},\mathbf{k}'} u_{\mathbf{k}'} v_{\mathbf{k}'} u_{\mathbf{k}} v_{\mathbf{k}} \tag{3.38}$$

(assuming already $u_{\mathbf{k}}$ and $v_{\mathbf{k}}$ to be real, as it could easily be proved just as for $T = 0$). In this expression, $u_{\mathbf{k}} v_{\mathbf{k}}$ is the contribution from the ground state. This now needs to be multiplied by the probability $1/Z$ to have the ground state. Regarding the excited pair state, we have seen in Section 3.1 that it is obtained from the ground state by the exchange $u_{\mathbf{k}} \to v_{\mathbf{k}}$ and $v_{\mathbf{k}} \to -u_{\mathbf{k}}$, so that it gives in the interaction energy a contribution $-u_{\mathbf{k}} v_{\mathbf{k}}$, instead of $u_{\mathbf{k}} v_{\mathbf{k}}$. This has to be multiplied by the probability $e^{-2\beta E_{\mathbf{k}}}/Z$ to have the excited pair state. Adding this contribution to the one coming from the ground state, we see that $u_{\mathbf{k}} v_{\mathbf{k}}$ in Eq. (3.38) has to be replaced by

$$u_{\mathbf{k}} v_{\mathbf{k}} \frac{1 - e^{-2\beta E_{\mathbf{k}}}}{(1 + e^{-\beta E_{\mathbf{k}}})^2} = u_{\mathbf{k}} v_{\mathbf{k}} (1 - 2f_k) \tag{3.39}$$

Once again, in terms of fermionic excitations, this result may just be seen as the ground state result, multiplied by the probability $1 - 2f_k$ to have no excitations. With the same arguments applied to the $u_{\mathbf{k}'}v_{\mathbf{k}'}$ term, we obtain finally for the interaction energy

$$E_{int} = -V \sum_{\mathbf{k},\mathbf{k}'} u_{\mathbf{k}}v_{\mathbf{k}}(1 - 2f_k)u_{\mathbf{k}'}v_{\mathbf{k}'}(1 - 2f_{k'}) \tag{3.40}$$

Finally the description of the excited states in terms of free fermionic excitations allows one to write the standard expression for the entropy corresponding to this case

$$S = -2k_B \sum_{\mathbf{k}} f_k \ln f_k + (1 - f_k)\ln(1 - f_k) \tag{3.41}$$

with the factor 2 accounting again for the spin of these excitations.

According to the procedure we have followed at $T = 0$, in order to obtain the explicit expression of $v_{\mathbf{k}}$, we should now minimize the free energy

$$F = E_c + E_{int} - TS \tag{3.42}$$

with respect to $v_{\mathbf{k}}$ (to which $u_{\mathbf{k}}$ is directly related by the normalization condition). This seems like a fairly complicated matter since we have to take now into account that f_k depends on E_k, which is itself related to $u_{\mathbf{k}}$ and $v_{\mathbf{k}}$. Fortunately, we can easily sidestep this problem by a formal argument. The expressions we have written above for F both depend on f_k and $v_{\mathbf{k}}$ (to which $u_{\mathbf{k}}$ is directly related); hence it is a functional $F(\{v_{\mathbf{k}}\}, \{f_k\})$ of $v_{\mathbf{k}}$ and f_k. Although we know actually, through Eq. (3.37), the actual expression of f_k, let us assume that we do not know this result and that we are looking for it. This is equivalent to say that we do not know the expression of the Boltzmann weight that should be attributed to each excited state. In the case of free fermions, it is well known in statistical physics that the explicit expression of the Fermi distribution Eq. (3.37) can be obtained by minimizing the expression of the free energy with respect to f_k. Here we follow the same path, with in addition the fact that $v_{\mathbf{k}}$ is also unknown. In order to obtain both $v_{\mathbf{k}}$ and f_k we minimize the free energy F with respect to $v_{\mathbf{k}}$ and f_k. This leads us to write that the two corresponding functional derivatives are zero:

$$\left.\frac{\delta F(\{v_{\mathbf{k}}\}, \{f_k\})}{\delta v_{\mathbf{k}}}\right|_{f_k} = 0 \qquad\qquad \left.\frac{\delta F(\{v_{\mathbf{k}}\}, \{f_k\})}{\delta f_k}\right|_{v_{\mathbf{k}}} = 0 \tag{3.43}$$

These functional derivatives are analogous to partial derivatives. Accordingly, when we vary with respect to $v_{\mathbf{k}}$, we have to keep f_k constant, and conversely when we vary with respect to f_k, we have to keep $v_{\mathbf{k}}$ constant.

The first variation is easily done since, from Eq. (3.41), the entropy S depends only on f_k, which must be kept constant. Hence this term disappears from the variation with respect to $v_{\mathbf{k}}$. Then, since the expressions for E_c and E_{int} are very similar to their $T = 0$ counterparts, Eq. (2.64) is generalized for real $u_{\mathbf{k}}$ and $v_{\mathbf{k}}$ into

$$\delta F = \sum_{\mathbf{k}} 4\xi_{\mathbf{k}} v_{\mathbf{k}}(1 - 2f_k)\delta v_{\mathbf{k}} - 2V \sum_{\mathbf{k},\mathbf{k}'} u_{\mathbf{k}'}v_{\mathbf{k}'}(1 - 2f_{k'})(1 - 2f_k)\delta(u_{\mathbf{k}}v_{\mathbf{k}}) \tag{3.44}$$

If we set

$$\Delta = V \sum_{\mathbf{k}}^{\omega_D} u_{\mathbf{k}} v_{\mathbf{k}} (1 - 2f_k) \tag{3.45}$$

which generalizes the $T = 0$ expression Eq. (2.66), we obtain from Eq. (3.44)

$$\delta F = 2 \sum_{\mathbf{k}} (1 - 2f_k) [2\xi_{\mathbf{k}} v_{\mathbf{k}} \delta v_{\mathbf{k}} - \Delta \delta(u_{\mathbf{k}} v_{\mathbf{k}})] \tag{3.46}$$

In order to write that the variation of F with respect to $v_{\mathbf{k}}$ is zero, we have to write that the factor of $\delta v_{\mathbf{k}}$ in the bracket under the summation is zero. But, by comparison with Eq. (2.67), we see that this is exactly what we have already done at $T = 0$. The temperature has completely disappeared at this stage of the calculation; it is only present in the definition Eq. (3.45) of Δ. Accordingly, the calculation following Eq. (2.67) is fully unchanged, and we end up for $u_{\mathbf{k}}$ and $v_{\mathbf{k}}$ with the same expressions Eq. (2.75) as for $T = 0$, which we repeat here for reference

$$u_{\mathbf{k}} = \sqrt{\frac{1}{2}\left(1 + \frac{\xi_{\mathbf{k}}}{E_{\mathbf{k}}}\right)} \qquad v_{\mathbf{k}} = \sqrt{\frac{1}{2}\left(1 - \frac{\xi_{\mathbf{k}}}{E_{\mathbf{k}}}\right)} \tag{3.47}$$

Let us finally check that the second variation condition in Eq. (3.43) provides, as we stated, the Fermi-Dirac distribution. Since f_k appears in a simple way in the expression of F, one easily obtains that this second condition amounts to requiring that

$$\frac{\delta F(\{v_{\mathbf{k}}\}, \{f_k\})}{\delta f_k}\bigg|_{v_{\mathbf{k}}} = 2\xi_{\mathbf{k}}[1 - 2v_{\mathbf{k}}^2] + 4Vu_{\mathbf{k}}v_{\mathbf{k}} \sum_{\mathbf{k}'} u_{\mathbf{k}'} v_{\mathbf{k}'} (1 - 2f_{k'}) + 2k_B T \ln \frac{f_k}{1 - f_k} \tag{3.48}$$

is zero. Making use of the definition Eq. (3.45) of Δ, together with Eq. (2.72) $u_{\mathbf{k}} v_{\mathbf{k}} = \Delta/2E_{\mathbf{k}}$ and the expression Eq. (3.47) of $v_{\mathbf{k}}$, this leads to

$$k_B T \ln \frac{f_k}{1 - f_k} = -\frac{\xi_{\mathbf{k}}^2}{E_{\mathbf{k}}} - \frac{\Delta^2}{E_{\mathbf{k}}} = -E_{\mathbf{k}} \tag{3.49}$$

in agreement with the Fermi-Dirac distribution Eq. (3.37).

Let us note finally that we have actually assumed from the very beginning in Section 3.1 that the ground state has the same structure as at zero temperature, except for a possible change for the value of the gap, which was left unspecified. The above discussion has shown that this is indeed the case and provides a justification for our starting hypothesis. Hence the whole BCS framework at nonzero temperature is a self-consistent one.

3.4 Gap Equation, Critical Temperature

Let us now focus on the gap equation Eq. (3.45). Making use of $u_{\mathbf{k}} v_{\mathbf{k}} = \Delta/2E_{\mathbf{k}}$, and noting that $1 - 2f_k = (e^{\beta E_{\mathbf{k}}} - 1)/(e^{\beta E_{\mathbf{k}}} + 1) = \tanh(\beta E_{\mathbf{k}}/2)$, it can be rewritten as

$$\Delta = V \sum_{\mathbf{k}}^{\omega_D} \frac{\Delta}{2E_{\mathbf{k}}} \tanh(\frac{\beta E_{\mathbf{k}}}{2}) \tag{3.50}$$

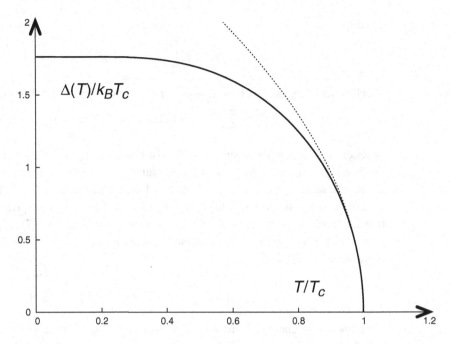

Fig. 3.6 Full line: temperature dependence of $\Delta(T)$ obtained by solving numerically Eq. (3.51). The excellent analytical approximation for the reciprocal function $x = y/\text{arctanh}\, y$, with $x = T(\Delta)/T_c$ and $y = \Delta/(1.76k_BT_c)$ is also plotted as a dashed line, but it is at this scale undistinguishable from the exact numerical result. The dotted line is the quadratic approximation Eq. (3.55) valid for T in the vicinity of T_c.

Discarding again the solution $\Delta = 0$, which corresponds to the normal state, and replacing the summation over $\mathbf{k}$ by an integration over $\xi_{\mathbf{k}}$ with the density of states assumed to be constant, equal to N_0, because the range of integration $|\xi_{\mathbf{k}}| < \hbar\omega_D$ is narrow, we obtain

$$\frac{1}{N_0 V} = \int_{-\hbar\omega_D}^{\hbar\omega_D} d\xi_{\mathbf{k}} \frac{1}{2E_{\mathbf{k}}} \tanh(\frac{\beta E_{\mathbf{k}}}{2}) \tag{3.51}$$

where the gap Δ still appears through $E_{\mathbf{k}} = \sqrt{\xi_{\mathbf{k}}^2 + \Delta^2}$. As long as this equation has a nonzero solution for Δ, we have a BCS-like superconducting state.

One can fairly easily see that when temperature T is increasing, the corresponding solution of Eq. (3.51) for Δ is decreasing. Obtaining $\Delta(T)$ for general T requires a simple numerical calculation from Eq. (3.51). The result is displayed[1] in Fig. (3.6). However, when due to this decrease, Δ reaches zero, we are as we have seen in the normal state, and so superconductivity has disappeared. Hence the temperature corresponding to $\Delta = 0$ is the critical temperature T_c. From Eq. (3.51), it satisfies

$$\frac{1}{N_0 V} = \int_0^{\hbar\omega_D} d\xi \frac{1}{\xi} \tanh(\frac{\beta_c \xi}{2}) \tag{3.52}$$

---

[1] The reciprocal function $T(\Delta)$ has the excellent analytical approximation $x = y/\text{arctanh}\, y$, where $x = T/T_c$ and $y = \Delta/(1.76k_BT_c)$. It differs from the numerical result at most by a few tenths of a percent. It is plotted in Fig. (3.6), but it is not distinguishable from the exact result at this scale.

where $\beta_c = 1/(k_B T_c)$, and we have taken advantage of the fact that in Eq. (3.51), the integrand is even in $\xi_{\mathbf{k}}$ to reduce the integration to the interval $[0, \hbar\omega_D]$. The dependence of T_c on $N_0 V$ is easily obtained by taking $x = \beta_c \xi/2$ as new variable in Eq. (3.52) and integrating by parts over x. This gives, with $W = \beta_c \hbar\omega_D/2$,

$$\frac{1}{N_0 V} = [\ln x \, \tanh x]_0^W - \int_0^W dx \, \frac{\ln x}{\cosh^2 x} \tag{3.53}$$

We can now make use of the weak coupling condition implying that $k_B T_c \ll \hbar\omega_D$, so $W \gg 1$. Hence in the first term we may replace $\tanh W$ by 1, and in the second term the upper bound may be set to $+\infty$, since the integral converges very rapidly due to the factor $1/\cosh^2 x$. This implies that this second term is just a constant. Hence one finds immediately $k_B T_c \propto \hbar\omega_D \, e^{-1/N_0 V}$. The proportionality constant may be obtained numerically, or analytically,[2] and one finds

$$k_B T_c = \frac{2e^C}{\pi} \hbar\omega_D \, e^{-\frac{1}{N_0 V}} \simeq 1.13 \, \hbar\omega_D \, e^{-\frac{1}{N_0 V}} \tag{3.54}$$

where $C = 0.577215...$ is the Euler constant.

This celebrated BCS formula for the critical temperature sets the important energy scale for T_c, namely a typical phonon energy $\hbar\omega_D$. Otherwise it is not an effective formula to obtain an accurate value for the critical temperature. Indeed $\hbar\omega_D$ does not have a very precise definition in the BCS model, and is barely more than an order of magnitude. However, the most important source of uncertainty is the evaluation of the interaction strength V, since it characterizes a very rough modeling of the attractive interaction, as we have already seen. Moreover, consistently with the weak coupling condition $k_B T_c \ll \hbar\omega_D$, $N_0 V$ is small in our regime of interest, which makes T_c extremely sensitive to the evaluation of V because of the exponential dependence. To take a specific example if $\hbar\omega_D/k_B = 300K$ and $N_0 V = 0.3$, we find $T_c \simeq 12K$. If $N_0 V = 0.37$, that is a 20% uncertainty which is difficult to avoid, we find $T_c \simeq 23K$, which is almost a factor of 2 higher than the preceding evaluation. This situation gets worse for smaller $N_0 V$. This is clearly not so satisfactory.

On the other hand, the functional dependence that appears on the right-hand side of Eq. (3.54) is the same as the one we had for the zero temperature gap $\Delta(0)$ found in Eq. (2.81). Taking the ratio between these two quantities, we have

---

[2] Starting from Eq. (3.51) with the boundaries extended to $[-\hbar\omega_D, \hbar\omega_D]$ and the change of variable $x = \beta_c \xi/2$, one has $2/N_0 V = \int_{-W}^{W} dx \, \tanh x/x$. One closes the integration contour in the upper complex plane by a semi-circle of radius W. The contribution of this semi-circle may be easily shown to be negligible for large W, since in this case $\tanh x = 1$ for $\mathrm{Re}\, x > 0$ while $\tanh x = -1$ for $\mathrm{Re}\, x < 0$. The contour integral is calculated by residues, making use of the pole expansion $\tanh x = \sum_{n=-\infty}^{\infty} 1/[x - i\pi(2n+1)/2]$. This leads to $1/N_0 V \simeq 2 \sum_{n=0}^{E[W/\pi]} 1/(2n+1) \simeq \ln(4W/\pi) + C$, where $E[W/\pi]$ is the integer part of W/π. This gives Eq. (3.54). The sum in the preceding equation is obtained from the standard definition of the Euler constant $\lim_{N\to\infty}(\sum_{n=1}^{N} 1/n - \ln N) = C$ by separating the contribution for even n, and noting that $\sum_{p=1}^{E[N/2]}(1/2p) \simeq (1/2)[\ln(N/2) + C]$ from the definition of C.

$$\frac{\Delta(0)}{k_B T_c} = \pi e^{-C} \simeq 1.76 \tag{3.55}$$

This ratio is a simple number; there is nothing adjustable in it, and all the parameters coming in the BCS model have disappeared. On the other hand, both $\Delta(0)$ and T_c are easy to measure very directly, as we have seen. It is therefore remarkable that for standard superconductors, the experimental result for this ratio falls within a few percent of the BCS theoretical result Eq. (3.55).

For temperature in the vicinity of T_c, $\Delta(T)$ is small. The precise behavior of $\Delta(T)$ can be extracted from Eq. (3.51). One has just to vary this equation with respect to Δ^2 and $\beta = 1/k_B T$. The coefficients turning out to be finite, this shows already the important result that Δ^2 is proportional to $1 - T/T_c$. One finds[3] precisely

$$\frac{\Delta(T)}{k_B T_c} = \pi \left[\frac{8}{7\zeta(3)} \left(1 - \frac{T}{T_c} \right) \right]^{1/2} \simeq 3.06 \left(1 - \frac{T}{T_c} \right)^{1/2} \tag{3.56}$$

where $\zeta(s) = \sum_{p=1}^{\infty}(1/p^s)$ is the Riemann function, with $\zeta(3) \simeq 1.20$.

One can see from Fig. (3.6) that this expansion Eq. (3.56) around $T = T_c$ has a fair range of validity. The other noticeable qualitative feature of $\Delta(T)$ in Fig. (3.6) is that it stays essentially equal to its $T = 0$ value for quite a large range, extending almost to $T = 0.4\, T_c$. This can be related to the existence of the gap itself. Indeed, since $1 - 2f_k = \tanh(\beta E_k/2)$, one sees that at low temperature $f_k = [e^{\beta E_k} + 1]^{-1} \leq [e^{\beta \Delta} + 1]^{-1}$ so that as long as $k_B T/\Delta$ stays fairly small, f_k is very small. There is an exponentially strong suppression of the fermionic excitations. Hence $\tanh(\beta E_k/2) \approx 1$ in Eq. (3.50), so that the result is essentially the same as at $T = 0$.

3.5 Specific Heat, Critical Field

Since we have found the free energy, we can obtain all the thermodynamic properties of the superconductor. One has just to be careful of the implicit temperature dependence carried by the gap $\Delta(T)$. We address first the specific heat. It is obtained from the entropy, given explicitly by Eq. (3.41). This leads to

$$C = T \frac{dS}{dT} = -2k_B T \sum_k \ln \frac{f_k}{1 - f_k} \frac{df_k}{dT} = 2 \sum_k E_k \frac{df_k}{dT} \tag{3.57}$$

---

[3] By differentiating Eq. (3.51) for $\Delta^2 \to 0$ and $\beta - \beta_c \to 0$, one finds the equation $\delta^2 \int_{-\infty}^{\infty} dx\, [d(\tanh x/x)/dx](1/2x)$
$= [(\beta_c - \beta)/\beta_c] \int_{-\infty}^{\infty} dx \cosh^{-2} x = 2(1 - \beta/\beta_c)$, where $\delta = \beta_c \Delta/2$. and $x = \beta_c \xi_k/2$ is taken as integration variable. The first integral is transformed into an integral over a contour $\mathcal{C}$ encircling the upper complex plane (avoiding the singularity at $x = 0$), which after by-part integration becomes $\int_{\mathcal{C}} dx\, \tanh x/x^3$. Just as in the preceding note for Eq. (3.54), an expression is found by residues integration using the pole expansion of $\tanh x$. This gives $\int_{\mathcal{C}} dx\, \tanh x/x^3 = -(16/\pi^2) \sum_{n=0}^{\infty}(2n + 1)^{-3}$. Again as in the preceding note, this last sum is easily related to $\zeta(3) = \sum_{p=1}^{\infty} 1/p^3$ by separating the terms with p odd or even, these last ones giving a contribution $(1/8)\zeta(3)$. As a result, one obtains $\sum_{n=0}^{\infty}(2n + 1)^{-3} = (7/8)\zeta(3)$. Gathering these results leads to Eq. (3.56).

by making use of the explicit expression Eq. (3.37) of the Fermi distribution. This expression implies also

$$\frac{df_k}{dT} = -\frac{e^{\beta E_k}}{(e^{\beta E_k} + 1)^2} \frac{d(\beta E_k)}{dT} \qquad \frac{d(\beta E_k)}{dT} = -\frac{E_k}{k_B T^2} + \frac{1}{k_B T} \frac{dE_k}{dT} \tag{3.58}$$

Introducing $-k_B T(\partial f_k / \partial E_k) = e^{\beta E_k} / (e^{\beta E_k} + 1)^2 = 1/4 \cosh^2(\beta E_k / 2)$, we can rewrite Eq. (3.57) as

$$C = 2 \sum_k \left(-\frac{\partial f_k}{\partial E_k} \right) \left[\frac{E_k^2}{T} - E_k \frac{dE_k}{dT} \right] \tag{3.59}$$

The relation $E_k^2 = \xi_k^2 + \Delta^2$ implies $E_k(dE_k/dT) = -\xi_k(d\mu/dT) + \Delta(d\Delta/dT)$. As usual, we transform the summation over $\mathbf{k}$ into an integral over ξ_k with constant density of states N_0, where the lower boundary can be extended to $-\infty$ with negligible error. The term proportional to ξ_k in $E_k(dE_k/dT)$ disappears by parity. We use the even parity of the remaining integrand to reduce the integration to the interval $[0, \infty]$, and we obtain finally

$$C = \frac{4N_0}{T} \int_0^\infty d\xi_k \left(-\frac{\partial f_k}{\partial E_k} \right) \left(E_k^2 - \frac{T}{2} \frac{d\Delta^2(T)}{dT} \right) \tag{3.60}$$

At $T = T_c$, we have $E_k = |\xi_k|$, and Eq. (3.60) becomes

$$C_s = \frac{4N_0}{T} \int_0^\infty d\xi_k \left(-\frac{\partial f_k}{\partial \xi_k} \right) \left(\xi_k^2 - \frac{T}{2} \frac{d\Delta^2(T)}{dT} \right) \tag{3.61}$$

The first term is what is obtained if we set $\Delta = 0$, so it is the normal state result

$$C_n = \frac{4N_0}{T} \int_0^\infty d\xi_k \left(-\frac{\partial f_k}{\partial \xi_k} \right) \xi_k^2 = \frac{2\pi^2}{3} N_0 k_B^2 T_c \equiv \gamma T_c \tag{3.62}$$

where we have made use of the normal state integral $\int_0^\infty dx \, (x^2 / \cosh^2 x) = \pi^2/12$, and $\gamma = 2\pi^2 N_0 k_B^2 / 3$ is the well-known Sommerfeld constant. Hence, at $T = T_c$, the difference $C_s - C_n$ is just given by the second term in Eq. (3.61). Since $\int_0^\infty d\xi_k(-\partial f_k/\partial \xi_k) = 1/2$, one finds the simple result

$$C_s(T_c) - C_n(T_c) = -N_0 \left. \frac{d\Delta^2(T)}{dT} \right|_{T_c} = \frac{8\pi^2}{7\zeta(3)} N_0 k_B^2 T_c \tag{3.63}$$

making use, in the last step, of the expansion Eq. (3.56) for $\Delta(T)$ in the vicinity of T_c. Compared to the normal state specific heat, this gives for the specific heat jump

$$\left. \frac{C_s - C_n}{C_n} \right|_{T_c} = \frac{12}{7\zeta(3)} \simeq 1.43 \tag{3.64}$$

This has to be compared with the result of 2, found in Eq. (1.19) in Chapter 1 by a phenomenological approach. The numbers are similar, but experimental results fall nearer to the BCS result Eq. (3.64), although there is some dispersion around this value. We can make about this result the same comment as we have made for Eq. (3.55) on the ratio $\Delta(0)/k_B T_c$. We have a pure number, with nothing adjustable for the theory. It is striking that the BCS result falls in fair vicinity of the experimental results.

The low temperature $T \to 0$ behavior can also be easily extracted from Eq. (3.60). As we have seen in Fig. (3.6), $d\Delta(T)/dT \to 0$ in this limit. Moreover, $\beta E_{\mathbf{k}}$ gets very large so $f_k \simeq e^{-\beta E_{\mathbf{k}}}$ and $-(\partial f_k/\partial E_{\mathbf{k}}) \simeq \beta e^{-\beta E_{\mathbf{k}}}$. Because this exponential is decreasing very rapidly at low T when $E_{\mathbf{k}}$ increases, only the values of $E_{\mathbf{k}}$ in the vicinity of its minimum Δ contribute significantly. This corresponds to small values of $\xi_{\mathbf{k}}$, and it is enough to retain for $E_{\mathbf{k}}$ the small $\xi_{\mathbf{k}}$ expansion $E_{\mathbf{k}} \simeq \Delta + \xi_{\mathbf{k}}^2/2\Delta$ in the exponential, and even take $E_{\mathbf{k}} \simeq \Delta$ in the prefactor. This leads to

$$C \simeq \frac{4N_0}{T} \int_0^\infty d\xi_{\mathbf{k}} \left(-\frac{\partial f_k}{\partial E_{\mathbf{k}}} \right) E_{\mathbf{k}}^2 \simeq \frac{4N_0\Delta^2}{k_B T^2} e^{-\Delta/k_B T} \int_0^\infty d\xi_{\mathbf{k}} \, e^{-\xi_{\mathbf{k}}^2/2k_B T\Delta} \qquad (3.65)$$

With $\int_0^\infty dx \, e^{-x^2} = \sqrt{\pi}/2$, one has finally in this limit

$$C(T) \simeq \frac{2N_0\Delta^2(0)}{k_B T^2} e^{-\frac{\Delta(0)}{k_B T}} \sqrt{2\pi k_B T \Delta(0)} = \sqrt{8\pi} N_0 k_B \frac{\Delta^{5/2}(0)}{(k_B T)^{3/2}} e^{-\frac{\Delta(0)}{k_B T}} \qquad (3.66)$$

where we have made explicit that one has naturally to use the $T = 0$ value $\Delta(0)$ for the gap.

The essential feature of this result for $C(T)$ is the exponential dependence $e^{-\Delta(0)/k_B T}$ which is obviously expected from the existence of the gap in the excitation spectrum, for example in analogy with semiconductor physics. Actually, since the possible presence of a gap in superconductors was a very important ingredient in the elaboration of the theoretical understanding of superconductivity, and since its existence should be clearly revealed by the low-temperature behavior of the specific heat, one may wonder why the experimental evidence has not been obtained earlier, whereas it has been found only about the time of the BCS paper publication. The answer is that a clean exponential behavior requires one to go to fairly low temperature compared to T_c in order to be seen clearly without any ambiguity, and this turned out to be fairly difficult to achieve. This is clearly seen in Fig. (3.7), where the BCS specific heat, calculated numerically from Eq. (3.60), is displayed. The existence of the gap is inferred typically from the temperature range $T \lesssim 0.1T_c$, where the specific heat is extremely small and accordingly difficult to measure with good precision.

Let us turn toward the explicit calculation of the free energy, which from Section 1.1.5 provides the critical field through Eq. (1.12)

$$F_n(T) - F_s(T) = \frac{1}{2}\mu_0 H_c^2(T) \qquad (3.67)$$

where we use the standard notation $F(T)$ for the thermodynamic potential for zero field $G(T, 0)$ considered in Section 1.1.5. First of all, at $T = 0$, we may easily obtain the critical field from the result for the condensation energy we have already found in Chapter 2, since the free energy difference is merely the energy difference obtained in Eq. (2.88). This gives

$$F_n(0) - F_s(0) = E_N - E_S = \frac{1}{2}N_0\Delta^2(0) = \frac{1}{2}\mu_0 H_c^2(0) \qquad (3.68)$$

For the ratio considered in Eq. (1.17) this leads to

$$\frac{\gamma T_c^2}{\mu_0 H_c^2(0)} = \frac{2\pi^2}{3} \left(\frac{k_B T_c}{\Delta(0)} \right)^2 \frac{N_0\Delta^2(0)}{\mu_0 H_c^2(0)} = \frac{2\pi^2}{3} \frac{e^{2C}}{\pi^2} = \frac{2e^{2C}}{3} \simeq 2.11 \qquad (3.69)$$

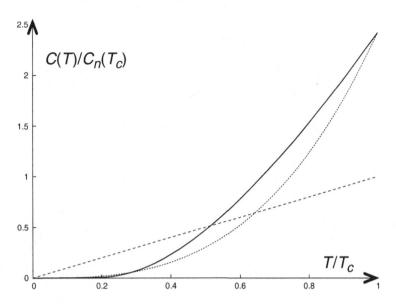

Fig. 3.7 Specific heat normalized to the normal state value $C_n(T_c)$ at T_c. Full line: BCS theory. Dashed line: extrapolation of the normal state specific heat. Dotted line: T^3 law, for comparison (see Chapter 1).

where we have used the explicit expression of the Sommerfeld constant $\gamma = 2\pi^2 N_0 k_B^2/3$, together with Eq. (3.55) and Eq. (3.68). This numerical result, where again nothing is adjustable, is remarkably close to the result of 2, which we have obtained in Chapter 1, in part from phenomenological considerations.

Next, we explore the case of a general temperature. Gathering the results from Eq. (3.36) for E_c and from Eq. (3.40) for E_{int}, using the gap definition Eq. (3.45), and replacing $u_\mathbf{k}$ and $v_\mathbf{k}$ by their explicit expressions Eq. (3.47), we obtain

$$E_c + E_{int} = \sum_\mathbf{k} \left\{ \left[\xi_\mathbf{k}(1 - \frac{\xi_\mathbf{k}}{E_\mathbf{k}}) - \frac{\Delta^2(T)}{2E_\mathbf{k}} \right] + 2f_k \left(\frac{\xi_\mathbf{k}^2}{E_\mathbf{k}} + \frac{\Delta^2(T)}{2E_\mathbf{k}} \right) \right\} \qquad (3.70)$$

In the normal state, obtained by setting $\Delta = 0$, the first term on the right-hand side reduces to $\sum_{k<k_F} 2\xi_\mathbf{k}$. This is the normal state free energy at zero temperature $F_n(T = 0)$, since in this case all the other contributions to the free energy are zero. If this expression $\sum_{k<k_F} 2\xi_\mathbf{k}$ is subtracted from the first term of Eq. (3.70), the result is identical to the expression we had in Chapter 2, at $T = 0$, on the right-hand side of Eq. (2.85), except naturally that the gap is now equal to its value $\Delta(T)$ at temperature T. Hence we can handle it in the same way, as we have done in Eq. (2.86) and Eq. (2.87). Accordingly, from Eq. (2.88), the result is $-N_0\Delta^2(T)/2$, so that the first term in Eq. (3.70) is equal to $F_n(T = 0) - N_0\Delta^2(T)/2$.

In order to obtain the free energy Eq. (3.42), we still have to subtract from $E_c + E_{int}$ the contribution $-TS$. The entropy S is obtained by inserting in its expression Eq. (3.41) the explicit expression Eq. (3.37) for the Fermi distribution. Gathering the terms containing $\ln(1 + e^{-\beta E_\mathbf{k}})$, it can be rewritten as

$$S = 2k_B \sum_\mathbf{k} \left[\ln(1 + e^{-\beta E_\mathbf{k}}) + \beta E_\mathbf{k} f_k \right] \qquad (3.71)$$

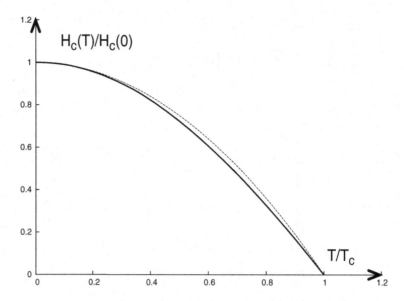

Full line: reduced critical field $H_c(T)/H_c(0)$ as a function of temperature T. The dashed line is the approximate parabolic law $H_c(T)/H_c(0) = 1 - (T/T_c)^2$.

We transform the first term by going as usual to an integration over $\xi_{\mathbf{k}}$ and then integrating by parts:

$$\frac{1}{N_0} \sum_{\mathbf{k}} \ln(1 + e^{-\beta E_{\mathbf{k}}}) = \int_{-\infty}^{\infty} d\xi_{\mathbf{k}} \ln(1 + e^{-\beta E_{\mathbf{k}}}) \tag{3.72}$$

$$= \left[\xi_{\mathbf{k}} \ln(1 + e^{-\beta E_{\mathbf{k}}})\right]_{-\infty}^{\infty} + \int_{-\infty}^{\infty} d\xi_{\mathbf{k}} \frac{\xi_{\mathbf{k}} e^{-\beta E_{\mathbf{k}}}}{1 + e^{-\beta E_{\mathbf{k}}}} \beta \frac{dE_{\mathbf{k}}}{d\xi_{\mathbf{k}}} = \beta \int_{-\infty}^{\infty} d\xi_{\mathbf{k}} f_k \frac{\xi_{\mathbf{k}}^2}{E_{\mathbf{k}}}$$

since the integrated term is zero. When the second term of Eq. (3.71) is added one obtains for TS an expression which turns out to be twice the second term on the right-hand side of Eq. (3.70). Gathering all the contributions, we have finally

$$F_s(T) - F_n(T = 0) = -N_0 \frac{\Delta^2(T)}{2} - 2N_0 \int_{-\infty}^{\infty} d\xi_{\mathbf{k}} f_k \left(\frac{\xi_{\mathbf{k}}^2}{E_{\mathbf{k}}} + \frac{\Delta^2(T)}{2E_{\mathbf{k}}}\right) \tag{3.73}$$

The integral is easily calculated numerically. On the other hand, the normal state specific heat $C_n(T) = \gamma T$ implies $F_n(T) - F_n(T = 0) = -\gamma T^2/2$, which is also obtained from Eq. (3.73) by setting $\Delta = 0$, with $\int_0^{\infty} dx \, x/(e^x + 1) = \pi^2/12$ and $\gamma = 2\pi^2 N_0 k_B^2/3$. Together with the thermodynamic relation Eq. (1.12), this yields finally for the critical field

$$\frac{1}{2} \mu_0 H_c^2(T) = N_0 \frac{\Delta^2(T)}{2} - \gamma \frac{T^2}{2} + 2N_0 \int_{-\infty}^{\infty} d\xi_{\mathbf{k}} f_k \left(\frac{\xi_{\mathbf{k}}^2}{E_{\mathbf{k}}} + \frac{\Delta^2(T)}{2E_{\mathbf{k}}}\right) \tag{3.74}$$

Naturally this reduces to Eq. (3.68) for $T = 0$, since $f_k \to 0$ in this limit.

The reduced critical field $H_c(T)/H_c(0)$ is plotted in Fig. (3.8), together with the parabolic experimental approximation $H_c(T)/H_c(0) = 1 - (T/T_c)^2$. It is remarkable that the BCS result, where again nothing is adjustable, comes so close to this experimental law. Nevertheless the two curves are not identical, and later more precise experiments on standard low-temperature superconductors have given results that are closer to the BCS result than to the parabolic law.

To summarize, the BCS theory provides a single parameter description of the superconductor: as soon as the critical temperature T_c is known (and it is naturally quite easy to measure, at least in principle), the gap is determined from Eq. (3.55) and everything else is determined by universal relations involving no further parameters. Hence if different superconductors follow BCS theory they follow identical laws, after proper rescaling of the critical temperature. There is a law of corresponding states between these different superconductors, a kind of universality. However, naturally all superconductors do not follow BCS theory, and we will return to this matter later.

3.6 Isotope Effect, Coulomb Repulsion

We have seen in Chapter 1 that the existence of an isotope effect on the critical temperature of mercury was quite important for drawing attention to the role played by lattice vibrations in the occurrence of superconductivity. We can now investigate quantitatively this effect with the BCS formula Eq. (3.54) for the critical temperature. In a single element superconductor like mercury, one expects a simple change for the phonon frequencies upon isotope substitution. Indeed lattice vibrations are analogous to the harmonic oscillations of a mass M attached to a spring with restoring force characterized by constant k. The frequency ω of these oscillations is given $\omega = \sqrt{k/M}$. If for fixed k, there is a small change δM in the mass value, there is a corresponding shift $\delta \omega$ in the frequency given by $\delta \omega / \omega = -(1/2)\delta M/M$, obtained by differentiating the frequency formula. When the mercury isotope is changed, the electronic properties are unchanged, so the restoring forces on the moving ions do not change. This is analogous to keeping k fixed for the spring. Since ω_D is some average of phonons frequencies, we expect it to follow the same law $\delta \omega_D / \omega_D = -(1/2)\delta M/M$ upon the mercury mass change δM due to isotope substitution. Since in Eq. (3.54) the interaction V is related to electron-ion and electron–electron interaction, it is a purely electronic property and should not change upon isotope substitution. As a result, differentiating Eq. (3.54), we obtain the critical temperature shift δT_c upon isotope substitution

$$\frac{\delta T_c}{T_c} = -\frac{1}{2}\frac{\delta M}{M} \tag{3.75}$$

This happens to be in quite good agreement with mercury experiments, which was an additional success of BCS theory.

In retrospect this very good agreement was somewhat coincidental, and actually it happens quite infrequently. This occurs, in particular, because Coulomb repulsion between electrons modifies the expected result for the isotope effect. Accordingly, it is worthwhile to try to investigate the modifications brought by this Coulomb repulsion, both for its own intrinsic interest and for its influence on the isotope effect. Unfortunately, for many reasons, including those examined in Section 2.2, this is basically an impossible task. Having a reliable description of the indirect attractive electron–electron interaction due to phonon exchange is already quite difficult, and above we have merely followed the simple BCS modeling. A description of the effective Coulomb repulsion is even more complex, in particular because there is no small parameter in this problem. Hence we will just investigate a simple generalization of the BCS modeling and examine its physical consequences.

As we have seen, the attractive interaction is ineffective beyond phonon frequencies, and this is what is modeled by the BCS cut-off at ω_D. This is because the ions are heavy so their motion is slow. In contrast, there is nothing of this kind with the repulsive electron–electron interaction, which has been fully ignored in the BCS model. Indeed electrons are light and move very rapidly. All the electrons are involved in this Coulomb repulsion. If we want to find something like a cut-off energy, it should be of the order of a plasma frequency or a Fermi energy. Hence we introduce for this repulsive interaction a cut-off ω_c which will be of the order of the Fermi energy. For our purpose, this is its essential feature. For the rest, we model this interaction in the BCS way, that is, by a constant interaction independent of energy.

Hence explicitly we consider the following generalization of the interaction in the BCS Hamiltonian

$$H_{int} = -V_p \sum_{\mathbf{k},\mathbf{k}'}^{\omega_D} c^\dagger_{\mathbf{k}'\uparrow} c^\dagger_{-\mathbf{k}'\downarrow} c_{-\mathbf{k}\downarrow} c_{\mathbf{k}\uparrow} + V_c \sum_{\mathbf{k},\mathbf{k}'}^{\omega_c} c^\dagger_{\mathbf{k}'\uparrow} c^\dagger_{-\mathbf{k}'\downarrow} c_{-\mathbf{k}\downarrow} c_{\mathbf{k}\uparrow} \tag{3.76}$$

The corresponding cut-offs, ω_D and ω_c, written above each sum, indicate that the summations are restricted respectively to $|\xi_{\mathbf{k}}|, |\xi_{\mathbf{k}'}| < \hbar\omega_D$, and $|\xi_{\mathbf{k}}|, |\xi_{\mathbf{k}'}| < \hbar\omega_c$. For the attractive part, we have used the notation V_p, instead of V, to remind that it is due to the electron–phonon interaction. The two interaction constants are assumed to be positive: $V_p, V_c > 0$.

The formal similarity between this interaction, Eq. (3.76), and the standard BCS one, Eq. (2.26), makes it easy to generalize the analysis we have done in the BCS case. Basically we have just, coming from the interaction, two similar terms instead of one. So we do not reproduce here explicitly all the steps we have been through. For example, the variation of the free energy Eq. (3.44) becomes now

$$\delta F = \sum_{\mathbf{k}} 4\xi_{\mathbf{k}} v_{\mathbf{k}}(1 - 2f_k)\delta v_{\mathbf{k}} - 2V_p \sum_{\mathbf{k},\mathbf{k}'}^{\omega_D} u_{\mathbf{k}'} v_{\mathbf{k}'}(1 - 2f_{k'})(1 - 2f_k)\delta(u_{\mathbf{k}} v_{\mathbf{k}}) \tag{3.77}$$

$$+ 2V_c \sum_{\mathbf{k},\mathbf{k}'}^{\omega_c} u_{\mathbf{k}'} v_{\mathbf{k}'}(1 - 2f_{k'})(1 - 2f_k)\delta(u_{\mathbf{k}} v_{\mathbf{k}})$$

where we assume again from the start that $v_{\mathbf{k}}$ is real. This introduces two gap-like quantities instead of one, namely

$$\Delta_p = V_p \sum_{k'}^{\omega_D} u_{\mathbf{k}'} v_{\mathbf{k}'}(1 - 2f_{k'}) \qquad \Delta_c = V_c \sum_{k'}^{\omega_c} u_{\mathbf{k}'} v_{\mathbf{k}'}(1 - 2f_{k'}) \tag{3.78}$$

Then canceling the coefficient of $\delta v_{\mathbf{k}}$ on the right-hand side of Eq. (3.77) leads, instead of Eq. (2.68), to

$$2\xi_{\mathbf{k}}u_{\mathbf{k}}v_{\mathbf{k}} = (u_{\mathbf{k}}^2 - v_{\mathbf{k}}^2)\left[\Delta_p\,\theta(\hbar\omega_D - |\xi_{\mathbf{k}}|) - \Delta_c\,\theta(\hbar\omega_c - |\xi_{\mathbf{k}}|)\right] \tag{3.79}$$

where the factors $(1 - 2f_k)$ have disappeared, just as in Eq. (3.46). The Heaviside step functions $\theta(\hbar\omega_D - |\xi_{\mathbf{k}}|)$ and $\theta(\hbar\omega_c - |\xi_{\mathbf{k}}|)$ indicate explicitly the presence of the cut-off.

For $|\xi_{\mathbf{k}}| < \hbar\omega_D$, the term in brackets in Eq. (3.79) is $\Delta_p - \Delta_c$, and we find from this equation exactly the same result as in the pure BCS case

$$u_{\mathbf{k}}v_{\mathbf{k}} = \frac{\Delta}{2E_{\mathbf{k}}} \tag{3.80}$$

provided we introduce $\Delta = \Delta_p - \Delta_c$ and $E_{\mathbf{k}} = \sqrt{\xi_{\mathbf{k}}^2 + \Delta^2}$. On the other hand, for $\hbar\omega_D < |\xi_{\mathbf{k}}| < \hbar\omega_c$, the Δ_p term in Eq. (3.79) disappears but otherwise the calculation is the same, so we obtain

$$u_{\mathbf{k}}v_{\mathbf{k}} = -\frac{\Delta_c}{2|\xi_{\mathbf{k}}|} \tag{3.81}$$

where we have written $\sqrt{\xi_{\mathbf{k}}^2 + \Delta_c^2} \simeq |\xi_{\mathbf{k}}|$, using the fact that in this range, from weak coupling, $|\xi_{\mathbf{k}}| > \hbar\omega_D \gg \Delta_c$ (which can be checked in the final result to be of the same order as the standard gap). We notice that in this range, $u_{\mathbf{k}}v_{\mathbf{k}}$ has changed its sign to adjust at best to the repulsive interaction.

We find equations for Δ_p and Δ_c by introducing these results in the definitions Eq. (3.78) for these quantities. As usual, to perform the $\mathbf{k}$ summations we will turn to an integration over $\xi_{\mathbf{k}}$. However, our standard assumption of a constant density of states is no longer valid for the summation with cut-off ω_c since it is comparable to the Fermi energy, instead of being small as for the cut-off ω_D. Nevertheless in order to get a simple result, we will assume a constant density of states N_0 even for this ω_c summation, because we are already in a situation of heavy modeling and it does not make sense to keep secondary complications. Anyway, one can see that the essential qualitative result, namely the appearance of a $\ln\omega_D$ dependence, appears also if one keeps an energy-dependent density of states. With this approximation, Eq. (3.78), together with Eq. (3.80) and Eq. (3.81), gives

$$\Delta_p = N_0V_p(\Delta_p - \Delta_c)I \tag{3.82}$$

$$\Delta_c = N_0V_c(\Delta_p - \Delta_c)I - N_0V_c\Delta_c\int_{\hbar\omega_D}^{\hbar\omega_c}d\xi_{\mathbf{k}}\,\frac{1}{\xi_{\mathbf{k}}} \tag{3.83}$$

where we have used as usual the even parity of the integrands and we have set

$$I = \int_0^{\hbar\omega_D}d\xi_{\mathbf{k}}\,\frac{\tanh(\beta E_{\mathbf{k}}/2)}{E_{\mathbf{k}}} \tag{3.84}$$

The explicit integral in Eq. (3.83) has been obtained by replacing $\tanh(\beta E_{\mathbf{k}}/2)$ with 1, since in the integration range $E_{\mathbf{k}} \simeq \xi_{\mathbf{k}} \gg k_BT$. This integral is equal to $\ln(\omega_c/\omega_D)$. Comparing the value of $(\Delta_p - \Delta_c)I$ from Eq. (3.82) and Eq. (3.83), we find

$$\ln\frac{\omega_c}{\omega_D} + \frac{1}{N_0V_c} = \frac{\Delta_p}{N_0V_p\Delta_c} \tag{3.85}$$

Carrying this ratio for Δ_p/Δ_c into Eq. (3.82) gives finally

$$\frac{1}{I} = N_0 V_p - \frac{1}{\ln \frac{\omega_c}{\omega_D} + \frac{1}{N_0 V_c}} \tag{3.86}$$

This result is interesting on several points. First of all, if the Coulomb repulsion V_c is zero so that the last term on the right-hand side is absent, it reduces naturally to the standard gap equation Eq. (3.51) $I = 1/N_0 V_p$. Otherwise this last term of the right-hand side is effectively reducing the attractive interaction V_p, so Coulomb repulsion is unfavorable to superconductivity, as expected. If V_c is small enough, V_p is just replaced by $V_p - V_c$. This comes, as we have seen in the above derivation, from the effect of the Coulomb repulsion in the low energy range $|\xi_{\mathbf{k}}| < \hbar \omega_D$.

However, we see that when V_c increases, the negative effect of Coulomb repulsion saturates since the last term on the right-hand side is always smaller than $1/\ln(\omega_c/\omega_D)$. Interestingly this upper bound is small when the ratio ω_c/ω_D is large. This is the quantitative manifestation that the retardation effect from the lattice vibrations is favorable to superconductivity. As we have noticed qualitatively, when ω_D is small, ions move slowly so the attractive interaction of two electrons through the lattice is a retarded interaction. As a consequence, the two interacting electrons are not at the same spatial location, which reduces effectively the strength of their direct Coulomb repulsion. And the slower the phonon frequency ω_D compared to the typical electronic frequency ω_c, the larger this effective reduction of Coulomb repulsion. This effect does not look very strong since it comes only through a logarithmic term. Nevertheless, since the ratio ω_c/ω_D is typically large, the effect is sizeable. For a typical value $\omega_c/\omega_D = 10^2$, $1/\ln(\omega_c/\omega_D) \simeq 0.2$. Hence, for $N_0 V_p > 0.2$, which is still a fairly weak coupling value, superconductivity cannot in our model be overwhelmed by Coulomb repulsion whatever its strength.

Finally we can extract the critical temperature from Eq. (3.86) since in this case both Δ_p and Δ_c go to zero. We have found the value of I in this specific case when we have obtained the BCS critical temperature Eq. (3.54). The corresponding value was $I = \ln(2e^C \hbar \omega_D/\pi k_B T_c) \simeq \ln(1.13 \, \hbar \omega_D/k_B T_c)$ so that the critical temperature is obtained from

$$\frac{1}{\ln \frac{1.13 \, \hbar \omega_D}{k_B T_c}} = N_0 V_p - \frac{1}{\ln \frac{\omega_c}{\omega_D} + \frac{1}{N_0 V_c}} \tag{3.87}$$

We can guess, from the result Eq. (3.87) of this model, how difficult it is to obtain a reliable microscopic result for the critical temperature. Not only the actual evaluation of the effective attractive interaction, equivalent to V_p, is quite difficult but one has to subtract an effective Coulomb repulsion, equivalent to the second term on the right-hand side, which is even more difficult to evaluate. Since these two terms are roughly of comparable magnitude, it is clear that there is a good deal of uncertainty in their difference. Moreover, this comes in a quantity that has to be exponentiated to obtain T_c; hence the sensitivity of the critical temperature to theoretical uncertainties is even worse.

This is all the more frustrating since obviously T_c is one of the most essential characteristics of a superconductor for applications. This inability of BCS theory to come with a reliable way to obtain T_c has often been pointed out as a weakness of the theory. Perhaps the proper conclusion is the opposite one: it is remarkable that despite the fundamental difficulty in predicting the critical temperature of a superconductor, the BCS theory comes out with so many simple and reliable results. One could very well dream of a terrible situation where many superconducting properties would be as difficult to predict as the critical temperature.

One last point can be extracted from Eq. (3.87), which regards the isotope effect. If we consider that we have a single element superconductor where, for example, we have increased ω_D by isotopic substitution, we see that T_c no longer rises proportionally to ω_D because the rise in ω_D also increases the Coulomb repulsion term on the right-hand side of Eq. (3.87), which goes in the direction of reducing the effective attractive interaction, and thus of reducing T_c. Hence the simple result Eq. (3.75) is no longer valid. More precisely, by differentiating Eq. (3.87), we find

$$\frac{\delta T_c}{T_c} = \frac{\delta \omega_D}{\omega_D} \left[1 - \left(\frac{\ln \frac{1.13 \hbar \omega_D}{k_B T_c}}{\ln \frac{\omega_c}{\omega_D} + \frac{1}{N_0 V_c}} \right)^2 \right] \tag{3.88}$$

Indeed we see that the second term in the bracket of the right-hand side comes to reduce the relative change of the critical temperature $\delta T_c / T_c$ with respect to the simple proportionality $\delta T_c / T_c = \delta \omega_D / \omega_D$. It is even possible in this simple model to have no isotope effect at all, that is no change of T_c while ω_D is changed. This occurs if the bracket is zero, that is, from Eq. (3.87),

$$\frac{1}{\ln \frac{\omega_c}{\omega_D} + \frac{1}{N_0 V_c}} = \frac{1}{2} N_0 V_p \tag{3.89}$$

Hence we see that the simple isotope effect Eq. (3.75) does not hold in general, in actual agreement with experiments, as indicated at the beginning of this section.

Actually, the situation with respect to the isotope effect in general is even more complex because clearly most superconductors are not made of a single element. However, it is possible to proceed to an isotopic substitution for only one of the elements of a multi-element compound. (Regardless, even if one tried to substitute several elements, the relative mass changes would not be the same for the various elements, so the situation would not simplify.) The resulting modification of the phonon frequencies becomes obviously much more involved and requires detailed calculations, instead of the simple proportionality rule we have for a single element superconductor. Clearly the isotope effect gets difficult to predict.

An additional complication comes from the experimental side. What one would really like to have in an isotopic substitution is exactly the same compound with only one isotope replaced by another for one atomic element. However, whatever the practical method followed to reach this goal, it is almost impossible to guaranty that all imperfections, such as defects or departure from correct stoichiometry, which all may affect T_c, are identical in the original and the substituted samples. Hence this provides an additional source for the change of critical temperature upon isotopic substitution. If these experimental uncertainties can be eliminated within a good accuracy, then the observation of an isotopic shift for T_c, whatever its value, is a clear indication that lattice vibrations play some role in the electron–electron interaction responsible for pairing. On the other hand, as we have just seen above, the lack of isotopic shift is not enough to eliminate the possibility that electron–phonon interaction plays some role. In summary, if experimental uncertainties can be eliminated, the existence of an isotope shift is a strong indication for a role of the electron–phonon interaction. But its quantitative interpretation is a complicated matter.

3.7 Electron-Phonon Interaction

In this section, we come back to the microscopic aspects of pairing, by considering a more precise description of the effective attractive electron–electron interaction through electron–phonon interaction.

3.7.1 Jellium Model

We consider first the "jellium" model. This is an oversimplification of the physical situation in a metal, but it allows one to obtain simple explicit results which are moreover quite reasonable physically. In this model the ionic lattice is replaced by a positively charged continuous background, the "jellium," which is uniform at equilibrium, with density n_{i0}. Accordingly, the electron gas is also uniform at equilibrium, with density n_0. In this way, one gets rid of the complexity linked to the discrete structure of the ionic lattice. We take for simplicity the case of a metal with a single ion of mass M, and charge $|e|$ ($e < 0$ is the electron charge), per unit cell of volume v_0. Hence the equilibrium jellium and electronic densities are equal $n_{i0} = n_0$.

This jellium has a natural vibration mode at its plasma frequency. Indeed any density fluctuation $\delta n_i(\mathbf{r}, t)$ of the jellium produces a corresponding charge fluctuation $\delta \rho_i(\mathbf{r}, t) = |e| \, \delta n_i(\mathbf{r}, t)$, and a resulting electric field $\mathbf{E}(\mathbf{r}, t)$ given by Gauss equation div $\mathbf{E} = \delta \rho_i / \epsilon_0$. This electric field produces a jellium motion that follows Newton's law $M d\mathbf{v}_i / dt = |e| \mathbf{E}$, where $\mathbf{v}_i$ is the jellium velocity. (We have taken an elemental volume of the jellium equal to the unit cell volume v_0.) Finally this jellium velocity is linked to the jellium density fluctuation by the continuity equation $\partial(\delta n_i)/\partial t + \text{div}\,(n_i \mathbf{v}_i) = 0$.

In the case of very small fluctuations, to which we restrict ourselves, the velocity $\mathbf{v}_i$ is small, so we can replace in the continuity equation the jellium density n_i by its equilibrium value $n_{i0} = 1/v_0$. This leads to linearized equations. Similarly, in this case of very small velocities, we can replace in Newton's law the total derivative $d\mathbf{v}_i/dt$ by the partial derivative $\partial \mathbf{v}_i/\partial t$, the additional contribution $(\mathbf{v}_i \cdot \nabla)\mathbf{v}_i$ being negligible. Going to Fourier transforms with the same notations as in Sections 1.3.1 and 1.3.2, which leads to the substitutions $\nabla \to i\mathbf{q}$ and $\partial/\partial t \to -i\omega$, these equations become respectively

$$ i\mathbf{q} \cdot \mathbf{E} = \frac{|e|\delta n_i}{\epsilon_0} \qquad -i\omega M\mathbf{v}_i = |e|\mathbf{E} \qquad -i\omega \delta n_i + in_{i0}\,\mathbf{q} \cdot \mathbf{v}_i = 0 \qquad (3.90) $$

where for clarity we have not written the variables $(\mathbf{q}, \omega)$ for all the various physical quantities $\mathbf{E}$, δn_i and $\mathbf{v}_i$. Proceeding to the simple elimination of these quantities between these three equations leads to the conclusion that a spontaneous jellium density fluctuation can exist provided the frequency ω is equal to the jellium plasma frequency ω_i given by

$$\omega_i^2 = \frac{n_{i0}e^2}{M\epsilon_0} \tag{3.91}$$

However, in this calculation, we have not taken into account the reaction of the electron gas which screens the jellium charge fluctuations. We consider now this screening by the electron gas.

In handling this screening, we first take into account that we are only interested in low frequencies of the order of the jellium plasma frequency ω_i. These are much smaller than the typical electronic frequencies, which are of the order of the electronic plasma frequency ω_p (since $\omega_i/\omega_p = (m/M)^{1/2}$), or the Fermi energy. Hence we may consider that the reaction of the electron gas to the jellium motion is instantaneous and can be evaluated in its static limit $\omega = 0$. Moreover, for simplicity, we will restrict ourselves to the Thomas-Fermi approximation for this evaluation. In this approximation, one considers the small region around $\mathbf{r}$, of extension $\delta\mathbf{r}$, to be macroscopic so it can be described by standard thermodynamics. Naturally this is valid only for small wavevectors corresponding to slow spatial variations of the various physical quantities. Nevertheless, when extrapolated to larger wavevectors, it leads to physically reasonable results, although it misses some important physics, such as Friedel oscillations for the electronic density, which are linked to the existence of the Fermi surface.

The reaction of the electron gas is characterized by its dielectric constant. If some electrostatic perturbation is created by "external" charges $\rho_{\text{ext}}(\mathbf{r})$, with the corresponding electric displacement field $\mathbf{D}(\mathbf{r})$ satisfying div $\mathbf{D}(\mathbf{r}) = \rho_{\text{ext}}(\mathbf{r})$, "induced" electronic charges $\rho_{\text{ind}}(\mathbf{r})$ appear. The total electronic charge density is $en_0 + \rho_{\text{ind}}(\mathbf{r})$, but the equilibrium electronic charge density en_0 is compensated by the equilibrium jellium charge density $|e|n_{i0}$, since with our assumptions $n_{i0} = n_0$. The electric field $\mathbf{E}(\mathbf{r})$ is linked to the "total" charge density $\rho_{\text{tot}}(\mathbf{r}) = \rho_{\text{ext}}(\mathbf{r}) + \rho_{\text{ind}}(\mathbf{r})$ by div $\mathbf{E}(\mathbf{r}) = \rho_{\text{tot}}(\mathbf{r})/\epsilon_0$. The dielectric constant $\epsilon(\mathbf{q}) = \epsilon_0\epsilon_r(\mathbf{q})$ is defined from the Fourier transforms $\mathbf{D}(\mathbf{q})$ and $\mathbf{E}(\mathbf{q})$ by $\mathbf{D}(\mathbf{q}) = \epsilon(\mathbf{q})\mathbf{E}(\mathbf{q})$. Since, going to Fourier transforms, we have $i\mathbf{q}\cdot\mathbf{D}(\mathbf{q}) = \rho_{\text{ext}}(\mathbf{q})$ and $i\mathbf{q}\cdot\mathbf{E}(\mathbf{q}) = \rho_{\text{tot}}(\mathbf{q})/\epsilon_0$, the relative dielectric constant $\epsilon_r(\mathbf{q})$ is obtained by

$$\epsilon_r(\mathbf{q}) = \frac{\rho_{\text{ext}}(\mathbf{q})}{\rho_{\text{tot}}(\mathbf{q})} \tag{3.92}$$

We evaluate the induced electronic charges by writing that in the static limit, in the Thomas-Fermi approximation, the chemical potential of the electron gas should be a constant independent of spatial position. However, the external charges $\rho_{\text{ext}}(\mathbf{r})$ create an additional space-dependent electric potential. This produces an increase for an electron energy, and accordingly raises its chemical potential by the same amount. This is compensated by an increase $\delta n(\mathbf{r}) = n_{\text{ind}}(\mathbf{r}) = \rho_{\text{ind}}(\mathbf{r})/e$ in the electronic density, raising the equilibrium chemical potential by $\delta\mu(\mathbf{r}) = (\partial\mu/\partial n)\delta n(\mathbf{r})$, where $(\partial n/\partial\mu) = 2N_0 = mk_F/(\pi^2\hbar^2)$ is the electronic density of states for both spin populations at the Fermi energy. On the other hand, to obtain the electric potential $V_{\text{tot}}(\mathbf{r})$ acting on an electron, and the resulting increase in chemical potential $eV_{\text{tot}}(\mathbf{r})$, we have to take into account all charges $\rho_{tot}(\mathbf{r}) = \rho_{ext}(\mathbf{r}) + \rho_{ind}(\mathbf{r})$. This potential is linked to the corresponding charges by Poisson's equation $\Delta V_{tot}(\mathbf{r}) = -\rho_{tot}(\mathbf{r})/\epsilon_0$. Hence in order to obtain a constant chemical potential we must have that its total variation is equal to zero, so

$$eV_{tot}(\mathbf{r}) + \frac{\rho_{ind}(\mathbf{r})}{2eN_0} = 0 \tag{3.93}$$

Going to Fourier transforms, for which Poisson's equation implies $\epsilon_0 q^2 V_{tot}(\mathbf{q}) = \rho_{tot}(\mathbf{q})$, we find

$$\epsilon_r(\mathbf{q}) = \frac{\rho_{tot}(\mathbf{q}) - \rho_{ind}(\mathbf{q})}{\rho_{tot}(\mathbf{q})} = 1 + \frac{k_s^2}{q^2} \qquad k_s^2 \equiv \frac{2e^2 N_0}{\epsilon_0} \tag{3.94}$$

This result is physically quite reasonable. It implies that in the limit of small wavevector $q \to 0$, the dielectric constant goes to infinity, which means that an external potential is fully screened at large distances. This is related to the well-known fact that for a metal in static conditions, the bulk has no charges at a macroscopic scale and any additional charges appear only at the surface. On the other hand, for large wavevectors, $\epsilon_r(\mathbf{q}) \simeq 1$, which means that at short distances screening is weak, as it is physically expected.

More quantitatively Eq. (3.92) implies, through Poisson's equation, that the same relation holds between the corresponding electric potentials $\epsilon_r(\mathbf{q}) = V_{ext}(\mathbf{q})/V_{tot}(\mathbf{q})$. If we take the example of an external charge Q, localized at the origin, so that $\rho_{ext}(\mathbf{r}) = Q\,\delta(\mathbf{r})$, the corresponding electric potential is $V_{ext}(\mathbf{r}) = Q/(4\pi\epsilon_0 r)$. In Fourier transform $\rho_{ext}(\mathbf{q}) = Q$ and $V_{ext}(\mathbf{q}) = Q/\epsilon_0 q^2$, so that from Eq. (3.94) the screened potential is $V_{tot}(\mathbf{q}) = V_{ext}(\mathbf{q})/\epsilon_r(\mathbf{q}) = Q/[\epsilon_0(k_s^2 + q^2)]$. The Fourier transform is easily obtained,[4] and one finds for the screened potential a Yukawa function

$$V_{tot}(\mathbf{r}) = \frac{Q}{4\pi\epsilon_0 r} e^{-k_s r} \tag{3.95}$$

This expression displays explicitly the fact that the Coulomb potential is screened over a distance of order $1/k_s$, which is called the screening length. In standard metals it is of the order of the inter-electronic distance or the Fermi wavelength.

Let us now come back to the plasma mode of the jellium. If we take into account the electronic screening, we have merely to replace the vacuum permittivity ϵ_0 by the electron gas permittivity $\epsilon_0 \epsilon_r(\mathbf{q})$, since this electron gas reacts essentially instantaneously to the jellium motion. This leads us to replace Eq. (3.91) with

$$\omega_q^2 = \frac{n_{i0}e^2}{M\epsilon_0 \epsilon_r(\mathbf{q})} = \frac{n_{i0}e^2}{M\epsilon_0} \frac{q^2}{k_s^2 + q^2} \tag{3.96}$$

We see that instead of a fixed frequency, we obtain now a dispersion relation with the frequency depending on the wavevector, as it occurs for the acoustic phonon dispersion relation in a standard metal. Moreover, for small wavevectors, we find a linear dispersion relation $\omega_q = c_s q$ with a sound velocity $c_s^2 = n_{i0}e^2/(M\epsilon_0 k_s^2)$. Taking into account

---

[4] It is even easier to check the inverse Fourier transform: $V_{tot}(\mathbf{q}) = \int d\mathbf{r}\, e^{-i\mathbf{q}\cdot\mathbf{r}} V_{tot}(\mathbf{r}) = (Q/4\pi\epsilon_0)2\pi \int_0^\infty dr\, r^2(e^{-k_s r}/r)\int_0^\pi d\theta \sin\theta \exp(-iqr\cos\theta)$ after azimuthal integration. The integration over the polar angle θ gives $\int_0^\pi d\theta \sin\theta \exp(-iqr\cos\theta) = 2\,\mathrm{Im}\,e^{iqr}/qr$ and the remaining integration is $\mathrm{Im}\int_0^\infty dr\, e^{-k_s r + iqr} = \mathrm{Im}\,[1/(k_s - iq)] = q/(k_s^2 + q^2)$, which leads indeed to $V_{tot}(\mathbf{q}) = Q/[\epsilon_0(k_s^2 + q^2)]$.

$n_{i0} = n_0$ and $2N_0 = 3n_0/2E_F$ for a free electron gas, one obtains from Eq. (3.94) $c_s = v_F(m/3M)^{1/2}$. For a Fermi velocity $v_F \sim 10^6$ m/s and a typical electron to ion mass ratio $m/M \sim 10^{-4}$, this gives a sound velocity $c_s \sim 10^4$ m/s, which is quite a reasonable estimate for the sound velocity in a standard metal. Moreover, the large wavevector value $\omega_q = \omega_i = \omega_p(m/M)^{1/2}$ gives an energy of order 0.1 eV, which is a reasonable order of magnitude for optical phonon. In conclusion, the jellium model provides a phonon spectrum that is physically quite satisfactory.

We finally consider the effective electron–electron interaction for the jellium model. The bare electron–electron interaction is naturally the Coulomb interaction $U_{Coul}(\mathbf{r}) = e^2/(4\pi\epsilon_0 r)$, corresponding to the bare electric potential $V_{ext}(\mathbf{r}) = e/(4\pi\epsilon_0 r)$ created by one electron and felt by the other electron. However, this potential is modified by the screening reaction of both the jellium and the electron gas. This results in a total electric potential $V_{tot}(\mathbf{r})$, which is felt by the other electron, so the effective interaction of the two electrons is $U_{eff}(\mathbf{r}) = eV_{tot}(\mathbf{r})$. Just as above, the ratio in the Fourier transform between V_{ext} and V_{tot} is the dielectric constant, except that we have now to take into account its frequency dependence since we are interested in frequencies of the order of the jellium plasma frequency ω_i. Similarly we obtain this dielectric constant, just as in Eq. (3.92), by finding the ratio between the total charges and the external ones. Hence

$$\epsilon_r(\mathbf{q}, \omega) = \frac{V_{ext}(\mathbf{q}, \omega)}{V_{tot}(\mathbf{q}, \omega)} = \frac{\rho_{ext}(\mathbf{q}, \omega)}{\rho_{tot}(\mathbf{q}, \omega)} \qquad (3.97)$$

To obtain the total charge $\rho_{tot}(\mathbf{q}, \omega)$, we have to take into account both the response of the electron gas, which we call again $\rho_{ind}(\mathbf{q}, \omega)$, and the response of the jellium, which we call $\rho_i(\mathbf{q}, \omega)$, so that

$$\rho_{tot}(\mathbf{q}, \omega) = \rho_{ext}(\mathbf{q}, \omega) + \rho_{ind}(\mathbf{q}, \omega) + \rho_i(\mathbf{q}, \omega) \qquad (3.98)$$

Both responses $\rho_{ind}(\mathbf{q}, \omega)$, and $\rho_i(\mathbf{q}, \omega)$ are produced by the total charge $\rho_{tot}(\mathbf{q}, \omega)$. Actually, since the electronic gas response is very fast and we are interested in low frequencies, its response is the static response, and it is still given by the result we have already found in Eq. (3.94), that is

$$\rho_{ind}(\mathbf{q}, \omega) = -\frac{k_s^2}{q^2} \rho_{tot}(\mathbf{q}, \omega) \qquad (3.99)$$

With respect to the jellium response, the last two equations in Eq. (3.90) are still valid (the first one is equivalent to Poisson's equation), but we have to take into account in the second one that $\mathbf{E}$ is now the total field so that $i\mathbf{q} \cdot \mathbf{E}(\mathbf{q}, \omega) = q^2 V_{tot}(\mathbf{q}, \omega) = \rho_{tot}(\mathbf{q}, \omega)/\epsilon_0$. Together with the third equation, this gives

$$\rho_i(\mathbf{q}, \omega) = |e|\delta n_i(\mathbf{q}, \omega) = \frac{n_{i0}e^2}{M\epsilon_0\omega^2} \rho_{tot}(\mathbf{q}, \omega) = \frac{\omega_i^2}{\omega^2} \rho_{tot}(\mathbf{q}, \omega) \qquad (3.100)$$

with the definition Eq. (3.91) for ω_i. Introducing Eq. (3.99) and Eq. (3.100) into Eq. (3.98) gives finally from Eq. (3.97) the dielectric constant

$$\epsilon_r(\mathbf{q}, \omega) = 1 + \frac{k_s^2}{q^2} - \frac{\omega_i^2}{\omega^2} \qquad (3.101)$$

We note that $\epsilon_r(\mathbf{q}, \omega) = 0$ gives the eigenmodes of the system since, from Eq. (3.97), it implies that a spontaneous finite charge fluctuation $\rho_{\text{tot}}(\mathbf{q}, \omega)$ can appear from a vanishingly small external excitation $\rho_{\text{ext}}(\mathbf{q}, \omega) \to 0$. In the present case we obtain that from Eq. (3.101), the eigenfrequencies are given by

$$\omega^2 = \omega_i^2 \frac{q^2}{k_s^2 + q^2} = \omega_q^2 \tag{3.102}$$

Hence we recover the jellium phonon modes ω_q in agreement with Eq. (3.96). We may factorize the electron gas screening in Eq. (3.101) to make this more obvious and write

$$\epsilon_r(\mathbf{q}, \omega) = [1 + \frac{k_s^2}{q^2}](1 - \frac{\omega_q^2}{\omega^2}) \tag{3.103}$$

From Eq. (3.103), we obtain finally for the effective electron–electron interaction

$$U_{\text{eff}}(\mathbf{q}, \omega) = \frac{U_{\text{Coul}}(\mathbf{q})}{\epsilon_r(\mathbf{q}, \omega)} = \frac{e^2}{\epsilon_0(k_s^2 + q^2)} \frac{\omega^2}{\omega^2 - \omega_q^2} = \frac{e^2}{\epsilon_0(k_s^2 + q^2)}[1 + \frac{\omega_q^2}{\omega^2 - \omega_q^2}] \tag{3.104}$$

The first term is the screened Coulomb interaction we had already found above. It is naturally repulsive, and it is this contribution which has been modeled by the V_c term in the Hamiltonian Eq. (3.76) in Section 3.6. The second term is the contribution of the jellium: it disappears if we let the ionic mass M go to infinity so that the ions cannot move and their plasma frequency ω_i goes to zero, as well as the phonon frequencies ω_q.

We see that at high frequency $\omega \gg \omega_i$, where the heavy ions cannot follow the excitation, this contribution is positive and mostly negligible. On the other hand, at small frequency $\omega \lesssim \omega_i$ (the details depending on the wavevector $\mathbf{q}$), it is negative and accordingly provides an attractive contribution to the effective interaction. This change of sign can be understood as an overscreening from the ions: because the frequency ω is in the vicinity of their resonance frequency, the ions motions go beyond the displacement necessary to merely screen the bare electron–electron interaction. This is the attractive interaction, which is merely modeled by the V_p term in the Hamiltonian Eq. (3.76) in Section 3.6, and the only one which is retained in the standard BCS theory. One notes that in Eq. (3.104), the effective interaction is zero at $\omega = 0$ so that there is a perfect compensation between the repulsive and the attractive contributions. This is a specific feature of this simple jellium model, and since electron gas and ionic screening are not in general related in a simple way, this cancellation has no reason to persist in more elaborated models.

3.7.2 Electron-Phonon Hamiltonian

We consider now this question of the indirect electron–electron attractive interaction in a somewhat more general and formal way by introducing the electron–phonon interaction term in the electronic Hamiltonian. In the preceding case of the jellium model the modification of the electronic energy due to the jellium displacement is $U_{ep} = eV_{ep}$, where V_{ep} is the electric potential created by the jellium charge fluctuations $\rho_i = |e|\delta n_i$. This potential is related to ρ_i by Poisson's equation $\Delta V_{ep} = -\rho_i/\epsilon_0$. On the other hand, the jellium density fluctuation δn_i is obtained from the jellium local displacement from equilibrium $\mathbf{u}_i(\mathbf{r}, t)$ by mass conservation $\delta n_i = -\text{div}\,(n_i \mathbf{u}_i) \simeq -n_{i0}\text{div}\,(\mathbf{u}_i)$, which is the same as the continuity equation integrated over time.

We write now the second quantized version of this term in the electronic Hamiltonian: $U_{ep}(\mathbf{r}) \rightarrow \sum_{\mathbf{k},\mathbf{k}'} U_{ep}(\mathbf{k}' - \mathbf{k})c_{\mathbf{k}'}^{\dagger}c_{\mathbf{k}}$, where $U_{ep}(\mathbf{k}' - \mathbf{k}) = \langle \mathbf{k}'|U_{ep}(\mathbf{r})|\mathbf{k}\rangle$ is the Fourier transform of $U_{ep}(\mathbf{r})$. It is related to the jellium charge fluctuations and displacement by Poisson's equation $V_{ep}(\mathbf{q}) = \rho_i(\mathbf{q})/(\epsilon_0 q^2) = -i|e|n_{i0}\mathbf{q} \cdot \mathbf{u}_i(\mathbf{q})/(\epsilon_0 q^2)$. We write the jellium displacement in terms of the phonon creation and annihilation operators $a_{\mathbf{q}}^{\dagger}$ and $a_{\mathbf{q}}$ for the mode with wavevector $\mathbf{q}$. This is essentially the same[5] as the expression of the position x of an harmonic oscillator of mass m and frequency ω in terms of the corresponding creation and annihilation operators $x = \sqrt{\hbar/2m\omega}\,(a + a^{\dagger})$. It reads

$$\mathbf{u}_i(\mathbf{q}) = \left(\frac{\hbar}{2n_{i0}M\omega_{\mathbf{q}}}\right)^{1/2}\vec{\epsilon}_{\mathbf{q}}\left(a_{\mathbf{q}} + a_{-\mathbf{q}}^{\dagger}\right) \tag{3.105}$$

where $\vec{\epsilon}_{\mathbf{q}}$ is the polarization unit vector of the corresponding mode. In the present case, the mode is longitudinal, parallel to the wavevector $\mathbf{q}$, since transverse oscillations do not produce any jellium density fluctuations. Hence $\mathbf{q} \cdot \vec{\epsilon}_{\mathbf{q}} = q$. In Eq. (3.105), n_{i0} is the number of unit cell in our sample with unit volume, and it is equal to electronic density $n_{i0} = n$ since we have assumed monovalent ions.

Gathering the various contributions we find for the second quantized expression of the electron–phonon interaction

$$H_{\text{ep}} = \sum_{\mathbf{k},\mathbf{k}',\mathbf{q}} g_{\mathbf{k}\,\mathbf{k}'\mathbf{q}}\,c_{\mathbf{k}'}^{\dagger}c_{\mathbf{k}}\left(a_{\mathbf{q}} + a_{-\mathbf{q}}^{\dagger}\right) \tag{3.106}$$

with

$$g_{\mathbf{k}\,\mathbf{k}'\mathbf{q}} = \left(\frac{\hbar}{2nM\omega_{\mathbf{q}}}\right)^{1/2}\frac{iqne^2}{\epsilon_0 q^2}\delta_{\mathbf{k}',\mathbf{k}+\mathbf{q}} \tag{3.107}$$

where the Kronecker delta $\delta_{\mathbf{k}',\mathbf{k}+\mathbf{q}}$ indicates explicitly that $\mathbf{k}' - \mathbf{k} = \mathbf{q}$, corresponding to momentum conservation. In Eq. (3.107) and our above argument, we have not taken into account the screening of the jellium charge fluctuations by the electron gas. When we take it into account we should replace the vacuum permittivity ϵ_0 by the corresponding value $\epsilon_0\epsilon_r(\mathbf{q})$ for the electron gas, with $\epsilon_r(\mathbf{q})$ given by Eq. (3.94) within the Thomas-Fermi approximation, as we have done it in Eq. (3.96). When this is done, the expression of the electron–phonon coupling $g_{\mathbf{k}\,\mathbf{k}'\mathbf{q}}$ is changed into

$$g_{\mathbf{k}\,\mathbf{k}'\mathbf{q}} = \left(\frac{\hbar}{2nM\omega_{\mathbf{q}}}\right)^{1/2}\frac{iqne^2}{\epsilon_0(k_s^2 + q^2)}\delta_{\mathbf{k}',\mathbf{k}+\mathbf{q}} \tag{3.108}$$

Naturally Eq. (3.106) is a completely general form for the electron–phonon interaction. Starting from this expression, we can write the corresponding effective electron–electron interaction. We treat H_{ep} as a perturbation since it is expected to bring a small change to the overall Fermi sea. We are looking for the effective matrix element $V_{\mathbf{k},\mathbf{k}'}$ of H_{int} in Eq. (2.23), which gives the amplitude for the scattering of an electron pair $(\mathbf{k}, -\mathbf{k})$ into an

---

[5] For details, see the Appendix on Second Quantization.

electron pair $(\mathbf{k}', -\mathbf{k}')$ due to the electron–phonon interaction. We note that there is no contribution of H_{ep} to first order in perturbation since we want to deal with a situation where no thermal phonons are present (this is a $T = 0$ situation with respect to the phonons). Hence we want no phonons in the initial or final state, while H_{ep} either creates or annihilates a phonon. Hence we have to go to second order in perturbation theory, where a (virtual) phonon can be created in a first process and then annihilated in a second process, as is shown in Fig. (2.2).

From standard second-order perturbation theory for nearly degenerate states, the effect of $H_{\rm ep}$ is to give to the Hamiltonian in the pair $(\mathbf{k}, -\mathbf{k})$ subspace a contribution:

$$H_{int}^{(2)} = \sum_{n} \frac{H_{\rm ep}|n\rangle \langle n|H_{\rm ep}}{E_0 - E_n} \tag{3.109}$$

where the sum extends over all excited states $|n\rangle$ having two electrons and a phonon. In the excited state $|n\rangle$, from Eq. (3.106), an electron has, for example, been scattered from state $\mathbf{k}$ to state $\mathbf{k}'$, while a phonon with momentum $-\mathbf{q} = \mathbf{k} - \mathbf{k}'$ and energy $\hbar\omega_{\mathbf{k}'-\mathbf{k}}$ has been created. In practice, coherently with the weak coupling approximation, we are only interested in states with small kinetic energies $\xi_{\mathbf{k}}$, comparable to the gap or the temperature, which are negligible compared to a typical phonon energy $\hbar\omega_{\mathbf{q}}$. Hence the difference between the excited and the unperturbed energy $E_n - E_0$ is essentially equal to $\hbar\omega_{\mathbf{q}}$. The corresponding matrix element $\langle n|H_{\rm ep}|(\mathbf{k}, -\mathbf{k})\rangle$ is $g_{\mathbf{k}\,\mathbf{k}'\mathbf{q}}$ from Eq. (3.106). Then the phonon is annihilated in the process where the remaining electron is scattered from state $-\mathbf{k}$ to state $-\mathbf{k}'$, with the corresponding matrix element $\langle(\mathbf{k}', -\mathbf{k}')|H_{\rm ep}|n\rangle$ being $g_{-\mathbf{k},-\mathbf{k}',-\mathbf{q}}$, from Eq. (3.106), with $g_{-\mathbf{k},-\mathbf{k}',-\mathbf{q}} = g^*_{\mathbf{k}\,\mathbf{k}'\mathbf{q}}$ from time reversal invariance. Hence the overall coefficient in the numerator of Eq. (3.109) is $|g_{\mathbf{k}\,\mathbf{k}'\mathbf{q}}|^2$. We have also to take into account that an equivalent possibility is that this is the electron $-\mathbf{k}$, which is scattered to state $-\mathbf{k}'$, with emission of a phonon $\mathbf{q}$; and then electron $\mathbf{k}$, which is scattered to state $\mathbf{k}'$ with annihilation of phonon $\mathbf{q}$. The corresponding coefficient is $g_{\mathbf{k}\,\mathbf{k}'\mathbf{q}} g_{-\mathbf{k},-\mathbf{k}',-\mathbf{q}} = |g_{\mathbf{k}\,\mathbf{k}'\mathbf{q}}|^2$, which is the same as for the first process we considered. Gathering these results, we obtain from Eq. (3.109) an effective electron–electron interaction

$$U_{\rm eff}(\mathbf{k}, \mathbf{k}') = -\frac{2|g_{\mathbf{k}\,\mathbf{k}'\mathbf{q}}|^2}{\hbar\omega_{\mathbf{q}}} \tag{3.110}$$

which is indeed an attractive interaction.

Coming back to the specific case of the jellium model by making use of Eq. (3.108), we find

$$U_{\rm eff}(\mathbf{k}, \mathbf{k}') = -\frac{2}{\hbar\omega_{\mathbf{q}}} \left(\frac{\hbar}{2nM\omega_{\mathbf{q}}}\right) \left(\frac{qne^2}{\epsilon_0(k_s^2 + q^2)}\right)^2 = -\frac{e^2}{\epsilon_0(k_s^2 + q^2)} \tag{3.111}$$

by making use of the expression Eq. (3.96) for $\omega_{\mathbf{q}}$. This result is in agreement with the jellium contribution (second term) in Eq. (3.104) in the static limit $\omega = 0$, appropriate for electrons with small kinetic energies $\xi_{\mathbf{k}}$ and $\xi_{\mathbf{k}'}$.

3.8 Other Pairing Symmetries and Mechanisms

The electron–phonon interaction mechanism for pair formation, which we have considered in detail in the preceding section, is clearly a very important ingredient in BCS theory since it provides the physical justification for the existence of Cooper pairs. Actually, quite often "BCS Theory" is meant to correspond not only to the BCS formalism but also to the specific electron–phonon interaction mechanism for pairing. This is perhaps too restrictive since we have seen that as soon as the pairing mechanism is found, the BCS formalism can be developed basically without any reference to the specific pairing mechanism. Anyway, almost as soon as BCS theory appeared, physicists realized that pair formation could also arise from other physical processes.

For example, Pitaevskii [27] suggested in 1959 that in liquid ^{3}He, long-ranged van der Waals attractive forces could lead to pair formation and to superfluidity in a way completely analogous to what happens in a BCS superconductor. The main difference with a superconductor is that ^{3}He atoms are not charged, although they are fermions giving rise in the involved temperature range to quantum degeneracy with the presence of ^{3}He atoms Fermi seas, with a Fermi energy equivalent to a temperature of order 1 K (depending on the specific definition and on pressure). More specifically, for a ^{3}He atom, there is no intrinsic angular momentum coming from the electrons, which form full shells. Rather there is an intrinsic angular momentum, corresponding to a spin $I = 1/2$, coming from the ^{3}He nucleus, with an associated nuclear magnetic moment. Hence the situation is in this respect quite analogous to the one found for electrons in a metal. However, the case of superfluid ^{3}He is a remarkable example of a BCS superfluid where the symmetry as well as the pairing mechanism are different from the ones occurring in the traditional BCS theory we have just described. For this reason, it is worth considering it in some detail.

3.8.1 Superfluid ^{3}He

A first interesting point is the determination of the critical temperature for the appearance of this superfluidity. The initial evaluations of this temperature were rather optimistic, with a T_c in the 0.1 K range, which could be reached fairly easily experimentally. When experiments failed to find the predicted superfluid phase more refined calculations rather led to a T_c in the 0.01 K range. This is harder to obtain, but these predictions led to considerable experimental progress in reaching lower temperatures. Nevertheless the expected superfluid phase did not appear in this temperature range. Further theoretical evaluations turned quite pessimistic, finding a critical temperature in an unreachably low range.

Afterward, experimentalists lost interest in looking for this unattainable superfluid phase. However, further experimental progress allowed to reach the mK range. The interest rather focused on solid ^{3}He, which arises in the low-temperature range under the pressure of typically 30 bars. Physicists were more specifically interested in the magnetic order, expected to arise in the solid at very low temperature. It was experimentally convenient to work on the melting curve of the solid. It was accidentally, in the course of these explorations, that the manifestation of superfluidity[6] in liquid ^{3}He was actually discovered [28] at 2.6 mK on this melting curve at 34 bars. The general conclusion to be drawn from these events is that the prediction of a critical temperature for strongly interacting systems is quite a difficult task.

Regarding the symmetry of the pair wave function, it was clear from the start that the s-wave symmetry of the standard BCS theory could not be the appropriate answer. Indeed for s-wave there is a nonzero probability to have the two pairing fermions at the same position, corresponding to a zero separation $\mathbf{r} = \mathbf{0}$. However, two ^{3}H atoms have at short distance an extremely strong hard-core repulsion, so the s-wave state has clearly a high positive contribution to its energy coming from this repulsion. Hence it cannot be the lowest energy solution. On the other hand, any non s-wave solution $\ell \neq 0$ will give a zero probability to have the two pairing atoms at the same place, so basically the hard-core repulsion does not give any contribution to the pair energy. However, there was no obvious answer from theory for the appropriate angular momentum. It was unclear if this was for example p-wave ($\ell = 1$) or d-wave ($\ell = 2$) pairing. Finally the analysis[7] of the nuclear magnetic resonance properties of the superfluid proved that the ^{3}He atoms formed p-wave pairs.

Coherent, with our discussion of the Cooper problem in Section 2.1, we have to retain the $\ell = 1$ component V_1 of the interaction potential, with wavevectors taken at the Fermi surface. The angular dependence of the three degenerate solutions for the pair wave function is given by the spherical harmonics $Y_{1,m}(\hat{\mathbf{k}})$. It is more convenient to use equivalently $\hat{\mathbf{k}}_x$, $\hat{\mathbf{k}}_y$, and $\hat{\mathbf{k}}_z$ for the angular dependence of these three orthogonal wave functions.

In addition to this orbital degeneracy, we have to take into account the presence of a spin degeneracy. Indeed we have seen that the s-wave symmetry of the standard BCS wave function implies that the spins of the two paired electrons must be in a singlet state, since the overall wave function must be odd with respect to the exchange of the variables corresponding to these two fermions. Now, in the case of superfluid ^{3}He, the $\ell = 1$ orbital part changes sign under the exchange of the atoms orbital variables (which corresponds to make $\hat{\mathbf{k}} \rightarrow -\hat{\mathbf{k}}$). The overall odd symmetry of the wave function implies then that the spin part of the wave function must be even under exchange, which means that the two spins are in a triplet state. Hence we can have any combination of $|\uparrow\uparrow\rangle$, $|\downarrow\downarrow\rangle$, and $(|\uparrow\downarrow\rangle + |\downarrow\uparrow\rangle)/\sqrt{2}$.

A triplet state corresponds physically to a state which can be represented by a vector. It is convenient to make this explicit in the pair wave function. Considering only the spin dependences, let us write $\psi_{\alpha\beta}$ the pair wave function, with α and β equal to $\uparrow$ or $\downarrow$. This is a 2×2 matrix. Since, together with the unit matrix, Pauli matrices σ_x, σ_y, and σ_z form a basis for 2×2 matrices, we can express $\psi_{\alpha\beta}$ in terms of these matrices. More precisely, since the singlet state is antisymmetric, we have $\psi_{\alpha\beta} = -\psi_{\beta\alpha}$ for its corresponding wave function. This is just the behavior of $(i\sigma_y)_{\alpha\beta}$. Hence the triplet state can be written in terms of the three orthogonal combinations, that is, $\psi_{\alpha\beta} = i(d_x\sigma_x + d_y\sigma_y + d_z\sigma_z)\sigma_y = i(\boldsymbol{\sigma} \cdot \mathbf{d})\sigma_y$. Indeed $i\sigma_x\sigma_y = -\sigma_z$ corresponds to $|\uparrow\uparrow\rangle - |\downarrow\downarrow\rangle$, $\sigma_y\sigma_y = 1$ to $|\uparrow\uparrow\rangle + |\downarrow\downarrow\rangle$, and $i\sigma_z\sigma_y = \sigma_x$ to $|\uparrow\downarrow\rangle + |\downarrow\uparrow\rangle$, so we find indeed a basis for the triplet state. Hence $\mathbf{d}$ is the vector we are looking for to describe the triplet state. Actually, although we have discussed pairing

---

[6] For this discovery D. D. Osheroff, R. C. Richardson, and D. M. Lee received the Nobel Prize in 1996.

[7] A. J. Leggett received the Nobel Prize in 2003 for the theoretical work leading to the proper identification of the superfluid phases of liquid ^{3}He.

symmetry in terms of the pair wave function, the usual way to do it is rather in terms of the corresponding order parameter, to be introduced below in Chapter 5 and mostly Chapter 6.

Another striking experimental discovery about superfluid ^{3}He is that while a single superfluid phase was expected, two different phases (in the absence of a magnetic field) were found. Indeed, at the melting pressure, in addition to the standard second-order transition from the normal state to a superfluid state found at 2.6 mK, a second (first-order) transition to another superfluid phase was found at 2 mK. The phase found right below 2.6 mK is called the A phase, while the second one below 2 mK is the B phase. Actually, their domain of existence depends strongly on pressure. The A phase domain decreases with pressure, and below a polycritical point around 21 bars, the A phase disappears and only the B phase is found. The normal-superfluid critical temperature decreases also with pressure to go below 1 mK at zero pressure.

The precise nature of the pairing has been mostly identified through the nuclear magnetic resonance properties. In the A phase pairs are formed between ↑ spins, and between ↓ spins. This means that the above $\mathbf{d}$ vector is in the $x - y$ plane, the z component being zero. However, we have not specified our choice of axes, and since ^{3}He is an isotropic liquid, any order obtained by rotation of these "spin axes" is equally valid. On the other hand, the common orbital dependence of the wave function corresponds merely to the $Y_{1,1}(\hat{\mathbf{k}})$ spherical harmonic, so it is proportional to $\hat{\mathbf{k}}_x + i\hat{\mathbf{k}}_y$. This is again for a specific choice of the z-axis, usually called the $\boldsymbol{\ell}$ direction. But this $\boldsymbol{\ell}$ vector can in principle be in any direction. The overall huge degeneracy is somewhat reduced, due to the quite weak dipole-dipole interaction between ^{3}He atoms. This creates a link between the "spin space" and the "orbital space," which makes it that $\mathbf{d}$ should be parallel (or antiparallel) to $\boldsymbol{\ell}$. Otherwise these directions are mainly determined by the boundary conditions at the surface of the vessel and by the possible applied magnetic field.

The order is quite different for the B phase. In contrast with the A phase, $\mathbf{d}$ depends on $\hat{\mathbf{k}}$. The simplest form of this relation is merely $\mathbf{d} = \hat{\mathbf{k}}$. However, just as we have indicated for the A phase, we can apply any rotation independently in the "spin space" and in the "orbital space." Hence the general form of the relation between $\mathbf{d}$ and $\hat{\mathbf{k}}$ in the B phase is $\mathbf{d} = R(\hat{\mathbf{k}})$, where R is any rotation. Actually, just as in the A phase, the weak dipole-dipole interaction breaks the independence between the "spin space" and the "orbital space." It turns out that the dipole-dipole interaction constrains the angle θ of the rotation R to be such that $\cos \theta = -1/4$, which gives $\theta \simeq 104^o$. However, the axis of the rotation is completely free, and its actual direction is mainly ruled by boundary conditions and magnetic field.

The B phase turns out to be the phase that was expected from BCS-like calculations, prior to the experimental discovery. On the other hand, although it had also been theoretically investigated beforehand, the existence of the A phase was a surprise. The existence of these two very different phases is a clear sign that the pairing interaction is not a simple one. The standard understanding is that in addition to a van der Waals-like attractive interaction, an additional contribution comes from an indirect interaction, somewhat analogous to the standard BCS phonon-mediated interaction. In the ^{3}He case, this interaction is through "spin fluctuations."

Indeed the ^{3}He susceptibility is much higher than the one corresponding to a non-interacting Fermi gas. This is as if ^{3}He was near a ferromagnetic transition, where all its nuclear spins would be aligned. In the case of a real ferromagnetic transition, the static susceptibility would diverge at the transition, and for small but nonzero frequency, the susceptibility would be very high. If the ferromagnetic transition is not reached but is close to happening, the susceptibility is high in the low-frequency domain. Physically this corresponds to a situation where the nuclear spins fluctuate, getting almost but not completely aligned. These low-frequency fluctuations are also called "paramagnons." This situation is quite analogous to the phonon case, where in Eq. (3.110) low-frequency phonons lead to a strong effective interaction. Actually, for ^{3}He one does not get so near a ferromagnetic transition. Nevertheless the static susceptibility increases markedly when pressure increases toward the melting curve. Since this is also in this higher pressure domain that the A phase appears, while at low pressure only the standard B phase is present, it is quite likely that the appearance of the A phase is linked to the increase in susceptibility and to the corresponding existence of low-frequency spin fluctuations.

3.8.2 Spin Fluctuations

A simple way to model this high susceptibility is to introduce in the Fermi gas an interaction H_{int}, which is unfavorable for antiparallel spins and accordingly favors spin alignment. A very simple model for this interaction is to take a very short-ranged repulsion, that is, with $I > 0$,

$$H_{int} = I \int d\mathbf{r}\, n_\uparrow(\mathbf{r}) n_\downarrow(\mathbf{r}) = I \sum_{\mathbf{k},\mathbf{k}',\mathbf{q}} c^\dagger_{\uparrow \mathbf{k}'+\mathbf{q}/2} c^\dagger_{\downarrow -\mathbf{k}'+\mathbf{q}/2} c_{\downarrow -\mathbf{k}+\mathbf{q}/2} c_{\uparrow \mathbf{k}+\mathbf{q}/2} \qquad (3.112)$$

where $n_\uparrow(\mathbf{r}) = \psi^\dagger_\uparrow(\mathbf{r})\psi_\uparrow(\mathbf{r})$ is the up spin density operator at $\mathbf{r}$, and similarly for $\downarrow$ spins. The last form in Eq. (3.112) is obtained by inserting the relations Eq. (2.30) and Eq. (2.31) between $\psi(\mathbf{r})$ and $c_\mathbf{k}$, and finding momentum conservation through $\int d\mathbf{r}\, e^{i\mathbf{K}\cdot\mathbf{r}} = \delta_{\mathbf{K},0}$.

We recall that the Pauli susceptibility χ_P of the non-interacting Fermi gas is $\chi_P = \mu_0\mu_f^2 N(0)$, where μ_f is the magnetic moment of the fermions (in our case, this is the nuclear magnetic moment μ_n of ^{3}He atoms) and $N(0)$ the density of states at the Fermi surface for both spin populations. Indeed, for a weak magnetic field $\mathbf{B}$ applied along the z-axis, the spin-up fermions get an additional energy $-\mu_f B$, while this is $+\mu_f B$ for the down-spins. This leads to a difference $\mu_\uparrow - \mu_\downarrow = -2\mu_f B$ between the $\uparrow$ and $\downarrow$ spins chemical potentials. In order to bring back the equilibrium condition $\mu_\uparrow = \mu_\downarrow$, down-spin fermions have to be converted into up-spins, to fill the energy interval of width $\mu_f B$ at the Fermi surface. Their number is $N_0\mu_f B$, where $N_0 = N(0)/2$ is the single spin density of states at the Fermi surface. Hence the resulting magnetic moment is $M = 2\mu_f(N_0\mu_f B) = \mu_0\mu_f^2 N(0)H$, leading to the above expression for the Pauli susceptibility.

Compared with Eq. (2.26), we see that because it is repulsive instead of attractive, interaction Eq. (3.112) is unfavorable for singlet pairing. On the other hand, if we consider triplet pairing, and more specifically the $|\uparrow\uparrow\rangle$ component of the triplet, we can see that the situation is different. Indeed, to second order in perturbation we have to consider the contribution analogous to Eq. (3.109), where H_{ep} is replaced with Eq. (3.112). Calling this contribution $H_{sf}^{(2)}$, it reads

$$H_{sf}^{(2)} = \sum_n \frac{H_{\text{int}}|n\rangle \langle n|H_{\text{int}}}{E_0 - E_n} \tag{3.113}$$

Physically, this contribution takes into account that a ↑ electron can polarize the surrounding ↓ electrons, that is, perturb the ↓ Fermi sea, and in this way interact indirectly with another ↑ electron. This is quite analogous with the screening of an electron charge by the Fermi sea, we have seen above in Eq. (3.92) through Eq. (3.94), which results in the modification through screening of the interaction between two electrons. Since this involves charges, this is characterized by the electron gas dielectric constant. In the present case, the indirect interaction comes through the magnetic properties and the related interaction H_{int}, so it involves the susceptibility of the electron gas.

Specifically, if we consider a ($\mathbf{k}$ ↑, $-\mathbf{k}$ ↑) pair, it can be scattered by $H_{sf}^{(2)}$, given by Eq. (3.113), into a ($\mathbf{k}'$ ↑, $-\mathbf{k}'$ ↑) pair in the following way. H_{int} can scatter $\mathbf{k}$ ↑ into $\mathbf{k}'$ ↑ by creating in the ↓ Fermi sea a particle-hole pair with a $\mathbf{k}_h$ hole below the Fermi surface and a particle $\mathbf{k}_h + \mathbf{q}$ above it (we assume for simplicity that temperature $T = 0$), with $\mathbf{q} = \mathbf{k} - \mathbf{k}'$, in order to satisfy momentum conservation. Then, again through the action of H_{int}, the ↓ particle-hole pair can be annihilated by transferring its momentum to the $-\mathbf{k}$ ↑ particle, which gets in this way a final momentum $-\mathbf{k} + \mathbf{q} = -\mathbf{k}'$, as it must be from total momentum conservation. This whole process is depicted in Fig. (3.9).

We obtain the scattering matrix element $J(\mathbf{k}', \mathbf{k}) = \langle \mathbf{k}' \uparrow, -\mathbf{k}' \uparrow |H_{sf}^{(2)}|\mathbf{k} \uparrow, -\mathbf{k} \uparrow \rangle$ corresponding to the above process by taking the intermediate state $|n\rangle$ as corresponding to the addition of the particle-hole pair, in analogy with Eq. (3.109), where it contains the additional phonon. The energy for creation of this particle-hole pair is $\xi_{\mathbf{k}_h+\mathbf{q}} - \xi_{\mathbf{k}_h}$. This leads to

$$J(\mathbf{k}', \mathbf{k}) = I^2 \sum_{\mathbf{k}_h} \frac{f(\xi_{\mathbf{k}_h}) - f(\xi_{\mathbf{k}_h+\mathbf{q}})}{-(\xi_{\mathbf{k}_h+\mathbf{q}} - \xi_{\mathbf{k}_h})} = -I^2 \sum_{\mathbf{k}_h} -\frac{f(\xi_{\mathbf{k}_h+\mathbf{q}}) - f(\xi_{\mathbf{k}_h})}{\xi_{\mathbf{k}_h+\mathbf{q}} - \xi_{\mathbf{k}_h}} \tag{3.114}$$

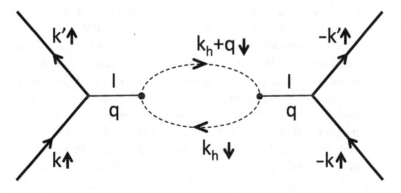

Fig. 3.9   Indirect interaction between two spin ↑ fermions (full lines) through the polarization of the spin ↓ Fermi sea, by the creation and annihilation of particle-hole pair ($\mathbf{k}_h$, $\mathbf{k}_h + \mathbf{q}$), represented by the dashed lines.

We have inserted the statistical factor[8] $f(\xi_{\mathbf{k}_h}) - f(\xi_{\mathbf{k}_h+\mathbf{Q}})$ (where f is the Fermi distribution) to ensure that the hole corresponds to a wavevector $\mathbf{k}_h$ below the Fermi surface (so $f(\xi_{\mathbf{k}_h}) = 1$) and the particle is above it (so $f(\xi_{\mathbf{k}_h+\mathbf{q}}) = 0$). In the last expression in Eq. (3.114), the sum over $\mathbf{k}_h$ reduces in the limit $\mathbf{q} \to \mathbf{0}$ to $\sum_{\mathbf{k}} -\partial f(\xi_{\mathbf{k}})/\partial \xi_{\mathbf{k}} = N_0$, so it is clearly a positive quantity. Hence $J(\mathbf{k}', \mathbf{k})$ is negative, which corresponds indeed to an attractive pairing interaction in the triplet state for these particles.

This factor of $-I^2$ in Eq. (3.114) is essentially the susceptibility of the Fermi gas, since as we have mentioned the indirect interaction between the $\uparrow$ spin particles comes through the magnetic polarization of the $\downarrow$ Fermi sea. More precisely, it is its zero frequency value. One can generalize by considering the Fermi sea susceptibility at nonzero frequency ω and wavevector $\mathbf{q}$. This introduces the reduced susceptibility

$$\bar{\chi}_0(\omega, \mathbf{q}) = \frac{1}{N_0} \sum_{\mathbf{k}} \frac{f(\xi_{\mathbf{k}+\mathbf{q}}) - f(\xi_{\mathbf{k}})}{\hbar\omega - (\xi_{\mathbf{k}+\mathbf{q}} - \xi_{\mathbf{k}})} \tag{3.115}$$

which is equal to 1 for $\omega = 0$ and $\mathbf{q} \to \mathbf{0}$. If we generalize the Pauli susceptibility to the case of nonzero ω and $\mathbf{q}$, the result is $\mu_0 \mu_f^2 N(0) \bar{\chi}_0(\omega, \mathbf{q})$. We do not go here in the details of this calculation since it is essentially considered specifically in Chapter 4, where one has to deal with the current-current response function, whereas we have here to calculate the spin-spin response in order to obtain the susceptibility, which makes basically no difference (compare Eq. (3.115) with Eq. (4.31)).

However, for the pairing interaction as well as for gas susceptibility, our handling is not quite satisfactory since our treatment is equivalent to consider that the Fermi sea itself is unaffected by the interaction Eq. (3.112). Hence this is equivalent to a Hartree-Fock approximation. It is possible to improve on this approximation by making it self-consistent, which leads to the RPA approximation. This is completely analogous to our handling of the electron gas dielectric constant in Eq. (3.92)–Eq. (3.94). If we take, for example, the case of the susceptibility, we want to take into account that H_{int} provides an equivalent internal field that acts in addition to the external field.

Let us start from a zero-field situation where $n_\uparrow(\mathbf{r}) = n_\downarrow(\mathbf{r}) = n_0$. Upon application of a field $B(\mathbf{r}) = B_{\mathbf{q}}e^{i\mathbf{q}\cdot\mathbf{r}}$, along the z-axis, the spin populations become $n_\uparrow(\mathbf{r}) = n_0 + \delta n_\uparrow(\mathbf{r})$ and $n_\downarrow(\mathbf{r}) = n_0 + \delta n_\downarrow(\mathbf{r})$ with $\delta n_\uparrow(\mathbf{r}) + \delta n_\downarrow(\mathbf{r}) = 0$, and the magnetization is $M(\mathbf{r}) = \mu_f(\delta n_\uparrow(\mathbf{r}) - \delta n_\downarrow(\mathbf{r})) = 2\mu_f \delta n_\uparrow(\mathbf{r}) = -2\mu_f \delta n_\downarrow(\mathbf{r})$. When these expressions for $n_\uparrow(\mathbf{r})$ and $n_\downarrow(\mathbf{r})$ are introduced in Eq. (3.112), we obtain a term $I \int d\mathbf{r}\, \delta n_\uparrow(\mathbf{r})\delta n_\downarrow(\mathbf{r})$. The $\uparrow$ spins get from the field an energy $-\mu_f \delta n_\uparrow(\mathbf{r})B(\mathbf{r})$. However, from the interaction with the $\downarrow$ spins, there is an additional contribution $I\delta n_\uparrow(\mathbf{r})\delta n_\downarrow(\mathbf{r})$, so the total $\uparrow$ spins energy term is $-\mu_f \delta n_\uparrow(\mathbf{r})[B(\mathbf{r}) - I\delta n_\downarrow(\mathbf{r})/\mu_f] = -\mu_f \delta n_\uparrow(\mathbf{r})[B(\mathbf{r}) + IM(\mathbf{r})/2\mu_f^2]$. The same argument works for the $\downarrow$ spins. Hence, because of the interactions, the external field $B(\mathbf{r})$ is replaced by $B(\mathbf{r}) + IM(\mathbf{r})/2\mu_f^2$. The above Pauli susceptibility relation $M = 2\mu_f(N_0\mu_f B)$, which corresponds to the response of the non-interacting Fermi sea, can be generalized for a nonzero wavevector $\mathbf{q}$ into $M_{\mathbf{q}} = 2\mu_f(N_0\mu_f \bar{\chi}_0(0, \mathbf{q})B_{\mathbf{q}})$. We can now take into account the

---

[8] We refer to Chapter 4 for a more detailed handling of these statistical factors, including the case of nonzero temperature. In this chapter, the response function $K(\mathbf{q}, \omega)$ is considered, which is quite analogous to the situation we consider here with the susceptibility, see for example Eq. (4.31).

interactions by replacing $B_{\mathbf{q}}$ by $B_{\mathbf{q}} + IM_{\mathbf{q}}/2\mu_f^2$. Generalizing this reasoning to a nonzero frequency situation, which does not create any problems, this leads for the susceptibility to

$$\chi(\omega, \mathbf{q}) = \frac{\chi_P \, \bar{\chi}_0(\omega, \mathbf{q})}{1 - \bar{I}\bar{\chi}_0(\omega, \mathbf{q})} \tag{3.116}$$

where we have set $\bar{I} = N_0 I$.

For itinerant electrons in metals, this result is known as the Stoner theory. In particular, if we think of increasing $\bar{I}$ starting from low values, we obtain a divergence for $\omega = 0$ and $\mathbf{q} = \mathbf{0}$ when $\bar{I} = 1$. This divergence corresponds to the instability of the electron gas toward ferromagnetism and $\bar{I} = 1$ is the Stoner criterion for this ferromagnetic transition. In our case, we are interested in situations where $\bar{I}$ is less than 1 but not so far from it, so that the susceptibility is large.

The above argument for the susceptibility can be extended to the pairing interaction. The details are slightly more complicated and we will not consider them. In this way, Eq. (3.114) gets essentially the same enhancement factor as the susceptibility in Eq. (3.116), with $\omega = 0$, and so it becomes within this RPA approximation

$$J(\mathbf{k}', \mathbf{k}) = -\frac{\bar{I}^2}{2N_0} \frac{\bar{\chi}_0(0, \mathbf{q})}{1 - \bar{I}\bar{\chi}_0(0, \mathbf{q})} \tag{3.117}$$

with $\mathbf{q} = \mathbf{k} - \mathbf{k}'$. The prefactor $1/2$ arises because density fluctuations are actually also coming in our argument, leading to Eq. (3.114). However, these density fluctuations do not get large in the vicinity of the ferromagnetic transition (they get an enhancement factor $1/[1 + \bar{I}\bar{\chi}_0(\omega, \mathbf{q})]$), so they are uninteresting for our purpose, and we have not written their contribution. Actually, in the vicinity of the ferromagnetic transition, $\bar{I}\bar{\chi}(0, \mathbf{q}) \simeq 1$, and we can write Eq. (3.117) in an even simpler way

$$J(\mathbf{k}', \mathbf{k}) = -\frac{\bar{I}}{2N_0} \frac{1}{1 - \bar{I}\bar{\chi}_0(0, \mathbf{q})} \tag{3.118}$$

Let us finally remark that we have written this indirect pairing interaction in the BCS way, assuming it to be essentially instantaneous, and taking accordingly a zero frequency in the susceptibility $\bar{\chi}_0$. This is actually inconsistent since polarizing the Fermi sea is not an instantaneous process, in the same way as moving ions is actually not instantaneous in BCS theory, and one has properly to take into account that the indirect interaction is retarded. Hence, in practical applications of the spin fluctuation mechanism, this has to be taken into account by making use of the frequency- dependent susceptibility Eq. (3.115). Then one has to treat this retarded interaction with a formalism similar to the one that is considered in Chapter 9 for handling the retarded nature of the phonon mechanism for pairing.

3.8.3 Other Pairing Symmetries and Mechanisms in Superconductors

Spin fluctuations in superfluid ^{3}He have provided a convincing example of the possibility of a pairing interaction mediated by a collective mode different from the phonons of the standard mechanism of BCS theory in superconductors. It is worth noting that this collective mode arises among the ^{3}He atoms themselves, which is the same system where pairing occurs. Nevertheless, one treats it as if it is coming from another system, in a way similar to the phonons, which are essentially a collective mode of the metallic ions. It is natural to wonder if such a mechanism can be found in superconductors, for example, arising from a collective mode of the electron gas. Actually, this consideration occurred even before the discovery of ^{3}He superfluidity, but rather as an interaction unfavorable for s-wave pairing. Indeed spin fluctuations have been pointed out as a possible explanation for the absence of superconductivity in some transition metals.

The discovery of cuprates superconductors, with their corresponding huge jump in critical temperature among known superconducting compounds, gave an enormous impulse to the investigation for new mechanisms and pairing symmetries, since it was fairly rapidly established that pairing occurred[9] in these compounds. The idea of a new mechanism is quite attractive since it introduces naturally new characteristic energies, analogous to the Debye temperature for phonons. Finding a new higher characteristic energy is a simple and appealing way to explain a large increase in critical temperature. Among these mechanisms, those linked to magnetic degrees of freedom, and in particular spin fluctuations, have been very frequently proposed since critical temperatures for magnetic transitions may be quite high. In the case of cuprates superconductors, this is a quite natural idea since closely related compounds are known to be indeed materials exhibiting magnetic transitions.

A very-well-known example is $YBa_2Cu_3O_{7-x}$, with its maximal critical temperature of 95 K. Indeed the compound $YBa_2Cu_3O_6$ is an antiferromagnetic insulator, with a critical temperature of 415 K. By increasing the oxygen content, this critical temperature decreases, and then the compound becomes a superconductor with a critical temperature increasing with oxygen content. Note that this increase in oxygen is conventionally called "doping," by analogy with the vocabulary of semiconductor physics, where "doping" means putting impurities, which modifies the density of charge carriers. Although the effect with oxygen is also to increase the charge carrier density, this wording may be somewhat misleading because the physics in the cuprates is quite different from the one in semiconductors.

A major problem with these cuprates materials is their high complexity, in particular, as compared to the standard elemental low-temperature superconducting metals. This makes both their experimental control quite difficult and their precise theoretical description quite complex. Hence they are almost systematically represented by more or less simple models, with the hope that these models contain the (most) important and relevant physical features of the real compound. In the case of cuprates, these materials are known to be very anisotropic compounds, almost two-dimensional. Their essential electronic components are CuO_2 planes, with the Cu atoms being the essential metallic ingredients. As shown in Fig. (3.10a), they form an essentially square lattice, with the O atoms making bridges between two nearby Cu atoms. Hence a simple model is just a square lattice of Cu atoms, with a distance a between two Cu neighbors. Moreover, a single band (for each electronic spin) is relevant and the electronic balance between all the atoms is such that this band is almost half-filled.

---

[9] This has been obtained from the frequency of the Josephon effect, see Chapter 5

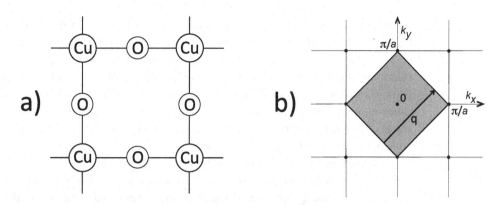

Fig. 3.10 a) Schematic representation of the CuO_2 planes, which are the essential part of $YBa_2Cu_3O_{7-x}$ with respect to the electronic properties. The distance between two neighboring Cu atoms is a. b) Reciprocal lattice of the CuO_2 planes, with the first Brillouin zone corresponding to the square $-\pi/a \le k_x \le \pi/a, -\pi/a \le k_y \le \pi/a$. The shaded square is the Fermi sea at half-filling, for the simple dispersion relation $\epsilon_{\mathbf{k}} = -2t[\cos(k_x a) + \cos(k_y a)]$. The vector $\mathbf{q}$ shifts one side of the Fermi 'surface' to another, displaying the perfect nesting of this simple Fermi surface.

Within the simple nearest neighbor tight-binding approximation with matrix element t, the electronic dispersion relation reads

$$\epsilon_{\mathbf{k}} = -2t[\cos(k_x a) + \cos(k_y a)] \tag{3.119}$$

If one adds to the corresponding non-interacting Hamiltonian an on-site repulsive interaction between opposite spins

$$H_{int} = U \sum_i n_{i\uparrow} n_{i\downarrow} \tag{3.120}$$

where the sum extends over all the sites $\mathbf{R}_i$ of the lattice, one obtains the simplest version of the Hubbard Hamiltonian, a very frequently used model for the cuprates superconductors. This interaction is the discrete equivalent of the repulsive interaction Eq. (3.112) we have considered above, and it has qualitatively the same effects.

However, one rather finds in the present case that this interaction favors an antiferromagnetic order, which is indeed found in $YBa_2Cu_3O_6$. One possible way to understand it is as follows. With the simple dispersion relation Eq. (3.119), the Fermi "surface" (which is actually a line in two dimensions) for a half-filled band corresponds to $\epsilon_{\mathbf{k}} = 0$, its shape being a simple square with the corners at $k_x = \pm\pi/a, k_y = 0$, and $k_x = 0, k_y = \pm\pi/a$. Its sides are the straight lines $\pm k_x \pm k_y = \pi/a$. Naturally this is an oversimplified situation, but the idea is that physical reality is not so far from it. This Fermi surface is shown in Fig. (3.10b).

Now if we calculate the $\omega = 0$ reduced susceptibility Eq. (3.115) and look for large contributions, they come from wavevectors $\mathbf{k}$ satisfying $\xi_{\mathbf{k+q}} \simeq \xi_{\mathbf{k}}$. On the other hand, to have a nonzero statistical factor $f(\xi_{\mathbf{k+q}}) - f(\xi_{\mathbf{k}})$, $\mathbf{k}$ and $\mathbf{k} + \mathbf{q}$ must be above and below the Fermi surface. Hence they are both essentially at the Fermi surface, $\xi_{\mathbf{k}}$ and $\xi_{\mathbf{k+q}}$ being close to zero. For a given $\mathbf{q}$, the corresponding $\mathbf{k}$ are found at the intersection of the Fermi surface by the same Fermi surface shifted by $-\mathbf{q}$. For a three-dimensional Fermi surface, this intersection is a line. In two dimensions the intersection is only made of few points. On the other hand, with our above square Fermi surface, we see that, for example, if we shift the side $-k_x - k_y = \pi/a$ by $\mathbf{q}$, with $q_x = q_y = \pi/a$, we find the whole side $k_x + k_y = \pi/a$. Hence the intersection is no longer a few points, but a whole line. Such a situation is called Fermi surface nesting. Even if we have considered for simplicity and clarity a quite extreme situation, it is clear that even an approximate nesting leads to an important increase in the susceptibility for the corresponding wavevector $\mathbf{q}$, because one has in Eq. (3.115) a large domain of $\mathbf{k}$ for which $\xi_{\mathbf{k+q}} - \xi_{\mathbf{k}}$ is small.

In the present case, the actual wavevectors for this strong increase are $q_x = \pm \pi/a$, $q_y = \pm \pi/a$. This is quite different from the ferromagnetic case, where the maximum susceptibility is found for $\mathbf{q} = \mathbf{0}$, corresponding to a uniform magnetization over the system. In the present case, the magnetization has a $e^{i\mathbf{q}\cdot\mathbf{R}_i}$ dependence. Hence $\mathbf{R}_i = \mathbf{0}$ and $\mathbf{R}_i = (a, a)$ have, for example, a maximum magnetization, since $\mathbf{q} \cdot \mathbf{R}_i$ is an integer multiple of 2π. On the other hand, for $\mathbf{R}_i = (\pm a, 0)$ or $\mathbf{R}_i = (0, \pm a)$, we have $e^{i\mathbf{q}\cdot\mathbf{R}_i} = -1$, so that the magnetization is opposite to the one found at $\mathbf{R}_i = \mathbf{0}$. This situation corresponds to an antiferromagnetic ordering. Moreover, just as in Eq. (3.116) for the ferromagnetic case, where interactions enhance the susceptibility and may lead to a ferromagnetic transition, this strong susceptibility for $\mathbf{q} = (\pm\pi/a, \pm\pi/a)$ is found to be enhanced by H_{int}, and this may lead to an antiferromagnetic transition. Hence the nesting of the Fermi surface together with a strong repulsive interaction between opposite spins provides a simple and coherent understanding of the antiferromagnetic order in $YBa_2Cu_3O_6$.

If we consider now the case where $YBa_2Cu_3O_6$ is doped toward $YBa_2Cu_3O_{7-x}$, the resulting modification in charge carrier density makes the Fermi surface go away from the nesting situation, which may explain the progressive disappearance of the antiferromagnetic order and the transformation into a metallic state. However, we can argue that although this metal is not antiferromagnetic, it is not so far from it so that a large susceptibility for $\mathbf{q} = (\pm\pi/a, \pm\pi/a)$ is still present. Hence the corresponding spin fluctuations may provide a pairing mechanism.

However, the magnetic properties of the cuprates have shown that this is singlet pairing[10] which occurs in these compounds. This raises the problem we have already mentioned that spin fluctuations lead to a repulsive contribution for pairing. Specifically in the continuous model with interaction Eq. (3.112), instead of Eq. (3.120), one finds for singlet pairing that spin fluctuations lead to a pairing interaction

$$J(\mathbf{k}', \mathbf{k}) = \frac{3\bar{I}^2}{2N_0} \frac{\bar{\chi}_0(0, \mathbf{q})}{1 - \bar{I}\bar{\chi}_0(0, \mathbf{q})} \tag{3.121}$$

However, this repulsive pairing interaction arises because BCS-like s-wave pairing was considered. But other possibilities are open.

In our discussion of the Cooper problem in Section 2.1, we have concluded that in the isotropic situation we have considered, pairing can in principle occur in any angular momentum ℓ, the actual pairing arising in the angular momentum corresponding to the strongest attraction. For a repulsive interaction, one might think that whatever the angular

---

[10] This has been found from the Knight shift, see Chapter 4.

momentum, the interaction is repulsive and no pairing is possible. This is certainly what happens if we take a repulsive contact interaction (for which only the $\ell = 0$ component is nonzero). But, with respect to the actual description of physical systems, such an interaction is an oversimplification. In general, a repulsive interaction will have an attractive component in some higher angular momentum.

Actually, Kohn and Luttinger [29] have shown that because of the peculiar response of the electronic gas due to the sharpness of the Fermi surface, a bare repulsive interaction will necessarily lead for the effective interaction to an attractive component for some angular momentum ℓ. However, the critical temperature resulting from this argument could be so low that the corresponding superconducting transition might be unobservable. In practice this means, for example, that some component of the Coulomb interaction could actually be attractive and lead to pairing. Such a purely electronic mechanism for pairing has been considered for superconductors, in particular, the high T_c cuprates. Naturally the appealing aspect of such a pairing mechanism is that the characteristic energy of Coulomb interaction is very high, typically of the order of the Fermi energy. On the other hand, just as in the Kohn and Luttinger argument, there is a strong reduction due to the fact that one has to find an attractive component out of a basically repulsive interaction. Hence the net result is not obvious.

The preceding discussion is for an isotropic system, but for a metallic superconductor we have naturally to consider the actual symmetry of its crystal. Nevertheless the basic principles are similar. Instead of being invariant by any rotation, the crystal is invariant under a smaller symmetry group, containing, for example, symmetry with respect to points, lines or planes, or rotations with specific axes and angles. Corresponding to the angular momentum classification of the pair wave function for rotationally invariant systems, the possible pairing wave functions are classified according to the way in which they are transformed when the crystal symmetries are applied. Actually, many of the possibilities compatible with crystal symmetries have been classified, but this does not tell which ones actually occur. The simplest case is the one where the pair wave function is invariant with respect to all the symmetries of the crystal. This is analogous to the s-wave case where the wave function is invariant under any rotation. These kinds of pairing are often called extended s-wave pairing.

Similarly, we may have pairing wave functions that transform under the crystal symmetries in the same way as k_x, k_y, or k_z. This is analogous to p-wave states that behave as the spherical harmonics $Y_{1,m}(\hat{\mathbf{k}})$. However, just as we have seen for superfluid ^3He, since these orbital parts are odd with respect to the exchange of the two electrons in the pair, the spin part must be even, which means that such pairing wave functions are possible only for triplet pairing.

Coming back to superconductivity in $\text{YBa}_2\text{Cu}_3\text{O}_{7-x}$, the crystal is tetragonal if we consider that the part which is doped with O atoms is unimportant for superconductivity. Forgetting about the third dimension, the corresponding symmetries are those which leave invariant the square corresponding to the unit cell, represented in Fig. (3.10). Since we are looking for singlet pairing, with accordingly an orbital part that is even under $\mathbf{k} \rightarrow -\mathbf{k}$, the next symmetry after the repulsive s-wave-like symmetry is the d-wave-like one. In particular, wave functions that transform in the same way as $k_x^2 - k_y^2$, analogous to the spherical harmonics $Y_{2,\pm 2}(\hat{\mathbf{k}})$, are of particular interest. This is called $d_{x^2-y^2}$ pairing. An example is a function proportional to $\cos(k_x a) - \cos(k_y a)$, but only the symmetry of this function is really relevant, not its very precise form.

In order to see if this symmetry can lead to pairing within the BCS framework, we have to look if the gap equation can have a solution. This is, in BCS theory, the analog of finding the wave function in the Cooper problem. However, we have to generalize this gap equation to the case where the interaction and the gap depend on the wavevector, as in Eq. (2.23). We consider only the $T = 0$ case for simplicity (the $T \neq 0$ follows along the same lines). This is easily done by following the derivation leading to Eq. (2.66), with the interaction $V_{\mathbf{k},\mathbf{k}'}$ staying inside the summations. One has to generalize the gap into the gap function $\Delta_{\mathbf{k}}$, defined by

$$\Delta_{\mathbf{k}} = \sum_{\mathbf{k}'}^{\omega_D} V_{\mathbf{k},\mathbf{k}'} u_{\mathbf{k}'} v_{\mathbf{k}'} \tag{3.122}$$

The rest of this derivation follows, with Δ replaced by $\Delta_{\mathbf{k}}$. This introduces $E_{\mathbf{k}} = \sqrt{\xi_{\mathbf{k}}^2 + \Delta_{\mathbf{k}}^2}$ and the gap equation becomes

$$\Delta_{\mathbf{k}} = \sum_{\mathbf{k}}^{\omega_D} V_{\mathbf{k},\mathbf{k}'} \frac{\Delta_{\mathbf{k}'}}{2E_{\mathbf{k}'}} \tag{3.123}$$

where in our case $V_{\mathbf{k},\mathbf{k}'}$ has to be replaced by $-J(\mathbf{k}', \mathbf{k})$ (since we have explicitly introduced a minus sign in the interaction for the attractive case in order to handle positive quantities). Hence we have to satisfy

$$\Delta_{\mathbf{k}} = - \sum_{\mathbf{k}}^{\omega_D} J(\mathbf{k}', \mathbf{k}) \frac{\Delta_{\mathbf{k}'}}{2E_{\mathbf{k}'}} \tag{3.124}$$

Without performing the detailed calculation one can see that this can be done with $d_{x^2-y^2}$ waves by considering an oversimplified case. We assume that the Fermi surface is not so different from the square Fermi surface considered above for half-filling and shown in Fig. (3.10). Then we take advantage of the fact that $J(\mathbf{k}', \mathbf{k})$ is large only for transfer momentum $\mathbf{q} = \mathbf{k} - \mathbf{k}' = (\pm \pi/a, \pm \pi/a)$ to conclude that the wavevectors $\mathbf{k}' = (\pm \pi/a, 0)$ and $\mathbf{k}' = (0, \pm \pi/a)$ give the dominant contribution in the summation in Eq. (3.124) (or more precisely the regions in the vicinity of these wavevectors). Now consider for example the case $\mathbf{k} = (\pi/a, 0)$, for which a gap function $\Delta_{\mathbf{k}}$, proportional to $\cos(k_x a) - \cos(k_y a)$, is negative. It is strongly connected by $J(\mathbf{k}', \mathbf{k})$ to $\mathbf{k}' = (0, \pm \pi/a)$, for which $\Delta_{\mathbf{k}'}$ is positive. Hence, due to the change of sign of the gap function linked to its d-wave symmetry, the signs on both sides of Eq. (3.124) are the same, which makes it possible to satisfy this equation, while this would be obviously impossible if $\Delta_{\mathbf{k}}$ had the same sign all over the Fermi surface. One checks easily that this argument works for all the values of $\mathbf{k} = (\pm \pi/a, 0)$ and $\mathbf{k} = (0, \pm \pi/a)$. In conclusion, we see that the strongly peaked repulsive interaction mediated by spin fluctuations, together with a gap function having a $d_{x^2-y^2}$ symmetry, can provide an unconventional mechanism and symmetry for pairing in $YBa_2Cu_3O_{7-x}$.

More generally, we have exposed BCS theory in the simplest physical situation, with a single isotropic electronic band having a free electron-like dispersion relation and a simple interaction. This does not correspond to any real metallic compound. Some of our simplistic assumptions are clearly not important. In particular, the physics of metals (with a large enough Fermi energy) comes from the immediate vicinity of the Fermi surface, so taking a free electron-like dispersion relation is not really a problem, and it is easy to reformulate the theory so that only properties at the Fermi surface are involved. Similarly nearly isotropic metals are quite infrequent, and one would expect an angular dependence of the gap function. However (see Chapter 4), one can see that the effect of non-magnetic impurities is to scatter electrons over the whole Fermi surface, and in this way produce by averaging a gap independent of the location on the Fermi surface. This may explain the somewhat surprising, and even puzzling, quantitative success of BCS theory in a number of metals.

Nevertheless the variety of physical situations to be found in condensed matter is such that departures from the simple BCS framework, that is from conventional superconductivity, are expected to occur. This may be due to complex features in the band structure (van Hove singularities, multiband situations which are actually quite frequent) or peculiar complicated normal state properties, such as found in "heavy fermions" metals. Strong spin-orbit coupling may also have important consequences. Hence the search for superconducting compounds with different pairing symmetries and mechanisms is naturally very active. The counterpart of all these interesting physical situations is that the corresponding compounds display quite often a more complex physics, as compared to the standard metals at low temperature. This makes all the analyses more complicated.

Response to an External Perturbation

4.1 Introduction

The purpose of this chapter is to examine the implications of BCS theory for the physical properties of a superconductor submitted to an external excitation. Actually, most of it is devoted to the case where this excitation is caused by an electromagnetic field. This corresponds to the calculation of the response function $K(\mathbf{q}, \omega)$, introduced in Section 1.3.2. In particular, we will show that BCS theory indeed describes a superconductor. Although this is likely from the physical ideas underlying this theory, it is obviously quite important to obtain this proof explicitly since the basic purpose of BCS theory is to provide an explanation for the very existence of superconductivity. However, we will also consider ultrasonic attenuation and nuclear magnetic resonance (NMR) because they display physically striking behaviors in the superconducting state, which are properly accounted for by BCS theory. Moreover, their theoretical description is close to the one for the electromagnetic response, so that consideration of these effects provides a complementary understanding.

In all these situations, the external perturbation produces an additional term V in the Hamiltonian, which describes the scattering of electrons by this perturbation, so it is of the general form

$$V = \sum_{\mathbf{k}\,\mathbf{k}'\sigma\sigma'} \mathcal{V}_{\mathbf{k}\,\mathbf{k}'\sigma\sigma'}(t)\, c^{\dagger}_{\mathbf{k}'\sigma'} c_{\mathbf{k}\sigma} \tag{4.1}$$

The matrix element $\mathcal{V}_{\mathbf{k}\,\mathbf{k}'\sigma\sigma'}(t)$ results from a semiclassical treatment of the perturbing field. For example, in the case of an electromagnetic perturbation, it is not necessary to quantize the electromagnetic field and make use of creation and annihilation photon operators, so we may treat it as a classical field. We first address the electromagnetic response.

4.2 Coupling to the Electromagnetic Field

We have first to obtain the specific expression of the coupling, Eq. (4.1), in the case of an electromagnetic perturbation. As we have seen in Chapter 1, for all cases of interest, in order to know the reaction of a superconductor to an electromagnetic perturbation, we only need to know how the electronic Hamiltonian is modified in the presence of a general vector potential $\mathbf{A}(\mathbf{r}, t)$. In order to handle a simple situation, we stay at the level used in Chapter 2 and ignore any band structure effect in the metal. In this case, as we have already seen, the electron kinetic energy is merely modified by replacing its momentum $\mathbf{p}$ by $\mathbf{p} - e\mathbf{A}(\mathbf{r}, t)$, so that for a single electron,

$$H_c = \frac{1}{2m}\,(\mathbf{p} - e\mathbf{A}(\mathbf{r}, t))^2 \simeq \frac{\mathbf{p}^2}{2m} - \frac{e}{2m}\,(\mathbf{p}\cdot\mathbf{A}(\mathbf{r}, t) + \mathbf{A}(\mathbf{r}, t)\cdot\mathbf{p}) = \frac{\mathbf{p}^2}{2m} - \frac{e}{m}\mathbf{p}\cdot\mathbf{A}(\mathbf{r}, t) \qquad (4.2)$$

In the first step, we merely have expanded the kinetic energy to first order in $\mathbf{A}$, since we are only interested in cases where the electromagnetic perturbation is small. The last step is obtained by noticing that $\mathbf{A}$ and $\mathbf{p}$ commute in our case, because for any wave function $\psi(\mathbf{r})$, we have

$$(\mathbf{p}\cdot\mathbf{A} - \mathbf{A}\cdot\mathbf{p})\psi = \frac{\hbar}{i}\,(\nabla\cdot(\mathbf{A}\psi) - \mathbf{A}\cdot\nabla\psi) = \frac{\hbar}{i}\psi\,\nabla\cdot\mathbf{A} = 0 \qquad (4.3)$$

since we will only consider vector potentials satisfying the London gauge condition $\nabla\cdot\mathbf{A} = 0$.

Summing this coupling term $-(e/m)\mathbf{p}\cdot\mathbf{A}(\mathbf{r}, t)$ over all the electrons, we have for the coupling term Eq. (4.1) in the Hamiltonian to the electromagnetic field the well-known result

$$V = -\frac{e}{m}\sum_i \mathbf{p_i}\cdot\mathbf{A}(\mathbf{r_i}, t) = -e\int d\mathbf{r}\,\mathbf{j}(\mathbf{r})\cdot\mathbf{A}(\mathbf{r}, t) \qquad (4.4)$$

where $\mathbf{j}(\mathbf{r}) = \sum_i \delta(\mathbf{r} - \mathbf{r}_i)\,\mathbf{p}_i/m$ is the total particle current at $\mathbf{r}$. We now have to write the second quantized version of this first quantized expression Eq. (4.4). Before doing it, we specialize $\mathbf{A}(\mathbf{r}, t)$ to our case of interest, namely when it corresponds to a well-defined wavevector $\mathbf{q}$, so that

$$\mathbf{A}(\mathbf{r}, t) = e^{i\mathbf{q}\cdot\mathbf{r}}\,\mathbf{A_q}(t) \qquad (4.5)$$

The coefficient of $c_{\mathbf{k}'\sigma'}^{\dagger}c_{\mathbf{k}\sigma}$ in Eq. (4.1) is the matrix element of $-(e/m)\mathbf{p}_i\cdot\mathbf{A}(\mathbf{r}_i, t)$ between plane-waves $\mathbf{k}'$ and $\mathbf{k}$, so we need to calculate

$$\langle\mathbf{k}'|\mathbf{p}_i\,e^{i\mathbf{q}\cdot\mathbf{r_i}}|\mathbf{k}\rangle = \frac{\hbar}{i}\int d\mathbf{r_i}\,e^{-i\mathbf{k}'\cdot\mathbf{r_i}}\nabla_i(e^{i\mathbf{q}\cdot\mathbf{r_i}}e^{i\mathbf{k}\cdot\mathbf{r_i}}) = \hbar\mathbf{k}'\,\delta_{\mathbf{k}',\mathbf{k}+\mathbf{q}} \qquad (4.6)$$

since we deal with a superconductor having unit volume. On the other hand, the momentum operator does not act on spin variables (which were not written explicitly in Eq. (4.4)), which imply a factor $\delta_{\sigma,\sigma'}$ for the matrix element. Hence we end up with

$$V = -\frac{e\mathbf{A_q}(t)}{m}\cdot\sum_{\mathbf{k}\sigma}\hbar\mathbf{k}\,c_{\mathbf{k}+\mathbf{q}/2,\sigma}^{\dagger}c_{\mathbf{k}-\mathbf{q}/2,\sigma} \equiv -e\mathbf{A_q}(t)\cdot\mathbf{j}_{-\mathbf{q}} \qquad (4.7)$$

where we have made the change of summation variable $\mathbf{k} \to \mathbf{k} - \mathbf{q}/2$, and made use of $\mathbf{q}\cdot\mathbf{A_q}(t) = 0$, which results from the gauge condition $\nabla\cdot\mathbf{A} = 0$ when Eq. (4.5) is taken into account. The last equality in Eq. (4.7), which comes out clearly from the last equality in

Eq. (4.4), expresses the result in terms of the Fourier transform $\mathbf{j_q} = \int d\mathbf{r}\, e^{-i\mathbf{q}\cdot\mathbf{r}}\, \mathbf{j(r)}$ of $\mathbf{j(r)}$

$$\mathbf{j_q} = \sum_{\mathbf{k}\sigma} \frac{\hbar\mathbf{k}}{m}\, c^{\dagger}_{\mathbf{k}-\mathbf{q}/2,\sigma}\, c_{\mathbf{k}+\mathbf{q}/2,\sigma} \equiv \sum_{\mathbf{k}\sigma} \mathbf{j_{q\,k\sigma}} \tag{4.8}$$

where $\mathbf{j_{q\,k\sigma}} = (\hbar\mathbf{k}/m)\, c^{\dagger}_{\mathbf{k}-\mathbf{q}/2,\sigma}\, c_{\mathbf{k}+\mathbf{q}/2,\sigma}$.

Finally, in order to obtain the response function $K(\mathbf{q}, \omega)$, we have to write the expression of the physical quantity we consider as resulting from the perturbation. From the definition Eq. (1.38) of $K(\mathbf{q}, \omega)$, this is the value of the charge current $\mathbf{J(r}, t)$ with its corresponding operator $e(\mathbf{p} - e\mathbf{A(r}, t))/m$ for a single electron, as we have seen in Eq. (1.31). Summing over the electrons, we have

$$\mathbf{J(r}, t) = \frac{e}{m}\left\langle \sum_i \delta(\mathbf{r} - \mathbf{r}_i)\left(\mathbf{p_i} - e\mathbf{A(r}_i, t)\right)\right\rangle \tag{4.9}$$

the average corresponding to the quantum statistical average, since we will deal also with nonzero temperature situations. The first term is directly proportional to the particle current $\mathbf{j(r)}$ introduced in Eq. (4.4).

On the other hand, after averaging, the last term is proportional to the local electronic density $n(\mathbf{r}, t)$ (since for a single electron with wave function $\psi(\mathbf{r_i})$, $\langle\delta(\mathbf{r}-\mathbf{r_i})\rangle = \int d\mathbf{r}_i\delta(\mathbf{r}-\mathbf{r_i})\psi^*(\mathbf{r_i})\psi(\mathbf{r_i}) = |\psi(\mathbf{r})|^2$). However, because electrons carry charges, any departure of the electronic density from its equilibrium value n implies the appearance of the very strong Coulomb forces, as we have seen when considering the plasma frequency in Section 3.7.1. This leads to the electronic plasma frequency being quite high, in the ultraviolet range, as already indicated in Chapter 1, which corresponds to energies of the order of 10 eV. However, the frequency range for ω that we will consider for the physical properties of superconductors will be much smaller, corresponding to typical energies around 1 meV. Hence these frequencies are unable to excite any sizeable electronic density fluctuation. (They should be comparable to the plasma frequency to do so.) As a result, it is an excellent approximation to take the electronic density constant, equal to its equilibrium value $n(\mathbf{r}, t) = n$. In this way, Eq. (4.9) reduces to

$$\mathbf{J(r}, t) = e\langle\mathbf{j(r)}\rangle - \frac{ne^2}{m}\mathbf{A(r}, t) \tag{4.10}$$

If we consider now a plane wave excitation given by Eq. (4.5), only the corresponding Fourier components will be nonzero. Taking the Fourier transform of Eq. (4.10) leads, for the Fourier transform $\mathbf{J_q}(t)$ of $\mathbf{J(r}, t)$, to

$$\mathbf{J_q}(t) = e\langle\mathbf{j_q}\rangle - \frac{ne^2}{m}\mathbf{A_q}(t) \tag{4.11}$$

4.3 Linear Response

We now have to obtain the current $\langle\mathbf{j_q}\rangle$ arising from the perturbation Eq. (4.7). Since this perturbation is small, this is merely achieved by making use of time-dependent first-order perturbation theory. The result is known as the linear response, and its derivation is a standard matter in statistical physics. However, the details are somewhat cumbersome, and we will not reproduce them here. Hence we may just refer the reader to standard textbooks in statistical physics. However, it is interesting to go around this problem and take another path[1] to reach the result. Indeed it is easy to find the imaginary part of the response function directly and then obtain the full response function by making use of general relations for the response function expressing causality. Let us first consider this last point.

The consequence (in our case, the charge current) appears after its cause (in our case, the pertubation V) arises; this is causality. The response function $K(\mathbf{q}, \omega)$ in Eq. (1.38) is the value of the charge current $\mathbf{j}_s(\mathbf{q}, \omega)$ when the pertubation $\mathbf{A}(\mathbf{q}, \omega)$ is actually a simple constant, independent of ω. Translated by Fourier transform to the time variable t, such a perturbation corresponds to a $\delta(t)$ peak, that is, a perturbation occurring just at time $t = 0$. To simplify notations, let us for the moment call merely $R(\omega)$, instead of $K(\mathbf{q}, \omega)$, the response function. When we use the time variable, the response function is the Fourier transform of $R(\omega)$: $R_F(t) = \int (d\omega/2\pi)e^{-i\omega t}R(\omega)$. Since $R_F(t)$ is the response to the $\delta(t)$ perturbation, from causality it is nonzero only for $t \geq 0$. Then a standard behavior for $R_F(t)$ as a function of time is a progressive decay for large t. A simple explicit example could be $R_F(t) = \theta(t)e^{-\alpha t}$, where $\theta(t) = 1$ for $t > 0$ and $\theta(t) = 0$ for $t < 0$ is the Heaviside function and α is a positive constant. In the inverse Fourier transform,

$$R(\omega) = \int_0^\infty dt\, e^{i\omega t} R_F(t) \tag{4.12}$$

the integral runs only over positive t because of causality, expressed by $\theta(t)$.

Now we consider the continuation of $R(\omega)$ for complex values of the frequency variable $\omega = \omega' + i\omega''$, where ω' and ω'' are respectively the real and imaginary part of ω. This can be obtained explicitly by making use of Eq. (4.12). If we continue $R(\omega)$ toward the upper complex plane, that is, we have $\omega'' > 0$, we see that this produces an additional factor $e^{-\omega'' t}$ in the integrand of the integral Eq. (4.12). This factor makes the integrand converge more rapidly to 0 for $t \to \infty$. Hence since the integral is already defined for $\omega'' = 0$, it will exist for any $\omega'' > 0$. If the lower boundary of the integral, instead of being 0, is extended to $-\infty$ where $e^{-\omega'' t}$ diverges, our argument would fail. Hence this result is directly linked to causality. We have in this way come to the conclusion that $R(\omega)$ has no singularity in the upper complex plane, that is, it is an analytical function of ω in this domain. On the other hand, if we try to continue $R(\omega)$ in the lower complex plane $\omega'' < 0$, the factor $e^{-\omega'' t}$ provides a divergent contribution, and if we go deep enough in this lower complex plane with large $|\omega''|$, the integral Eq. (4.12) will diverge at some stage, so we have singularities of $R(\omega)$ in this domain. To take our above simple explicit example, performing the Fourier transform Eq. (4.12), we find $R(\omega) = 1/(\alpha - i\omega)$, which indeed has a pole in the lower complex plane at $\omega = -i\alpha$ and is analytical in the upper complex plane.

---

[1] This approach is directly related to Kramers–Kronig relations.

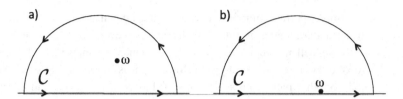

Fig. 4.1 (a) General integration contour $\mathcal{C}$ in Eq. (4.14) for the Cauchy formula. (b) Special case, where ω goes extremely close to the real axis, which is used in Eq. (4.15) to relate the real part of $R(\omega)$ to its imaginary part.

Conversely, we can make use of this analyticity of $R(\omega)$ in the upper complex plane to calculate the inverse Fourier transform,

$$R_F(t) = \frac{1}{2\pi} \int_{-\infty}^{\infty} d\omega\, e^{-i\omega t} R(\omega) \qquad (4.13)$$

If we consider negative values of time $t < 0$, the exponential factor provides a convergent factor $e^{-\omega''|t|}$ if we go into the upper complex plane for ω. Hence if we go on the semicircle at infinity in this upper complex plane, where ω'' is infinitely large, the integrand will be zero (one can see from Eq. (4.12) that on this semicircle $R(\omega)$ goes to zero, so it cannot counter the effect of the exponential factor). So we can add to the integration range $[-\infty, \infty]$ in Eq. (4.13) this semicircle at infinity, in the counterclockwise direction, since its contribution is zero. The result is an integral over a closed contour encircling the upper complex plane. However, the integrand in this domain is analytical since both $R(\omega)$ and $e^{-i\omega t}$ are analytical. Accordingly, this integral is zero, and we find $R_F(t) = 0$ for $t < 0$, which is causality. Hence we see that conversely analyticity of $R(\omega)$ in the upper complex plane implies causality.

This analytical property can be used to obtain $R(\omega)$ anywhere in the upper complex plane from its knowledge on the real ω axis by Cauchy's integral formula,

$$R(\omega) = \frac{1}{2i\pi} \int_{\mathcal{C}} dz\, \frac{R(z)}{z - \omega} = \frac{1}{2i\pi} \int_{-\infty}^{\infty} dx\, \frac{R(x)}{x - \omega} \qquad (4.14)$$

In Eq. (4.14), contour $\mathcal{C}$, which is displayed in Fig. (4.1a), is just the closed contour we considered in the preceding paragraph, going on the real axis from $-\infty$ to $+\infty$ and closed counterclockwise by the semicircle at infinity. The first equality (Cauchy's formula) is due to the fact that the integral is unchanged if the integration contour is deformed, provided it stays in a domain where the integrand is analytical. Since $R(\omega)$ is analytical in the whole upper complex plane, the only singularity is the explicit pole at $z = \omega$. Accordingly we may, without changing the integral, deform contour $\mathcal{C}$ into a very small circle around this pole. Since z on this new contour is everywhere very near ω, we may just write $R(z) \simeq R(\omega)$ and take this constant out of the integral. The standard remaining integral along the small circle $\oint dz/(z - \omega)$ is easily calculated by the change of variable $z - \omega = \epsilon e^{i\theta}$ and is equal to $i \int_0^{2\pi} d\theta = 2i\pi$. This justifies the first equality in Eq. (4.14). The last one merely results from the fact that $R(\omega)$ goes to zero on the semicircle at infinity, so that the only remaining contribution is the integral on the real frequency axis.

Now we make use of this general formula by letting ω be extremely close to the real axis $\omega = \omega' + i\epsilon$, with $\epsilon \to 0_+$, and ω' real, as shown in Fig. (4.1b). We make use of

$1/(x - \omega' - i\epsilon) = 1/(x - \omega') + i\pi\delta(x - \omega')$, where $\delta(x)$ is the Dirac delta function. We explicitly write the real R' and imaginary R'' parts of $R(x) = R'(x) + iR''(x)$. Equating the real parts of both members of this Eq. (4.14) gives

$$\text{Re}[R(\omega' + i\epsilon)] = \frac{1}{2\pi} \int_{-\infty}^{\infty} dx \, \frac{R''(x)}{x - \omega'} + \frac{1}{2}R'(\omega') \tag{4.15}$$

The integral in Eq. (4.15) has to be understood as a principal part. Letting ϵ go to zero, we have $\text{Re}[R(\omega')] = R'(\omega')$, so that Eq. (4.15) reduces to

$$R'(\omega') = \frac{1}{\pi} \int_{-\infty}^{\infty} dx \, \frac{R''(x)}{x - \omega'} \tag{4.16}$$

In this way, we have been able to express the real part of R in terms of its imaginary part. Adding the obvious equality $iR''(\omega') = (1/\pi) \int_{-\infty}^{\infty} dx \, R''(x) \, i\pi\delta(x - \omega')$ to this equation, and making use again of $1/(x - \omega') + i\pi\delta(x - \omega') = 1/(x - \omega' - i\epsilon)$, we finally express $R(\omega)$ in terms of its imaginary part

$$R(\omega) = \frac{1}{\pi} \int_{-\infty}^{\infty} dx \, \frac{R''(x)}{x - \omega - i\epsilon} \tag{4.17}$$

where we switch back to the notation ω, instead of ω', since we assume from now on ω to be real. For example, in our simple case $R(\omega) = 1/(\alpha - i\omega)$ considered above, we have $R''(\omega) = \omega/(\alpha^2 + \omega^2)$ and we can check that

$$\frac{1}{\alpha - i\omega} = \frac{1}{\pi} \int_{-\infty}^{\infty} dx \, \frac{x}{(x^2 + \alpha^2)(x - \omega - i\epsilon)} \tag{4.18}$$

for $\epsilon \to 0$, the integral being most easily calculated by residue, by closing the contour in the lower half complex plane where the only pole is for $x = -i\alpha$. Finally, it is interesting to note that the analyticity of $R(\omega)$ in the upper complex plane is explicit in Eq. (4.17) since the poles are for $\omega = x - i\epsilon$, which implies that they are all located in the lower complex plane.

Coming back to our problem of finding the response to an electromagnetic perturbation, we can easily relate the imaginary part $K''(\mathbf{q}, \omega)$ of $K(\mathbf{q}, \omega)$ to dissipation. Indeed, in simple terms, if we consider in Eq. (4.5) a vector potential with specific frequency ω,

$$\mathbf{A}(\mathbf{r}, t) = \mathbf{A_q}(t)e^{i\mathbf{q}\cdot\mathbf{r}} = \mathbf{A_{q,\omega}} \, e^{i\mathbf{q}\cdot\mathbf{r}} \, e^{-i\omega t} \tag{4.19}$$

the conductivity is obtained from $K(\mathbf{q}, \omega)$ by generalizing Eq. (1.39) to any wavevector $\sigma(\mathbf{q}, \omega) = i K(\mathbf{q}, \omega)/\omega$. In the static case, power dissipation per unit volume P is given by $P = \mathbf{j} \cdot \mathbf{E} = \sigma E^2$, leading to the Joule effect. Generalizing to nonzero frequency, dissipation is linked to the in-phase part of the current $\mathbf{j}$ with respect to the field $\mathbf{E}$ – that is, to the real part $\sigma'(\omega)$ of the conductivity.

More precisely, when we deal with complex perturbations, we actually have in mind for physical quantities the real part of the complex quantity. Hence, instead of Eq. (4.19), we take specifically

$$\mathbf{A}(\mathbf{r}, t) = \mathbf{A_{q,\omega}} \left(e^{i(\mathbf{q}\cdot\mathbf{r}-\omega t)} + e^{-i(\mathbf{q}\cdot\mathbf{r}-\omega t)} \right) = 2 \, \mathbf{A_{q,\omega}} \, \cos(\mathbf{q} \cdot \mathbf{r} - \omega t) \tag{4.20}$$

with $\omega > 0$. The corresponding electric field is $\mathbf{E}(\mathbf{r}, t) = -\partial \mathbf{A}(\mathbf{r}, t)/\partial t = -2\omega \mathbf{A}_{\mathbf{q},\omega} \times \sin(\mathbf{q} \cdot \mathbf{r} - \omega t)$, and the in-phase part of the current is $\sigma'(\omega)\mathbf{E}(\mathbf{r}, t)$. Hence the power dissipation has a $\sin^2(\mathbf{q} \cdot \mathbf{r} - \omega t)$ dependence, which after averaging over time gives a factor of $1/2$. In this way, we obtain for the power dissipation

$$P = \frac{1}{2}[-\frac{K''(\mathbf{q}, \omega)}{\omega}][2\omega A_{\mathbf{q},\omega}]^2 = -2\omega A_{\mathbf{q},\omega}^2 K''(\mathbf{q}, \omega) \qquad (4.21)$$

On the other hand, dissipation can be obtained from Fermi golden rule, which states that under time-dependent perturbation V with frequency ω, the probability per unit time W to have the system excited from initial state $|i\rangle$, with energy E_i, to final state $|f\rangle$, with energy E_f, is given by

$$W = \frac{2\pi}{\hbar}|\langle f|V|i\rangle|^2 \delta(E_f - E_i - \hbar\omega) \qquad (4.22)$$

Coherently with Eq. (4.20), since we have also introduced the wavevector $-\mathbf{q}$, we have instead of Eq. (4.7) to write now the perturbation as

$$V = -e\mathbf{A}_{\mathbf{q},\omega} \cdot \left(e^{-i\omega t}\mathbf{j}_{-\mathbf{q}} + e^{i\omega t}\mathbf{j}_{\mathbf{q}}\right) \qquad (4.23)$$

We begin with the contribution from the first term of Eq. (4.23). Taking an initial state $|n\rangle$ and a final state $|m\rangle$, the absorption of the electromagnetic wave provides to the system an energy $\hbar\omega = E_m - E_n > 0$, so the power absorbed by the system is $P = \hbar\omega W$. To obtain the total absorbed power, we have to sum over all possible initial and final states. We also have to take into account that at temperature T, the probability for the system to be in state $|n\rangle$ is given by the Boltzmann factor $e^{-\beta E_n}/Z$, where $\beta = 1/k_B T$, with k_B being the Boltzmann constant. Here $Z = \sum_n e^{-\beta E_n}$ is the partition function. Hence to have the absorbed power, we have to multiply by the probability $e^{-\beta E_n}/Z$ to have state $|n\rangle$ and sum over all possible $|n\rangle$ and $|m\rangle$. This gives to the absorbed power a contribution

$$P_1 = 2\pi\,\omega\,e^2[A_{\mathbf{q},\omega}^{(x)}]^2 \sum_{m,n} \frac{e^{-\beta E_n}}{Z}|\langle m|j_{-\mathbf{q}}^{(x)}|n\rangle|^2\,\delta(E_m - E_n - \hbar\omega) \qquad (4.24)$$

where, to be definite, we have taken the vector potential along the x-axis so that only the component $A_{\mathbf{q},\omega}^{(x)}$ is nonzero. (We write the cartesian component x as a superscript for clarity.)

The contribution from the second term of Eq. (4.23) is handled in a similar way, by changing ω to $-\omega$ and $\mathbf{q}$ to $-\mathbf{q}$. However, in this case, $E_m - E_n = -\hbar\omega < 0$ and the system emits power by lowering its energy from E_n to E_m. This leads to a negative contribution to the absorbed power

$$P_2 = -2\pi\,\omega\,e^2[A_{\mathbf{q},\omega}^{(x)}]^2 \sum_{m,n} \frac{e^{-\beta E_n}}{Z}|\langle m|j_{\mathbf{q}}^{(x)}|n\rangle|^2\,\delta(E_m - E_n + \hbar\omega) \qquad (4.25)$$

Exchanging the summation variables n and m in Eq. (4.25), and taking into account that since $\mathbf{j}_{\mathbf{q}}^{\dagger} = \mathbf{j}_{-\mathbf{q}}$ from its expression Eq. (4.8), we have $\langle n|j_{\mathbf{q}}^{(x)}|m\rangle = (\langle m|j_{-\mathbf{q}}^{(x)}|n\rangle)^*$, we can combine the two results to obtain the total absorbed power $P = P_1 + P_2$:

$$P = 2\pi\,\omega\,e^2 [A^{(x)}_{\mathbf{q},\omega}]^2 \sum_{m,n} \frac{e^{-\beta E_n} - e^{-\beta E_m}}{Z} |\langle m|j^{(x)}_{-\mathbf{q}}|n\rangle|^2\,\delta(E_m - E_n - \hbar\omega) \qquad (4.26)$$

Comparing with Eq. (4.21), we find the general expression for $K''(\mathbf{q}, \omega)$ we were looking for:

$$K''(\mathbf{q}, \omega) = -\pi e^2 \sum_{m,n} \frac{e^{-\beta E_n} - e^{-\beta E_m}}{Z} |\langle m|j^{(x)}_{-\mathbf{q}}|n\rangle|^2\,\delta(E_m - E_n - \hbar\omega) \qquad (4.27)$$

In order to obtain our final result, we have to make use of Eq. (4.17) to obtain the complete response function from its imaginary part. Moreover, we have to realize that the above considerations were devoted to obtaining in Eq. (4.11) the response corresponding to the first term. However, the second term[2] provides a trivial additional contribution ne^2/m to the global response function. Taking this term into account gives our final expression:

$$K(\mathbf{q}, \omega) = \frac{ne^2}{m} + \frac{e^2}{\hbar} \sum_{m,n} \frac{e^{-\beta E_n} - e^{-\beta E_m}}{Z} \frac{|\langle m|j^{(x)}_{-\mathbf{q}}|n\rangle|^2}{\omega - \omega_{mn} + i\epsilon} \qquad (4.28)$$

where we have set $\omega_{mn} = (E_m - E_n)/\hbar$. Since, as we have seen, $\langle m|j^{(x)}_{-\mathbf{q}}|n\rangle^* = \langle n|j^{(x)}_{\mathbf{q}}|m\rangle$, we may also write $|\langle m|j^{(x)}_{-\mathbf{q}}|n\rangle|^2 = \langle n|j^{(x)}_{\mathbf{q}}|m\rangle\langle m|j^{(x)}_{-\mathbf{q}}|n\rangle$.

4.4 Coherence Factors

Expression Eq. (4.28) is valid for any Hamiltonian for the considered system, which could, in particular, involve strong interactions. Fortunately we do not have to deal with such complicated situations. Rather we will first consider the simplest case, namely the non-interacting case where the Hamiltonian contains only the kinetic energy of the electrons. This will be a first step for the calculation of the BCS case, which will be rather similar.

4.4.1 Free Electrons

We first write explicitly in Eq. (4.28) the operator $j^{(x)}_{\mathbf{q}}$, as in Eq. (4.8), which leads to

$$K(\mathbf{q}, \omega) - \frac{ne^2}{m} = \frac{e^2}{\hbar} \sum_{\mathbf{k}\,\mathbf{k}'\sigma\sigma'} \sum_{m,n} \frac{e^{-\beta E_n} - e^{-\beta E_m}}{Z} \frac{\langle n|j^{(x)}_{\mathbf{q}\,\mathbf{k}'\sigma'}|m\rangle\langle m|j^{(x)}_{-\mathbf{q}\,\mathbf{k}\sigma}|n\rangle}{\omega - \omega_{mn} + i\epsilon} \qquad (4.29)$$

The eigenstates $|n\rangle$ for our free electrons are merely the tensorial product $\Pi_i \otimes |\mathbf{k}_i\rangle$ of all the plane wave states $\mathbf{k}_i$, corresponding to single-electron eigenstates of the kinetic energy. We naturally work in the grand canonical ensemble, in order to decouple essentially what happens in the different plane wave states. Accordingly, the kinetic energies are taken from

---

[2] Including this term in our general considerations would require us to take into account much higher frequencies, of the order of the plasma frequency. This would make matters much more complicated; it is much easier to handle this term separately.

the chemical potential μ, and the energies of the eigenstates are $E_n = \sum_{\mathbf{k}_i}(\epsilon_{\mathbf{k}_i} - \mu) = \sum_{\mathbf{k}_i} \xi_{\mathbf{k}_i}$, where the sum runs over all occupied plane wave states.

Operator $j^{(x)}_{-\mathbf{q}\,\mathbf{k}\sigma} = (\hbar k_x/m)\, c^{\dagger}_{\mathbf{k}+\mathbf{q}/2,\sigma} c_{\mathbf{k}-\mathbf{q}/2,\sigma}$ destroys an electron in the plane wave state $\mathbf{k} - \mathbf{q}/2$ (with spin σ) and creates an electron in plane wave $\mathbf{k} + \mathbf{q}/2$. Hence, in order to have a nonzero matrix element $\langle m|j^{(x)}_{-\mathbf{q}\,\mathbf{k}\sigma}|n\rangle$, $|n\rangle$ must have a plane wave $\mathbf{k} - \mathbf{q}/2$ occupied by an electron, and a plane wave $\mathbf{k} + \mathbf{q}/2$ must be empty. Conversely, $|m\rangle$ must have plane wave $\mathbf{k} - \mathbf{q}/2$ empty and plane wave $\mathbf{k} + \mathbf{q}/2$ occupied. Otherwise $|n\rangle$ and $|m\rangle$ must have, for all the other wavevectors, identical occupations by electrons. Hence, as soon as $|n\rangle$ is known, $|m\rangle$ is also determined, so that the summation over m disappears in Eq. (4.29). In this case, $\langle m|c^{\dagger}_{\mathbf{k}+\mathbf{q}/2,\sigma} c_{\mathbf{k}-\mathbf{q}/2,\sigma}|n\rangle = 1$. Then, in the same way, in order to be have matrix element $\langle n|j^{(x)}_{\mathbf{q}\,\mathbf{k}'\sigma'}|m\rangle$ nonzero, operator $j^{(x)}_{\mathbf{q}\,\mathbf{k}'\sigma'} = (\hbar k'_x/m)\, c^{\dagger}_{\mathbf{k}'-\mathbf{q}/2,\sigma'} c_{\mathbf{k}'+\mathbf{q}/2,\sigma'}$ must transform back $|m\rangle$ into $|n\rangle$. This implies that we must have $\mathbf{k}' = \mathbf{k}$ and $\sigma' = \sigma$, so the summations over $\mathbf{k}'$ and σ' disappear in Eq. (4.29). This leads for our nonzero matrix element to $\langle n|j^{(x)}_{\mathbf{q}\,\mathbf{k}\sigma}|m\rangle\langle m|j^{(x)}_{-\mathbf{q}\,\mathbf{k}\sigma}|n\rangle = (\hbar k_x/m)^2$.

Regarding the energies of our eigenstates, since only the occupations of $\mathbf{k}-\mathbf{q}/2$ and $\mathbf{k}+\mathbf{q}/2$ differ between $|n\rangle$ and $|m\rangle$, we have $\hbar\omega_{mn} = (E_m - E_n) = (\xi_{\mathbf{k}+\mathbf{q}/2} - \xi_{\mathbf{k}-\mathbf{q}/2})$. More precisely, we have $E_n = \xi_{\mathbf{k}-\mathbf{q}/2} + \sum_{\mathbf{k}_{i\neq}} \xi_{\mathbf{k}_i}$ and $E_m = \xi_{\mathbf{k}+\mathbf{q}/2} + \sum_{\mathbf{k}_{i\neq}} \xi_{\mathbf{k}_i}$, where $\mathbf{k}_{i\neq}$ stands for all the plane wave states occupied in $|m\rangle$ and $|n\rangle$ other than $\mathbf{k}-\mathbf{q}/2$ and $\mathbf{k}+\mathbf{q}/2$. In Eq. (4.29) we can, in $e^{-\beta E_n} - e^{-\beta E_m}$, factorize all the contributions $\mathbf{k}_{i\neq}$ and write $e^{-\beta E_n} - e^{-\beta E_m} = (e^{-\beta\xi_{\mathbf{k}-\mathbf{q}/2}} - e^{-\beta\xi_{\mathbf{k}+\mathbf{q}/2}})\Pi_{\mathbf{k}_{i\neq}} e^{-\beta\xi_{\mathbf{k}_i}}$. For a specific $\mathbf{k}_i$, when we sum over all possible $|n\rangle$, the plane wave $\mathbf{k}_i$ will be either occupied by an electron (which gives a factor $e^{-\beta\xi_{\mathbf{k}_i}}$) or empty (which gives a factor 1). Hence, summing over these two possibilities gives a factor of $1 + e^{-\beta\xi_{\mathbf{k}_i}}$ from this $\mathbf{k}_i$. This leads to $\sum_n \Pi_{\mathbf{k}_{i\neq}} e^{-\beta\xi_{\mathbf{k}_i}} = \Pi_{\mathbf{k}_{i\neq}}(1 + e^{-\beta\xi_{\mathbf{k}_i}})$. However, the same factors appear in the partition function, which has exactly the same expression, except that there is no restriction on the wavevectors so that the values $\mathbf{k}_i = \mathbf{k} - \mathbf{q}/2$ and $\mathbf{k}_i = \mathbf{k} + \mathbf{q}/2$ are also included in Z. Hence the factor $\Pi_{\mathbf{k}_{i\neq}}(1 + e^{-\beta\xi_{\mathbf{k}_i}})$ drops out in the ratio $(e^{-\beta E_n} - e^{-\beta E_m})/Z$. In this way, we are left with

$$\sum_n \frac{e^{-\beta E_n} - e^{-\beta E_m}}{Z} = \frac{e^{-\beta\xi_{\mathbf{k}-\mathbf{q}/2}} - e^{-\beta\xi_{\mathbf{k}+\mathbf{q}/2}}}{(1 + e^{-\beta\xi_{\mathbf{k}-\mathbf{q}/2}})(1 + e^{-\beta\xi_{\mathbf{k}+\mathbf{q}/2}})} \qquad (4.30)$$
$$= f(\xi_{\mathbf{k}-\mathbf{q}/2})\left(1 - f(\xi_{\mathbf{k}+\mathbf{q}/2})\right) - f(\xi_{\mathbf{k}+\mathbf{q}/2})\left(1 - f(\xi_{\mathbf{k}-\mathbf{q}/2})\right) = f(\xi_{\mathbf{k}-\mathbf{q}/2}) - f(\xi_{\mathbf{k}+\mathbf{q}/2})$$

In the last equalities of Eq. (4.30), we have expressed the result in terms of the Fermi distribution $f(\xi) = 1/(e^{\beta\xi} + 1)$. This makes clear the physical interpretation of this statistical factor. Indeed this term is due to the transfer of an electron from plane wave $\mathbf{k} - \mathbf{q}/2$ to plane wave $\mathbf{k} + \mathbf{q}/2$, due to the electromagnetic excitation. For this process to occur, plane wave $\mathbf{k} - \mathbf{q}/2$ must be occupied in the initial state (probability $f(\xi_{\mathbf{k}-\mathbf{q}/2})$) and plane wave $\mathbf{k} + \mathbf{q}/2$ must be empty (probability $1 - f(\xi_{\mathbf{k}+\mathbf{q}/2})$), which explains the presence of the first term in the last line in Eq. (4.30). The second term is similarly due to the reverse process of de-excitation from plane wave $\mathbf{k} + \mathbf{q}/2$ to plane wave $\mathbf{k} - \mathbf{q}/2$, as we have already seen at the level of Eq. (4.27).

Carrying all these results into Eq. (4.29) leads finally for our free electrons to

$$K(\mathbf{q}, \omega) - \frac{ne^2}{m} = -2\frac{e^2\hbar}{m^2} \sum_{\mathbf{k}} k_x^2 \frac{f(\xi_{\mathbf{k}+\mathbf{q}/2}) - f(\xi_{\mathbf{k}-\mathbf{q}/2})}{\omega - (\xi_{\mathbf{k}+\mathbf{q}/2} - \xi_{\mathbf{k}-\mathbf{q}/2})/\hbar + i\epsilon}$$ (4.31)

where the factor of 2 comes from the summation over spin σ. We will explore below the physical consequences of this result, but for the moment we rather switch immediately to the BCS case. Note, however, that Eq. (4.31) displays explicitly the singular behavior for $\mathbf{q} \to \mathbf{0}$ and $\omega \to 0$, as mentioned in Section 1.3.2. Indeed, if we set first $\mathbf{q} = \mathbf{0}$, with $\omega \neq 0$, the argument of the sum on the right-hand side is zero. On the other hand, if we set first $\omega = 0$, this argument is proportional to $[f(\xi_{\mathbf{k}+\mathbf{q}/2}) - f(\xi_{\mathbf{k}-\mathbf{q}/2})]/[\xi_{\mathbf{k}+\mathbf{q}/2} - \xi_{\mathbf{k}-\mathbf{q}/2}]$, which has a nonzero limit when $\mathbf{q} \to \mathbf{0}$, as we will see in more detail below.

4.4.2 BCS Case

Indeed, since in the BCS theory the excited states behave as free fermions, the calculation of the response function is quite similar to the above one for free electrons. The essential difference is that while the coupling to the electromagnetic field Eq. (4.7) has a simple expression in terms of electronic operators, this expression is not convenient for dealing with the free fermionic excitations of BCS theory, namely the bogolons. In order to recover the simplicity we have found in dealing with free electrons, we have to convert this expression of the coupling into an expression involving the bogolons creation and annihilation operators $\gamma_{\mathbf{k}}^{\dagger}$ and $\gamma_{\mathbf{k}}$. This is merely done by carrying into Eq. (4.7) the expressions Eq. (3.16) of electronic operators in terms of the bogolon operators. This is most easily done by first rewriting the expression Eq. (4.8) of the current operator $\mathbf{j}_{-\mathbf{q}}$ as

$$\mathbf{j}_{-\mathbf{q}} = \sum_{\mathbf{k}\sigma} \frac{\hbar\mathbf{k}}{m} c_{\mathbf{k}_+,\sigma}^{\dagger} c_{\mathbf{k}_-,\sigma} = \sum_{\mathbf{k}} \frac{\hbar\mathbf{k}}{m} \left(c_{\mathbf{k}_+\uparrow}^{\dagger} c_{\mathbf{k}_-\uparrow} + c_{-\mathbf{k}_+\downarrow} c_{-\mathbf{k}_-\downarrow}^{\dagger} \right)$$ (4.32)

$$= \sum_{\mathbf{k}} \frac{\hbar\mathbf{k}}{m} \left(c_{\mathbf{k}_+\uparrow}^{\dagger} \quad c_{-\mathbf{k}_+\downarrow} \right) \begin{pmatrix} c_{\mathbf{k}_-\uparrow} \\ c_{-\mathbf{k}_-\downarrow}^{\dagger} \end{pmatrix}$$

where we have used the compact notations $\mathbf{k}_+ \equiv \mathbf{k} + \mathbf{q}/2$ and $\mathbf{k}_- \equiv \mathbf{k} - \mathbf{q}/2$. Moreover, the last term for $\downarrow$ spins has been obtained by changing $\mathbf{k}$ into $-\mathbf{k}$ in the summation over $\mathbf{k}$ and making use of the anticommutation of the fermionic operators for different wavevectors. Inserting the transformation Eq. (3.16), we obtain

$$\mathbf{j}_{-\mathbf{q}} = \sum_{\mathbf{k}} \frac{\hbar\mathbf{k}}{m} \left(\gamma_{\mathbf{k}_+\uparrow}^{\dagger} \quad \gamma_{-\mathbf{k}_+\downarrow} \right) U_{\mathbf{k}_+}^{\mathsf{T}} U_{\mathbf{k}_-} \begin{pmatrix} \gamma_{\mathbf{k}_-\uparrow} \\ \gamma_{-\mathbf{k}_-\downarrow}^{\dagger} \end{pmatrix}$$ (4.33)

with explicitly, from the definition Eq. (3.15),

$$U_{\mathbf{k}_+}^{\mathsf{T}} U_{\mathbf{k}_-} = \begin{pmatrix} C_s(\mathbf{k}_+, \mathbf{k}_-) & C_{ca}(\mathbf{k}_+, \mathbf{k}_-) \\ -C_{ca}(\mathbf{k}_+, \mathbf{k}_-) & C_s(\mathbf{k}_+, \mathbf{k}_-) \end{pmatrix}$$ (4.34)

where we have defined

$$C_s(\mathbf{k}_+, \mathbf{k}_-) \equiv u_{\mathbf{k}_+} u_{\mathbf{k}_-} + v_{\mathbf{k}_+} v_{\mathbf{k}_-} \qquad C_{ca}(\mathbf{k}_+, \mathbf{k}_-) \equiv u_{\mathbf{k}_+} v_{\mathbf{k}_-} - v_{\mathbf{k}_+} u_{\mathbf{k}_-} \quad (4.35)$$

As a result, we obtain for the current operator the following expression in terms of bogolon operators

$$\mathbf{j}_{-\mathbf{q}} = \sum_{\mathbf{k}} \frac{\hbar \mathbf{k}}{m} \left\{ C_s(\mathbf{k}_+, \mathbf{k}_-) \left(\gamma_{\mathbf{k}_+\uparrow}^\dagger \gamma_{\mathbf{k}_-\uparrow} - \gamma_{-\mathbf{k}_-\downarrow}^\dagger \gamma_{-\mathbf{k}_+\downarrow} \right) \right. \tag{4.36}$$
$$\left. + C_{ca}(\mathbf{k}_+, \mathbf{k}_-) \left(\gamma_{\mathbf{k}_+\uparrow}^\dagger \gamma_{-\mathbf{k}_-\downarrow}^\dagger + \gamma_{\mathbf{k}_-\uparrow} \gamma_{-\mathbf{k}_+\downarrow} \right) \right\}$$

making use again of the anticommutation of the fermionic operators.

The coefficients $C_s(\mathbf{k}_+, \mathbf{k}_-)$ and $C_{ca}(\mathbf{k}_+, \mathbf{k}_-)$ we have just introduced in Eq. (4.35) are called "coherence factors." Their expressions are fairly easy to understand. The coefficient $C_s(\mathbf{k}_+, \mathbf{k}_-)$ comes in the first term on the right-hand side of Eq. (4.36), which describes the scattering of an $\uparrow$ spin bogolon with momentum $\mathbf{k}_- = \mathbf{k} - \mathbf{q/2}$ into a bogolon with momentum $\mathbf{k}_+ = \mathbf{k} + \mathbf{q/2}$. However, from the definition Eq. (3.11) of the operator $\gamma_{\mathbf{k}}^\dagger$, this $\mathbf{k}_-$ bogolon is just, with probability amplitude u_{k_-}, an electron with momentum $\mathbf{k}_-$. This electron can naturally be scattered by the electromagnetic interaction into an electron with momentum $\mathbf{k}_+$. Since the $\mathbf{k}_+$ bogolon can just be this electron, with probability amplitude u_{k_+}, it is natural that the scattering of bogolon $\mathbf{k}_-$ into $\mathbf{k}_+$ (corresponding to the scattering of the underlying electron) comes with a probability amplitude $u_{k_-} u_{k_+}$. On the other hand, from Eq. (3.11), the $\mathbf{k}_-$ bogolon is as well, with probability amplitude $-v_{k_-}$, a hole with momentum $-\mathbf{k}_-$. This hole can be scattered into a $-\mathbf{k}_+$ hole, and bogolon $\mathbf{k}_+$ is such a hole with probability amplitude $-v_{k_+}$. This provides another way, with probability amplitude $v_{k_-} v_{k_+}$, in which bogolon $\mathbf{k}_-$ scatters into bogolon $\mathbf{k}_+$. Because a bogolon is a coherent combination of an electron and a hole, these two ways for bogolon scattering combine coherently, so that the probability amplitude for the whole process is the sum of the respective amplitudes, namely $u_{k_-} u_{k_+} + v_{k_-} v_{k_+} = C_s(\mathbf{k}_+, \mathbf{k}_-)$. This coherent combination is naturally responsible for the name of this coefficient. Repeating the same argument for the scattering of a $\downarrow$ spin bogolon with momentum $-\mathbf{k}_+ = -\mathbf{k} - \mathbf{q/2}$ into a bogolon with momentum $-\mathbf{k}_- = -\mathbf{k} + \mathbf{q/2}$ leads to the same probability amplitude $C_s(\mathbf{k}_+, \mathbf{k}_-)$.

The last two terms on the right-hand side of Eq. (4.36) are qualitatively new since they correspond respectively to the creation and annihilation of a pair of bogolons, which has no equivalent in Eq. (4.8). Naturally, electrons are conserved in these processes, and the bogolon pair creation corresponds actually to the creation of an electron-hole pair. For example, the $\uparrow$ spin bogolon $\mathbf{k}_+$ is a $\mathbf{k}_+ \uparrow$ electron, with probability amplitude u_{k_+}, while the $\downarrow$ spin bogolon $-\mathbf{k}_-$ is a $\mathbf{k}_- \uparrow$ hole, with probability amplitude v_{k_-} from Eq. (3.13), which leads to an overall amplitude $u_{k_+} v_{k_-}$. On the other hand, the $\uparrow$ spin bogolon $\mathbf{k}_+$ may as well be a $-\mathbf{k}_+ \downarrow$ hole, with probability amplitude $-v_{k_+}$, while the $\downarrow$ spin bogolon $-\mathbf{k}_-$ may be a $-\mathbf{k}_- \downarrow$ electron, with probability amplitude u_{k_-}, leading to an overall amplitude $-v_{k_+} u_{k_-}$. The coherent combination of these two possibilities gives an overall amplitude $u_{k_+} v_{k_-} - v_{k_+} u_{k_-} = C_{ca}(\mathbf{k}_+, \mathbf{k}_-)$. The same kind of reasoning leads, in the case of bogolon pair annihilation, to the same coherence factor $C_{ca}(\mathbf{k}_+, \mathbf{k}_-)$.

4.5 BCS Response Function

With the above expression, Eq. (4.36), for the current operator, we turn now to the calculation of the response function for BCS theory. First of all, if we proceed as we have done in Eq. (4.29) by introducing explicitly the $\mathbf{k}$ summation in Eq. (4.36), we see that starting from a specific eigenstate $|n\rangle$, the four different terms on the right-hand side of Eq. (4.36) connect $|n\rangle$ to different eigenstates $|m\rangle$ because either the spins or the number of bogolons in $|m\rangle$ are different. Hence there is no interference between the four different kind of processes, and their respective contributions to $K(\mathbf{q}, \omega)$ will merely add up.

Then, since the bogolons are not interacting, the situation for the scattering terms is almost unchanged with respect to our above calculation for free electrons. More specifically, the coherence factor $C_s(\mathbf{k}_+, \mathbf{k}_-)$ will naturally produce an additional $C_s^2(\mathbf{k}_+, \mathbf{k}_-)$ factor in the result. Otherwise we have merely to take into account that the energy of a $\mathbf{k}$ bogolon is $E_{\mathbf{k}}$, instead of $\xi_{\mathbf{k}}$ for free electrons. This leads, instead of the right-hand side of Eq. (4.31), to a contribution to $K(\mathbf{q}, \omega)$ equal to

$$-\frac{e^2 \hbar}{m^2} \sum_{\mathbf{k}} k_x^2 \, C_s^2(\mathbf{k}_+, \mathbf{k}_-) \, \frac{f(E_{\mathbf{k}+\mathbf{q}/2}) - f(E_{\mathbf{k}-\mathbf{q}/2})}{\omega - (E_{\mathbf{k}+\mathbf{q}/2} - E_{\mathbf{k}-\mathbf{q}/2})/\hbar + i\epsilon} \tag{4.37}$$

Next, the second scattering term in Eq. (4.36) can be transformed into the first one, except for the spin, which is $\downarrow$ instead of $\uparrow$. This is done by changing $\mathbf{k}$ into $-\mathbf{k}$ in the summation, and noticing that $C_s(\mathbf{k}_+, \mathbf{k}_-)$ is unchanged in this transformation since both $u_{\mathbf{k}}$ and $v_{\mathbf{k}}$ are even in $\mathbf{k}$. As a result, the contribution of this second scattering term is identical to the one Eq. (4.37) of the first scattering term, and just as in Eq. (4.31), the overall result gets a factor of 2 due to summation over spin.

For the contribution of the third term, we have to take into account that since this term creates a bogolon pair, the energy difference between states $|n\rangle$ and $|m\rangle$ becomes $E_m - E_n = E_{\mathbf{k}+\mathbf{q}/2} + E_{\mathbf{k}-\mathbf{q}/2}$, instead of $E_m - E_n = \xi_{\mathbf{k}+\mathbf{q}/2} - \xi_{\mathbf{k}-\mathbf{q}/2}$. Formally, this is obtained by changing $E_{\mathbf{k}-\mathbf{q}/2}$ into $-E_{\mathbf{k}-\mathbf{q}/2}$, after the substitution of $\xi_{\mathbf{k}}$ by $E_{\mathbf{k}}$. Since the coherence factor is now $C_{ca}(\mathbf{k}_+, \mathbf{k}_-)$, we obtain from bogolon pair creation a contribution

$$\frac{e^2 \hbar}{m^2} \sum_{\mathbf{k}} k_x^2 \, C_{ca}^2(\mathbf{k}_+, \mathbf{k}_-) \, \frac{1 - f(E_{\mathbf{k}+\mathbf{q}/2}) - f(E_{\mathbf{k}-\mathbf{q}/2})}{\omega - (E_{\mathbf{k}+\mathbf{q}/2} + E_{\mathbf{k}-\mathbf{q}/2})/\hbar + i\epsilon} \tag{4.38}$$

where we have made use of $f(-E) = 1 - f(E)$. The result for the statistical factor gets clearer if we follow the argument already made for free electrons. To create a bogolon pair with wavevectors $\mathbf{k} + \mathbf{q}/2$ and $\mathbf{k} - \mathbf{q}/2$, these states must be initially empty. The corresponding probabilities are $1 - f(E_{\mathbf{k}+\mathbf{q}/2})$ and $1 - f(E_{\mathbf{k}-\mathbf{q}/2})$, leading to an overall probability $(1 - f(E_{\mathbf{k}+\mathbf{q}/2}))(1 - f(E_{\mathbf{k}-\mathbf{q}/2}))$. In addition, we have to subtract, as we have done for free electrons, the contribution of the reverse process where the two bogolons $\mathbf{k} + \mathbf{q}/2$ and $\mathbf{k} - \mathbf{q}/2$ are annihilated. The probability of having these two states initially occupied is $f(E_{\mathbf{k}+\mathbf{q}/2})f(E_{\mathbf{k}-\mathbf{q}/2})$. Accordingly, the global statistical factor is $(1 - f(E_{\mathbf{k}+\mathbf{q}/2}))(1 - f(E_{\mathbf{k}-\mathbf{q}/2})) - f(E_{\mathbf{k}+\mathbf{q}/2})f(E_{\mathbf{k}-\mathbf{q}/2}) = 1 - f(E_{\mathbf{k}+\mathbf{q}/2}) - f(E_{\mathbf{k}-\mathbf{q}/2})$, in agreement with Eq. (4.38).

Finally we proceed in much the same way to obtain the contribution from bogolon pair annihilation. Here, the energy difference is rather $E_m - E_n = -E_{\mathbf{k}+\mathbf{q}/2} - E_{\mathbf{k}-\mathbf{q}/2}$ since the bogolons are present in state $|n\rangle$ and absent in state $|m\rangle$. Formally this is accounted by the change of $E_{\mathbf{k}+\mathbf{q}/2}$ into $-E_{\mathbf{k}+\mathbf{q}/2}$, after the substitution of $\xi_{\mathbf{k}}$ by $E_{\mathbf{k}}$. Since the coherence factor is again $C_{ca}(\mathbf{k}_+, \mathbf{k}_-)$, this leads to the contribution

$$-\frac{e^2 \hbar}{m^2} \sum_{\mathbf{k}} k_x^2 \, C_{ca}^2(\mathbf{k}_+, \mathbf{k}_-) \, \frac{1 - f(E_{\mathbf{k}+\mathbf{q}/2}) - f(E_{\mathbf{k}-\mathbf{q}/2})}{\omega + (E_{\mathbf{k}+\mathbf{q}/2} + E_{\mathbf{k}-\mathbf{q}/2})/\hbar + i\epsilon} \tag{4.39}$$

The statistical factor is actually opposite to the contribution from bogolon pair creation, as it is obvious by repeating in the present case the above derivation of this factor. The sum of contributions Eq. (4.38) and Eq. (4.39) are conveniently combined into a single term. This leads to our final result for the BCS response:

$$K(\mathbf{q}, \omega) = \frac{ne^2}{m} - \frac{2e^2 \hbar}{m^2} \sum_{\mathbf{k}} k_x^2 \Bigg[\, C_s^2(\mathbf{k}_+, \mathbf{k}_-) \, \frac{f(E_{\mathbf{k}_+}) - f(E_{\mathbf{k}_-})}{\omega - (E_{\mathbf{k}_+} - E_{\mathbf{k}_-})/\hbar + i\epsilon} \tag{4.40}$$
$$- \, C_{ca}^2(\mathbf{k}_+, \mathbf{k}_-) \, \frac{(1 - f(E_{\mathbf{k}_+}) - f(E_{\mathbf{k}_-})) \, (E_{\mathbf{k}_+} + E_{\mathbf{k}_-})/\hbar}{(\omega + i\epsilon)^2 - (E_{\mathbf{k}_+} + E_{\mathbf{k}_-})^2/\hbar^2} \Bigg]$$

where in this long final formula, we have uniformly used the same compact notation $\mathbf{k}_\pm \equiv \mathbf{k} \pm \mathbf{q}/2$ as above.

The first term on the right-hand side is often called the "diamagnetic" term since, if it were the only one, it would lead essentially to the result of London theory and to the diamagnetic properties of superconductors. The second term, which turns out to have the opposite sign, is called the "paramagnetic" term. There is plenty of information in this result Eq. (4.40), but we will only consider limiting cases. However, before proceeding for the superconducting case, it is important as a benchmark to know what happens in the normal state, and first of all to examine how the result Eq. (4.31) is recovered when we take the normal state limit of Eq. (4.40).

As we have seen, the normal state is recovered when one takes the limit $\Delta \to 0$. However, one must be careful that in this limit E_k reduces to $|\xi_k|$ (since it is an excitation energy, it must be positive), and not to ξ_k, so that one does not recover straightforwardly Eq. (4.31). Correspondingly, from their expressions Eq. (2.75), u_k has the limiting value 1 for $\xi_k > 0$, that is, $k > k_F$, and 0 in the opposite case $k < k_F$. For v_k, these two situations are exchanged, so that one merely has $v_k = 1 - u_k$. From these results one sees that if both $|\mathbf{k}_-|$ and $|\mathbf{k}_+|$ are larger or smaller than k_F, one has from their definitions Eq. (4.35) $C_s(\mathbf{k}_+, \mathbf{k}_-) = 1$ and $C_{ca}(\mathbf{k}_+, \mathbf{k}_-) = 0$. As a result, one survives the first term, corresponding to bogolon scattering, in the large bracket on the right-hand side of Eq. (4.40). The second term is zero. For $\xi_{k_+} > 0$ and $\xi_{k_-} > 0$, one recovers in this way directly the expression appearing in Eq. (4.31). In the case where both $\xi_{k_+} < 0$ and $\xi_{k_-} < 0$, one has $|\xi_{k_+}| = -\xi_{k_+}$ and $|\xi_{k_-}| = -\xi_{k_-}$, so that $E_{\mathbf{k}_+} - E_{\mathbf{k}_-} = \xi_{\mathbf{k}_-} - \xi_{\mathbf{k}_+}$. Moreover, making use of $f(-\xi) = 1 - f(\xi)$, one has also $f(-\xi_{\mathbf{k}_+}) - f(-\xi_{\mathbf{k}_-}) = f(\xi_{\mathbf{k}_-}) - f(\xi_{\mathbf{k}_+})$. However, changing $\mathbf{k}$ into $-\mathbf{k}$ in the summation over $\mathbf{k}$ exchanges the roles of $\mathbf{k}_+$ and $\mathbf{k}_-$, because $\xi_{\mathbf{k}}$ is even in $\mathbf{k}$, so that one recovers again Eq. (4.31).

In the remaining case where $|\mathbf{k}_+|$ is larger than k_F while $|\mathbf{k}_-|$ is smaller, one finds from Eq. (4.35) that $C_s(\mathbf{k}_+, \mathbf{k}_-) = 0$ and $C_{ca}(\mathbf{k}_+, \mathbf{k}_-) = 1$. Hence the normal state result comes now from the second term in the large bracket on the right-hand side of Eq. (4.40), while the first term is zero. This is at first surprising since this term describes the contribution of bogolon pair creation and annihilation, which sounds specific to the superconducting state. However, this is a matter of vocabulary, since the creation in the normal state of an electron-hole pair corresponds in our superconducting language to the creation of two bogolons because both the electron creation and the hole creation correspond to the creation of an excited state, which means that they are limiting cases of bogolon creation. In this case, as above, one has $E_{\mathbf{k}_+} + E_{\mathbf{k}_-} = \xi_{\mathbf{k}_+} - \xi_{\mathbf{k}_-}$, and $1 - f(E_{\mathbf{k}_+}) - f(E_{\mathbf{k}_-}) = 1 - f(\xi_{\mathbf{k}_+}) - f(-\xi_{\mathbf{k}_-}) = f(\xi_{\mathbf{k}_-}) - f(\xi_{\mathbf{k}_+})$, so that the second term in the large bracket leads indeed to the normal state result (in Eq. (4.31), one has to combine with the expression obtained by $\mathbf{k} \to -\mathbf{k}$). Finally one finds naturally the same result for the inverse case of $|\mathbf{k}_+| < k_F$, and $|\mathbf{k}_-| > k_F$.

We are now ready to investigate the conductivity and the Meissner effect in the BCS state. However, it is again quite useful, for the sake of comparison, to look first at the normal state results for these two basic properties. From Eq. (1.39) we have, for the conductivity, to deal with $K(\mathbf{0}, \omega)$. In this case, where $\mathbf{q} = \mathbf{0}$, the result of Eq. (4.31) is quite simple, because the term on the right-hand side is zero. Indeed, in the sum over $\mathbf{k}$, we have $f(\xi_{\mathbf{k}+\mathbf{q}/2}) - f(\xi_{\mathbf{k}-\mathbf{q}/2}) = 0$, while the denominator is nonzero because $\omega \neq 0$. Hence we merely have $K(\mathbf{0}, \omega) = ne^2/m$, which leads for the conductivity $\sigma(\omega)$ to

$$\sigma(\omega) = i \frac{K(\mathbf{0}, \omega)}{\omega} = i \frac{ne^2}{m\omega} \tag{4.41}$$

This is the expected result for free electron dynamics since, from Newton's law $m\, d\mathbf{v}/dt = e\mathbf{E}$ for a single electron, we have for an electric field $\mathbf{E}(t) = \mathbf{E}_0\, e^{-i\omega t}$ at frequency ω, $\mathbf{v}(t) = \mathbf{v}_0\, e^{-i\omega t}$ with $-i\omega m\mathbf{v}_0 = e\mathbf{E}_0$. For the current $\mathbf{j} = ne\mathbf{v}$, this leads to $\mathbf{j} = i(ne^2/m\omega)\mathbf{E}$, in agreement with Eq. (4.41) for the conductivity. Naturally, we find that in the limit $\omega \to 0$, the conductivity diverges and the free electron gas has, as expected, an infinite conductivity.

It is convenient to recall at this stage the simple Drude phenomenology for metallic electrons, since it will be useful in the following. To account for the finite conductivity of metals, Drude introduced in Newton's law an additional term that slows down electrons and can compensate the accelerating force due to the electric field. He took for this purpose a simple viscous term, proportional to the electron velocity, so that Newton's law reads with it

$$m \frac{d\mathbf{v}}{dt} = e\mathbf{E} - \frac{m}{\tau}\mathbf{v} \tag{4.42}$$

Physically, this slowing down of the electric current is due to electron scattering on impurities or any other kind of defect which counters the free propagation of electrons in metals. The simple Drude form for this term gives, for the evolution of the velocity in the absence of an electric field, $m\, d\mathbf{v}/dt = -m\mathbf{v}/\tau$, which results in an exponential decay $\mathbf{v}(t) = \mathbf{v}_0\, e^{-t/\tau}$ for the velocity of an electron with initial velocity $\mathbf{v}_0$. Hence the relaxation time τ is the characteristic time for this loss of velocity due to impurity scattering. In the presence of an electric field $\mathbf{E}(t) = \mathbf{E}_0 \, e^{-i\omega t}$, the above expression for the velocity is modified into $(1/\tau - i\omega)m\mathbf{v}_0 = e\mathbf{E}_0$, which leads for the frequency-dependent conductivity to

$$\sigma(\omega) = \sigma_0 \frac{1}{1 - i\omega\tau} \qquad \sigma_0 = \frac{ne^2\tau}{m} \qquad (4.43)$$

Hence this Drude model provides a finite static conductivity σ_0, in contrast with the free-electron case. The Drude phenomenology looks very simplistic for the description of all the complicated processes responsible for the finite conductivity of metals. However, while it cannot quantitatively provide reliable results, it has turned out to be a physically reasonable and sound description of metallic conductivity, interpolating properly between the low-frequency dissipative regime and the high frequency domain dominated by the reactive inertial electronic response.

Having considered the conductivity, we turn now to the Meissner effect, where we need to consider the zero-frequency value $K(\mathbf{q}, 0)$ of the response function. More specifically, we are interested in the zero wavevector limit $\mathbf{q} \to \mathbf{0}$ of this quantity. On the right-hand side of Eq. (4.31), we find in this case the ratio $[f(\xi_{\mathbf{k}+\mathbf{q}/2}) - f(\xi_{\mathbf{k}-\mathbf{q}/2})]/[\xi_{\mathbf{k}+\mathbf{q}/2} - \xi_{\mathbf{k}-\mathbf{q}/2}]$ which, in the $\mathbf{q} \to \mathbf{0}$ limit, is equal to the derivative $\partial f(\xi_{\mathbf{k}})/\partial \xi_{\mathbf{k}}$ of the Fermi distribution with respect to energy. Since at $T = 0$ the Fermi distribution is a step function $f(\xi) = 1 - \theta(\xi)$, its derivative is a $-\delta(\xi_{\mathbf{k}})$ peak so that $|\mathbf{k}|$ is constrained to be equal to k_F. At low temperature compared to E_F, $\partial f(\xi_{\mathbf{k}})/\partial \xi_{\mathbf{k}}$ is a narrow peak with a width of order $k_B T$ around $\xi_{\mathbf{k}} = 0$, and we find the same constraint $|\mathbf{k}| \simeq k_F$. Transforming the sum over $\mathbf{k}$ into an integral, we see that the only quantity in the integrand which depends on the direction of $\mathbf{k}$ is k_x^2. Just as in the calculation leading to Eq. (1.33), the angular average gives a factor $1/3$. Since $|\mathbf{k}| = k_F$, k_x^2 merely produces a factor $k_F^2/3$. Hence we are left with an integration over $|\mathbf{k}|$, so we can introduce the single spin density of states $N(\xi_k)$ and integrate over ξ_k. Since we have $|\mathbf{k}| = k_F$, we have $N(\xi_k) = N(0) \equiv N_0$, where N_0 is the single spin density of states at the Fermi energy. We are left with $\int_{-E_F}^{\infty} d\xi_k \, \partial f(\xi_{\mathbf{k}})/\partial \xi_{\mathbf{k}} = f(\infty) - f(-E_F) = -1$, since $f(-E_F) - 1 \sim e^{-E_F/k_B T}$ is exponentially negligible at low T. To summarize, we have

$$\sum_{\mathbf{k}} k_x^2 \frac{f(\xi_{\mathbf{k}_+}) - f(\xi_{\mathbf{k}_-})}{\xi_{\mathbf{k}_+} - \xi_{\mathbf{k}_-}} = \frac{k_F^2}{3} \sum_{\mathbf{k}} \frac{\partial f(\xi_{\mathbf{k}})}{\partial \xi_{\mathbf{k}}} = -\frac{k_F^2}{3} N_0 = -\frac{k_F^2}{3} \frac{3n}{4E_F} = -\frac{nm}{2\hbar^2} \qquad (4.44)$$

where we have used the free electrons value $N_0 = mk_F/(2\pi^2\hbar^2) = 3n/4E_F$, already seen below Eq. (3.92). Taking into account the additional coefficient $-2e^2\hbar^2/m^2$ coming on the right-hand side of Eq. (4.31), we find that this right-hand side is equal to $-ne^2/m$. Accordingly, the paramagnetic term cancels exactly the diamagnetic term, and we obtain finally $\lim_{q\to 0} K(\mathbf{q}, 0) = 0$. So we obtain that the free electron gas does not display any Meissner effect, which was naturally the expected result. Note that from our calculation of the conductivity and the Meissner effect, we have found explicitly that in the normal state, $ne^2/m = \lim_{\omega\to 0} K(\mathbf{0}, \omega) \neq \lim_{q\to 0} K(\mathbf{q}, 0) = 0$, as stated in Section 1.3.2.

We notice finally that for $q \neq 0$, the cancellation between the paramagnetic and diamagnetic terms is not perfect, so that in this case $K(\mathbf{q}, 0) \neq 0$. The actual result can be seen to be proportional to q^2. This corresponds to the weak orbital diamagnetic response of the electron gas, also known as Landau diamagnetism.

4.6 Conductivity and Meissner Effect for the BCS State

- Conductivity

We now calculate the conductivity and the Meissner effect for the BCS state. Addressing first the conductivity, we consider again the response function at zerowave vector $\mathbf{q} = \mathbf{0}$ and nonzero frequency, but this time we make use of the appropriate Eq. (4.40) for $K(\mathbf{q}, \omega)$. For $\mathbf{q} = \mathbf{0}$, we merely have in this formula $\mathbf{k}_+ = \mathbf{k} + \mathbf{q}/2 = \mathbf{k} = \mathbf{k} - \mathbf{q}/2 = \mathbf{k}_-$. This implies from Eq. (4.35) for the coherence factors $C_s(\mathbf{k}_+, \mathbf{k}_-) = u_{\mathbf{k}} u_{\mathbf{k}} + v_{\mathbf{k}} v_{\mathbf{k}} = 1$ from normalization, and $C_{ca}(\mathbf{k}_+, \mathbf{k}_-) = u_{\mathbf{k}} v_{\mathbf{k}} - v_{\mathbf{k}} u_{\mathbf{k}} = 0$. Hence the bogolon creation and annihilation contribution disappears in the paramagnetic term. But the bogolon scattering term is also zero, since we have in the numerator $f(E_{\mathbf{k}}) - f(E_{\mathbf{k}}) = 0$. So in this limit, the whole paramagnetic term is zero, just as for free electrons. Accordingly, we find in the same way for the conductivity

$$\sigma(\omega) = i\,\frac{K(\mathbf{0}, \omega)}{\omega} = i\,\frac{ne^2}{m\omega} \tag{4.45}$$

In the limit $\omega \to 0$, we have $\sigma(\omega) \to \infty$. So we might believe we have shown that the BCS state displays superconductivity.

However, this is an empty result because this result is just the same as Eq. (4.41) for the corresponding normal state of the free electron gas. Physically, the whole point about superconductivity is that the normal state conductivity is finite while in the superconducting state it becomes infinite. Therefore, in order to prove meaningfully that the BCS state describes indeed a superconducting state, we must have a corresponding normal state that displays finite conductivity. However, as we have just seen with the Drude model, this is the result of the presence of impurities (or anything equivalent). Hence, in order to prove anything about the superconducting properties of the BCS state, we have to include the effect of impurities in this state, which we have not done yet. We will address this extension in the following section.

- Meissner Effect

Turning now to the Meissner effect, we have again to consider the BCS response function Eq. (4.40) for zero frequency in the limit $\mathbf{q} \to \mathbf{0}$. Just as above for the conductivity, we have in this limit $C_s(\mathbf{k}_+, \mathbf{k}_-) = 1$ and $C_{ca}(\mathbf{k}_+, \mathbf{k}_-) = 0$, so that the bogolon creation and annihilation term disappears again. On the other hand, for the bogolon scattering term we find, much as in the free-electron case, the ratio $[f(E_{\mathbf{k}_+}) - f(E_{\mathbf{k}_-}]/[E_{\mathbf{k}_+} - E_{\mathbf{k}_-}]$, which in the limit $\mathbf{q} \to \mathbf{0}$ becomes the derivative $\partial f(E_{\mathbf{k}})/\partial E_{\mathbf{k}}$ of the Fermi distribution with respect to the energy. The important difference is that this energy is $E_{\mathbf{k}}$ and no longer $\xi_{\mathbf{k}}$. Otherwise we can follow the same steps as in the normal state. Indeed $\partial f(E_{\mathbf{k}})/\partial E_{\mathbf{k}}$ is sharply peaked around $k = k_F$ because, as soon as $|\xi_{\mathbf{k}}|$ gets large compared to Δ (but nevertheless being small compared to E_F), we have $E_{\mathbf{k}} = |\xi_{\mathbf{k}}|$, and $f(E_{\mathbf{k}})$, as well as $\partial f(E_{\mathbf{k}})/\partial E_{\mathbf{k}}$, becomes exponentially small as soon as $|\xi_{\mathbf{k}}| \gg k_B T$. Since the angular average gives again a factor $1/3$, we see that k_x^2 gives again a factor $k_F^2/3$. We go again from the sum over $\mathbf{k}$ to an integration over ξ_k. Since the integrand is strongly peaked around $\xi_k = 0$, the density of

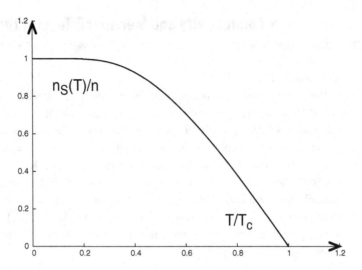

Fig. 4.2   Reduced superfluid density $n_S(T)/n$ as a function of temperature T.

states can again be replaced by its value N_0 at the Fermi energy. Taking again into account, as in Eq. (4.44), $N_0 k_F^2/3 = nm/2\hbar^2$, we have

$$\sum_{\mathbf{k}} k_x^2 \frac{f(E_{\mathbf{k}_+}) - f(E_{\mathbf{k}_-})}{E_{\mathbf{k}_+} - E_{\mathbf{k}_-}} = \frac{k_F^2}{3} \sum_{\mathbf{k}} \frac{\partial f(E_{\mathbf{k}})}{\partial E_{\mathbf{k}}} = \frac{nm}{2\hbar^2} \int_{-\infty}^{\infty} d\xi_k \frac{\partial f(E_{\mathbf{k}})}{\partial E_{\mathbf{k}}} \tag{4.46}$$

where we have extended the lower boundary of the integral to $-\infty$ since for large E_k the integrand is exponentially small. This leads finally to

$$\lim_{q \to 0} K(\mathbf{q}, 0) = \frac{n_s(T)e^2}{m} \qquad \frac{n_s(T)}{n} \equiv 1 - \int_{-\infty}^{\infty} d\xi_k \left(-\frac{\partial f(E_{\mathbf{k}})}{\partial E_{\mathbf{k}}} \right) \tag{4.47}$$

where we introduced the "superfluid density" $n_s(T)$ (note that $\partial f(E_{\mathbf{k}})/\partial E_{\mathbf{k}} < 0$).

Since at $T = 0$ we have $f(E_k) = 0$ because the existence of the gap makes $E_k \geq \Delta > 0$, the derivative $\partial f(E_{\mathbf{k}})/\partial E_{\mathbf{k}}$ is also zero and $n_s(0) = n$. On the other hand, when $T \to T_c$, we have $E_k \to |\xi_k|$, and by making the change $\xi_k \to -\xi_k$ in the part of the integral going from $-\infty$ to 0, we obtain that the integral is equal to $2 \int_0^{\infty} d\xi_k (-\partial f(\xi_{\mathbf{k}})/\partial \xi_{\mathbf{k}}) = 2f(\xi_k = 0) = 1$. Hence we recover the normal state result for this integral. As a result, at T_c, the superfluid density goes to zero. In the vicinity of T_c one finds[3] $n_s(T)/n = 2(1 - T/T_c)$. Hence $n_s(T)/n$ goes smoothly from 1 to 0 when T goes from 0 to T_c, as it is displayed in Fig. (4.2).

---

[3] With $x = \beta_c \xi_k/2$ as integration variable, we have from Eq. (4.47) $1 - n_s(T)/n = \int_0^{\infty} dx \cosh^{-2} y$ with $y^2 = x^2 + \delta^2$, and $\delta = \beta \Delta/2$. Differentiating this relation for small δ^2 gives $n_s(T)/n = -\delta^2 \int_0^{\infty} dx \, [d(\cosh^{-2} x)/dx](1/2x) = (\delta^2/2) \int_0^{\infty} dx \, (1 - \cosh^{-2} x)/x^2]$ after by-parts integration. On the other hand, we have found in the note related to Eq. (3.56), where $\Delta(T)$ is obtained in the vicinity of T_c, $\delta^2 \int_{-\infty}^{\infty} dx \, [d(\tanh x/x)/dx](1/2x) = 2(1 - \beta/\beta_c) \simeq -2(1 - T/T_c)$. The integrand in this last integral is even in x, so by writing explicitly the x derivative $\int_{-\infty}^{\infty} dx \, [d(\tanh x/x)/dx](1/2x) = \int_0^{\infty} dx \, [(x - \tanh x)/x^3 - (1 - \cosh^{-2} x)/x^2]$. However, integrating by parts $\int_0^{\infty} dx \, (x - \tanh x)/x^3 = (1/2) \int_0^{\infty} dx \, (1 - \cosh^{-2} x)/x^2$, so that finally $\int_{-\infty}^{\infty} dx \, [d(\tanh x/x)/dx](1/2x) = -(1/2) \int_0^{\infty} dx \, (1 - \cosh^{-2} x)/x^2$. Comparing with the above expression for $n_s(T)/n$ leads immediately to the result given in the text.

Result Eq. (4.47) coincides formally with the expression we have found in London theory. Hence we find quite satisfactorily that BCS theory leads to a state displaying the Meissner effect. However, in contrast with the situation in London theory, n_s is now a well-defined quantity. In particular, it depends on temperature. This is physically reasonable since nonzero temperature gives rise to excited states, and one does not expect these states to fully participate in the superfluid screening currents responsible for the Meissner effect. Correspondingly, BCS theory provides an explicit temperature dependence for the penetration depth

$$\lambda_L(T) = \left(\frac{m}{\mu_0 n_s(T) e^2} \right)^{1/2} \tag{4.48}$$

with $n_s(T)$ given by Eq. (4.47). In particular, in the vicinity of the critical temperature, the penetration depth diverges as $\lambda_L(T) \sim (1 - T/T_c)^{-1/2}$. This is again physically quite reasonable, since one expects the superconductor to be less and less able to screen out the magnetic field in the vicinity of T_c, where it is not so far from being a normal metal.

Finally, it is interesting to take a closer look at the reasons for which the Meissner effect is present in the BCS state and not in the normal state. The simplest situation is found at $T = 0$. In the normal state, for $\xi_{k_+} > 0$ and $\xi_{k_-} > 0$ for example, we have seen that $[f(\xi_{k_+}) - f(\xi_{k_-})]/[\xi_{k_+} - \xi_{k_-}]$ has a nonzero limit when $\mathbf{q} \to \mathbf{0}$. This gives a nonzero contribution from the first term of the paramagnetic term in Eq. (4.40), which cancels exactly the diamagnetic contribution, even if it gets concentrated at $\xi = 0$ when the temperature goes to zero. On the other hand, in the BCS state, whatever ξ, the corresponding contribution from $[f(E_{k_+}) - f(E_{k_-})]/[E_{k_+} - E_{k_-}]$ disappears exponentially rapidly when T goes to 0, due to the existence of the gap Δ. If we consider the second term of the paramagnetic term in Eq. (4.40), it disappears because the coherence factor $C_{ca}^2(\mathbf{k}_+, \mathbf{k}_-)$ goes to zero when $\mathbf{q} \to \mathbf{0}$. However, in contrast with what may happen in the normal state, this cannot be balanced by a zero denominator because $E_{\mathbf{k}_+} + E_{\mathbf{k}_-} \geq 2\Delta$, due to the presence of the gap. Hence we see that the disappearance of the paramagnetic term, leaving the diamagnetic term unbalanced and leading to the Meissner effect, is due to the existence of the gap. This comes as a remarkable justification of London's intuitive argument that the electronic wave function is frozen due to the existence of a gap.

Actually, as is often the case, the real situation with superconductors is more complicated. There are known examples of superconductors where the gap for excitations is zero. They arise, for example, when magnetic impurities are introduced in a standard superconductor. These impurities may be shown to have negative effects on the formation of Cooper pairs. However, just before the number of impurities is high enough to fully destroy superconductivity, one finds a superconductor with a gap equal to zero. Such a zero gap may also possibly arise in known cases of strongly anisotropic superconductors. In these cases, one still gets a superconductor because, in order to obtain a Meissner effect, it is enough to have the diamagnetic term not exactly compensated by the paramagnetic one. Clearly, for this purpose, it is enough to have $E_k \neq |\xi_k|$ for some wavevectors; it is not at all necessary to have it for all wavevectors. Hence these cases come as counterexamples to London's argument.

4.7 Impurities Effect

We now come to the effect of impurities on the electromagnetic response in BCS theory. As we have seen in the preceding section, this inclusion is necessary if we want to properly show that BCS theory describes indeed a superconductor.

Although our purpose is only to deal with the effect of impurities on the response function, we nevertheless make two general physical remarks on the effect of impurities on superconductivity. First, we may very well worry that the presence of impurities completely spoils Cooper pair formation since, because of scattering of electrons by impurities, the electronic momentum $\mathbf{k}$ is no longer a conserved quantity, and the formation of $(\mathbf{k} \uparrow, -\mathbf{k} \downarrow)$ Cooper pairs becomes meaningless. However, it has been rapidly realized [30] that if one considers the electronic eigenstates in the presence of impurities, one can still form Cooper pairs with such electronic states related to each other by time-reversal transformation, which generalizes the fact that $(-\mathbf{k} \downarrow)$ is obtained from $(\mathbf{k} \uparrow)$ by time reversal. One can then think of building a BCS-like theory from these pairs. Since the density of electronic states should not be strongly affected by standard non-magnetic impurities, one does not expect drastic modifications of the thermodynamic properties with respect to standard BCS theory, and this agrees reasonably well with experimental facts. On the other hand, the situation is completely different with magnetic impurities, because in contrast with the preceding situation, the electronic spin is flipped in the electronic scattering on the impurity so that it is no longer possible to have the standard $(\uparrow, \downarrow)$ time-reversed pairing. Indeed it is known that magnetic impurities strongly affect standard superconductors, and that increasing their concentration ultimately leads to the destruction of superconductivity.

Another important effect due to impurities is related to anisotropy. The standard situation regarding metals is that they display some important anisotropy, with possibly complicated Fermi surfaces. In these cases, one would expect the gap itself to vary depending on the location on the Fermi surface and so present some anisotropy. This is actually what is found experimentally for very pure superconductors. However, in the presence of impurities, an electron with wavevector $\mathbf{k}$ near the Fermi surface can be scattered to some other wavevector $\mathbf{k}'$, also in the vicinity of the Fermi surface. If there are enough impurities, this process is repeated fairly rapidly, so that the electron is in practice "visiting" the whole Fermi surface. As a result, it is sensitive not to the gap at some specific location at the Fermi surface, but to some kind of average of this gap over the whole Fermi surface. This average is naturally the same for all the electrons on the Fermi surface, so that the resulting gap becomes in this way effectively isotropic. This impurity effect allows one to understand to an important extent the surprising quantitative success of simple BCS theory, which assumes from the start an isotropic situation for simplicity. Experimentally one finds indeed that adding impurities to a pure anisotropic superconductor makes it evolve rapidly toward isotropy, with little change when the concentration of impurities is further increased.

We go now to our handling of the effect of impurities on the electromagnetic response. Translational invariance[4] is broken due to the presence of impurities, this is their essential effect which we want to describe here phenomenologically, following the method of Mattis and Bardeen [31]. Indeed translational invariance is responsible for momentum conservation, which implies that in Eq. (4.7), electron with momentum $\hbar(\mathbf{k} - \mathbf{q}/2)$ is scattered into electron with momentum $\hbar(\mathbf{k} + \mathbf{q}/2)$ by the electromagnetic perturbation, which provides the additional momentum $\hbar\mathbf{q}$. To make the following discussion clearer let us rewrite in a compact form the paramagnetic term in the response function Eq. (4.40) as

$$\sum_{\mathbf{k}} k_x^2 \tilde{F}(\mathbf{k}_-, \mathbf{k}_+) = \frac{1}{(2\pi)^3} \int d\mathbf{k}_1 \, d\mathbf{k}_2 \, k_{1,x} k_{2,x} \, \tilde{F}(\mathbf{k}_1, \mathbf{k}_2) \, \delta(\mathbf{k}_2 - \mathbf{k}_1 - \mathbf{q}) \qquad (4.49)$$

where we have omitted the prefactor, and set

$$\tilde{F}(\mathbf{k}_1, \mathbf{k}_2) = C_s^2(\mathbf{k}_2, \mathbf{k}_1) \frac{f(E_{\mathbf{k}_2}) - f(E_{\mathbf{k}_1})}{\omega - (E_{\mathbf{k}_2} - E_{\mathbf{k}_1})/\hbar + i\epsilon} \qquad (4.50)$$
$$- C_{ca}^2(\mathbf{k}_2, \mathbf{k}_1) \frac{\left(1 - f(E_{\mathbf{k}_1}) - f(E_{\mathbf{k}_2})\right)(E_{\mathbf{k}_1} + E_{\mathbf{k}_2})/\hbar}{(\omega + i\epsilon)^2 - (E_{\mathbf{k}_1} + E_{\mathbf{k}_2})^2/\hbar^2}$$

In writing Eq. (4.49), we have replaced $\mathbf{k}_-$ by $\mathbf{k}_1$, and $\mathbf{k}_+$ by $\mathbf{k}_2$, but we have expressed momentum conservation $\mathbf{k}_2 - \mathbf{k}_1 = \mathbf{q}$ explicitly by introducing the Dirac δ function $\delta(\mathbf{k}_2 - \mathbf{k}_1 - \mathbf{q})$. Integrating over $\mathbf{k}_2$ on the right-hand side and making the change of variable $\mathbf{k}_1 \to \mathbf{k} - \mathbf{q}/2$, we recover the left-hand side by $(1/(2\pi)^3 \int d\mathbf{k} \to \sum_{\mathbf{k}}$ and taking into account, as we have seen, that $q_x = 0$ so that $\mathbf{k}_{-,x} = \mathbf{k}_{+,x} = \mathbf{k}_x$.

In the presence of impurities, momentum conservation does not hold anymore. After propagating over a typical distance of order of the mean free path ℓ, an electron is scattered by an impurity and changes its momentum. As a result, this electron does not have a well-defined momentum and accordingly it gets a wavevector uncertainty δk. This uncertainty is directly related to the size, of order ℓ, of the region over which the electron has a well-defined plane wave behavior. This relation is just the standard uncertainty principle $\delta k \cdot \ell \sim 1$.

We will account phenomenologically for the effect of this mean free path by analogy with the well-known corresponding situation in the time-frequency domain. Indeed, if an excited state has a finite lifetime τ, its energy gets a corresponding uncertainty $\delta\omega$ related to τ by the uncertainty principle $\delta\omega \cdot \tau \sim 1$ (we assume $\hbar = 1$ in this paragraph). Consequently, the width of the optical line corresponding to the transition from this excited state to the ground state is no longer infinitely narrow, it gets instead a width of order $1/\tau$. Instead of having an optical emission line proportional, for example, to $\delta(\omega - \omega_0)$, we have a widening for which a standard phenomenology is a Lorentzian shape, with width $1/\tau$, that is, $(1/\pi\tau)/((\omega - \omega_0)^2 + 1/\tau^2)$ if we take a normalization over frequency equal to 1,

---

[4] In a real compound, we have to take into account the presence of the lattice, and the translational invariance is only by the discrete translations that leave the lattice invariant. But this does not change the net result that there is no longer momentum conservation due to impurity scattering.

just as for the $\delta(\omega - \omega_0)$ function. Hence, due to the finite lifetime, we have for the spectral shape

$$\delta(\omega - \omega_0) \longrightarrow \frac{1}{\pi} \frac{1/\tau}{(\omega - \omega_0)^2 + 1/\tau^2} = \frac{1}{2\pi} \int_{-\infty}^{\infty} dt \, e^{i(\omega - \omega_0)t - |t|/\tau} \qquad (4.51)$$

The Fourier transform representation of the Lorentzian, which is written in the last equality, shows quite clearly that this shape is directly related to the fact that instead of having an oscillation with a well-defined frequency $e^{-i\omega_0 t}$ which would produce a $\delta(\omega - \omega_0)$, a limitation of order τ in the time during which this oscillation occurs, as produced by the exponentially decreasing factor $e^{-|t|/\tau}$, induces a linewidth of order $1/\tau$.

Coming back to our impurity problem, we will similarly take into account the effect of the finite mean free path ℓ on momentum conservation by the substitution

$$(2\pi)^3 \delta(\mathbf{k_2} - \mathbf{k_1} - \mathbf{q}) \longrightarrow \int d\mathbf{R} \, e^{i(\mathbf{k_2} - \mathbf{k_1} - \mathbf{q}) \cdot \mathbf{R}} e^{-R/\ell} \qquad (4.52)$$

analogous to Eq. (4.51). This produces an uncertainty of order $1/\ell$ for wavevector conservation, and gives back the proper δ function in the limit $\ell \to \infty$. As a result, including the prefactor, the paramagnetic term becomes in the presence of impurities

$$K(\mathbf{q}, \omega) - \frac{ne^2}{m} = -\frac{2e^2\hbar}{m^2(2\pi)^6} \int d\mathbf{k_1} \, d\mathbf{k_2} \, d\mathbf{R} \, e^{i(\mathbf{k_2} - \mathbf{k_1} - \mathbf{q}) \cdot \mathbf{R}} e^{-R/\ell} k_{1,x} k_{2,x} \tilde{F}(\mathbf{k_1}, \mathbf{k_2}) \qquad (4.53)$$

Since $\tilde{F}(\mathbf{k_1}, \mathbf{k_2})$ actually depends only on $k_1 = |\mathbf{k_1}|$ and $k_2 = |\mathbf{k_2}|$, the angular averages over $\mathbf{k_1}$ and $\mathbf{k_2}$ can easily be performed. This is done conveniently by writing $k_{1,x} e^{-i\mathbf{k_1} \cdot \mathbf{R}} = i \partial(e^{-i\mathbf{k_1} \cdot \mathbf{R}})/\partial R_x$, and similarly for the $\mathbf{k_2}$ factor. The angular average of $e^{-i\mathbf{k_1} \cdot \mathbf{R}}$ is obtained as in Eq. (2.93) and gives in the same way $\int d\Omega_{\mathbf{k_1}} e^{-i\mathbf{k_1} \cdot \mathbf{R}} = 4\pi \sin(k_1 R)/(k_1 R)$.

A further simplifying feature in this calculation is that the variables $k_1 R$ and $k_2 R$ are essentially always very large. Indeed, just as in the preceding section, $\mathbf{k_1}$ is constrained to be in the vicinity of the Fermi surface in order to have a sizeable value for $\tilde{F}(\mathbf{k_1}, \mathbf{k_2})$ (this is clearer below with explicit expressions), and the same is true for $\mathbf{k_2}$. Hence we have $k_1 \simeq k_F \simeq k_2$. On the other hand, from the exponential factor $e^{-R/\ell}$ in Eq. (4.53), R is constrained to be at most of order of the mean free path ℓ. This implies that in essentially all the integration range for R, $k_1 R \simeq k_F \ell \gg 1$. The last inequality results from a physical constraint. Indeed we want the electronic mean free path to be much larger than a typical electronic wavelength $\lambda_F \sim 1/k_F$ in order to have locally (at a scale small compared to ℓ) free electronic behavior with electronic wavelength λ_F corresponding to metallic properties; hence we want physically to restrict our considerations to situations where $\ell \gg \lambda_F$.

In taking the derivative of $\sin(k_1 R)/(k_1 R)$ with respect to R_x, with $\partial/\partial R_x = (R_x/R)\partial/\partial R$, we have merely $\partial[\sin(k_1 R)/(k_1 R)]/\partial R \simeq \cos(k_1 R)/R$ when we take into account that $k_1 R \gg 1$. Taking similarly into account the $\mathbf{k_2}$ factor, we obtain finally

$$\int \frac{d\Omega_{\mathbf{k_1}}}{4\pi} \frac{d\Omega_{\mathbf{k_2}}}{4\pi} e^{i(\mathbf{k_2} - \mathbf{k_1}) \cdot \mathbf{R}} k_{1,x} k_{2,x} = \frac{R_x^2}{R^4} \cos(k_1 R) \cos(k_2 R) \qquad (4.54)$$

$$\simeq \frac{R_x^2}{2R^4} \cos[(k_1 - k_2)R]$$

The last step results from the known trigonometric identity $2\cos(k_1 R)\cos(k_2 R) = \cos[(k_1 - k_2)R)] + \cos[(k_1 + k_2)R]$. The contribution from $\cos[(k_1 + k_2)R]$ can be omitted because, in the R integration in Eq. (4.53), it behaves essentially as $\cos(2k_F R)$, with very rapid oscillations as a function of R, leading to an almost zero contribution by destructive interference. On the other hand, since $k_1 \simeq k_F \simeq k_2$, the oscillations are much slower in the first term $\cos[(k_1 - k_2)R]$ and its contribution dominates.

The final integration over $\mathbf{R}$ can be performed in Eq. (4.53) in the general case. However, the general result is lengthy and we have no specific use for it, since we want to concentrate on the $\mathbf{q} \to \mathbf{0}$ limit relevant both for the Meissner effect and the conductivity. Before proceeding, we want to point out that in Eq. (4.53) the limit $\lim_{\omega \to 0, q \to 0} K(\mathbf{q}, \omega)$ is regular and the order in which we take ω and $\mathbf{q}$ going to zero does not matter anymore, in contrast with the situation we had in the absence of impurities. This occurs because ω appears in $\tilde{F}(\mathbf{k_1}, \mathbf{k_2})$, while $\mathbf{q}$ appears in the factor $e^{-i\mathbf{q}\cdot\mathbf{R}}$ in Eq. (4.53), so they are no longer compared directly.

For $\mathbf{q} = \mathbf{0}$, the $\mathbf{R}$ integration in Eq. (4.53) is easily performed. Just as above Eq. (1.33), the angular integration of R_x^2 gives a $4\pi/3$ factor, and one is left with calculating $\text{Re}[\int_0^\infty dR\, e^{-R/\ell}\, e^{i(k_1 - k_2)R}] = \ell/[1 + (k_1 - k_2)^2 \ell^2]$. This leads to

$$\int \frac{d\Omega_{\mathbf{k_1}}}{4\pi} \frac{d\Omega_{\mathbf{k_2}}}{4\pi} \, d\mathbf{R}\, e^{i(\mathbf{k_2} - \mathbf{k_1})\cdot\mathbf{R}} e^{-R/\ell}\, k_{1,x} k_{2,x} = \frac{2\pi}{3} \frac{\ell}{1 + (k_1 - k_2)^2 \ell^2} \qquad (4.55)$$

It is more convenient to express the wavevector difference $k_1 - k_2$ in terms of the corresponding kinetic energy difference $\xi_1 - \xi_2$ as $k_1 - k_2 = (\xi_1 - \xi_2)/(\hbar v_F)$. Since the typical order of magnitude of $\xi_1 - \xi_2$ is the gap Δ, one sees that the result in Eq. (4.55) involves the ratio ℓ/ξ_0 of the mean free path ℓ to the BCS coherence length $\xi_0 = \hbar v_F/(\pi \Delta)$ (see Eq. (2.89)). In the case where the mean free path ℓ goes to infinity, the right-hand side of Eq. (4.55) becomes proportional to $\delta(k_1 - k_2)$, just as for $\tau \to \infty$ in Eq. (4.51), so one recovers momentum conservation (for $\mathbf{q} = \mathbf{0}$) as it should be for a superconductor in the absence of impurities.

In the other limiting case, where the mean free path is much shorter than the coherence length $\ell \ll \xi_0$, there is no constraint anymore relating the two kinetic energies ξ_1 and ξ_2, since the right-hand side of Eq. (4.55) is independent of them, so these variables are completely decoupled. Hence this so-called dirty limit is very simple. It is suitable to describe the physical situation in superconductors, which have a fair quantity of impurities implying a small mean free path ℓ, and for which the critical temperature is low, so that Δ is small and the coherence length ξ_0 is large. This kind of situation is fairly frequent.

Our last step in rewriting the response function is to replace as usual the integration over k_1 and k_2 by an integration over ξ_1 and ξ_2 through the introduction of the density of states $\int d\mathbf{k_1}/(2\pi)^3 \to N_0 \int d\xi_1$. Since the contributions from large $|\xi_1|$ and $|\xi_2|$ are vanishingly small, the integration boundaries can be extended to $\pm\infty$. This leads us to

$$K(\mathbf{0}, \omega) = \frac{ne^2}{m} + \int_{-\infty}^{\infty} \int_{-\infty}^{\infty} d\xi_1\, d\xi_2\, g(\xi_1 - \xi_2) F(\xi_1, \xi_2) \qquad (4.56)$$

with

$$g(\xi_1 - \xi_2) = -\frac{4\pi}{3} \frac{e^2 N_0^2 \hbar}{m^2} \frac{\ell}{1 + (\frac{\xi_1 - \xi_2}{\hbar v_F})^2 \ell^2} \tag{4.57}$$

and

$$F(\xi_1, \xi_2) = (u_1 u_2 + v_1 v_2)^2 \frac{f(E_2) - f(E_1)}{\omega - (E_2 - E_1)/\hbar + i\epsilon} \tag{4.58}$$

$$- (u_1 v_2 - v_1 u_2)^2 \frac{(1 - f(E_1) - f(E_2))(E_1 + E_2)/\hbar}{(\omega + i\epsilon)^2 - (E_1 + E_2)^2/\hbar^2}$$

is just $\tilde{F}(\mathbf{k_1}, \mathbf{k_2})$ rewritten with the simpler notations $\xi_{1,2} \equiv \xi_{\mathbf{k_1}, \mathbf{k_2}}$ and $E_{1,2} \equiv E_{\mathbf{k_1}, \mathbf{k_2}}$. We have made use of the explicit expressions Eq. (4.35) of the coherence factors, with $u_{1,2} \equiv u_{\mathbf{k_1}, \mathbf{k_2}}$ and $v_{1,2} \equiv v_{\mathbf{k_1}, \mathbf{k_2}}$.

4.7.1 Normal State

Before proceeding to the calculation in the BCS state, we naturally want to check that our phenomenological procedure gives an appropriate description of the effect of impurities in the normal state. Accordingly, we first evaluate Eq. (4.56) in the normal state, for which $F(\xi_1, \xi_2)$ reduces to

$$F(\xi_1, \xi_2) = \frac{f(\xi_2) - f(\xi_1)}{\omega - (\xi_2 - \xi_1)/\hbar + i\epsilon} \tag{4.59}$$

corresponding to Eq. (4.31), as we have already seen. This calculation can be performed, but one has to deal with not so interesting technical details. Hence, in order to avoid as much as possible these details, we calculate from Eq. (4.56) only the imaginary part $K''(0, \omega)$ of the response function. Then we rely on Eq. (4.17), as we have already done for the general case, to obtain the whole response function. This imaginary part comes from the imaginary part $F''(\xi_1, \xi_2) = -\pi [f(\xi_2) - f(\xi_1)]\delta(\omega - (\xi_2 - \xi_1)/\hbar)$ of $F(\xi_1, \xi_2)$, so it is given by

$$K''(0, \omega) = -\pi \int_{-\infty}^{\infty} \int_{-\infty}^{\infty} d\xi_1 \, d\xi_2 \, g(\xi_1 - \xi_2)[f(\xi_2) - f(\xi_1)] \, \delta(\omega - (\xi_2 - \xi_1)/\hbar) \tag{4.60}$$

We perform this integration by making the change of variables $x = \xi_2 - \xi_1$ and $y = (\xi_1 + \xi_2)/2$. We have $\int d\xi_1 \, d\xi_2 = \int dx \, dy$. The δ function makes $x = \hbar \omega$, so we are left with

$$K''(0, \omega) = -\pi \hbar \, g(\hbar \omega) \int_{-\infty}^{\infty} dy \, [f(y + \frac{x}{2}) - f(y - \frac{x}{2})] \tag{4.61}$$

The remaining integral over y of the difference between two Fermi distributions, shifted by the energy x, is merely equal to $-x = -\hbar \omega$. This is obvious at $T = 0$, where this difference between Fermi distributions is equal to -1 for $-x/2 < y < x/2$, and zero otherwise. But it is easy to see[5] that this result is valid at any temperature.

---

[5] Instead of the boundaries $-\infty$ and ∞, we introduce equivalently boundaries $-A$ and A, with A large enough so that $f(-A) = 1$ and $f(A) = 0$. Making the appropriate changes of variable, we have then $\int_{-\infty}^{\infty} dy \, [f(y + \frac{x}{2})$ $- f(y - \frac{x}{2})] = \int_{-A}^{A} dy \, [f(y + \frac{x}{2}) - f(y - \frac{x}{2})] = \int_{-A+x/2}^{A+x/2} dz \, f(z) - \int_{-A-x/2}^{A-x/2} dz \, f(z) = -\int_{-A-x/2}^{-A+x/2} dz \, f(z) +$ $\int_{A-x/2}^{A+x/2} dz \, f(z) = -x$, since $f(z) = 1$ for $-A - x/2 < z < -A + x/2$, and $f(z) = 0$ for $A - x/2 < z < A + x/2$.

Introducing the explicit expression Eq. (4.57) for $g(\hbar\omega)$, we obtain

$$K''(\mathbf{0}, \omega) = -\pi\hbar^2\omega \frac{4\pi}{3} \frac{e^2 N_0^2 \hbar}{m^2} \frac{v_F \tau}{1 + (\omega\tau)^2} = -\frac{ne^2}{m} \frac{\omega\tau}{1 + (\omega\tau)^2} \tag{4.62}$$

where we have defined the relaxation time τ by $\ell = v_F \tau$, which is physically quite a natural definition, since it is the time required for an electron at the Fermi surface, with velocity v_F, to propagate over the mean free path ℓ. In the last step, we used $v_F = \hbar k_F/m$, $N_0 = mk_F/(2\pi^2\hbar^2)$, and $n = k_F^3/(3\pi^2)$. From this last expression, we see that going to the conductivity $\sigma(\omega)$ by Eq. (4.41), we obtain for its real part a result that is in exact agreement with the Drude phenomenological theory Eq. (4.43).

Finally we can, as we indicated above, obtain the whole response function from its imaginary part. Actually, we have already done this calculation, since it corresponds to the simple example, with a single pole in the lower complex plane, which we have handled in Eq. (4.18). However, this calculation is even not necessary since we know that there is a unique response function given by Eq. (4.17) as soon as we know its imaginary part. Since, for the imaginary part, our result coincides with Drude theory, the agreement extends necessarily to the whole response function. In conclusion, our phenomenological description of the effect of impurities is quite satisfactory since, in the normal state, it leads exactly to physically sound Drude theory. We can now address in the same framework the effect of impurities on the BCS response.

4.7.2 Infrared Absorption

For simplicity, we will restrict our investigation of the superconducting state to the dirty limit, where the mean free path ℓ is very small compared to the coherence length ξ_0. Similarly, because we are interested in frequencies ω that are at most of the order of a few gap $\hbar\omega \sim \Delta$, we have $\omega\tau \sim \Delta\ell/\hbar v_F \sim \ell/\xi_0 \ll 1$, so we assume also coherently $\omega\tau \ll 1$. In this dirty limit, the energy variables ξ_1 and ξ_2 of the two electrons are completely decoupled, since from Eq. (4.57) $g(\xi_1 - \xi_2)$ is merely a constant $g(\xi_1 - \xi_2) = -4\pi e^2 N_0^2 \hbar\ell/(3m^2) = -ne^2\tau/(\pi\hbar^2 m) = -\sigma_0/(\pi\hbar^2)$, following our evaluation Eq. (4.62) in the normal state. In our case, the normal state conductivity Eq. (4.43) is just a constant equal to σ_0, since we assume $\omega\tau \ll 1$.

Instead of $K(\mathbf{0}, \omega)$, we write directly the conductivity $\sigma(\omega) = iK(\mathbf{0}, \omega)/\omega$, which is here our physical quantity of interest. From Eq. (4.56–4.58), we have

$$\frac{\sigma(\omega)}{\sigma_0} = \frac{i}{\omega\tau} - \frac{i}{\pi\hbar^2\omega} \int_{-\infty}^{\infty} \int_{-\infty}^{\infty} d\xi_1 \, d\xi_2 \, F(\xi_1, \xi_2) \tag{4.63}$$

In performing the integrations in Eq. (4.63), only the parts of the integrand that are even with respect to ξ_1 and ξ_2 contribute; the odd parts give vanishing contributions. Since E_1 and E_2 are even, this simplifies to a large extent the relevant expressions of the coherence factors. Indeed, in Eq. (4.58),

$$(u_1 u_2 + v_1 v_2)^2 = u_1^2 u_2^2 + 2u_1 v_1 u_2 v_2 + v_1^2 v_2^2 = \tag{4.64}$$

$$\frac{1}{4}(1 + \frac{\xi_1}{E_1})(1 + \frac{\xi_2}{E_2}) + \frac{1}{2}\frac{\Delta^2}{E_1 E_2} + \frac{1}{4}(1 - \frac{\xi_1}{E_1})(1 - \frac{\xi_2}{E_2}) \longrightarrow \frac{1}{2}(1 + \frac{\Delta^2}{E_1 E_2})$$

by making use of the explicit expressions Eq. (2.75) for $u_{\mathbf{k}}$ and $v_{\mathbf{k}}$, and of $u_{\mathbf{k}}v_{\mathbf{k}} = \Delta/2E_{\mathbf{k}}$. Eliminating the parts odd with respect to ξ_1 and ξ_2, we see that only a contribution $(1/2)(1 + \Delta^2/E_1 E_2)$ survives from this coherence factor. Similarly for the creation-annihilation coherence factor,

$$(u_1 v_2 - v_1 u_2)^2 = u_1^2 v_2^2 - 2u_1 v_1 u_2 v_2 + v_1^2 u_2^2 = \tag{4.65}$$

$$\frac{1}{4}(1 + \frac{\xi_1}{E_1})(1 - \frac{\xi_2}{E_2}) - \frac{1}{2}\frac{\Delta^2}{E_1 E_2} + \frac{1}{4}(1 - \frac{\xi_1}{E_1})(1 + \frac{\xi_2}{E_2}) \longrightarrow \frac{1}{2}(1 - \frac{\Delta^2}{E_1 E_2})$$

and only a contribution $(1/2)(1 - \Delta^2/E_1 E_2)$ survives. In this way, Eq. (4.63) reduces to

$$\frac{\sigma(\omega)}{\sigma_0} = \frac{i}{\omega\tau} - \frac{i}{2\pi\hbar^2\omega}\int_{-\infty}^{\infty}\int_{-\infty}^{\infty} d\xi_1\, d\xi_2 \left[(1 + \frac{\Delta^2}{E_1 E_2})\frac{f(E_2) - f(E_1)}{\omega - (E_2 - E_1)/\hbar + i\epsilon} \right. \tag{4.66}$$

$$\left. - (1 - \frac{\Delta^2}{E_1 E_2})\frac{(1 - f(E_1) - f(E_2))(E_1 + E_2)/\hbar}{(\omega + i\epsilon)^2 - (E_1 + E_2)^2/\hbar^2}\right]$$

We now make use of this general expression to obtain the electromagnetic absorption, which is given by the real part of $\sigma(\omega)$ (that is, the imaginary part of $K(\mathbf{0}, \omega)$). Moreover, in order to be as simple as possible, we restrict ourselves for the explicit expressions to the $T = 0$ limit. In this case, there are no thermally excited bogolons and $f(E_1) = f(E_2) = 0$. The first term in the integral is zero since it corresponds to bogolon scattering, but there are no bogolons to be scattered. The only contribution to the real part of $\sigma(\omega)$ comes from the second term in the integral, which describes bogolon pair creation. Taking $\omega > 0$, the real part comes from the pole at $\hbar\omega = E_1 + E_2 - i\epsilon$, with $\mathrm{Im}(1/[\omega + i\epsilon - (E_1 + E_2)/\hbar]) = -\pi\delta(\omega - (E_1 + E_2)/\hbar)$. This leads to

$$\frac{\mathrm{Re}\,\sigma(\omega)}{\sigma_0} = \frac{1}{4\hbar^2\omega}\int_{-\infty}^{\infty}\int_{-\infty}^{\infty} d\xi_1\, d\xi_2\, (1 - \frac{\Delta^2}{E_1 E_2})\delta(\omega - (E_1 + E_2)/\hbar) \tag{4.67}$$

Since the integrand is even with respect to ξ_1 and ξ_2, we have $\int_{-\infty}^{\infty} d\xi_1 = 2\int_0^{\infty} d\xi_1$, and similarly for ξ_2. Then, instead of ξ_1 and ξ_2, we go to the variables E_1 and E_2 by introducing the reduced BCS density of states $d\xi/dE = n(E) = E/\sqrt{E^2 - \Delta^2}$, just as in Eq. (3.30). This gives

$$\frac{\mathrm{Re}\,\sigma(\omega)}{\sigma_0} = \frac{1}{\hbar^2\omega}\int_{\Delta}^{\infty}\int_{\Delta}^{\infty} dE_1\, dE_2\, \frac{E_1 E_2 - \Delta^2}{\sqrt{(E_1^2 - \Delta^2)(E_2^2 - \Delta^2)}}\delta(\omega - (E_1 + E_2)/\hbar) \tag{4.68}$$

Naturally, we can make use of the δ function to reduce this expression to an integral over a single variable.[6] But the resulting expression does not lend itself to a simple integration (the result can be expressed in terms of elliptic integrals). Actually, the behavior of $\mathrm{Re}\,\sigma(\omega)$ is fairly simple and its qualitative features are easily obtained directly from Eq. (4.68).

---

[6] We set $\hbar = 1$ for simplicity. Eliminating E_2 in favor of E_1 through the δ function, and then taking $x = 2E_1 - \omega$ as a variable to obtain a more symmetrical expression, one finds $\mathrm{Re}\,\sigma(\omega)/\sigma_0 = (1/\omega)\int_0^{\omega - 2\Delta} dx\, (\omega^2 - 4\Delta^2 - x^2)/\sqrt{[(\omega - 2\Delta)^2 - x^2][(\omega + 2\Delta)^2 - x^2]}$. The numerical evaluation of this integral is used to plot Fig. (4.3). For $\omega \simeq 2\Delta$, one gets $\mathrm{Re}\,\sigma(\omega)/\sigma_0 \simeq (1/2\Delta)\int_0^{\omega - 2\Delta} dx\,(\omega - 2\Delta)/\sqrt{[(\omega - 2\Delta)^2 - x^2]} = \pi(\omega - 2\Delta)/(4\Delta)$.

First of all, the δ function in Eq. (4.68) implies $\hbar\omega = E_1 + E_2$. Since from their range of integration we have $E_1, E_2 \geq \Delta$, we must have $\hbar\omega \geq 2\Delta$ in order to have a nonzero $\mathrm{Re}\,\sigma(\omega)$. The physical interpretation is clear. The absorbed electromagnetic radiation (we may think of it as corresponding to a single photon) provides an energy $\hbar\omega$ to the superconductor. It is used to create two excited states, that is two bogolons. Since the creation of each bogolon requires at least an energy equal to the gap Δ, the photon must have at least an energy $\hbar\omega = 2\Delta$ for this process to be allowed by energy conservation.

It is interesting to contrast this situation with the one found in tunneling experiments in Section 3.2 when we considered the transfer of an electron from a normal metal to a superconductor. Here a single electron is transferred to the superconductor, which implies the creation of a single bogolon, and accordingly the energy eV provided by the voltage has to be larger than the gap Δ to have this process feasible. In the present case of electromagnetic absorption, we can see the process as the breaking of a Cooper pair by the absorbed photon. However, the number of electrons in the superconductor is conserved in this process. Hence the breaking of the Cooper pair leaves the two electrons of the pair present in the superconductor. Each one requires to present an energy at least equal to the gap Δ. Accordingly, this pair breaking process is allowed only for a photon energy $\hbar\omega$ larger than 2Δ. In the same way, it is interesting to note that except for the coherence factor, the integral in Eq. (4.68) is essentially the same as in Eq. (3.30).

Experimentally, for standard low-temperature superconductors, this threshold 2Δ for electromagnetic absorption falls in the far-infrared domain, with a typical millimeter range wavelength and frequency around 100–1000 GHz. This is in practice a very inconvenient domain for experiments, and it is probably in part for this reason that the manifestation of the threshold in infrared absorption has been first observed [32] fairly late, roughly at the time of the elaboration of BCS theory.

In the opposite limit of large frequency, the constraint $\hbar\omega = E_1 + E_2$ implies that in most of the integration range, E_1 and E_2 are of order $\hbar\omega$, that is very large compared to the gap Δ. In this case, the integrand in Eq. (4.68) reduces to unity, and making use of the δ function to eliminate the E_2 integration, we are left with $\int_\Delta^{\hbar\omega-\Delta} dE_1 \simeq \hbar\omega$, since $E_2 = \hbar\omega - E_1 > \Delta$ implies that $E_1 < \hbar\omega - \Delta$. Hence one finds for large frequencies that $\mathrm{Re}\,\sigma(\omega) = \sigma_0$, that is, one recovers as expected the normal state property. The numerical calculation of Eq. (4.68) shows that $\mathrm{Re}\,\sigma(\omega)/\sigma_0$ goes smoothly from 0 to 1 when $\hbar\omega$ goes from 2Δ to infinity, as can be seen in Fig. (4.3) where the result is displayed. At $\hbar\omega = 2\Delta$ the slope is finite, since one finds easily from Eq. (4.68) that $\mathrm{Re}\,\sigma(\omega)/\sigma_0 \simeq \pi(\hbar\omega - 2\Delta)/(4\Delta)$. This Mattis–Bardeen result Eq. (4.68) is found to be in fair agreement with experiments.

At nonzero temperature the absorption becomes smooth, without the discontinuous sharp rise which occurs right at threshold in Fig. (4.3). It is nonzero at any frequency, with a progressive rise occurring around the threshold at $\omega = 2\Delta(T)$. Indeed, at nonzero temperature, there is, in addition, the possibility of thermally excited bogolons being scattered by the electromagnetic excitation, as described by the first term in Eq. (4.66). There is no frequency threshold for these processes, so they give in particular, contributions for $\omega < 2\Delta(T)$ and in this way, the infrared absorption becomes nonzero below the threshold at $\omega = 2\Delta(T)$ in Fig. (4.3). Since the number of excited bogolons grows when temperature is raised, these contributions progressively fill the whole frequency region below threshold.

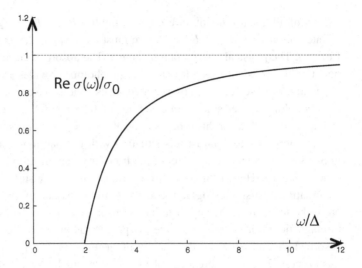

Fig. 4.3 Real part of the conductivity $\mathrm{Re}\,\sigma(\omega)$ at $T = 0$, corresponding physically to the absorption of the electromagnetic radiation by the superconductor. This result, valid in the dirty limit $\ell \ll \xi_0$ of BCS theory, is normalized by the normal state conductivity $\sigma_0 = ne^2\tau/m$. It displays for this infrared absorption a threshold at 2Δ, corresponding physically to the fact that the photon is absorbed by breaking a Cooper pair, which leaves two electrons in excited states, each one having at least an energy equal to the gap Δ.

In addition, there is a peculiar behavior at low frequency, which we will discuss when we consider nuclear-spin relaxation in the following section.

4.7.3 Superconductivity

We now turn to our main purpose, which is to show that BCS theory indeed describes a superconductor when the effect of impurities is taken into account. This is done by obtaining the imaginary part of the conductivity, or equivalently the real part of the response function[7] $K'(\omega) = \omega\,\mathrm{Im}\,\sigma(\omega)$. With this aim, we can start[8] from its imaginary part $K''(\omega)$, which is obtained from Eq. (4.66) by $K''(\omega) = -\omega\,\mathrm{Re}\,\sigma(\omega)$. Considering a general temperature, we have (for simplicity, we set $\hbar = 1$ in this calculation)

$$\frac{K''(\omega)}{\sigma_0} = \frac{1}{2}\int_{-\infty}^{\infty}\int_{-\infty}^{\infty} d\xi_1\,d\xi_2\left[(1+\frac{\Delta^2}{E_1E_2})\,(f(E_2){-}f(E_1))\,\delta(\omega{+}E_1{-}E_2)\right. \tag{4.69}$$

$$\left.-\frac{1}{2}(1-\frac{\Delta^2}{E_1E_2})\,(1{-}f(E_1){-}f(E_2))\,(\delta(\omega{-}E_1{-}E_2) - \delta(\omega{+}E_1{+}E_2))\right]$$

In the second term on the right-hand side, in the part which has the factor $f(E_2)$, we exchange the variables ξ_1 and ξ_2. Since the factor of $f(E_2)$ is invariant in this exchange,

---

[7] From this point forward, we drop the $\mathbf{q}$ variable and write $K(\omega)$ instead of $K(\mathbf{q} = \mathbf{0}, \omega)$.

[8] Starting the calculation of $K'(\omega)$ directly from Eq. (4.66) would involve dealing with divergent behavior for large ξ_1 and ξ_2. These are eliminated by subtracting the corresponding result for the normal state, which is equal to zero in our case. This allows one, at the same time, to handle the diamagnetic contribution in Eq. (4.66). Our procedure avoids these mathematical complications.

we have merely $f(E_2)$ changed into $f(E_1)$, so that this second term gets a general statistical factor $1 - 2f(E_1)$. We make this same factor appearing in the first term by writing $f(E_2)-f(E_1) = (1/2)[(1-2f(E_1))-(1-2f(E_2))]$, and by exchanging the variables ξ_1 and ξ_2 in the integral containing $1-2f(E_2)$. This leads to

$$\frac{K''(\omega)}{\sigma_0} = \frac{1}{4} \int_{-\infty}^{\infty} \int_{-\infty}^{\infty} d\xi_1 \, d\xi_2 \left[(1+\frac{\Delta^2}{E_1 E_2}) \, (\delta(\omega+E_1-E_2)-\delta(\omega+E_2-E_1)) \right. \tag{4.70}$$
$$\left. - (1-\frac{\Delta^2}{E_1 E_2}) \, (\delta(\omega-E_1-E_2)-\delta(\omega+E_1+E_2)) \right] \, (1-2f(E_1))$$

To get $K'(\omega)$, we insert Eq. (4.70) in Eq. (4.16), making use of $\int_{-\infty}^{\infty} dx \, \delta(x+E_1-E_2)$ $/(x-\omega) = -1/(\omega+E_1-E_2)$, and so on. Gathering the terms with $1/(\omega+E_1\pm E_2)$ and with $1/(\omega-E_1\pm E_2)$ gives us

$$\frac{K'(\omega)}{\sigma_0} = -\frac{1}{2\pi} \int_{-\infty}^{\infty} \int_{-\infty}^{\infty} d\xi_1 \, d\xi_2 \left[\frac{\omega+E_1+\frac{\Delta^2}{E_1}}{(\omega+E_1)^2-E_2^2} - \frac{\omega-E_1-\frac{\Delta^2}{E_1}}{(\omega-E_1)^2-E_2^2} \right] (1-2f(E_1)) \tag{4.71}$$

As $E_2^2 = \xi_2^2 + \Delta^2$, the integration over ξ_2 is easily performed because, for $a^2 > 0$, $\int_{-\infty}^{\infty} d\xi_2/(a^2 - \xi_2^2) = (1/2a) \int_{-\infty}^{\infty} d\xi_2[1/(a - \xi_2) + 1/(a + \xi_2)] = 0$, since the integrals must be understood as principal parts, just as in Eq. (4.16). The only case where the integral is nonzero is for $a^2 = -b^2 < 0$. In this case, $\int_{-\infty}^{\infty} d\xi_2/(b^2 + \xi_2^2) = \pi/b$. Considering only the case $\omega > 0$, this cannot occur with the denominator $(\omega + E_1)^2 - E_2^2$, since in this case $a^2 = (\omega + E_1)^2 - \Delta^2 > 0$, due to $E_1 \geq \Delta$. Only the second term in Eq. (4.71) gives a nonzero contribution, but only for a limited range of E_1 values.

We now concentrate on the case which is of most interest for us, namely $\omega \to 0$, for which a simple result can be obtained. Otherwise the end of the calculation has to be done numerically. In this case of very small ω, we must have $E_1 \simeq \Delta$ to have $a^2 = (E_1 - \omega)^2 - \Delta^2 < 0$; otherwise for larger E_1, we have clearly $a^2 > 0$. This simplifies a great deal our expression since we can take $1-2f(E_1) \simeq 1-2f(\Delta)$ out of the ξ_1 integration. Similarly the numerator is merely $\omega - E_1 - \Delta^2/E_1 \simeq -2\Delta$. Performing the ξ_2 integration as indicated above with $b^2 = \Delta^2 - (E_1 - \omega)^2 = 2\omega E_1 - \xi_1^2 + \omega^2 \simeq 2\omega\Delta - \xi_1^2$ (which requires $|\xi_1| < \sqrt{2\omega\Delta}$ to be positive), we find

$$\frac{K'(\omega)}{\sigma_0} = \Delta \, (1 - 2f(\Delta)) \int_{-\sqrt{2\omega\Delta}}^{\sqrt{2\omega\Delta}} d\xi_1 \frac{1}{\sqrt{2\omega\Delta - \xi_1^2}} \tag{4.72}$$

The remaining integral is equal to π. Writing $(1 - 2f(\Delta)) = \tanh(\Delta/2k_B T)$ we have finally the very simple result

$$\frac{K'(0)}{\sigma_0} = \pi \, \Delta \, \tanh \frac{\Delta}{2k_B T} \tag{4.73}$$

and, restoring $\hbar$, the corresponding value for the conductivity

$$\frac{\text{Im} \, \sigma(\omega)}{\sigma_0} = \frac{\pi \, \Delta}{\hbar\omega} \, \tanh \frac{\Delta}{2k_B T} \tag{4.74}$$

with even simpler forms at zero temperature and in the vicinity of the critical temperature

$$T = 0 \quad \frac{\text{Im}\,\sigma(\omega)}{\sigma_0} = \frac{\pi\,\Delta}{\hbar\omega} \qquad\qquad T \simeq T_c \quad \frac{\text{Im}\,\sigma(\omega)}{\sigma_0} = \frac{\pi\,\Delta^2}{2\hbar\omega k_B T_c} \qquad (4.75)$$

Eq. (4.74) shows explicitly that the conductivity goes to infinity when the frequency goes to zero, and accordingly that despite the presence of impurities that limit the normal state conductivity, the state corresponding to BCS theory behaves indeed as a superconductor. In principle, to complete this proof, we should also handle the effect of phonons since it is well known that lattice vibrations, in addition to lattice defects, are another source for the limitation of conductivity in normal metals. Hence we should prove that even the presence of thermally excited phonons does not spoil the infinite conductivity we have just found for the BCS state. This has indeed been done, but this is quite a complicated matter, much beyond our scope.

Finally, it is interesting to consider quantitatively the Meissner effect corresponding to our result Eq. (4.73). Writing explicitly $\sigma_0 = ne^2\tau/m$, the corresponding penetration depth λ_L is given by

$$\frac{1}{\mu_0\lambda_L^2} = K'(0) = \frac{ne^2}{m}\frac{\pi\tau\Delta}{\hbar}\tanh\frac{\Delta}{2k_B T} \qquad (4.76)$$

which should be compared to Eqs. (4.47–4.48). We first remark that if we still want to write $K'(0)$ as $n_s(T)e^2/m$, as we have done at the level of London theory, the physical interpretation of the $n_s(T)$ corresponding to Eq. (4.76) becomes far from obvious, due to the presence of impurities. We next note that in the vicinity of T_c, from Eq. (4.75), the penetration depth has still the divergent behavior $\lambda_L(T) \sim (1 - T/T_c)^{-1/2}$, just as in Eq. (4.48).

On the other hand, the magnitude of the penetration depth is strongly modified. Taking for simplicity the $T = 0$ case, we have to compare $ne^2\pi\tau\Delta/m\hbar$ to the result ne^2/m in the absence of impurities. The ratio between these two results is $\pi\tau\Delta/\hbar = v_F\tau\pi\Delta/(\hbar v_F) = \ell/\xi_0 \ll 1$, where the last inequality comes from the fact that we have been working in the dirty limit $\ell \ll \xi_0$. Note that this modification is in agreement with the result of the phenomenological Pippard theory, where the coherence length ξ is equal to the mean free path ℓ for a dirty superconductor and the response function (see Eq. (1.33)) is indeed reduced by a factor $\xi/\xi_0 = \ell/\xi_0$. We see that correspondingly the presence of impurities has made the penetration depth much larger, which means that the superconductor is far less able to resist to the penetration of the magnetic field. This weakening of superconductivity by the presence of impurities is physically quite reasonable and expected.

Finally it is worth mentioning that this effect of impurities has a quite interesting and surprising consequence. Indeed we have seen that superconductors with a short penetration depth compared to the coherence length are type I superconductors. However, we have just seen that a large increase in the penetration depth can be obtained by merely putting a good deal of impurities in the involved superconductor. If, as a result of this increase the penetration depth gets larger than the coherence length, the superconductor will be a type II superconductor. Hence, as a result, it is possible to transform a type I superconductor into a type II superconductor by merely putting enough impurities in the involved super-conductor. This surprising effect, corresponding to a strong modification of the magnetic properties, has indeed been obtained experimentally.

4.8 Other Responses

4.8.1 Ultrasonic Attenuation

Another interesting response of the superconductor to an external perturbation is obtained by investigating how sound waves are attenuated when they travel through the super-conductor. The convenient range of frequencies for these experiments is the MHz-GHz domain, corresponding to wavelengths in the μm-mm range. These sound waves produce local compression and dilatation in the metal. A compression may raise, for example, the energy of an electron of the metal. If the energy increase is $\mathcal{V}$, independent of the electron momentum, the corresponding coupling Hamiltonian is merely $V = \mathcal{V} \sum_{\mathbf{k}} c_{\mathbf{k}}^{\dagger} c_{\mathbf{k}}$. If we take into account the possible dependence on electron momentum $\mathbf{k}$, we have for the coupling Hamiltonian, corresponding to Eq. (4.1)

$$V = \sum_{\mathbf{k}\,\mathbf{k}'\sigma} \mathcal{V}_{\mathbf{k}\,\mathbf{k}'}(t)\, c_{\mathbf{k}'\sigma}^{\dagger} c_{\mathbf{k}\sigma} \tag{4.77}$$

where we have taken into account that the effect of the sound wave does not depend on the electronic spin, so that the electronic spin σ is conserved.

The coupling $\mathcal{V}_{\mathbf{k}\,\mathbf{k}'}(t)$ oscillates at the frequency ω of the sound wave. Momentum con-servation occurring in a perfect metal would imply $\mathbf{k}' - \mathbf{k} = \mathbf{q}$, where $\mathbf{q}$ is the sound wave momentum. However, because of the small frequency of the sound wave, the wavevector q is quite small and the small wavevector condition $q\ell \ll 1$ is usually satisfied. Moreover we assume in the following that the dirty limit is satisfied, so that the kinetic energies of the wavevectors $\mathbf{k}$ and $\mathbf{k}'$ are decoupled. As usual, the involved electrons are in the vicinity of the Fermi surface and $k \simeq k_F \simeq k'$. We have still to deal with the dependence of $\mathcal{V}_{\mathbf{k}\,\mathbf{k}'}$ on the angular orientation of $\mathbf{k}$ and $\mathbf{k}'$. For simplicity, we assume that $\mathcal{V}_{\mathbf{k}\,\mathbf{k}'}$ does not depend on the directions of $\mathbf{k}$ and $\mathbf{k}'$, so we merely take it as a constant $\mathcal{V}$ (otherwise we would end up with an angular average of $|\mathcal{V}_{\mathbf{k}\,\mathbf{k}'}|^2$).

The ultrasonic attenuation is, for sound waves, the physical quantity analogous to the infrared absorption for electromagnetic waves. Accordingly, its calculation is similar to the one leading to Eqs. (4.66–4.67). However, there is one crucial difference. In obtaining the coherence factors in Eqs. (4.35–4.36), our first step has been to rewrite in Eq. (4.32) the interaction of the electrons with electromagnetic perturbation. In the present case, the corresponding step is to write for the interaction Eq. (4.77)

$$\sum_{\mathbf{k}\,\mathbf{k}'\sigma} c_{\mathbf{k}',\sigma}^{\dagger} c_{\mathbf{k},\sigma} = \sum_{\mathbf{k}\,\mathbf{k}'} \left(c_{\mathbf{k}'\uparrow}^{\dagger} c_{\mathbf{k}\uparrow} - c_{-\mathbf{k}'\downarrow} c_{-\mathbf{k}\downarrow}^{\dagger} \right) \tag{4.78}$$

In contrast with Eq. (4.32), a minus sign appears on the right-hand side in front of the $\downarrow$ spin term, because the fermionic operators anticommute, and we have changed their order.

In Eq. (4.32), this sign change was compensated by the fact that $\mathbf{k}$ is also changed into $-\mathbf{k}$, along with the presence of a factor $\mathbf{k}$ in the interaction. In the present case, there is no factor $\mathbf{k}$ – hence the sign difference. When we carry out the Bogoliubov–Valatin transformation, as in Eq. (4.33), we have now to replace $U_{\mathbf{k}}^{\mathsf{T}} U_{\mathbf{k}}$ with

$$
U_{\mathbf{k}'}\begin{pmatrix} 1 & 0 \\ 0 & -1 \end{pmatrix}{}^{\mathsf{T}}U_{\mathbf{k}} = \begin{pmatrix} C_s'(\mathbf{k}',\mathbf{k}) & C_{ca}'(\mathbf{k}',\mathbf{k}) \\ C_{ca}'(\mathbf{k}',\mathbf{k}) & -C_s'(\mathbf{k}',\mathbf{k}) \end{pmatrix} \tag{4.79}
$$

with the new definitions

$$
C_s'(\mathbf{k}',\mathbf{k}) \equiv u_{\mathbf{k}'}u_{\mathbf{k}} - v_{\mathbf{k}'}v_{\mathbf{k}} \qquad C_{ca}'(\mathbf{k}',\mathbf{k}) \equiv u_{\mathbf{k}'}v_{\mathbf{k}} + v_{\mathbf{k}'}u_{\mathbf{k}} \tag{4.80}
$$

In this way, the sum coming in the interaction reads in terms of bogolon operators

$$
\sum_{\mathbf{k}\,\mathbf{k}'\sigma} c_{\mathbf{k}',\sigma}^\dagger c_{\mathbf{k},\sigma} = \sum_{\mathbf{k}\mathbf{k}'} C_s'(\mathbf{k}',\mathbf{k})\left(\gamma_{\mathbf{k}'\uparrow}^\dagger \gamma_{\mathbf{k}\uparrow} + \gamma_{-\mathbf{k}\downarrow}^\dagger \gamma_{-\mathbf{k}'\downarrow}\right) \tag{4.81}
$$

$$
+ C_{ca}'(\mathbf{k}',\mathbf{k})\left(\gamma_{\mathbf{k}'\uparrow}^\dagger \gamma_{-\mathbf{k}\downarrow}^\dagger - \gamma_{\mathbf{k}\uparrow}\gamma_{-\mathbf{k}'\downarrow}\right)
$$

using again anticommutation. In conclusion, we see that for sound attenuation the expressions of the coherence factors are slightly modified by sign changes with respect to the expressions we had for infrared absorption.

Coming specifically to sound attenuation we are interested, as we have seen, in quite low frequencies that are much smaller than the superconducting gap Δ. Accordingly, $\hbar\omega$ is far too small to allow Cooper pair breaking, in contrast to the physical situation we have found for infrared absorption. Thus, we do not have to consider the processes of bogolon pair creation and annihilation described by the second term on the right-hand side of Eq. (4.81). On the other hand, bogolon scattering is allowed, but we need to have them thermally excited, so we will deal naturally with the nonzero temperature case. These processes are described by the first term on the right-hand side of Eq. (4.81). Accordingly, the sound attenuation coefficient $\alpha_S(\omega)$ in the superconducting state is given by an expression essentially equal, within a multiplicative constant, to the imaginary part $K''(\mathbf{0},\omega)$ of the response function in Eq. (4.56), retaining only the first term of $F(\xi_1,\xi_2)$ in Eq. (4.58). Explicitly we have

$$
\alpha_S(\omega) = \frac{K_0}{2}\int_{-\infty}^{\infty}\int_{-\infty}^{\infty} d\xi_1\,d\xi_2\,(u_1u_2 - v_1v_2)^2\,(f(E_1) - f(E_2))\,\delta(\hbar\omega - (E_2 - E_1)) \tag{4.82}
$$

where we have used the appropriate expression $C_s'(\mathbf{k_1},\mathbf{k_2})$ for the coherence factor, given in Eq. (4.80). The coefficient K_0 we have introduced in Eq. (4.82) contains all the ingredients needed to obtain specifically the sound attenuation α_S from its precise definition, but there is no need to write them explicitly. We note that Eq. (4.82) could be obtained directly from Fermi golden rule, making use of the expression Eq. (4.81) for the perturbation. This would be quite analogous to the procedure we have followed to handle tunneling in Section 3.2.

Now we can follow the same steps as for infrared absorption after Eq. (4.63). First, the coherence factor reads explicitly

$$
(u_1u_2 - v_1v_2)^2 = u_1^2u_2^2 - 2u_1v_1u_2v_2 + v_1^2v_2^2 = \tag{4.83}
$$

$$
\frac{1}{4}\left(1 + \frac{\xi_1}{E_1}\right)\left(1 + \frac{\xi_2}{E_2}\right) - \frac{1}{2}\frac{\Delta^2}{E_1E_2} + \frac{1}{4}\left(1 - \frac{\xi_1}{E_1}\right)\left(1 - \frac{\xi_2}{E_2}\right) \longrightarrow \frac{1}{2}\left(1 - \frac{\Delta^2}{E_1E_2}\right)
$$

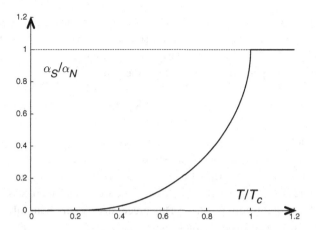

Fig. 4.4 Ultrasonic attenuation α_S in the superconducting state from BCS theory as a function of temperature T. This attenuation is normalized by its value α_N in the normal state just above T_c.

with all the terms odd with respect to ξ_1 or ξ_2 disappearing in the integrations, so that only the contribution $(1/2)(1 - \Delta^2/E_1 E_2)$ survives. Then we go to the variables E_1 or E_2, instead of ξ_1 or ξ_2, introducing in this way the BCS density of states $n(E) = E/\sqrt{E^2 - \Delta^2}$. This gives

$$\alpha_S(\omega) = K_0 \int_\Delta^\infty \int_\Delta^\infty dE_1\, dE_2\, \frac{(E_1 E_2 - \Delta^2)(f(E_1) - f(E_2))}{\sqrt{(E_1^2 - \Delta^2)(E_2^2 - \Delta^2)}}\, \delta(\hbar\omega - (E_2 - E_1)) \qquad (4.84)$$

The sound frequencies we are interested in are so low compared to the gap that we can evaluate Eq. (4.84) in the limit $\omega \to 0$. In this case, the delta function implies $E_2 = E_1$. Inserting this in the integrand, we see that there is a complete cancellation between the coherence factor and the BCS density of states. On the other hand, $E_2 - E_1 = \hbar\omega$ gives, to first order in ω, $f(E_1) - f(E_2) = -\hbar\omega\, \partial f(E_1)/\partial E_1$. This leads to the simple result

$$\alpha_S(\omega) = -K_0 \hbar\omega \int_\Delta^\infty dE_1 \frac{\partial f(E_1)}{\partial E_1} = K_0 \hbar\omega f(\Delta) \qquad (4.85)$$

On the other hand, it is convenient, both theoretically and experimentally, to compare this result to the ultrasonic attenuation $\alpha_N(\omega)$ in the normal state right above the critical temperature. This is merely obtained by setting $\Delta = 0$ in Eq. (4.85), so one has $\alpha_N(\omega) = K_0 \hbar\omega/2$. Hence the ratio is frequency independent and one finds, for this ratio α_S/α_N between ultrasonic attenuation in the superconducting and in the normal state, the extremely simple result

$$\frac{\alpha_S}{\alpha_N} = 2f(\Delta) = \frac{2}{e^{\frac{\Delta(T)}{k_B T}} + 1} \qquad (4.86)$$

which is plotted in Fig. (4.4). The physical interpretation for this behavior of the ultrasonic attenuation in the superconducting state is quite simple: the electronic thermal excitations responsible for sound attenuation disappear rapidly when the superconducting gap opens and are no longer present when $T \to 0$. Hence $\alpha_S(\omega)$ goes to zero, since this source for sound attenuation is suppressed. The above formula is in quite good agreement with experiments. It is a particularly nice feature that sound attenuation gives such direct access to the temperature dependence of the gap.

4.8.2 Nuclear Relaxation

Nuclear spins are weakly coupled to electronic spins, which gives them the very interesting role of close passive observers of nearby electronic properties. This is usually done by applying a static magnetic field[9] to polarize the nuclear spins. A transverse radiofrequency field is applied, which, in a classical view excites the precession of the nuclear spins around their equilibrium position. This precession occurs naturally at the Larmor frequency, and there is a resonance when the frequency of the transverse field is equal to the Larmor frequency. In the quantum picture, the radiofrequency field produces transitions of the nuclear spins between their ground state (corresponding to their magnetic moment along the static magnetic field) and some excited states, with the resonance occurring when the radiofrequency matches the energy $\hbar\omega$ difference between the ground state and the excited state. Nuclear spins can relax to their ground state by transferring their angular momentum to electronic spins. This gives rise to a finite lifetime, called T_1, for nuclear spins in their excited states. This is, for example, measured by looking at the width of the radiofrequency resonance.

Details of the specific coupling between electronic spins and nuclear spins are unfortunately fairly complicated. Nevertheless it is basically due to the hyperfine coupling between the electron and the nucleus, and for our purposes, the relevant term is proportional to the scalar product $\mathbf{I} \cdot \mathbf{S}$ of the nuclear spin $\mathbf{I}$ and the electronic spin $\mathbf{S}$. With the z-axis along the applied field, the term in this scalar product responsible for nuclear relaxation is proportional to $I_x S_x + I_y S_y = (I^- S^+ + I^+ S^-)/2$, where $I^{\pm} = I_x \pm I_y$ and $S^{\pm} = S_x \pm S_y$ are the operators flipping the electronic and nuclear spins from being parallel to antiparallel to the applied field, or conversely. The term $I^- S^+$ is responsible for flipping the electronic spin from $\downarrow$ to $\uparrow$, while the nuclear spin goes from $\uparrow$ to $\downarrow$, and conversely for the $I^+ S^-$ term.

When we write the perturbation on the electrons due to the nuclear spin in the general form Eq. (4.1), flipping the electronic spin from $\downarrow$ to $\uparrow$ corresponds to a contribution proportional to $c_{\uparrow}^{\dagger} c_{\downarrow}$. As we have indicated for Eq. (4.1), there is no need to quantize the nuclear spins and we can treat these degrees of freedom classically, considering that the dynamics of the nuclear spin are equivalent to the presence of an internal transverse radiofrequency field at frequency ω, which produces electronic spin flips. This is completely analogous to what we have done for the case of the electromagnetic perturbation. With respect to the wavevector dependence of the perturbation, we have to take into account that the hyperfine coupling is essentially local, the electron needs to be on the nucleus, or quite near, in order to have a sizeable hyperfine interaction. If we had a strictly local interaction proportional to $\delta(\mathbf{r})$, with $\mathbf{r}$ being the position of the electron with

---

[9] In this case, one has to deal with the fact that except in the penetration depth, the field is canceled in the bulk of the superconductor due to the Meissner effect. One way out of this problem is to work with fine powders of small size compared to the penetration depth.

respect to the nucleus, we would have for the interaction a matrix element proportional to $\int d\mathbf{r}\, e^{-i\mathbf{k}'\cdot\mathbf{r}}\delta(\mathbf{r})e^{i\mathbf{k}\cdot\mathbf{r}} = 1$, taking plane waves for electronic eigenstates. This is independent of the wavevectors $\mathbf{k}$ and $\mathbf{k}'$. Although this is not a realistic result, it is not too different from reality, and we will assume for simplicity that the matrix element $\mathcal{V}$ is independent of wavevectors. Accordingly, we write the perturbation Hamiltonian

$$V = \mathcal{V}(t)\sum_{\mathbf{k}\,\mathbf{k}'} c^{\dagger}_{\mathbf{k}'\uparrow}c_{\mathbf{k}\downarrow} + c^{\dagger}_{\mathbf{k}'\downarrow}c_{\mathbf{k}\uparrow} \tag{4.87}$$

where the second term is the Hermitian conjugate of the first one, and describes the opposite spin-flip process.

The physical situation we have to deal with is completely similar to the one we have met in the preceding section for sound attenuation. Indeed the frequencies involved for the nuclear magnetic resonance are very small compared to the gap, so that again the only processes which can occur are bogolon scatterings. Hence in expressing the perturbation Eq. (4.87) in terms of bogolon operators, we are only interested in these scattering terms. The simplest way to obtain them is to substitute in the first term in Eq. (4.87), after having made the change $\mathbf{k} \to -\mathbf{k}$, the expressions from Eq. (3.16) for $c^{\dagger}_{\mathbf{k}'\uparrow} = u_{\mathbf{k}'}\gamma^{\dagger}_{\mathbf{k}'\uparrow} + v_{\mathbf{k}'}\gamma_{-\mathbf{k}'\downarrow}$ and $c_{-\mathbf{k}\downarrow} = u_{\mathbf{k}}\gamma_{-\mathbf{k}\downarrow} - v_{\mathbf{k}}\gamma^{\dagger}_{\mathbf{k}\uparrow}$. This gives the scattering terms $u_{\mathbf{k}}u_{\mathbf{k}'}\gamma^{\dagger}_{\mathbf{k}'\uparrow}\gamma_{-\mathbf{k}\downarrow}$ and $-v_{\mathbf{k}}v_{\mathbf{k}'}\gamma_{-\mathbf{k}'\downarrow}\gamma^{\dagger}_{\mathbf{k}\uparrow} = v_{\mathbf{k}}v_{\mathbf{k}'}\gamma^{\dagger}_{\mathbf{k}\uparrow}\gamma_{-\mathbf{k}'\downarrow}$. After exchanging $\mathbf{k}$ and $\mathbf{k}'$ in this last term, we thus find a scattering term $(u_{\mathbf{k}}u_{\mathbf{k}'} + v_{\mathbf{k}}v_{\mathbf{k}'})\gamma^{\dagger}_{\mathbf{k}'\uparrow}\gamma_{-\mathbf{k}\downarrow}$. Proceeding in the same way with the second term of Eq. (4.87), we obtain that V gives rise to the scattering terms

$$V \longrightarrow (u_{\mathbf{k}}u_{\mathbf{k}'} + v_{\mathbf{k}}v_{\mathbf{k}'})(\gamma^{\dagger}_{\mathbf{k}'\uparrow}\gamma_{-\mathbf{k}\downarrow} + \gamma^{\dagger}_{\mathbf{k}'\downarrow}\gamma_{-\mathbf{k}\uparrow}) \tag{4.88}$$

We see that the coherence factor $(u_{\mathbf{k}}u_{\mathbf{k}'} + v_{\mathbf{k}}v_{\mathbf{k}'})$ associated with these processes is the same as for electromagnetic perturbation Eq. (4.35), and has, for the $v_{\mathbf{k}}v_{\mathbf{k}'}$ term, a sign opposite to what we found for ultrasonic attenuation Eq. (4.80).

The inverse nuclear relaxation time $1/T_1$ is the exact analog of the sound attenuation α_S. The nuclear magnetization relaxes by transferring magnetization to the electronic system due to spin-flip processes. The rate of this transfer can be obtained from Fermi golden rule, in the same as we have done for sound attenuation in Eq. (4.82). The only difference is the change of sign in the coherence factor. Taking the same notations as in Eq. (4.82), and introducing in the same way a constant K'_0 containing all the specifics of the definition of the relaxation time, we have

$$1/T_1 = \frac{K'_0}{2}\int_{-\infty}^{\infty}\int_{-\infty}^{\infty}d\xi_1\,d\xi_2\,(u_1u_2+v_1v_2)^2\,(f(E_1)-f(E_2))\,\delta(\hbar\omega-(E_2-E_1)) \tag{4.89}$$

We follow the same steps as for sound attenuation. The only difference comes from the change of sign in the coherence factor, so that after elimination of the terms odd with respect to ξ_1 and ξ_2, we have

$$(u_1u_2 + v_1v_2)^2 \longrightarrow \frac{1}{2}\left(1 + \frac{\Delta^2}{E_1E_2}\right) \tag{4.90}$$

just as in Eq. (4.69) for the electromagnetic perturbation. Going again to the variables E_1 and E_2 leads to

$$1/T_1 = K_0' \int_\Delta^\infty \int_\Delta^\infty dE_1 \, dE_2 \, \frac{(E_1 E_2 + \Delta^2)(f(E_1) - f(E_2))}{\sqrt{(E_1^2 - \Delta^2)(E_2^2 - \Delta^2)}} \, \delta(\hbar\omega - (E_2 - E_1)) \qquad (4.91)$$

If we consider the normal state situation where $\Delta = 0$, the difference with the case of ultrasonic attenuation disappears completely, so we obtain exactly the same result $(1/T_1)_N = K_0' \hbar\omega/2$. Hence we have for the ratio of the inverse relaxation time $(T_1^{-1})_S$ in the superconducting state to its normal state value $(T_1^{-1})_N$

$$\frac{(T_1^{-1})_S}{(T_1^{-1})_N} = \frac{2}{\hbar\omega} \int_\Delta^\infty dE_1 \, \frac{(E_1(E_1 + \hbar\omega) + \Delta^2)(f(E_1) - f(E_1 + \hbar\omega))}{\sqrt{(E_1^2 - \Delta^2)((E_1 + \hbar\omega)^2 - \Delta^2)}} \qquad (4.92)$$

where we have again used $E_2 = E_1 + \hbar\omega$ from the δ function.

In contrast with the situation for ultrasonic attenuation, we cannot argue that $\hbar\omega$ is very small compared to the gap to set $\hbar\omega = 0$ in the factor of $f(E_1) - f(E_1 + \hbar\omega)$, because this would lead to a factor $(E_1^2 + \Delta^2)/(E_1^2 - \Delta^2)$ that gives a divergence in the integral at the lower boundary $E_1 = \Delta$. This divergence comes from the coalescence, when $E_2 \to E_1$, of the divergences of the BCS density of states $n(E_1) = E_1/\sqrt{E_1^2 - \Delta^2}$ and $n(E_2) = E_2/\sqrt{E_2^2 - \Delta^2}$, which are present in Eq. (4.91). In the case of ultrasonic attenuation, this divergence is not present because it is exactly canceled by the coherence factor. In the case of nuclear relaxation, the coherence factor has a sign change and the cancellation does not occur.

Actually, in practice, we do not physically expect the density of states to have the divergence given by the BCS formula. The first reason for this is because real superconductors are not perfectly isotropic, in contrast with simple BCS modeling. Accordingly, the gap should depend somewhat on the wavevector position on the Fermi surface, and this will spread the location of the divergence over a range of energy values. Another physical reason is that we do not expect in a real superconductor excited states to have an infinite lifetime, in contrast with the simple BCS description. For example, they may emit phonons or absorb thermally excited phonons. Corresponding to this finite lifetime, an uncertainty should exist for the energy of these excitations, and the energy location of the divergence in the BCS density of states should also be affected by this uncertainty. This would again lead to a smearing of this divergence. Taking into account these various reasons for the departure of the excitation density of states $n(E)$ from its ideal value given by the BCS result, we may take the limit $\hbar\omega \to 0$ in the integrand in Eq. (4.92). On the other hand, just as for sound attenuation, we write $f(E_1) - f(E_1 + \hbar\omega) = -\hbar\omega \, \partial f(E_1)/\partial E_1$. This leads finally to

$$\frac{(T_1^{-1})_S}{(T_1^{-1})_N} = 2 \int_0^\infty dE \, n^2(E) \left(1 + \frac{\Delta^2}{E^2}\right) \left(-\frac{\partial f(E)}{\partial E}\right) \qquad (4.93)$$

where the lower boundary is 0, and no longer Δ, since we take the density of states $n(E)$ different from the BCS result.

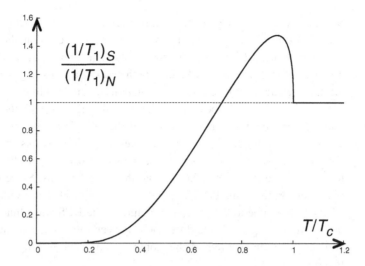

Fig. 4.5 Nuclear relaxation $(1/T_1)_S$ in the superconducting state from BCS theory as a function of temperature T, normalized by its normal state value $(1/T_1)_N$. It is obtained from Eq. (4.93) by introducing an imaginary part for the energy E, to take phenomenologically into account the finite lifetime of the electronic excited states (see the text for details).

For low temperature $T \to 0$, $-\partial f(E)/\partial E = \beta/(4\cosh^2(\beta E/2)) \simeq \beta e^{-\beta E} \leq \beta e^{-\beta \Delta}$ decreases extremely rapidly with temperature, just as for sound attenuation. Again, physically, nuclear relaxation disappears in this regime because the electronic excitations in the superconductor responsible for this relaxation are themselves disappearing. On the other hand, for $T \lesssim T_c$, even if the integral in Eq. (4.93) does not fully diverge as we have explained, it is not so far from diverging, so that it takes fairly large values. Hence, starting from its value equal to 1 at T_c, $(T_1^{-1})_S/(T_1^{-1})_N$ increases when T is lowered below T_c. Since it decreases at low temperature as we have seen, this means that $(T_1^{-1})_S/(T_1^{-1})_N$ has a peak just below T_c. An example of this behavior of $(T_1^{-1})_S/(T_1^{-1})_N$ as a function of temperature is given in Fig. (4.5). Here, we have taken the case where $n(E)$ is limited by lifetime effects. A simple phenomenology to account for this lifetime is to retain the BCS form for $n^2(E)(1 + \Delta^2/E^2) = [E/\sqrt{E^2 - \Delta^2}]^2 + [\Delta/\sqrt{E^2 - \Delta^2}]^2$, but to give a small imaginary part i/τ to the energy variable E. Then one goes back to real quantities by taking the real part of each contribution $E/\sqrt{E^2 - \Delta^2}$ and $\Delta/\sqrt{E^2 - \Delta^2}$, and one adds the squares of these real parts to obtain the factor of $-\partial f(E)/\partial E$ in Eq. (4.93). Specifically, we have taken an inverse lifetime proportional to energy since on fairly general ground excitations with zero energy have an infinite lifetime, as in Fermi liquid theory. Fig. (4.5) has been obtained with the choice $1/\tau = 0.1E$.

The existence of such a peak in nuclear relaxation just below T_c has been observed early on by Hebel and Slichter [33] in superconducting aluminum. The presence of such a Hebel–Slichter peak and the agreement in this respect with theory is a striking success of BCS theory. Indeed there is a strong contrast between Fig. (4.5) for nuclear relaxation and Fig. (4.4) for sound attenuation. This is at first quite surprising since in both cases one deals with the rate of an energy transfer from an "external" system (sound waves or nuclear spins) to the electronic excitations of a superconductor. If one had, instead of

BCS theory, a more phenomenological description merely linking superconductivity to the opening of a gap in the excitation spectrum, this description should give essentially the same temperature dependence for sound attenuation and nuclear relaxation. This is indeed the case at low temperature where, in both cases, the vanishing relaxation is clearly due to the disappearance of the electronic excitations due indeed to gap opening.

But just below T_c the behavior is completely different, with the presence of the Hebel–Slichter peak for nuclear relaxation and the complete absence of a similar feature for sound attenuation. Actually, we have seen the origin of this difference, since we have noticed that the coherence factors for the two processes are different, and that this difference plays a crucial role in the final theoretical results. Since these coherence factors have their origin in the detailed structures of the electronic excitations, which are themselves directly due to the detailed structure of the BCS wave function Eq. (2.44), with its factors $u_\mathbf{k} + v_\mathbf{k} c_{\mathbf{k}\uparrow}^\dagger c_{-\mathbf{k}\downarrow}^\dagger$, the difference between ultrasonic attenuation and nuclear relaxation can be seen as a remarkable confirmation of the validity of the detailed BCS wave function.

We notice finally that the Hebel–Slichter peak is not at all systematically present in standard superconductors. This can be easily understood as the result of the various physical ingredients, which are responsible, as we have seen, for the absence of an actual divergence in Eq. (4.93). If their effect is strong enough, this weakening of the divergence may lead to a complete washing out of the peak itself (as can be seen in our example by taking a shorter lifetime for excitations). Hence it is the presence of the Hebel–Slichter peak that is a strong argument in favor of the validity of BCS theory; it is much more difficult to argue that its absence is a sign that BCS theory is not valid.

To conclude, we mention another very interesting effect that occurs with nuclear spins. These spins can be polarized due to their magnetic environment which acts globally as an effective field. In particular, due to hyperfine coupling with the electrons, the nuclear magnetization is sensitive to the electronic polarization. In the presence of an applied magnetic field, these electronic spins are polarized. This is felt by the nuclear spins as an additional effective field, and as a result the frequency location of the nuclear magnetic resonance is shifted compared to its value in the absence of electronic polarization. This is known as the "Knight shift."

Through this shift one can measure the spin susceptibility of the electrons, which corresponds in the normal metal to Pauli paramagnetism with susceptibility χ_N. In the superconducting state, one cannot directly measure this spin susceptibility because the magnetization due to the orbital electronic currents present due to the Meissner effect is much stronger than the spin magnetization. Hence the Knight shift is quite interesting since it provides an indirect access to this electronic spin susceptibility χ_S in the superconducting state. A particularly interesting situation is the $T \to 0$ limit. In this case, there are no thermally excited bogolons to be polarized by an applied magnetic field; all the electrons belong to Cooper pairs. However, since a Cooper pair is made of a $\uparrow$ electron and a $\downarrow$ electron, it has no global magnetic moment, and accordingly it cannot be polarized by the field (provided this field is not strong enough to break a Cooper pair, but this would require very high fields). As a result, at zero temperature $\chi_S = 0$. More generally, only the electronic excitations of the superconductor can be polarized by the field, in a way that is analogous to electrons in the normal state. Precisely one can show that $\chi_S(T)/\chi_N = 1 - n_S(T)/n$, where $n_S(T)$ is the superfluid density introduced in Eq. (4.47). This gives the fraction of electronic excitations available for polarization by the field. This behavior of the Knight shift is quite interesting since it gives direct evidence that in a standard superconductor Cooper pairs are made of $(\uparrow, \downarrow)$ electronic pairs. Conversely, it provides for a general superconductor direct information on the spin structure of the Cooper pairs responsible for superconductivity. For example, one can conclude from the Knight shift that in the cuprate superconductor $YBa_2Cu_3O_{7-x}$, singlet pairing occurs.

Macroscopic Effects

In this chapter, we deal with striking effects that occur at a macroscopic scale. They rely on the basic physical understanding of superconductivity, but not so much on the detailed microscopic description of BCS theory that we have seen in the preceding chapters. Hence the results are actually independent of this theory (except for pairing). Nevertheless we will make use of it to introduce in a convenient and concrete way some physical ideas.

5.1 Gauge Invariance

Since we deal with phenomena occurring for large scales, the typical lengths we deal with are much larger than the typical microscopic scales coming in the description of the super-conducting state. Accordingly, the corresponding wavevectors are very small compared to microscopic wavevectors such as k_F, or the coherence length, so that we deal actually with the limit of vanishing wavevectors $\mathbf{q} \to 0$, that is, the London limit. So we do not need to take care of a possible non-local response of the superconducting current to the vector potential, and we can make use of the simple London equation Eq. (1.20), which we rewrite here for convenience

$$\mathbf{j}_s(\mathbf{r}) = -\frac{n_s e^2}{m}\mathbf{A}(\mathbf{r}) \tag{5.1}$$

As we have seen, this equation has been written in a specific gauge. However, physical laws should not depend on a particular gauge, and it must be possible to write them in a gauge-invariant way since the vector potential is not really physical, only the magnetic field having a physical meaning. We will now look for this gauge-invariant formulation.

As it happens, we have also in presenting the BCS theory chosen a specific gauge. Indeed it has been convenient to make a choice corresponding to take a real pair wave function $\Phi(\mathbf{r}_1 - \mathbf{r}_2)$. This occurred when we have decided to choose Δ real positive, just below Eq. (2.66). We have done it by arguing precisely that we could use the freedom in the gauge choice to achieve this. However, this was just a matter of convenience, and clearly the calculation could have been performed in any gauge, with the same result. Specifically we have shown that this gauge choice led to the replacement of Δ by $|\Delta|$. Hence, for example, for a general gauge Eq. (2.71) reads $E_\mathbf{k} = \sqrt{\xi_\mathbf{k}^2 + |\Delta|^2}$.

Instead of picking this choice, we are free to multiply the pair wave function by a phase factor $e^{i\varphi}$, so that $\Phi(\mathbf{r}_1 - \mathbf{r}_2)$ becomes $e^{i\varphi}\,\Phi(\mathbf{r}_1 - \mathbf{r}_2)$. Correspondingly, in the Fourier transform, $\Phi_{\mathbf{k}}$ becomes $e^{i\varphi}\Phi_{\mathbf{k}}$. In the BCS theory, $u_{\mathbf{k}}$ is just introduced to achieve a convenient normalization, and there is no reason to modify our simple choice to take it as a real positive. Hence, upon multiplication of the wave function by $e^{i\varphi}$, $v_{\mathbf{k}} = u_{\mathbf{k}}\Phi_{\mathbf{k}}$ becomes $e^{i\varphi}v_{\mathbf{k}}$. Finally as a result $\Delta = V\sum_{\mathbf{k}}^{\omega_D} u_{\mathbf{k}}v_{\mathbf{k}}$ is also multiplied by this same phase factor, which is independent of the wavevector, and becomes $e^{i\varphi}\Delta$. Hence Δ is physically a complex quantity, and when we chose it previously as real, this was just a matter of convenience, without deep physical implications.

In this chapter, we will consider inhomogeneous situations where Δ, or more precisely its phase φ, will depend the position $\mathbf{r}$ in the superconducting sample, so that we deal with $\Delta(\mathbf{r})$. The general situation where the modulus $|\Delta|$ depends on position will be considered in Chapter 6, where Ginzburg–Landau theory is addressed. At this stage, anticipating this concept which is at the root of Ginzburg–Landau theory, it is worthwhile to note that in superconductivity Δ also plays the role of an order parameter, in addition to giving the gap in the spectrum of the elementary excitations. This quantity is analogous to the magnetization in the phase transition to ferromagnetic order. It has to be zero in the high-temperature disordered phase, and to be nonzero in the low-temperature ordered phase. Moreover, it goes to zero continuously at the critical temperature T_c. As it is seen in Fig. (3.6), Δ displays all these specific features. As we have just stressed this order parameter Δ is physically a complex quantity, which comes directly from the fact that the wave function is also a complex quantity, manifesting immediately in this way its quantum nature. This complex nature of the order parameter is again a central feature of Ginzburg–Landau theory, as will be seen in Chapter 6.

Regarding the nature of BCS theory for the description of the superconducting state, it should also be noted that it has all the features of a mean-field theory. This is immediately apparent in the expression for the reduced Hamiltonian Eq. (2.26), or upon looking at the BCS wave function Eq. (2.44) itself, where it is seen that the various wavevectors $\mathbf{k}$ are decoupled. One sees that what happens to a specific pair $(\mathbf{k}_\uparrow, -\mathbf{k}_\downarrow)$ comes only through the average effect of all the other pairs that manifest themselves only through Δ, which from its very definition Eq. (2.66) is an average over all pairs and accordingly expresses their mean effect. This is quite analogous to the mean-field theory of ferromagnetism, where one considers that a specific spin is subject to a magnetic field that is obtained from an average over all the other spins.

We come back now to the problem of expressing the current in a gauge-invariant way. The general expression of gauge invariance in quantum mechanics implies that when the gauge is modified, a corresponding modification must be made on the wave function. Let us, for example, go from gauge $\mathbf{A}(\mathbf{r})$ to gauge $\mathbf{A}'(\mathbf{r})$ by

$$\mathbf{A}(\mathbf{r}) = \mathbf{A}'(\mathbf{r}) - \frac{\hbar}{e}\nabla\chi(\mathbf{r}) \tag{5.2}$$

where $\chi(\mathbf{r})$ is a single-valued function, so that the field $\mathbf{B}(\mathbf{r}) = \mathrm{curl}\,\mathbf{A}(\mathbf{r}) = \mathrm{curl}\,\mathbf{A}'(\mathbf{r})$ does not depend on the gauge choice. In order to have a gauge-invariant Schrödinger equation, we have at the same time to transform any wave function $\psi(\mathbf{r})$ according to

$$\psi(\mathbf{r}) = \psi'(\mathbf{r})e^{-i\chi(\mathbf{r})} \tag{5.3}$$

This is for a single-particle wave function. For a many-particle wave function, the same phase factor should apply for each particle so that the transform is

$$\psi(\mathbf{r}_1, \mathbf{r}_2, \cdots) = \psi'(\mathbf{r}_1, \mathbf{r}_2, \cdots) \, e^{-i\sum_i \chi(\mathbf{r}_i)} \tag{5.4}$$

Indeed, in this case, one has

$$(\mathbf{p} - e\mathbf{A}(\mathbf{r})) \, \psi(\mathbf{r}) = \left(\frac{\hbar}{i} \nabla - e\mathbf{A}(\mathbf{r}) \right) \psi(\mathbf{r}) \tag{5.5}$$

$$= \left(\frac{\hbar}{i} \nabla - e\mathbf{A}'(\mathbf{r}) + \hbar \nabla \chi(\mathbf{r}) \right) \psi'(\mathbf{r}) e^{-i\chi(\mathbf{r})} = e^{-i\chi(\mathbf{r})} \left(\frac{\hbar}{i} \nabla - e\mathbf{A}'(\mathbf{r}) \right) \psi'(\mathbf{r})$$

The same calculation can be made for the term $(\mathbf{p} - e\mathbf{A}(\mathbf{r}))^2 \psi(\mathbf{r})$ in the Schrödinger equation. The phase factor $e^{-i\chi(\mathbf{r})}$ can be factorized in all the terms of the Schrödinger equation, so that it drops out, and the resulting Schrödinger equation contains $(\mathbf{p} - e\mathbf{A}'(\mathbf{r}))^2 \psi'(\mathbf{r})$, instead of $(\mathbf{p} - e\mathbf{A}(\mathbf{r}))^2 \psi(\mathbf{r})$. Hence it does not depends on the choice of the gauge, as required by gauge invariance.

If we carry the gauge transformation Eq. (5.2) into the current expression Eq. (5.1), we find

$$\mathbf{j}_s(\mathbf{r}) = \frac{n_s e \hbar}{m} \nabla \chi(\mathbf{r}) - \frac{n_s e^2}{m} \mathbf{A}'(\mathbf{r}) \tag{5.6}$$

On the other hand, in the new gauge, the pair wave function $\Phi(\mathbf{r}_1 - \mathbf{r}_2)$ becomes, from Eq. (5.4) applied to the case of two particles, $\Phi'(\mathbf{r}_1, \mathbf{r}_2) = e^{i[\chi(\mathbf{r}_1) + \chi(\mathbf{r}_2)]} \Phi(\mathbf{r}_1 - \mathbf{r}_2)$, which generalizes our above considerations to the case where the phase factor has a spatial dependence. Since the positions $\mathbf{r}_1$ and $\mathbf{r}_2$ of the two electrons in the pair are very close, and we are only interested in long wavelength situations where phases vary very slowly on a microscopic scale, such as the size of the pair wave function, we have $\chi(\mathbf{r}_1) \simeq \chi(\mathbf{r}_2) \simeq \chi(\mathbf{r})$, where $\mathbf{r} = (\mathbf{r}_1 + \mathbf{r}_2)/2$ is the mass center of the pair. This allows one to more simply rewrite the transform as $\Phi'(\mathbf{r}_1, \mathbf{r}_2) = e^{2i\chi(\mathbf{r})} \Phi(\mathbf{r}_1 - \mathbf{r}_2)$. By the same arguments as above, this same phase factor comes in the transform of the order parameter Δ, so that we have

$$\Delta'(\mathbf{r}) = |\Delta| \, e^{i\varphi(\mathbf{r})} \qquad\qquad \varphi(\mathbf{r}) = 2\,\chi(\mathbf{r}) \tag{5.7}$$

Making use of the phase $\varphi(\mathbf{r})$ of the order parameter rather than $\chi(\mathbf{r})$, which has a less obvious physical meaning, we can rewrite the expression Eq. (5.6) for the current in a general gauge as

$$\mathbf{j}_s(\mathbf{r}) = \frac{e n_s}{2m} \left(\hbar \nabla \varphi(\mathbf{r}) - 2e\mathbf{A}(\mathbf{r}) \right) \tag{5.8}$$

where we have dropped the "prime" in the vector potential $\mathbf{A}(\mathbf{r})$ to make it clear that this expression is the general form we were looking for. Recalling London's general ideas about the link between superconductivity and quantum mechanics, it is worthwhile to note the strong similarity between this expression and the standard formula in quantum mechanics for the probability current in terms of its wave function $\psi(\mathbf{r})$ for a single particle with mass m and charge e

$$\mathbf{j}_s(\mathbf{r}) = \frac{e}{m}\left[\frac{\hbar}{2i}\left(\psi^*(\mathbf{r})\nabla\psi(\mathbf{r}) - \psi(\mathbf{r})\nabla\psi^*(\mathbf{r})\right) - e\mathbf{A}(\mathbf{r})|\psi(\mathbf{r})|^2\right] \quad (5.9)$$

If we apply this formula to the case of a wave function $\psi(\mathbf{r}) = |\psi|e^{i\varphi(\mathbf{r})}$ with constant modulus $|\psi(\mathbf{r})|$, which corresponds to the case we are considering for the order parameter $\Delta(\mathbf{r})$, we obtain exactly Eq. (5.8), provided that the presence probability $|\psi|^2$ is identified with the density of superconducting electrons n_s. The only difference is that the charge e has to be replaced by $2e$, and similarly the mass m has to be replaced by $2m$. This is quite reasonable since we are actually dealing with electronic pairs, and not a single electron; hence the factors 2 for the charge and the mass.

In Eq. (5.8) there is nothing new in the last term, proportional to $\mathbf{A}(\mathbf{r})$, since this contribution is identical to the one in our starting point Eq. (5.1). On the other hand, the first term is physically quite interesting. It tells us that a space-dependent phase $\varphi(\mathbf{r})$ of the order parameter implies the existence of a supercurrent given by

$$\mathbf{j}_s(\mathbf{r}) = en_s\frac{\hbar}{2m}\nabla\varphi(\mathbf{r}) = n_s e\,\mathbf{v}_s(\mathbf{r}) \quad (5.10)$$

In the last expression, we introduced the superfluid velocity $\mathbf{v}_s(\mathbf{r})$, defined by

$$\mathbf{v}_s(\mathbf{r}) = \frac{\hbar}{2m}\nabla\varphi(\mathbf{r}) \quad (5.11)$$

It gives an obvious physical meaning to the expression of the supercurrent due to a phase variation of the order parameter. Indeed it is just the current density carried by a charged fluid, with particle density n_s, charge e per particle, moving at the velocity $\mathbf{v}_s$.

The existence of a velocity $\mathbf{v}_s$ associated with the space dependence of the phase $\varphi(\mathbf{r})$ is easy to understand physically if we consider the particular case where, starting with an immobile electron gas, we make it moving by giving to each electron a velocity boost $\mathbf{V}$. This corresponds to an additional momentum $\mathbf{p} = m\mathbf{V}$, and this gives rise to an additional phase factor $e^{i\mathbf{p}\cdot\mathbf{r}_i/\hbar}$ for each electron, just as for a free particle. This is a particular case of the situation considered above where we take $\chi(\mathbf{r}) = \mathbf{p}\cdot\mathbf{r}/\hbar$. In this case, the velocity $\mathbf{V}$ is indeed obtained by $\mathbf{V} = \mathbf{p}/m = \hbar\nabla\chi(\mathbf{r})/m = \hbar\nabla\varphi(\mathbf{r})/2m$, which coincides in this particular case with the definition Eq. (5.11). Hence we have to interpret physically the superfluid velocity as the (local) drift velocity of all the Cooper pairs, in other words as the drift velocity of the condensate (described for example by the BCS wave function Eq. (2.44)). In setting up BCS theory, we were dealing with Cooper pairs having a fixed center of mass. Now we end up with a physical situation where these Cooper pairs are moving as a whole.

5.2 Flux Quantization

While the above considerations are clearly of very high interest, leading to a physical understanding of the space dependence of the order parameter, it might seem that one could also proceed in the opposite direction. Facing a situation with a space-dependent phase, one could make an appropriate gauge transformation, such that the resulting phase of the order parameter is zero and all the physics appears through the appropriate vector potential. This seems possible by merely choosing the function coming in the gauge transform as $\chi(\mathbf{r}) = \varphi(\mathbf{r})/2$. However, although this can be done in many cases, this is not possible systematically. The basic reason is that $\chi(\mathbf{r})$ and $\varphi(\mathbf{r})$ have to satisfy slightly different mathematical conditions. As we have seen, $\chi(\mathbf{r})$ must be a single-valued function of $\mathbf{r}$. But the same condition does not apply exactly to $\varphi(\mathbf{r})$. What is required physically is that the order parameter $\Delta(\mathbf{r})$, which describes the ordered phase at $\mathbf{r}$, is a single-valued function of $\mathbf{r}$. This condition defines the phase $\varphi(\mathbf{r})$ only within the addition of any multiple of 2π. This means that if starting from $\mathbf{r}$, we go around on some closed contour, with the phase evolving continuously on the contour, we should find upon going back to $\mathbf{r}$ the same phase only within a multiple of 2π. This has striking consequences if we consider a superconductor which has the shape of a torus, or any geometrical form which may be continuously deformed into a torus – in other words, a superconductor with the topology of a torus.

Let us specifically consider the annulus drawn in Fig. (5.1), which has this appropriate geometry. We take a contour C going around the annulus and staying in the superconducting region. If we want the flux Φ of the applied magnetic field $\mathbf{B}$ through the surface S enclosed by this contour we have, making use of $\mathbf{B} = \operatorname{curl}\mathbf{A}$,

$$\Phi = \int_S d\mathbf{S}\cdot\mathbf{B}(\mathbf{r}) = \oint_C d\boldsymbol{\ell}\cdot\mathbf{A}(\mathbf{r}) \tag{5.12}$$

$$= \oint_C d\boldsymbol{\ell}\cdot\mathbf{A}'(\mathbf{r}) - \frac{\hbar}{e}\oint_C d\boldsymbol{\ell}\cdot\nabla\chi(\mathbf{r}) = \oint_C d\boldsymbol{\ell}\cdot\mathbf{A}'(\mathbf{r}) - \frac{\hbar}{e}[\chi(\mathbf{r})]$$

where we have made the gauge transformation Eq. (5.2), and $[\chi(\mathbf{r})]$ is the variation of $\chi(\mathbf{r})$ when one goes around contour C. Naturally, since $\chi(\mathbf{r})$ is single-valued, we have $[\chi(\mathbf{r})] = 0$, but we see that this property is essential in order to have an expression for the flux in terms of $\mathbf{A}$ which is gauge-invariant: $\Phi = \oint_C d\boldsymbol{\ell}\cdot\mathbf{A} = \oint_C d\boldsymbol{\ell}\cdot\mathbf{A}'$.

On the other hand, if we had replaced $\chi(\mathbf{r})$ by $\varphi(\mathbf{r})/2$, this gauge invariance would not hold since, as we have seen above $[\varphi(\mathbf{r})] = 2\pi n$, where n is an integer that may be different from zero. For example, in Fig. (5.1), assuming that there is no magnetic field, we can very well take the choice $\varphi(\mathbf{r}) = n\theta$, where θ is the angle of the position $\mathbf{r}$ with respect to a reference direction, as indicated on the figure. This is physically perfectly acceptable, since the order parameter is single-valued with this choice $[\Delta(\mathbf{r})] = |\Delta|[e^{in\theta}] = 0$. Hence such a physical situation with $n \neq 0$ cannot be undone by a gauge transformation. This is physically quite clear since, with this order parameter that we are considering, corresponding to the superfluid velocity $\mathbf{v}_s = (\hbar/2m)\nabla\varphi$, we would have real supercurrents going around the annulus that cannot be removed by a gauge transform.

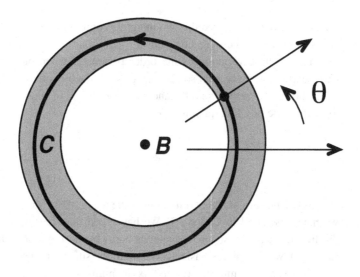

Superconducting annulus (shaded area) with an applied magnetic field *B*. A closed contour *C* goes around this annulus, staying inside the superconducting region, with the angle θ indicating the position of the running point on the contour.

It is interesting to note that such a phase variation, when one goes around such a contour, has topological stability. This means that it cannot change under continuous transformations of the contour or of the order parameter itself, merely because $\left[\varphi(\mathbf{r})/2\pi\right]$ is an integer and there is no way to go continuously from one integer to a different one. Such a change implies necessarily a jump that cannot be the result of a continuous evolution. This is a property of the phase itself. Actually, this stability is physically only a metastability because it is possible to go from an integer to another one by destroying superconductivity and going to the normal state, so that the order parameter goes to zero and does not have a well-defined phase. One can then go to a superconducting state corresponding to a different integer. Naturally, there is a strong energy cost to go to the normal state, so the barrier corresponding to this metastability is very high. However, it is possible to lower this energy cost to a large extent by making normal only a section of the superconducting annulus, so that at some places around contour *C* the phase is no longer well defined. An even more efficient way to go from one integer to another is to have a small normal region crossing some section of the annulus, so that the energy cost corresponds only to the creation of this small normal region. This can occur by having a vortex (vortices will be discussed in Chapter 6) crossing the annulus.

Let us now come back to the specific physical situation depicted in Fig. (5.1) where the superconducting annulus is in the presence of an applied magnetic field **B**. In this case, we know that the supercurrents are present only near the surface of the superconductor, typically within a few penetration depths. If we take expression Eq. (5.8) for the current and calculate the circulation of this current on contour *C*, we have

$$\oint_C d\boldsymbol{\ell} \cdot \mathbf{j}_s(\mathbf{r}) = \frac{en_s}{2m} \oint_C d\boldsymbol{\ell} \cdot (\hbar\nabla\varphi(\mathbf{r}) - 2e\mathbf{A}(\mathbf{r})) \qquad (5.13)$$

We have already discussed the two integrals coming on the right-hand side. $\Phi = \oint_C d\boldsymbol{\ell} \cdot \mathbf{A}$ is just the flux of the magnetic field through contour C, and $\oint_C d\boldsymbol{\ell} \cdot \nabla\varphi(\mathbf{r}) = [\varphi(\mathbf{r})] = 2\pi n$ is the phase variation of the order parameter when one goes once around contour C. In particular, if we choose contour C to be deep enough in the superconductor where supercurrents are zero, the left-hand side is zero and we find the striking result that Φ has to be multiple of a flux quantum Φ_0

$$\Phi = n\Phi_0 \qquad\qquad \Phi_0 = \frac{h}{2e} \qquad\qquad (5.14)$$

This can be seen as a generalization of the result of the Meissner–Ochsenfeld experiment we have discussed in Chapter 1. We have seen that due to the screening supercurrents, the flux through the superconductor is zero. In the more complex geometry depicted in Fig. (5.1), where there is basically a hole in the superconductor, the flux is not necessarily zero, it may be any multiple of the flux quantum Φ_0.

The existence of this flux quantization is quite a remarkable phenomenon. Indeed the superconducting annulus we consider is a macroscopic object (it could be as large as we like). Nevertheless a physical quantity attached to it, namely the magnetic flux through it, is quantized. In quantum mechanics, we are used to dealing with physical quantities which are quantized, but they apply to microscopic objects such as elementary particles, atoms, or molecules. Here we deal with a macroscopic system, and a physical quantity, the magnetic flux, which appears currently when one deals with the electrodynamics of macroscopic electric systems. Hence we have a manifestation of quantum mechanics at a macroscopic scale. This is a clear vindication of London's intuition, who was considering superconductivity itself as such a manifestation.

Actually, it was London [34] who realized that the magnetic flux should be quantized. However, he gave h/e as the value for the flux quantum, since the pairing theory did not exist at the time. This shows quite clearly that the presence of the pair charge $2e$ in the denominator of Eq. (5.14) is a direct signature that superconductivity is due to electron pairs. Hence the direct experimental observation [35] of the flux quantum Eq. (5.14) gave an additional very direct support to pairing theory.

The existence of the flux quantum is not only of major fundamental interest, it is also of very high practical importance. Indeed its numerical value

$$\Phi_0 = \frac{h}{2e} \simeq 2. \times 10^{-15}\,\mathrm{Wb} = 2. \times 10^{-7}\mathrm{gauss}\ \mathrm{cm}^2 \qquad\qquad (5.15)$$

is at the same time quite small, but nevertheless quite accessible experimentally with macroscopic objects (experiments [35] were performed with superconducting cylinders having typically 10^{-3} cm diameters, obtained by metallic deposits on thin copper wires). Hence the flux quantum provides a very small yet quite convenient unit. This translates into the possibility of high precision measurements of magnetic fields, down to very small values; hence its very high metrological interest. The flux quantum will appear again in the following on many occasions, in particular, in the next section dealing with the Josephson effect.

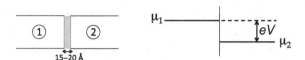

Fig. 5.2 Left: Schematic geometry of a tunnel junction (not to scale); the shaded area describes the very thin insulating domain, of typical thickness $15-20$ Å, separating the two metallic regions 1 and 2. Right: Relative position of the chemical potentials μ_1 and μ_2 in the two metals; their difference is linked to the applied voltage V by $\mu_1 - \mu_2 = eV$ (the figure is drawn assuming the electronic charge $e > 0$).

5.3 Josephson Effects

The Josephson effect[1] was first predicted [36] and observed in the tunnel junctions we have already considered in Chapter 3. For convenience, we reproduce here in Fig. (5.2) the corresponding figure. In Chapter 3, we considered the possibility of single electrons tunneling from one superconductor to the other. However, the possibility of coherent tunneling of Cooper pairs was not considered. In this case, the phase difference between the two superconductors is physically meaningful, since one can follow the phase of the pair wave function when one goes from one superconductor to the other. More specifically, we have in mind the phases φ_1 and φ_2 of the superconductors in the immediate vicinity of the tunneling junction. Actually, the Josephson effect has also been observed in other devices, different from the tunnel junction Fig. (5.2). They also allow one to follow the phase from one superconductor to the other, but superconductivity in the region between the two superconductors is not present or is much weaker than in the bulk of the two superconductors. All these devices provide a "weak link" between the two superconductors.

5.3.1 First Josephson Equation

In the preceding section, we saw that a phase variation of the order parameter in a superconductor leads to a current, corresponding physically to a Cooper pairs motion. For the same reason, we can conclude that the existence of a phase difference $\gamma = \varphi_2 - \varphi_1$ between the two superconductors gives rise to a current I flowing from superconductor 1 to superconductor 2, corresponding physically to Cooper pairs tunneling from one superconductor to the other. In general, the calculation of the current $I = f(\gamma)$ in terms of the phase difference γ is a complicated problem[2] with a result depending clearly on the specific weak link. Nevertheless it must have some properties resulting from general considerations. Indeed we have seen that, for example, superconductor 1 is fully characterized by its order parameter $\Delta_1 = |\Delta_1| e^{i\varphi_1}$, and that it is unchanged if we replace φ_1 by $\varphi_1 + 2\pi n$.

---

[1] Josephson received the Nobel Prize in 1973 for the prediction of this effect.

[2] More precisely we should also take into account the contribution from the last term in Eq. (5.8), in order to have a gauge-invariant phase difference and obtain a gauge-invariant expression for the current. For simplicity, we assume that this term is zero, or negligible, as it happens if the magnetic field is zero or weak enough. However, for a strong enough field, the critical current depends on the field and this term should be taken into account.

This means that there is no physical difference for the superconductor if we use φ_1 or $\varphi_1 + 2\pi n$. The same is valid for superconductor 2. Hence we conclude in the same way that there is no physical difference between situations described by the phase difference γ or $\gamma + 2\pi n$. This implies that the same current I should flow for these phase differences, so $f(\gamma + 2\pi n) = f(\gamma)$. This means that $f(\gamma)$ is a periodic function of γ with period 2π. Another property merely results from the fact that if there is no phase difference, there is physically no reason to have a current flow, so that $f(\gamma = 0) = 0$.

Otherwise there is no way in general to restrict $f(\gamma)$, and indeed one finds experimentally a variety of dependences of the current as a function of the phase difference, depending on the specific weak link. Nevertheless we will restrict ourselves to the particular case where the coupling between the two superconductors is so weak that it can be treated as a perturbation. This is the case, for example, of a tunneling junction with a wide enough spacing between the superconductors. In this case, if one considers the general Fourier series expansion of the periodic function $f(\gamma) = \sum_n f_n e^{in\gamma}$, one can show that only the lowest-order components $e^{i\gamma}$ and $e^{-i\gamma}$ are relevant for the current. Taking into account the condition $f(0) = 0$, this implies that only the combination $e^{i\gamma} - e^{-i\gamma} = 2i\sin\gamma$ can enter. So in this limiting case, we have merely for the current

$$I = I_c \sin\gamma \qquad (5.16)$$

where I_c is called the critical current of the junction. Indeed from Eq. (5.16) we have $I \leqslant I_c$.

5.3.2 Second Josephson Equation

The second Josephson equation is a general relation between the applied voltage V and the time derivative of the phase difference γ. It can be seen in a fairly concrete way in taking up the BCS framework. Coming back to our general expression Eq. (2.36) for the BCS wave function in terms of the wave functions $|\Psi\rangle_N$ at fixed number N of electrons, we have not written any time-dependent factor. This was actually a proper way to proceed since, at any given time t, this factor is just a phase factor that can be omitted because of gauge invariance, as we have seen at the beginning of this chapter. However, associated with the energy E_N of the state $|\Psi\rangle_N$, there is a phase factor $\exp(-iE_N t/\hbar)$, which we write now explicitly, so Eq. (2.36) becomes

$$|\Psi\rangle = \sum_{N=0}^{\infty} \frac{1}{(N/2)!} |\Psi\rangle_N \, e^{-iE_N t/\hbar} \qquad (5.17)$$

We have seen that this sum is actually very strongly peaked around the mean value $\bar{N}$ of the electron number, so that in practice only the terms with $N \simeq \bar{N}$ are relevant in Eq. (5.17). Since by definition the chemical potential μ is the derivative of the energy E_N with respect to N, we have in the vicinity of $\bar{N}$

$$E_N \simeq E_{\bar{N}} + \mu(N - \bar{N}) \qquad (5.18)$$

Hence when we go, for example, from N to $N + 2$, the phase factor in Eq. (5.17) goes from $\exp(-iE_N t/\hbar)$ to $\exp(-iE_{N+2}t/\hbar) = \exp(-iE_N t/\hbar)\exp(-2i\mu t/\hbar)$. So going from N to $N + 2$, that is, adding an electronic pair, introduces an additional time-dependent phase factor $\exp(-2i\mu t/\hbar)$. This is equivalent, in Eq. (2.37), to multiplying the pair wave function $\Phi_{\mathbf{k}}$ by $\exp(-2i\mu t/\hbar)$, so this equation becomes

$$|\Psi\rangle = \sum_{N=0}^{\infty} \frac{1}{(N/2)!} \left(\sum_{\mathbf{k}} \exp(-2i\mu t/\hbar)\Phi_{\mathbf{k}} c_{\mathbf{k}}^{\dagger} c_{-\mathbf{k}}^{\dagger} \right)^{N/2} |0\rangle \tag{5.19}$$

It is natural, since we write explicitly the time-dependent factors for the wave functions, that the pair wave function gets such a phase factor, but we see that the additional energy introduced by an additional pair is precisely twice the chemical potential 2μ. As we have seen at the beginning of this chapter, this phase factor $\exp(-2i\mu t/\hbar)$ for $\Phi_{\mathbf{k}}$ induces also the same phase factor for $v_{\mathbf{k}}$ and for Δ. But as we mentioned above, this phase factor could very well be omitted.

The situation changes drastically when we have to consider two superconductors with different chemical potentials. In this case, for superconductor 1, we have $\Delta_1(t) = |\Delta_1|\exp(i\varphi_1(t)) = |\Delta_1|\exp(-2i\mu_1 t/\hbar)$, while similarly $\Delta_2(t) = |\Delta_2|\exp(i\varphi_2(t)) = |\Delta_2|\exp(-2i\mu_2 t/\hbar)$ for superconductor 2. In other words, the phases for the two superconductors are now evolving with time at different speeds. This implies for the time-dependent phase difference $\gamma(t)$

$$\gamma(t) = \varphi_2(t) - \varphi_1(t) = -\frac{2(\mu_2 - \mu_1)t}{\hbar} \tag{5.20}$$

This is more often written by taking the time derivative as

$$\frac{\hbar}{2}\frac{d\gamma}{dt} = \mu_1 - \mu_2 = eV \tag{5.21}$$

where in the last step we have used the relation between the difference in chemical potential between the two superconductors and the voltage V applied at the junction. Introducing the flux quantum Φ_0 from Eq. (5.14), this can also be put under the simple form

$$\frac{d\gamma}{dt} = \frac{2\pi}{\Phi_0}V \tag{5.22}$$

One can get some additional intuition with Eq. (5.21) by noting the following relation. We apply this equation by taking points 1 and 2 in close vicinity, in the bulk of a superconductor. We have $\mathbf{r}_2 - \mathbf{r}_1 = d\boldsymbol{\ell}$, and $d\gamma = \varphi_2 - \varphi_1 = \nabla\varphi \cdot d\boldsymbol{\ell}$ with $\nabla\varphi = 2m\mathbf{v}_s/\hbar$ from the definition Eq. (5.11) of the superfluid velocity. On the other hand, $\mu_1 - \mu_2 = -\nabla\mu \cdot d\boldsymbol{\ell}$. Since the resulting equation is valid for any $d\boldsymbol{\ell}$, it implies

$$m\frac{d\mathbf{v}_s}{dt} = -\nabla\mu \tag{5.23}$$

This is just the standard Newton acceleration law for an electronic superfluid, which behaves as a perfect fluid. The force which produces this acceleration (this could be a gravitational field) derives from a potential energy that induces a variation in chemical potential, and a corresponding generalized force $-\nabla\mu$. Hence Eq. (5.21) is closely related to the hydrodynamic equations describing the electronic superfluid.

From Eq. (5.16) and Eq. (5.21), we can obtain the energy $E_J(\gamma)$ of the Josephson junction in terms of the phase difference γ. Indeed we have an electric current I given by Eq. (5.16) flowing with an applied voltage V given by Eq. (5.21). This implies that during an infinitesimal lapse of time dt, the energy dE_J provided to the Josephson junction is

$$dE_J = VI\,dt = I_c\,\sin\gamma\,\frac{\hbar}{2e}\frac{d\gamma}{dt}\,dt = \frac{h}{2e}\frac{I_c}{2\pi}\sin\gamma\,d\gamma \tag{5.24}$$

Integrating this relation over γ, we obtain

$$E_J(\gamma) = \frac{I_c\Phi_0}{2\pi}\,(1 - \cos\gamma) \tag{5.25}$$

where we choose the case $\gamma = 0$ to correspond to zero energy.

Conversely, it is interesting to note that starting from this expression for the energy of the Josephson junction, we can recover Eq. (5.16) and Eq. (5.21) by making use of the fact that particle number and phase are quantum-mechanically conjugate variables, just as particle position x and momentum p_x. Going to the classical limit, where in the present case, particle number and phase are still relevant variables, we can write the corresponding Hamilton equations. If q and p are conjugate variables, Hamilton equations read

$$\dot{q} \equiv \frac{dq}{dt} = \frac{\partial\mathcal{H}(q,p)}{\partial p} \qquad\qquad \dot{p} \equiv \frac{dp}{dt} = -\frac{\partial\mathcal{H}(q,p)}{\partial q} \tag{5.26}$$

where $\mathcal{H}(q,p)$ is the Hamiltonian expressed in terms of these variables.

In our case, the conjugate variables are more precisely particle number and phase multiplied by $\hbar$. We can, for example, think in terms of Cooper pairs and take as conjugate variables the pair number $n_{p2} = N_2/2$ in superconductor 2 and its phase $\hbar\varphi_2$. Since only the phase difference is relevant, we can for simplicity set $\varphi_1 = 0$ so that $\varphi_2 = \gamma$. On the other hand, if we consider that superconductor 1 and superconductor 2 form an isolated system with a total number of electrons $N = N_1 + N_2$ fixed, we have $N_1 = N - 2n_{p2}$. If $E_{02}(N_2)$ is the energy of isolated superconductor 2, as a function of its electron number N_2, and similarly $E_{01}(N_1)$ for superconductor 1, we have to add to these two energies the energy of the Josephson junction Eq. (5.25) to have the total energy of the coupled superconductors, which expressed in terms of n_{p2} and $\hbar\gamma$ gives us the expression of the Hamiltonian relevant for our problem. This leads to

$$\mathcal{H}(n_{p2}, \hbar\gamma) = E_{01}(N - 2n_{p2}) + E_{02}(2n_{p2}) + \frac{I_c\Phi_0}{2\pi}\,(1 - \cos\gamma) \tag{5.27}$$

The first Hamilton equation in Eq. (5.26) gives

$$\frac{dn_{p2}}{dt} = \frac{\partial\mathcal{H}}{\partial(\hbar\gamma)} = \frac{I_c\Phi_0}{2\pi\hbar}\sin\gamma = \frac{I_c}{2e}\sin\gamma \tag{5.28}$$

Since each Cooper pair carries a charge $2e$, the electric current $I = 2e\,(dn_{p2}/dt)$ flowing into superconductor 2 coincides indeed with the result of the first Josephson equation (5.16). On the other hand, the second Hamilton equation Eq. (5.26) leads to

$$\hbar\frac{d\gamma}{dt} = -\frac{\partial \mathcal{H}}{\partial n_{p2}} = -2(\mu_2 - \mu_1) \tag{5.29}$$

since $\mu_1 = dE_{01}(N_1)/dN_1$ and $\mu_2 = dE_{02}(N_2)/dN_2$. The result is in agreement with Eq. (5.16). One interest of this derivation of Josephson equations is to show that they result from very general considerations, and are clearly independent from a specific microscopic theory of superconductivity.

Finally it is convenient to show that if tunneling between the two superconductors is treated at the lowest order in perturbation, one ends up with Eq. (5.25) for the energy of the Josephson junction, which implies Eq. (5.16) as we have just seen and justifies that it is valid at lowest order, as we have stated. We are only interested in the tunneling of Cooper pairs through the junction and, just as we have done in Chapter 3 for single electron tunneling with Eq. (3.24), we can model this process by a pair tunneling Hamiltonian

$$H_T^{(2)} = \sum_{\mathbf{k}_1 \mathbf{k}_2} \mathcal{T}_{\mathbf{k}_1 \mathbf{k}_2}^{(2)} c_{\mathbf{k}_2 \uparrow}^{\dagger} c_{-\mathbf{k}_2 \downarrow}^{\dagger} c_{-\mathbf{k}_1 \downarrow} c_{\mathbf{k}_1 \uparrow} + h.c. \tag{5.30}$$

Actually, this Cooper pair transfer can be obtained by transferring each electron of the pair through the single electron tunneling Hamiltonian H_T given by Eq. (3.24), and accordingly can be obtained from a treatment of H_T to second order in perturbation (hence our superscript (2) to make this clear). However, the specific expression of $\mathcal{T}_{\mathbf{k}_1 \mathbf{k}_2}^{(2)}$ in terms of $\mathcal{T}_{\mathbf{k}_1 \mathbf{k}_2}$ is of no interest for us, so we do not need to make explicit this relation. We only need to use the fact that from this calculation, $\mathcal{T}_{\mathbf{k}_1 \mathbf{k}_2}^{(2)}$ is real.

Treating the coupling Eq. (5.30) to lowest order in perturbation we merely obtain $E_J(\gamma)$ by taking the average $\langle H_T^{(2)} \rangle$ of $H_T^{(2)}$ in the ground state corresponding to the uncoupled superconductors, each one being in its BCS ground state given by Eq. (2.44) (we consider only the zero temperature case). As above, we take $\varphi_1 = 0$ so that $\varphi_2 = \gamma$. Hence all terms with the pair wave function in superconductor 2 have a phase factor $e^{i\gamma}$, which we write explicitly. This gives for the unperturbed ground state of the two superconductors

$$|\Psi\rangle_1 \oplus |\Psi\rangle_2 = \prod_{\mathbf{k}_1, \mathbf{k}_2} \left(u_{\mathbf{k}_1} + v_{\mathbf{k}_1} c_{\mathbf{k}_1 \uparrow}^{\dagger} c_{-\mathbf{k}_1 \downarrow}^{\dagger} \right) |0\rangle_1 \oplus \left(u_{\mathbf{k}_2} + e^{i\gamma} v_{\mathbf{k}_2} c_{\mathbf{k}_2 \uparrow}^{\dagger} c_{-\mathbf{k}_2 \downarrow}^{\dagger} \right) |0\rangle_2 \tag{5.31}$$

Then the calculation of this average is completely analogous to the one we have already performed in Chapter 2 for the average of the interaction in the BCS ground state starting from Eq. (2.58). For a specific term $c_{\mathbf{k}_2 \uparrow}^{\dagger} c_{-\mathbf{k}_2 \downarrow}^{\dagger} c_{-\mathbf{k}_1 \downarrow} c_{\mathbf{k}_1 \uparrow}$ in Eq. (5.30), we have

$$\langle c_{\mathbf{k}_2 \uparrow}^{\dagger} c_{-\mathbf{k}_2 \downarrow}^{\dagger} c_{-\mathbf{k}_1 \downarrow} c_{\mathbf{k}_1 \uparrow} \rangle = \langle c_{\mathbf{k}_2 \uparrow}^{\dagger} c_{-\mathbf{k}_2 \downarrow}^{\dagger} \rangle \langle c_{-\mathbf{k}_1 \downarrow} c_{\mathbf{k}_1 \uparrow} \rangle \tag{5.32}$$

and, in the same way as Eq. (2.61) and Eq. (2.62), taking into account that $u_{\mathbf{k}}$ and $v_{\mathbf{k}}$ are real since we have written explicitly the phase factor,

$$\langle c_{-\mathbf{k}_1 \downarrow} c_{\mathbf{k}_1 \uparrow} \rangle = u_{\mathbf{k}_1} v_{\mathbf{k}_1} \qquad \langle c_{\mathbf{k}_2 \uparrow}^{\dagger} c_{-\mathbf{k}_2 \downarrow}^{\dagger} \rangle = u_{\mathbf{k}_2} (e^{i\gamma} v_{\mathbf{k}_2})^* = e^{-i\gamma} u_{\mathbf{k}_2} v_{\mathbf{k}_2} \tag{5.33}$$

Hence we see that in the presence of the phase difference γ, a phase term $e^{-i\gamma}$ can be factorized for all the contributions from $c_{\mathbf{k}_2 \uparrow}^{\dagger} c_{-\mathbf{k}_2 \downarrow}^{\dagger} c_{-\mathbf{k}_1 \downarrow} c_{\mathbf{k}_1 \uparrow}$ terms. Similarly for all the contributions from $c_{\mathbf{k}_1 \uparrow}^{\dagger} c_{-\mathbf{k}_1 \downarrow}^{\dagger} c_{-\mathbf{k}_2 \downarrow} c_{\mathbf{k}_2 \uparrow}$ terms, which come from the Hermitian conjugate term in Eq. (5.30), a phase term $e^{i\gamma}$ can be factorized. Otherwise, for $\gamma = 0$, the result would clearly be the same from these two parts of $H_T^{(2)}$ in Eq. (5.30). In this way, we see that the result is proportional to $e^{-i\gamma} + e^{i\gamma} = 2\cos\gamma$, in agreement with Eq. (5.25) (the constant in Eq. (5.25) results from our convention that $E_J(\gamma) = 0$ for $\gamma = 0$).

Finally it is worthwhile to note that from this evaluation, the prefactor of $\cos\gamma$ in the energy $E_J(\gamma)$, which is in practice the critical current I_c, is of order $\mathcal{T}^{(2)}$, that is, $[\mathcal{T}]^2$. Hence this pair tunneling current is of the same order as the standard tunneling current, due to the transfer of fermionic excitations from one superconductor to the other, which we considered in Chapter 3 in Eq. (3.30), for example. This is not so obvious, and at first it was thought that since transferring single electrons gives a current proportional to $[\mathcal{T}]^2$, transferring a pair, that is, two electrons, would have a probability proportional to the square of the probability for a single electron, that is, $[\mathcal{T}]^4$. Since $\mathcal{T}$ is small for a tunneling junction, it was thought that such a current would be much too small to be observable in practice. As a consequence, pair tunneling was at first overlooked.

5.3.3 Various Josephson Effects

Josephson equations Eq. (5.16) and Eq. (5.21) rule respectively the current I and the voltage V of the junction, but they do not form a closed set of equations because we do not have any information on how I and V are related. This is provided by electric organization set up around the junction. Since many setups can be considered, there is a considerable variety of possible effects.

The simplest one corresponds to the case where the voltage is set to zero. From the second Josephson equation, $V = 0$ implies that the phase difference γ does not depend on time. But there is no reason to have it equal to zero, and it is quite possible to have $\gamma \neq 0$. In this case, the first equation tells that there is a current flowing through the junction, even if there is no voltage applied to it. This is called the dc, or direct current, Josephson effect. This is a very surprising and counterintuitive effect, when one has in mind standard electronics. It is fairly difficult to observe, in particular, because the critical current I_c is small (a typical value is 1 mA). When the current flowing through the junction is larger than I_c a nonzero voltage appears, and such a current may already be produced by mere electric noise.

Another simple situation is found when the junction is left isolated so that the relation between current and voltage results from the simple capacitance C of the junction. With our conventions this leads to the relation $I = -dQ/dt = -CdV/dt$, where Q is the electric charge carried by the junction. Taking the time derivative of Eq. (5.22), this leads to

$$\frac{d^2\gamma}{dt^2} = \frac{2\pi}{\Phi_0}\frac{dV}{dt} = -\frac{2\pi}{\Phi_0}\frac{I}{C} = -\frac{2\pi I_c}{\Phi_0 C}\sin\gamma \tag{5.34}$$

This non-linear equation is identical to the equation of motion of a classical pendulum, moving under the action of gravity, γ being in this case the angle of the pendulum with the vertical direction. This equation can be integrated exactly and one has anyway a good intuitive physical understanding of the solution, which can give rise to various interesting situations. In the case of small oscillations where $\gamma \ll 1$, this equation reduces to a simple harmonic oscillator with frequency ω_J given by

$$\omega_J = \left(\frac{2\pi I_c}{\Phi_0 C}\right)^{1/2} \tag{5.35}$$

which is called the Josephson plasma frequency. Indeed the corresponding charge oscillation in the Josephson junction is analogous to the spontaneous charge oscillation which occurs at frequency ω_p in an isolated plasma (see Section 3.7.1). Note that Eq. (5.35) is the resonance frequency $1/\sqrt{LC}$ of an electric circuit with a capacitance C and an inductance $L_J = \Phi_0/(2\pi I_c)$, the Josephson junction acting in this small γ regime as an inductance L_J.

Finally another very simple situation is the case where the voltage V is kept constant. This gives rise, from Eq. (5.21), to a phase which grows linearly with time, and from Eq. (5.16) to a corresponding alternating current

$$I = I_c \sin(\omega t) \qquad\qquad \omega = \frac{2eV}{\hbar} = \frac{2\pi V}{\Phi_0} \tag{5.36}$$

This is the ac, or alternating current, Josephson effect. This is again a very surprising effect, since we are rather more used to constant voltage giving rise to direct, not alternating, current.

Comments on this ac Josephson effect are analogous to the ones for flux quantization. It is first of fundamental interest since it can again be seen as a manifestation of quantum mechanics at the macroscopic scale. Indeed it is completely analogous to the Rabi oscillation in a two-level microscopic system, with an energy difference δE between the two levels. When these two levels are coupled, the wave function of the system may oscillate with frequency $\omega = \delta E/\hbar$ between two extreme situations where one level or the other is occupied. In the present case, the role of the two levels is played by the two superconductors and, due to the coupling between them, Cooper pairs can oscillate from one to the other, giving rise to an alternating current. The energy difference for a Cooper pair being in one superconductor or the other is $(2e)V$, and accordingly the oscillation occurs with frequency $\omega = 2eV/\hbar$, in agreement with Eq. (5.36).

It is worth stressing that just as for the flux quantum, it is the pair charge $2e$ that is coming in this energy difference – not the charge e of a single electron. This shows quite directly that these are Cooper pairs which are the elementary charges for this alternating current, and accordingly that the formation of Cooper pairs is responsible for superconductivity. This direct link was used, for example, soon after the discovery of cuprates, to show that Cooper pairs were present in these superconductors. Indeed since the critical temperature was so high in these cuprates, it was not clear at all at the beginning that these new superconductors had anything to do with standard superconductivity and that superconductivity in these compounds was due to the formation of Cooper pairs. The observation of the ac Josephson effect, with a frequency involving $2e$, instead of e, allowed to settle definitely this question.

On the other hand, regarding possible applications, the ac Josephson effect is of utmost interest for metrology, since it allows one to relate extremely directly from an experimental point of view a frequency measurement to a voltage measurement. This is of very high interest since frequencies are among the physical quantities which are measured with the highest precision, while voltages, and more generally electrical quantities, are much harder to measure precisely. This state of affairs is such that very recently a fundamental change has been decided in the way basic physical units are defined. This follows a recent evolution where, instead of having material references for fundamental physical units, and measuring from them the values of fundamental constants, the opposite point of view has been taken because it turns out by now to be more convenient for precision measurements. It has been decided to give to some fundamental physical constants specific values, which serve as definitions for physical units. For example, it was decided in 1983 that the velocity of light in vacuum is exactly $c = 299\ 792\ 458$ m/s. This now serves as a definition for the length unit in the international system, the meter. Since frequencies are by now measured with a precision of order 10^{-16}, this provides in principle the same precision to length measurements. In the same way, in 2018, it was decided that Planck's constant has the exact value $h = 6.626\ 070\ 15 \times 10^{-34}$ m^2 kg/s, which serves as a definition for the mass unit. Similarly, from 2019, the electronic charge is exactly $|e| = 1.602\ 176\ 634 \times 10^{-19}$ C. The Josephson effect, together with the quantum Hall effect, plays an essential role in the experimental devices allowing one to translate this definition into practical units for current intensity and voltage.

5.3.4 SQUID and Applications

The last setup involving Josephson junctions we will consider can also be considered as displaying quantum mechanics at the macroscopic scale, since it provides evidence of the interference between two currents linked to two Josephson junctions. This device is shown schematically in Fig. (5.3). A loop made of bulk superconducting material is interrupted by two Josephson junctions. A total current I_t is fed to this loop and splits into two super-currents I and I', each one going respectively through one of the Josephson junctions. The two currents are collected on the other side of the loop into the total current I_t. The super-conducting loop is in the presence of a perpendicular magnetic field B. We assume for simplicity that the two Josephson junctions are identical, although this is not easily done in practice. However, small differences between the two junctions do not change qualitatively the results.

Calling as above φ_1 and φ_2 the phases of the two superconductors at the first Josephson junction, we have for the current I going through this junction the first Josephson equation, Eq. (5.16), $I = I_c \sin(\varphi_2 - \varphi_1)$. Similarly we have for the current I' going through the second Josephson junction $I' = I_c \sin(\varphi_2' - \varphi_1')$, where φ_1' and φ_2' are the phases for the two superconductors at the second Josephson junction. Accordingly, the total current going through the loop is

$$I_t = I + I' = I_c \left[\sin(\varphi_2 - \varphi_1) + \sin(\varphi_2' - \varphi_1') \right] \tag{5.37}$$

$$= 2I_c \sin \frac{(\varphi_2 - \varphi_1) + (\varphi_2' - \varphi_1')}{2} \cos \frac{(\varphi_2 - \varphi_1) - (\varphi_2' - \varphi_1')}{2}$$

where we have used in the last step a standard trigonometric identity.

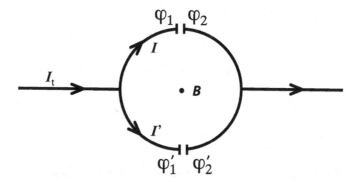

Fig. 5.3   Superconducting loop interrupted by two Josephson junction with an applied magnetic field B perpendicular to the loop. The loop is fed by a current I_t, which splits into currents I and I' going through either of the Josephson junctions.

The argument of the cosine in Eq. (5.37) can be related to the flux Φ of the magnetic field through the loop. Indeed, as we have seen in Section 5.2, in the bulk of the superconducting material making up the loop the supercurrent is zero, which implies from Eq. (5.8)

$$\nabla\varphi(\mathbf{r}) = \frac{2e}{\hbar}\mathbf{A}(\mathbf{r}) = \frac{2\pi}{\Phi_0}\mathbf{A}(\mathbf{r}) \tag{5.38}$$

Integrating this relation along the part of the superconducting loop where the total current enters, we obtain for the phase difference in the superconductors located on the left-side of the Josephson junction in Fig. 5.3

$$\varphi_1' - \varphi_1 = \int_1^{1'} d\boldsymbol{\ell}\cdot\nabla\varphi = \frac{2\pi}{\Phi_0}\int_1^{1'} d\boldsymbol{\ell}\cdot\mathbf{A} \tag{5.39}$$

where 1 and $1'$ indicate the locations in the superconductor right on the left side of the first and the second Josephson junction, respectively. Similarly we have on the right side of the loop, where the total current leaves,

$$\varphi_2 - \varphi_2' = \int_{2'}^{2} d\boldsymbol{\ell}\cdot\nabla\varphi = \frac{2\pi}{\Phi_0}\int_{2'}^{2} d\boldsymbol{\ell}\cdot\mathbf{A} \tag{5.40}$$

Together these expressions give for the phase difference coming in the argument of the cosine

$$\varphi_2 - \varphi_2' + \varphi_1' - \varphi_1 = \frac{2\pi}{\Phi_0}\left(\int_1^{1'} + \int_{2'}^{2}\right) d\boldsymbol{\ell}\cdot\mathbf{A} \tag{5.41}$$

However, in the Josephson junctions, the spacings $1-2$ and $1'-2'$ between the two superconductors are extremely small. Hence we may consider $\int_2^1 d\boldsymbol{\ell}\cdot\mathbf{A}$ and $\int_{1'}^{2'} d\boldsymbol{\ell}\cdot\mathbf{A}$ as negligibly small if the field is not too strong (this is coherent with our neglecting the last term in Eq. (5.8); see footnote 2). Taking this into account, the integrals on the right-hand side of Eq. (5.41) reduce to an integral around the complete loop $\oint d\boldsymbol{\ell}\cdot\mathbf{A} = \Phi$, equal to the magnetic flux Φ through the loop. Since in Eq. (5.37) the modulus of $\sin((\varphi_2-\varphi_1)/2 + (\varphi_2'-\varphi_1')/2)$ is less than 1, we see that the total current is bounded by

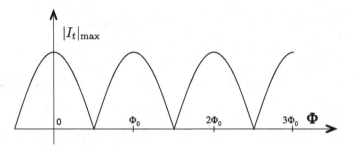

Critical current $|I_t|_{\max}$ for the loop represented in Fig. (5.3) as a function of the magnetic flux Φ through the loop, according to Eq. (5.42).

$$|I_t|_{\max} = 2I_c \left| \cos\left(\pi \frac{\Phi}{\Phi_0} \right) \right| \tag{5.42}$$

We arrive at the striking conclusion that the critical current through this device is a periodic function of the flux Φ through the loop, with period Φ_0, as it is represented in Fig. (5.4). In particular, for $\Phi = (n + 1/2)\Phi_0$, there is a complete destructive interference between the currents flowing through the two Josephson junctions, and the total current is zero. Note that the result Eq. (5.42) is gauge-invariant, as it should be.

The setup we have just considered is the starting point for devices known as SQUIDs, for Superconducting QUantum Interference Device. The unit scale for flux detection is the very small flux quantum Φ_0. But it is actually quite possible to point, on the curve represented in Fig. (5.4), the actual flux with a typical precision of order $10^{-3}\Phi_0$.

The actual SQUID devices come in two species, the dc SQUID or the rf SQUID. The dc SQUID is close to the setup we have considered. It has two Josephson junctions as in Fig. (5.3), but it is rather operated with a current I above the critical current, so that there is nonzero voltage V. Nevertheless the resulting voltage has a dependence on flux analogous to Fig. (5.4), which allows the flux to be obtained from the voltage measurement.

The rf SQUID is somewhat different since it contains a superconducting loop with a single Josephson junction. This loop is magnetically coupled to a resonating circuit, with a resonance rf frequency of order a few 10 MHz. The resonance features of the circuit are measured, and they can be shown to be modulated by the flux Φ through the superconducting loop, with period Φ_0. This rf SQUID is less sensitive than the dc SQUID, but it has been used early on for practical reasons. In both cases, the magnetic field under investigation is not directly measured from its flux through the superconducting loop, because one wants as much freedom as possible to optimize the characteristics of this loop. The field is rather measured by a pickup coil, which is magnetically coupled to the superconducting loop.

The most spectacular application of SQUIDs for the general public is the measurement of the extremely weak magnetic fields, of order of 10^{-9} gauss, generated by brain activity. This is the basis of magnetoencephalography (MEG). These fields are of interest because they can provide information on the neuronal activity located deep inside the brain. In order to eliminate as much as possible the signals resulting from much stronger fields coming from the environment, one rather measures the difference between the fields seen by nearby coils located near the skull. In this way, the perturbing fields coming from the environment, which are essentially uniform, are mostly removed. The only signals that are seen are the ones coming from the nearby sources in the brain, which are viewed differently by the various coils.

6   Ginzburg–Landau Theory

This chapter does not come in quite logical order. Indeed the Ginzburg–Landau theory we want to address here historically came before the BCS theory we have considered in the preceding chapters. However, we preferred to begin with the BCS theory because we wanted to go directly at our present deep physical understanding of the standard superconductivity phenomenon. There is also a pedagogical advantage to proceeding in this way. Having studied BCS theory, we are completely used to the fact that we have to deal with a complex order parameter, because it is directly linked to the fact that quantum mechanics tells us that the pair wave function, which is at the heart of BCS theory, is a complex object. By contrast, Ginzburg–Landau theory is a phenomenological theory that does not try to find the physical origin of superconductivity. It merely takes as a starting point the fact that the order parameter is a complex object, and proceeds from this hypothesis to describe the physical properties of the corresponding systems. This is a fairly abstract intellectual approach, which is easier to accept once one knows the specific physics that is behind superconductivity.

Ginzburg–Landau theory is all the more remarkable since, in its range of validity, it happens to be in full agreement with BCS theory. Indeed almost immediately after BCS theory came out, Gor'kov [37] showed that the equations obtained by Ginzburg and Landau could be derived exactly from BCS theory. This is of major interest not only as a matter of principle, but also in practice. It is indeed much easier to manipulate Ginzburg–Landau equations than to address the same problem starting from the microscopic BCS theory, taking into account that the relevant problems are those where one deals with inhomogeneous situations. Hence these equations can be considered as a very efficient tool to deal with the microscopic theory in these relevant situations. In particular, the vortex structure, which is crucial for the physical properties of type II superconductors, comes out quite naturally from these equations, as we will see.

Finally another interest of Ginzburg–Landau theory is that while it is fully consistent with BCS theory, it does not rely on BCS theory. It starts indeed from simpler and more general hypotheses, although this leads to a restricted range of validity compared to the microscopic theory. This independence is very useful if we want to describe the physical properties of a metal that is apparently a superconductor, but for which it is not clear at all that BCS theory provides a proper microscopic theory. This situation has in practice arisen in a number of cases, in particular with high T_c superconductors that display a fair deal of unusual and puzzling physical properties.

6.1 Simple Ginzburg–Landau Theory

This Ginzburg–Landau theory [38] is a specific application and extension to superconductivity of the general approach proposed by Landau [39] for the theory of "second-order" phase transitions. This Landau's theory deals with phase transitions where the symmetry of the system changes continuously when it goes from a high-temperature disordered phase, with high symmetry, existing above a critical temperature T_c, to a low-temperature ordered phase, with lower symmetry, existing below the critical temperature. Hence, in the low-temperature phase, the symmetry of the high-temperature phase is broken.

An essential idea of Landau's approach is that the order in the low-temperature ordered phase is quantitatively characterized by an "order parameter." This can actually be any mathematical quantity such as a simple real number, but also a vector or a matrix. For example, in the simple case of a transition toward a ferromagnetic order, the order parameter is the real vector $\mathbf{M}$ corresponding to the magnetization. Quite generally, this order parameter depends on temperature in the low-temperature phase, but it goes to zero at the critical temperature T_c and it is equal to zero in the high-temperature phase, where there is no order (in the ferromagnetic case there is no macroscopic magnetization above the critical temperature, so $\mathbf{M}=\mathbf{0}$).

For superconductivity, Ginzburg and Landau assumed [38] that this order parameter is a complex number, starting with the idea that it should represent some wave function of the superconducting electrons. Naturally, we have been made familiar with this idea by investigating BCS theory in detail. But this was not such an obvious idea at the time. They noted ψ for this quantity, but following our preceding notations, we will rather use the symbol Δ for this order parameter.[1]

The Ginzburg–Landau analysis starts from the free energy $F_s(T, \Delta)$ of the superconductor, which depends not only on temperature (we ignore, as above, any pressure dependence), but also on the order parameter Δ. In the case of a ferromagnet, it is clear that we can consider the free energy $F_s(T, \mathbf{M})$ because we can think of controlling the magnetization $\mathbf{M}$ via an external magnetic field, for example. In the case of a superconductor, it is not so clear how we could in practice control Δ, but in principle the general idea is just the same as for the ferromagnet.

Naturally, there is not so much to be said in general on $F_s(T, \Delta)$ without a specific microscopic theory. However, Ginzburg and Landau showed that a very fruitful analysis can nevertheless be made when the order parameter Δ is small. Since by definition the order parameter is zero at the critical temperature T_c, this occurs, in particular, in the vicinity of this critical temperature, and accordingly we will limit our considerations to this temperature domain.

The essential step made by Ginzburg and Landau was to assume that for small Δ, the free energy has a regular expansion in powers of Δ. Although this hypothesis seems innocuous, and almost obvious, theoretical research in the field of phase transitions showed in the

---

[1] This avoids introducing new notations, but one should be careful that our choice may introduce some confusion since we keep below the same notations, α, β, and γ, as in the original Ginzburg–Landau theory.

following decades that it is actually incorrect, in general. We will be back at the end of this chapter to this question but, as we will see, the assumption made by Ginzburg and Landau turns out to be excellent in the specific case of superconductivity. Hence, in this case, the assumption of a regular power expansion is justified.

If we had a single real variable x instead of Δ, we would have terms in x, x^2, x^3, and so on in our power expansion of the free energy. If we had two real variables x and y, we would have terms in x, y, x^2, xy, y^2, and so on in the expansion. Here the situation is similar since the complex order parameter Δ is equivalent to the knowledge of two real variables $\text{Re}\,\Delta$ and $\text{Im}\,\Delta$. However, it is more convenient to use instead Δ and Δ^*, which are just two independent linear combinations of $\text{Re}\,\Delta$ and $\text{Im}\,\Delta$.

The last essential ingredient in Ginzburg–Landau theory is that the free energy expansion must satisfy the symmetry requirements of the low-temperature phase and its order parameter. For example, in the case of an isotropic ferromagnet, the magnetization $\mathbf{M}$ is equivalent to the knowledge of the components $\{M_x, M_y, M_z\}$. The free energy is a simple real number, a scalar, which should not depend on the direction of the magnetization $\mathbf{M}$ for an isotropic ferromagnet. Hence it is not allowed to have a first-order term, such as M_x, in the expansion because such a term is not invariant by rotation of the magnetization: the result would depend on the direction of the magnetization with respect to the x-axis. More generally and briefly, there is no way to obtain a scalar (the free energy) from a term that depends linearly on a vector (the magnetization).

Going to second-order terms in the expansion, only the combination $M_x^2 + M_y^2 + M_z^2 = \mathbf{M}^2$ is invariant by rotation of the magnetization, and therefore acceptable in the free energy expansion, since any other quadratic form in the components would imply preferred directions incompatible with rotational invariance. Note that a first-order term such as $\sqrt{M_x^2 + M_y^2 + M_z^2} = |\mathbf{M}|$ is not acceptable in the expansion since, because of the square root, this term is not regular in M_x, M_y, and M_z, the resulting expansion is not analytic in these variables at $M_x = M_y = M_z = 0$. Quite generally the possible transformations of the order parameter form mathematically a group. The combinations which, at each order, produce a scalar that may belong to the free energy expansion can be systematically obtained from group theory.

Coming back to the case of a general superconductor, there are in principle many possibilities for its symmetry. One of the major interests of Ginzburg–Landau theory is that it can be adapted to any of these possibilities. For our purposes, we will only consider the case originally investigated by Ginzburg and Landau, which corresponds also to the standard BCS superconductor, namely the case of an isotropic s-wave superconductor. In this case, the order parameter is a simple complex number, and the only symmetry we have to consider is gauge invariance. The analysis is similar to that in the preceding paragraphs for the isotropic ferromagnet.

To first order, we have to consider terms proportional to Δ or Δ^*. But in a gauge transformation, which multiplies Δ by a phase factor, such linear terms will be multiplied by this phase factor (or its complex conjugate), and accordingly they will be modified by a gauge transformation. On the other hand, the free energy should not depend on the gauge. Hence these linear terms are incompatible with gauge invariance, and they should be rejected, just as the linear terms in the magnetization are rejected for the isotropic ferromagnet.

To second order, we have the three terms: $\Delta\Delta$, $\Delta\Delta^*$, and $\Delta^*\Delta^*$. Again the first and the third one are modified in a gauge transformation. On the other hand, the second one $\Delta\Delta^* = |\Delta|^2$ is not changed in such a gauge transformation, and it is the only one that survives at this order to the gauge invariance requirement. This is analogous to the $\mathbf{M}^2$ term for the ferromagnet. Similar considerations show that to third order, there is no way to write a gauge-invariant combination. Finally, to fourth order, only the term $\Delta^2\Delta^{*2} = |\Delta|^4$ is gauge-invariant and is allowed to appear in the expansion. Quite generally, it is clear that only the combination $|\Delta|^2$ can enter a gauge-invariant expansion. Finally the zeroth-order term $F_s(T, 0)$ in the expansion is identical to the free energy of the high-temperature phase, which, in the case of superconductivity is the normal phase, since we have seen that by definition $\Delta = 0$ in this high-temperature phase. Hence we merely write this term as $F_n(T)$ (below T_c it is the extrapolation of the normal state free energy to such temperatures).

Gathering all the terms we have kept, we end up for the superconductor free energy (for a unit volume) with the following expansion in powers of the order parameter:

$$F_s(T, \Delta) - F_n(T) = \alpha|\Delta|^2 + \frac{\beta}{2}|\Delta|^4 \tag{6.1}$$

where α and β are two undetermined phenomenological parameters which we discuss below. Naturally, we may wonder why we stop the expansion specifically at fourth order. The reason is that if we kept only the second-order term, we would end up with an incoherent theory as we will see. On the other hand, it can be checked that if we go to sixth order, for example, this does not bring any improvement because, if we stay consistently in the vicinity of T_c, the corrections brought by including the sixth-order term are negligible. These statements are valid only for our specific case. There are other situations where a Ginzburg–Landau expansion can be used, but where one needs to go to higher orders and, for example, include a sixth-order term. However, the general principle is the same. One goes to the order necessary to have a coherent theory, but one stops the expansion when additional terms would bring only negligible corrections.

We have written the free energy for any (small) value of the order parameter. However, in general, the superconductor is free to choose its order parameter, and at equilibrium the superconducting system goes to the order parameter $\Delta(T)$ which minimizes the free energy Eq. (6.1) at temperature T. This immediately implies conditions on the parameters coming in the expansion Eq. (6.1). Indeed we must have $\beta > 0$; otherwise, if we had $\beta < 0$, we could minimize $F_s(T, \Delta)$ by taking a very large value for $|\Delta|$, for which the fourth-order term would be dominant. But in such a case, we could indefinitely lower the free energy by taking ever-larger values of $|\Delta|$. This means that the equilibrium value of $|\Delta|$ would be very large, in contradiction with the starting assumption of the Ginzburg–Landau theory of a small order parameter (and naturally for very large $|\Delta|$ the expansion Eq. (6.1) is no longer valid). Accordingly, in order to have a consistent theory where the equilibrium order parameter is small, we must have $\beta > 0$.

Let us now consider the sign of α. For $\alpha > 0$, the two terms on the right-hand side of Eq. (6.1) are positive, and the minimum free energy is obviously obtained for $|\Delta| = 0$. However, by definition, this means that the system is in the normal state. The typical shape of the free energy difference Eq. (6.1) as a function of Δ is shown on the left side of Fig. (6.1) in this case. On the other hand, for $\alpha < 0$, it is clear that the minimum of $F_s(T, \Delta)$

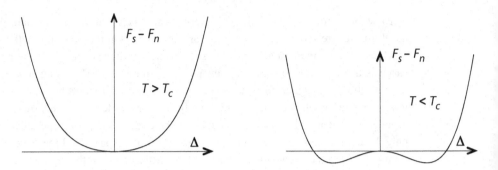

Qualitative behavior, as a function of the order parameter Δ, of the free energy difference $F_s(T, \Delta) - F_n(T)$ in Eq. (6.1) for $T > T_c$ on the left, and for $T < T_c$ on the right.

in Eq. (6.1) is obtained for $|\Delta| \neq 0$, since for very small $|\Delta|$ the second-order term is the dominant one on the right-hand side of Eq. (6.1), and one can lower the free energy by increasing $|\Delta|$. Hence for $\alpha < 0$, $|\Delta| \neq 0$ and the system is in the superconducting state. In this case, the typical shape of the free energy difference Eq. (6.1) as a function of Δ is shown on the right side of Fig. (6.1); it is often referred to as a "Mexican Hat" because of its shape.

Since by lowering the temperature from above T_c (where the system is normal, so $\alpha > 0$) to below T_c (where the system is superconducting, so $\alpha < 0$) α changes sign, this implies that $\alpha(T)$ depends on temperature T and changes sign at T_c. Now, just as for the free energy expansion, Ginzburg and Landau make for this temperature dependence a regularity assumption: at T_c, $\alpha(T)$ does not have any singular behavior, it merely changes sign, and since we are always in the close vicinity of this critical temperature, we can take for $\alpha(T)$ a linear approximation and write

$$\alpha(T) = \alpha_0(T - T_c) \tag{6.2}$$

where α_0 is a positive constant, independent of temperature. Coming back to β, it has also in principle a temperature dependence. However, since we stay in the close vicinity of T_c, this dependence is weak and it can be seen that taking it into account brings only negligible contributions. Hence we can, in practice, take β independent of temperature.

We can now calculate specifically the value of the order parameter $\Delta_0(T)$ in the superconducting state, corresponding to the minimum of $F_s(T, \Delta) - F_n(T)$ on the right-side of Fig. (6.1). Minimizing the right-hand side of Eq. (6.1) with respect to $|\Delta|$, or rather $|\Delta|^2$, we have immediately

$$\frac{\partial F_s(T, \Delta)}{\partial |\Delta|^2} = -|\alpha(T)| + \beta|\Delta|^2 = 0 \tag{6.3}$$

where we have made explicit the fact that $\alpha(T)$ is negative for $T < T_c$. Taking the real positive determination of the order parameter, this leads to

$$\Delta_0(T) = \left(\frac{|\alpha(T)|}{\beta}\right)^{1/2} = \left(\frac{\alpha_0}{\beta}\right)^{1/2} (T_c - T)^{1/2} \tag{6.4}$$

The temperature dependence in $(T_c - T)^{1/2}$ is in perfect agreement with the BCS result Eq. (3.56) found in Chapter 3, which is fairly remarkable since the paths we have taken to reach this result are completely different. Nevertheless, both BCS and Ginzburg–Landau theory are mean-field theories, so it is natural that they end up with the $1/2$ exponent which is the proper one for mean-field theory in this case.

Substituting Eq. (6.3) into Eq. (6.1) we obtain the equilibrium free energy $F_s(T)$ of the superconductor

$$F_s(T) - F_n(T) = -\frac{\alpha^2(T)}{2\beta} = -\frac{\alpha_0^2}{2\beta}(T - T_c)^2 \tag{6.5}$$

and, in the same way as in Chapter 1, Section 1.1.5, we find the entropy difference between the superconducting and the normal state

$$S_s(T) - S_n(T) = -\frac{\partial(F_s - F_n)}{\partial T} = \frac{\alpha_0^2}{\beta}(T - T_c) \tag{6.6}$$

and again, from $L = T(S_n - S_s)$, we obtain that there is no latent heat at T_c. Finally, we find the specific heat jump at T_c

$$C_s(T_c) - C_n(T_c) = T_c\frac{\partial(S_s - S_n)}{\partial T} = \frac{\alpha_0^2 T_c}{\beta} \tag{6.7}$$

Ginzburg–Landau theory does not provide any way to determine the phenomenological parameters α_0 and β, which must be obtained from a microscopic theory. The BCS theory, which agrees with all these results, provides precisely such a theory. Comparison of Eq. (6.4) with Eq. (3.56) gives the ratio α_0/β, while comparing the specific heat jumps Eq. (6.7) and Eq. (3.63) leads to α_0^2/β. In this way, one finds

$$\alpha_0 = \frac{N_0}{T_c} \qquad \beta = \frac{7\zeta(3)}{8\pi^2}\frac{N_0}{(k_B T_c)^2} \simeq 0.1\frac{N_0}{(k_B T_c)^2} \tag{6.8}$$

6.2 Magnetic Field Effect

Although it provides a nice way to obtain very directly the temperature dependence Eq. (6.4) of the order parameter in the vicinity of T_c, the Ginzburg–Landau theory would not be so interesting if it was limited to the above considerations and did not have an extension to a space-dependent order parameter in the presence of a magnetic field.

Let us first consider the extension to a space-dependent $\Delta(\mathbf{r})$ order parameter without magnetic field. This extension is valid only for an order parameter that varies very slowly with position $\mathbf{r}$. This restriction is somewhat similar to the limitation to small order parameter Δ, and it allows an analogous expansion in powers of the wavevector corresponding to the order parameter oscillations. Hence we consider only the large wavelength regime for the variations of the order parameter. In this situation, we are almost in the thermodynamic limit where the whole superconductor is made of its different parts with different order parameters $\Delta(\mathbf{r})$. In such a case it is natural to write, as in thermodynamics, the total free energy as the sum of the free energies of the various parts with different $\Delta(\mathbf{r})$. Taking for each part the expression Eq. (6.1) corresponding to its order parameter, this leads to generalize the free energy difference as $\int d\mathbf{r}\,[\alpha|\Delta(\mathbf{r})|^2 + (\beta/2)|\Delta(\mathbf{r})|^4]$.

However, such an expression treats the different parts of the superconductor as completely independent, it misses the physical fact that at equilibrium the order parameter tends to be uniform in the whole superconductor. Hence there must be an energy cost to have a non-uniform order parameter. This is analogous to the existence of an interface energy between parts of the superconductor with different order parameters. Now the spatial variations of the order parameter are quantitatively characterized by its spatial derivatives, that is, its gradient $\nabla\Delta(\mathbf{r})$, and accordingly the free energy depends in general not only on $\Delta(\mathbf{r})$, but also on all its spatial derivatives. For the slowly varying order parameter to which we restrict ourselves, we proceed to an expansion in powers of these derivatives and retain only the lowest-order, higher-order derivatives leading to negligible contributions. If we would go to Fourier transform, this would correspond to keep only the lowest powers of the wavevector $\mathbf{q}$ corresponding to the variations of the order parameter.

With respect to the order parameter itself, just as in our above expansion Eq. (6.1), the lowest-order term must contain Δ and Δ^* in order to satisfy gauge invariance. With respect to the spatial derivatives, we must build out of the ∇ operator, which is a vector, a scalar quantity since the free energy is such a scalar. The only possibility at lowest order is the scalar product $\nabla \cdot \nabla$. Taking these requirements together leads, for our isotropic superconductor, to $\nabla\Delta(\mathbf{r}) \cdot \nabla\Delta^*(\mathbf{r}) = |\nabla\Delta(\mathbf{r})|^2$ as the only possible lowest-order term[2] for this gradient expansion. Just as above, we introduce for this term a phenomenological coefficient γ, which must be positive in order that the inhomogeneous superconductor has a higher free energy than the uniform one. Just as β, γ can be considered as a constant in the vicinity of T_c. This leads to the following expression for the free energy

$$F_s(T, \{\Delta(\mathbf{r})\}) - F_n(T) = \int d\mathbf{r}\, \left(\alpha(T)\,|\Delta(\mathbf{r})|^2 + \frac{\beta}{2}\,|\Delta(\mathbf{r})|^4 + \gamma\,|\nabla\Delta(\mathbf{r})|^2 \right) \quad (6.9)$$

where we have specified that the free energy is now a functional of the whole order parameter field $\Delta(\mathbf{r})$.

We come now to the final generalization of the Ginzburg–Landau free energy, which includes the effect of a magnetic field $\mathbf{B}(\mathbf{r})$, with its corresponding vector potential $\mathbf{A}(\mathbf{r})$. We have seen in the preceding section that gauge invariance plays an essential role in selecting the terms which can enter the free energy expression. In the presence of a magnetic field, the gauge transformations which leave $\mathbf{B}(\mathbf{r})$ invariant are for the vector potential

$$\mathbf{A}(\mathbf{r}) = \mathbf{A}'(\mathbf{r}) - \frac{\hbar}{e}\,\nabla\chi(\mathbf{r}) \quad (6.10)$$

---

[2] Actually, we could also think of $\Delta^*(\mathbf{r})\nabla^2\Delta(\mathbf{r}) = \Delta^*(\mathbf{r})\sum_i(\partial^2\Delta(\mathbf{r})/\partial x_i^2)$, or $\Delta(\mathbf{r})\nabla^2\Delta^*(\mathbf{r})$, as other possibilities. However, after integration over $\mathbf{r}$ and by-parts integration, they reduce to the $|\nabla\Delta(\mathbf{r})|^2$ term we have in the text. The only difference, resulting from the by-parts integration, is a term involving the surface of the superconductor, which is irrelevant for our purpose.

As we have seen in Chapter 5, in this gauge transformation the corresponding change for the order parameter is

$$\Delta(\mathbf{r}) = e^{-2i\chi(\mathbf{r})}\Delta'(\mathbf{r}) \tag{6.11}$$

In this transformation $|\Delta(\mathbf{r})|^2$ is unchanged as before, so the first two terms of Eq. (6.9) are still gauge-invariant. On the other hand, this is not the case for the third term $|\nabla\Delta(\mathbf{r})|^2$, as is clear from Eq. (6.10) and Eq. (6.11). But, again we have already considered this question in Chapter 5, and we have seen that just as in the Schrödinger equation, the proper gauge-invariant combination is not $(\hbar/i)\nabla\Delta(\mathbf{r})$, but $[(\hbar/i)\nabla-2e\mathbf{A}(\mathbf{r})]\Delta(\mathbf{r})$, which, in the transformations Eq. (6.10) and Eq. (6.11) becomes appropriately $e^{-2i\chi(\mathbf{r})}[(\hbar/i)\nabla-2e\mathbf{A}'(\mathbf{r})]\Delta'(\mathbf{r})$. Hence, in order to obtain a gauge-invariant quantity, we have to make the same substitution in the third term of Eq. (6.9). This leads to the following generalization of the Ginzburg–Landau free energy in the presence of a magnetic field

$$F_s\left(T,\{\Delta(\mathbf{r})\},\{\mathbf{A}(\mathbf{r})\}\right) - F_n(T) = \tag{6.12}$$

$$\int d\mathbf{r}\left(\alpha(T)|\Delta(\mathbf{r})|^2 + \frac{\beta}{2}|\Delta(\mathbf{r})|^4 + \gamma\left|\left(\frac{1}{i}\nabla - \frac{2e}{\hbar}\mathbf{A}(\mathbf{r})\right)\Delta(\mathbf{r})\right|^2 + \frac{\mathbf{B}^2(\mathbf{r})}{2\mu_0}\right)$$

The last term in Eq. (6.12) is the contribution of the magnetic field $\mathbf{B}(\mathbf{r}) = \operatorname{curl}\mathbf{A}(\mathbf{r})$ to electromagnetic energy. This term must be included because, when we consider the equilibrium situation, the superconductor modifies the magnetic field through the currents it produces.

6.2.1 Ginzburg–Landau equations

Now, we want to find the equilibrium $\Delta(\mathbf{r})$ and $\mathbf{A}(\mathbf{r})$ by minimizing this free energy. In principle, this is analogous to what we have done in Eq. (6.3), except that the minimization is not with respect to the single variable Δ, but with respect to the fields $\Delta(\mathbf{r})$ and $\mathbf{A}(\mathbf{r})$. Hence we have to write that the corresponding functional derivatives are zero

$$\left.\frac{\delta F_s\left(T,\{\Delta(\mathbf{r})\},\{\mathbf{A}(\mathbf{r})\}\right)}{\delta\Delta(\mathbf{r})}\right|_{\mathbf{A}(\mathbf{r})} = 0 \qquad \left.\frac{\delta F_s\left(T,\{\Delta(\mathbf{r})\},\{\mathbf{A}(\mathbf{r})\}\right)}{\delta\mathbf{A}(\mathbf{r})}\right|_{\Delta(\mathbf{r})} = 0 \tag{6.13}$$

If we think that $\mathbf{r}$ takes only discrete values, this is quite analogous to the calculation we had to perform at the level of Eq. (3.43) in Chapter 3. In the same way, in the first equation of Eq. (6.13), we have, actually, to minimize with respect to both Re $\Delta(\mathbf{r})$ and Im $\Delta(\mathbf{r})$. As we have already done, it is more convenient to consider $\Delta(\mathbf{r})$ and $\Delta^*(\mathbf{r})$ as independent variables. Hence we consider first the (first-order) variation of $F_s\left(T,\{\Delta(\mathbf{r})\},\{\mathbf{A}(\mathbf{r})\}\right)$ with respect to $\Delta^*(\mathbf{r})$, at fixed $\Delta(\mathbf{r})$ and $\mathbf{A}(\mathbf{r})$.

When $\Delta^*(\mathbf{r})$ has a small variation $\delta\Delta^*(\mathbf{r})$, the contribution to the corresponding variation $\delta F_s\left(T,\{\Delta(\mathbf{r})\},\{\mathbf{A}(\mathbf{r})\}\right)$ of the free energy from the first two terms on the right-hand side of Eq. (6.12) is easily obtained as

$$\delta\left[\int d\mathbf{r}\left(\alpha(T)|\Delta(\mathbf{r})|^2 + \frac{\beta}{2}|\Delta(\mathbf{r})|^4\right)\right] = \int d\mathbf{r}\left(\alpha(T)+\beta|\Delta(\mathbf{r})|^2\right)\Delta(\mathbf{r})\delta\Delta^*(\mathbf{r}) \tag{6.14}$$

On the other hand, the variation of the third term of Eq. (6.12) is more complicated because one has

$$\delta \left[\int d\mathbf{r} \left| \left(\frac{1}{i} \nabla - \frac{2e}{\hbar} \mathbf{A}(\mathbf{r}) \right) \Delta(\mathbf{r}) \right|^2 \right] = \qquad (6.15)$$

$$\int d\mathbf{r} \left[\left(\frac{1}{i} \nabla - \frac{2e}{\hbar} \mathbf{A}(\mathbf{r}) \right) \Delta(\mathbf{r}) \right] \left[\left(-\frac{1}{i} \nabla - \frac{2e}{\hbar} \mathbf{A}(\mathbf{r}) \right) \delta \Delta^*(\mathbf{r}) \right]$$

and one has to deal with a $\nabla \delta \Delta^*(\mathbf{r})$ term. However, in order to obtain $\delta F_s / \delta \Delta^*(\mathbf{r})$, one must obtain this variation as an integral over $\mathbf{r}$ of an integrand where $\delta \Delta^*(\mathbf{r})$ appears as a general factor, as it does in Eq. (6.14), so that the factor of $\delta \Delta^*(\mathbf{r})$ can be identified with $\delta F_s / \delta \Delta^*(\mathbf{r})$. This difficulty is solved by integrating by parts the term with $\nabla \delta \Delta^*(\mathbf{r})$, so that after this integration the operator ∇ acts on the first factor $[\nabla / i - (2e/\hbar) \mathbf{A}(\mathbf{r})] \Delta(\mathbf{r})$, and no longer on $\delta \Delta^*(\mathbf{r})$. In this transformation the all integrated term corresponds to a surface contribution and is irrelevant for our purpose. One might, for example, assume that $\delta \Delta^*(\mathbf{r})$ is zero at the surface, so that the surface term is zero. In this way, one obtains

$$\delta \left[\int d\mathbf{r} \left| \left(\frac{1}{i} \nabla - \frac{2e}{\hbar} \mathbf{A}(\mathbf{r}) \right) \Delta(\mathbf{r}) \right|^2 \right] = \int d\mathbf{r} \, \delta \Delta^*(\mathbf{r}) \left(\frac{1}{i} \nabla - \frac{2e}{\hbar} \mathbf{A}(\mathbf{r}) \right)^2 \Delta(\mathbf{r}) \qquad (6.16)$$

Finally, the last term on the right-hand side of Eq. (6.12) does not contain $\Delta^*(\mathbf{r})$, so it does not contribute. Identifying to $\delta F_s / \delta \Delta^*(\mathbf{r})$ the overall factor of $\delta \Delta^*(\mathbf{r})$ in the integral giving δF_s, and setting it to zero according to Eq. (6.13), we obtain the first Ginzburg–Landau equation

$$\alpha(T) \, \Delta(\mathbf{r}) + \beta \, |\Delta(\mathbf{r})|^2 \, \Delta(\mathbf{r}) + \gamma \left(\frac{1}{i} \nabla - \frac{2e}{\hbar} \mathbf{A}(\mathbf{r}) \right)^2 \Delta(\mathbf{r}) = 0 \qquad (6.17)$$

which is the appropriate generalization of Eq. (6.3). It would be natural next to carry out the same calculation by varying $\Delta(\mathbf{r})$, instead of $\Delta^*(\mathbf{r})$. However, it can be checked that varying $F_s(T, \{\Delta(\mathbf{r})\}, \{\mathbf{A}(\mathbf{r})\})$ with respect to $\Delta(\mathbf{r})$, at fixed $\Delta^*(\mathbf{r})$ and $\mathbf{A}(\mathbf{r})$, does not provide any new result, one merely obtains the complex conjugate of the preceding equation. Hence we are left with a similar calculation of the variation of F_s with respect to $\mathbf{A}(\mathbf{r})$ at fixed $\Delta(\mathbf{r})$ and $\Delta^*(\mathbf{r})$.

The first two terms of Eq. (6.12) are independent of $\mathbf{A}(\mathbf{r})$, so we are left with the last two terms. The variation of the third term is easily obtained

$$\delta \left[\int d\mathbf{r} \left| \left(\frac{1}{i} \nabla - \frac{2e}{\hbar} \mathbf{A}(\mathbf{r}) \right) \Delta(\mathbf{r}) \right|^2 \right] = \qquad (6.18)$$

$$\int d\mathbf{r} \left(-\frac{2e}{\hbar} \delta \mathbf{A}(\mathbf{r}) \right) \left[\Delta^*(\mathbf{r}) \left(\frac{1}{i} \nabla - \frac{2e}{\hbar} \mathbf{A}(\mathbf{r}) \right) \Delta(\mathbf{r}) + \Delta(\mathbf{r}) \left(-\frac{1}{i} \nabla - \frac{2e}{\hbar} \mathbf{A}(\mathbf{r}) \right) \Delta^*(\mathbf{r}) \right]$$

On the other hand, for the variation of the last term, we have also to perform a by-parts integration, which is more easily done with tensorial notations at the appropriate stage. With $\delta \mathbf{B}(\mathbf{r}) = \text{curl} \, \delta \mathbf{A}(\mathbf{r})$, and omitting for clarity the dependence on $\mathbf{r}$ of the various fields

$$\delta \int dr \, \frac{\mathbf{B}^2}{2} = \int dr \, \mathbf{B} \cdot \delta \mathbf{B} = \int dr \, \mathbf{B} \cdot \text{curl} \, \delta \mathbf{A} = \epsilon_{ijk} \int dr \, B_i \, \frac{\partial \delta A_k}{\partial x_j} \qquad (6.19)$$

$$= -\epsilon_{ijk} \int dr \, \delta A_k \, \frac{\partial B_i}{\partial x_j} = \int dr \, \delta \mathbf{A} \cdot \text{curl} \, \mathbf{B}$$

where x_j are cartesian coordinates, ϵ_{ijk} is the completely antisymmetric unit tensor and summation over repeated indices is understood. Since Maxwell's equation $\text{curl} \, \mathbf{B}(\mathbf{r}) = \mu_0 \mathbf{j}(\mathbf{r})$ relates the magnetic field to the current density, writing $\delta F_s / \delta \mathbf{A}(\mathbf{r}) = 0$, we obtain finally from Eq. (6.18) and Eq. (6.19) the second Ginzburg–Landau equation

$$\mathbf{j}(\mathbf{r}) = \gamma \frac{2e}{\hbar} \left[\frac{1}{i} \left(\Delta^*(\mathbf{r}) \nabla \Delta(\mathbf{r}) - \Delta(\mathbf{r}) \nabla \Delta^*(\mathbf{r}) \right) - \frac{4e}{\hbar} \mathbf{A}(\mathbf{r}) |\Delta(\mathbf{r})|^2 \right] \qquad (6.20)$$

This last equation is remarkably similar to the standard expression of the probability current in quantum mechanics if we make the change of notation $\Delta(\mathbf{r}) \to \psi(\mathbf{r})$, to use the standard notation $\psi(\mathbf{r})$ for the wave function in quantum mechanics. Similarly the first equation is remarkably similar to the one-particle Schrödinger equation, except for the non-linear term in $|\Delta(\mathbf{r})|^2 \, \Delta(\mathbf{r})$. This is actually not so astonishing if one realizes that the expression Eq. (6.12) for the free energy itself (except for the $|\Delta(\mathbf{r})|^4$ term) is quite similar to the average energy in quantum mechanics for a charged particle with wave function $\psi(\mathbf{r})$. Since it is known that the Schrödinger equation can be obtained from a variational calculation by requiring that this energy is minimal for the actual wave function, it is not so surprising that by performing a similar variational calculation, we end up with a Schrödinger-like equation.

These striking similarities are sometimes made even more apparent by changing notations, and setting $e^* = 2e$ and $\gamma = \hbar^2/(2m^*)$, so that the two Ginzburg–Landau equations can be rewritten[3] as

$$\begin{cases} \dfrac{1}{2m^*} \left(\dfrac{\hbar}{i} \nabla - e^* \mathbf{A}(\mathbf{r}) \right)^2 \Delta(\mathbf{r}) + \alpha(T) \, \Delta(\mathbf{r}) + \beta \, |\Delta(\mathbf{r})|^2 \, \Delta(\mathbf{r}) = 0 \\[4mm] \mathbf{j}(\mathbf{r}) = \dfrac{e^*}{m^*} \dfrac{\hbar}{2i} \left(\Delta^*(\mathbf{r}) \nabla \Delta(\mathbf{r}) - \Delta(\mathbf{r}) \nabla \Delta^*(\mathbf{r}) \right) - \dfrac{e^{*2}}{m^*} \mathbf{A}(\mathbf{r}) |\Delta(\mathbf{r})|^2 \end{cases} \qquad (6.21)$$

6.2.2 Simple Cases

As a first simple application of Ginzburg–Landau equations, we note that one recovers the London equation when one takes the case of an order parameter which is space-independent $\Delta(\mathbf{r}) = \Delta_0$. The second Ginzburg–Landau equation gives

$$\mathbf{j}(\mathbf{r}) = -\frac{e^{*2}}{m^*} \mathbf{A}(\mathbf{r}) |\Delta_0|^2 \qquad (6.22)$$

which is indeed identical with Eq. (1.20) if we identify $(e^{*2}/m^*)|\Delta_0|^2$ with $n_s e^2/m$. On the other hand, from the first Ginzburg–Landau equation, assuming the field weak enough

---

[3] It is interesting to note that Ginzburg and Landau, who naturally did not know about pairs but realized that e^* should be a basic charge characterizing electrons, made the "natural" choice $e^* = e$, instead of $e^* = 2e$.

to make the first term negligible,[4] we have $|\Delta_0|^2 = |\alpha(T)|/\beta$, which is just what we had in Eq. (6.4). Since the London penetration depth λ_L is directly related to $n_s e^2/m$ by Eq. (1.27), this allows one to relate the phenomenological parameters α, β, and γ coming in the Ginzburg–Landau theory to the physical quantity $\lambda_L(T)$, which gets from this theory a specific temperature dependence. One has

$$\frac{1}{\lambda_L^2(T)} = \mu_0 \frac{e^{*2}}{m^*} \frac{|\alpha(T)|}{\beta} = \mu_0 \frac{8e^2}{\hbar^2} \frac{|\alpha(T)|\gamma}{\beta} \tag{6.23}$$

In the following discussion, we merely note $\lambda(T)$ this penetration depth occurring in Ginzburg–Landau theory.

We consider next a situation without a magnetic field so that $\mathbf{A}(\mathbf{r}) = \mathbf{0}$. We assume that the space-dependent order parameter $\Delta(x)$ depends only on the x coordinate, so we have a one-dimensional problem. Furthermore, we assume $\Delta(x)$ to be real. Physically, the domain $x > 0$ is supposed to be superconducting, while the $x < 0$ domain is in the normal metallic state, so that $\Delta(x) = 0$ in this region. By continuity we have $\Delta(x = 0) = 0$, which provides a boundary condition in the superconducting domain. With these hypotheses, it comes from the second Ginzburg–Landau equation Eq. (6.21) that there is no current $\mathbf{j}(\mathbf{r}) = \mathbf{0}$. The first Ginzburg–Landau equation reduces to

$$-\frac{\hbar^2}{2m^*} \frac{d^2 \Delta(x)}{dx^2} - |\alpha(T)| \, \Delta(x) + \beta \, \Delta^3(x) = 0 \tag{6.24}$$

It is more convenient to rescale the order parameter to its equilibrium value $\Delta_0^2 = |\alpha(T)|/\beta$ by setting $\Delta(x) = \Delta_0 f(x)$. This leads to the simpler equation

$$-\xi^2(T) \frac{d^2 f(x)}{dx^2} - f(x) + f^3(x) = 0 \tag{6.25}$$

where we have introduced the new characteristic length $\xi(T)$ for the superconductor, given by

$$\xi^2(T) = \frac{\hbar^2}{2m^* |\alpha(T)|} = \frac{\gamma}{|\alpha(T)|} \tag{6.26}$$

This length is called the Ginzburg–Landau coherence length. From Eq. (6.25), it is the typical length over which the amplitude of the superconducting order parameter varies. Qualitatively it is a quantity similar to the Pippard coherence length and the BCS coherence length we have already seen, although quantitatively from its specific definition Eq. (6.26) it is clear that there are differences.

---

[4] This is naturally not systematically the case. A very interesting and beautiful experiment by Little and Parks, where this approximation is not valid, is considered in the "Further Reading" section at the end of this chapter.

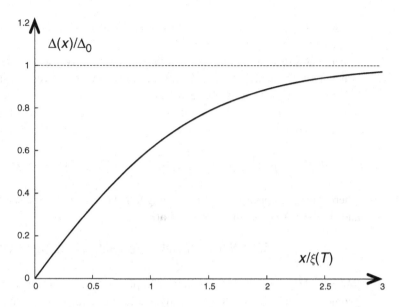

Fig. 6.2 Spatial dependence, given by Eq. (6.27), of the order parameter $\Delta(x)$ at a normal-superconductor contact.

It is easy[5] to integrate explicitly Eq. (6.25) into

$$\frac{\Delta(x)}{\Delta_0} \equiv f(x) = \tanh \frac{x}{\sqrt{2}\,\xi(T)} \tag{6.27}$$

where we have required physically that deep in the superconducting domain $x \to \infty$, the order parameter goes to its equilibrium value Δ_0. The resulting $\Delta(x)$ is displayed in Fig. (6.2).

The two length scales, which we have seen coming in the two above examples, characterize the physical properties of the superconductor, $\lambda(T)$ with respect to the spatial variations of the magnetic field, and $\xi(T)$ with respect to the spatial variations of the order parameter. We note that $\xi(T)$, as well as $\lambda(T)$, are proportional to $|\alpha(T)|^{-1/2}$. Hence they both diverge as $(T_c - T)^{-1/2}$ when T goes to T_c. On the other hand, their ratio $\lambda(T)/\xi(T)$ is independent of temperature. This constant ratio

$$\kappa = \frac{\lambda(T)}{\xi(T)} \tag{6.28}$$

is known as the Ginzburg–Landau parameter. If both λ and ξ change, but with this ratio κ kept fixed, the physical properties of the superconductor do not change, except for an

---

[5] With the variable $y = x/\xi(T)$, Eq. (6.25) reads $-f'' - f + f^3 = 0$. Multiplying this equation by f' yields by integration $-f'^2/2 - f^2/2 + f^4/4 = -1/4$, where the value of the integration constant is obtained by requiring that $f(y) = 1, f'(y) = 0$ satisfies the equation (this corresponds to the physical situation for large y where $\Delta(x) = \Delta_0$). This is rewritten more simply $df/dy = (1 - f^2)/\sqrt{2}$, which gives by integration $y = (1/\sqrt{2})\ln[(1+f)/(1-f)]$, taking into account the boundary condition $f = 0$ for $y = 0$. This is equivalent to Eq. (6.27).

overall change in length scale. Hence κ is the parameter that characterizes the physical properties of the superconductor.

It is interesting to display the crucial role of these two length scales by rewriting the Ginzburg–Landau free energy Eq. (6.12) in terms of reduced quantities. For the order parameter, we write it as above by taking $\Delta_0 = (|\alpha(T)|/\beta)^{1/2}$ as the unit and setting accordingly $\Delta(\mathbf{r}) = \Delta_0 f(\mathbf{r})$. Similarly we take $\hbar/(e^*\xi)$ as the unit for the vector potential $\mathbf{A}(\mathbf{r})$ and write $\mathbf{A}(\mathbf{r}) = \mathbf{a}(\mathbf{r})\,\hbar/(e^*\xi)$. Carrying these changes into Eq. (6.12), we obtain

$$F_s - F_n = \frac{\alpha^2}{\beta} \int d\mathbf{r} \left[-|f(\mathbf{r})|^2 + \frac{1}{2}\,|f(\mathbf{r})|^4 + \left|\left(\frac{\xi}{i}\nabla - \mathbf{a}(\mathbf{r})\right)f(\mathbf{r})\right|^2 + \lambda^2(\text{curl }\mathbf{a}(\mathbf{r}))^2 \right] \quad (6.29)$$

where it clearly appears that ξ is the length that comes in for the spatial variations of $f(\mathbf{r})$, and similarly λ comes in for those of $\mathbf{a}(\mathbf{r})$.

6.2.3 Normal Metal-Superconductor Surface Energy

We consider now a somewhat more complicated situation. Physically, this is the same geometry as the one we have investigated above, with a normal metal in the half-space $x < 0$ and a superconductor in the half-space $x > 0$, but we assume now that there is in addition an applied uniform magnetic field, parallel to the interface, the z-axis being, for example, along this field. Hence qualitatively the situation with respect to the field is analogous to the one represented in Fig. (1.6). With respect to the order parameter, the situation is similar to the one seen above in Fig. (6.2). However, quantitatively the results for $\Delta(x)$ and $B_z(x)$ are not the same as in these figures because the order parameter and the field are coupled in the Ginzburg–Landau equations.

We are specifically interested in the energy cost of the existence of this normal metal-superconductor interface. The evaluation of this energy has to be done by comparison with the situation that would exist if we had a bulk superconductor. In this last case, the order parameter would have its bulk value all over the superconducting region (we would have $f(x) = 1$ for any $x > 0$ in Fig. (6.2)). Similarly, we would have for the field $B_z(x) = 0$ for $x > 0$ in Fig. (1.6). Finding this surface energy quantitatively requires that we solve the Ginzburg–Landau equations for $\Delta(x)$ and $B_z(x)$, which is a fairly complicated matter in which we will not go. Nevertheless we can find a very important feature of the result by a simple qualitative discussion.

Indeed it is clear that the existence of the interface depresses the order parameter, qualitatively as in Fig. (6.2). Hence in the vicinity of the interface it is no longer equal to Δ_0 that makes the free energy minimum, as we have seen in Eq. (6.4). Accordingly, the free energy is raised because of this depression of the order parameter. Since this depression occurs over a depth of the order of the coherence length ξ, we see from Eq. (6.29) that the increase in free energy is of order $\xi(\alpha^2/\beta)$, for a unit surface of the interface.

On the other hand, the penetration of the field $B_z(x)$ in the superconductor, qualitatively as in Fig. (1.6), lowers the free energy. Indeed we have seen that there is an energy cost for the superconductor in expelling the magnetic field, which is compensated by the gain in energy due to the phase transition from the normal to the superconducting state. Technically this occurs because, in the presence of an external applied magnetic field $\mathbf{H}$, in order to find the equilibrium situation, we have to minimize $G = F - \mathbf{H} \cdot \int d\mathbf{r}\,\mathbf{B}(\mathbf{r})$, rather than the free energy F, as we have done in Section 1.1.5. (It is easy to check that this last term does not affect our derivation of the Ginzburg–Landau equations.) In the normal state where $F = \int d\mathbf{r}\, \mathbf{B}^2(\mathbf{r})/(2\mu_0)$, minimizing G leads to $\mathbf{B}(\mathbf{r}) = \mu_0 \mathbf{H}$, as it should, and $G = -\mu_0 \mathbf{H}^2/2$ per unit volume. On the other hand, we have $\mathbf{B}(\mathbf{r}) = 0$ if the magnetic induction is expelled, and G is indeed raised to $G = 0$. Since the penetration of the field occurs on a typical length λ, from Eq. (6.29) the energy lowering due to its penetration is of order $-\lambda(\alpha^2/\beta)$ per unit surface.

The net result is that if $\xi \gg \lambda$, that is, if $\kappa \ll 1$, the energy increase due to the depression of the order parameter wins, and the presence of the interface leads to an energy increase, a fairly conventional situation physically. This corresponds to the situation found in type I superconductors. On the other hand, if $\lambda \gg \xi$, that is, if $\kappa \gg 1$, this is the energy decrease due to the field penetration which wins, and the existence of the interface produces a decrease in energy. This is physically a surprising situation, and naturally as a result the superconductor will tend to multiply normal-superconducting interfaces as much as possible to decrease the total energy. This is the situation found in type II superconductors. A precise calculation of $\Delta(x)$ and $B_z(x)$ and of the surface energy is required to know when this energy is zero, going in this way from positive to negative values. It is found[6] that this occurs for $\kappa = 1/\sqrt{2}$. Hence we have specifically for type I and II superconductors

$$
\begin{cases}
\kappa < \dfrac{1}{\sqrt{2}} & \text{type \ I \ superconductor} \\[2mm]
\kappa > \dfrac{1}{\sqrt{2}} & \text{type \ II \ superconductor}
\end{cases}
\tag{6.30}
$$

In order to understand more specifically what happens, it is useful to investigate the upper critical field within Ginzburg–Landau theory, which we now consider.

6.2.4 Upper Critical Field

The condition for the appearance of the superconducting phase from the normal phase is quite easy to obtain since the order parameter goes to zero when this second-order transition occurs. In this case, in the free energy, the fourth-order term with coefficient β becomes negligible compared to the second-order one. Accordingly, we can write, for example, from Eq. (6.1) $F_s - F_n = \alpha(T)|\Delta|^2$, and the superconducting phase appears for $F_s = F_n$, that is, $\alpha(T) = 0$, as we have already seen. We can wonder what happens to this transition in the presence of a magnetic field. To answer this question, we have to take the first Ginzburg–Landau equation Eq. (6.21) and neglect again the $\beta|\Delta(\mathbf{r})|^2\,\Delta(\mathbf{r})$ term, since it is negligible compared to the linear term $\alpha(T)\,\Delta(\mathbf{r})$ when $\Delta(\mathbf{r})$ goes to zero.

It is obvious that the resulting equation has a solution since, as we have noticed, in the absence of the cubic term, it reduces to a Schrödinger equation. Hence we have to deal with the Schrödinger equation of a particle in the presence of a magnetic field, and the solutions correspond physically to the well-known Landau levels. If we take the uniform field $\mathbf{B}$ along the z-axis, and choose the corresponding gauge $A_y = xB$, $A_x = A_z = 0$, this first Ginzburg–Landau equation becomes explicitly

---

[6] This value already appears in the original paper [38] of Ginzburg and Landau.

$$-\frac{\hbar^2}{2m^*}\left(\frac{\partial^2}{\partial x^2}+\frac{\partial^2}{\partial z^2}\right)\Delta(\mathbf{r})+\frac{1}{2m^*}\left(\frac{\hbar}{i}\frac{\partial}{\partial y}-e^*Bx\right)^2\Delta(\mathbf{r})=|\alpha(T)|\,\Delta(\mathbf{r}) \qquad (6.31)$$

This is indeed a Schrödinger equation for a particle with mass m^* and charge e^*, if we identify $|\alpha(T)|$ with the energy E of the eigenstates. Translational invariance in the y and z directions allows us to take solutions with a plane wave dependence

$$\Delta(\mathbf{r})=g(x)\,e^{ik_y y}e^{ik_z z} \qquad (6.32)$$

and Eq. (6.31) becomes

$$-\frac{\hbar^2}{2m^*}\frac{d^2}{dx^2}g(x)+\frac{1}{2m^*}\left(\hbar k_y-e^*Bx\right)^2 g(x)=\left(E-\frac{\hbar^2 k_z^2}{2m^*}\right)g(x) \qquad (6.33)$$

This is the Schrödinger equation of a harmonic oscillator with mass m^* and frequency $\omega^*=e^*B/m^*$, centered on $x_0=\hbar k_y/(e^*B)$ (we assume for simplicity $e^*>0$, which does not change the final electric currents). The corresponding possible values for the energy of this harmonic oscillator are $\hbar(e^*B/m^*)(n+1/2)$. The characteristic length scale is $\ell=\sqrt{\hbar/m^*\omega^*}$ and $g(x)$ is proportional to $\exp\left[-(x-x_0)^2/2\ell^2\right]$.

We have now to replace E in Eq. (6.33) by $|\alpha(T)|$, which is fixed for a given temperature T. For each value of the harmonic oscillator quantum number n, and each value of k_z, we obtain a field value for which the Ginzburg–Landau equation has a solution, namely

$$B=\frac{m^*}{\hbar e^*}\frac{|\alpha(T)|-\frac{\hbar^2 k_z^2}{2m^*}}{n+\frac{1}{2}} \qquad (6.34)$$

Starting from a very high field, for which we know that we have the normal state (see Fig. (1.4)), and lowering progressively the field, we are interested in finding the first field for which the instability toward the superconducting state occurs. This is the highest possible field given by Eq. (6.34). It is obtained for $k_z=0$ and $n=0$. The other solutions corresponding to lower fields are physically irrelevant since the metal is already in the superconducting state when these fields are reached. The field we find is the upper critical field $B_{c2}=\mu_0 H_{c2}$ represented in Fig. (1.4), and it is given by

$$B_{c2}=\mu_0 H_{c2}=\frac{2m^*|\alpha(T)|}{\hbar e^*}=\frac{\Phi_0}{2\pi\xi^2(T)} \qquad (6.35)$$

where $\Phi_0=h/2e$ is the flux quantum. For this value of B, the harmonic oscillator length is merely $\ell=\sqrt{\hbar/e^*B_{c2}}=\sqrt{\hbar^2/(2m^*|\alpha(T)|)}=\xi$.

It is interesting to compare this field to the thermodynamic critical field $H_c(T)$. Since from Eq. (1.12), it is related to the free energy difference between the normal and the superconducting phase, and we have found in Eq. (6.5) this difference to be $\alpha^2(T)/2\beta$ in Ginzburg–Landau theory, we have $\mu_0 H_c^2(T)=\alpha^2(T)/\beta$. Making use of $\kappa^2=2(m^*/\hbar e^*)^2\beta/\mu_0$ from Eq. (6.23), Eq. (6.26), and Eq. (6.28), we find the simple result

$$\frac{H_{c2}}{H_c}=\sqrt{2}\,\kappa \qquad (6.36)$$

This result makes it clear that $\kappa=1/\sqrt{2}$ is the boundary between type I and type II superconductivity. Indeed, if $\kappa>1/\sqrt{2}$, we have $H_{c2}>H_c$ and starting from high fields, we meet the second-order superconducting instability Eq. (6.35) before entering the conventional superconducting phase, as represented in Fig. (1.4) for type II superconductivity. In the opposite case, $\kappa < 1/\sqrt{2}$, we have $H_{c2} < H_c$ and the first-order phase transition to superconducting order at $H_c(T)$ is met before the H_{c2} instability may arise (H_{c2} may still appear as a supercooling field below which the normal state is absolutely unstable). This corresponds to type I superconductivity, as shown in Fig. (1.3).

6.3 Vortices

In the preceding section, we saw that the superconducting phase appears at the upper critical field H_{c2}, but we have not yet clarified the precise nature of the superconducting order parameter. Indeed we have found solutions with the form given by Eq. (6.32), but they are inhomogeneous, and since the parameter k_y can take any value, we have a high degeneracy so that any linear combination of these solutions is also a possible solution. On the other hand, we face a physical situation with translational invariance in the $x-y$ plane. A uniform order parameter is not a solution of the Ginzburg–Landau equations as we have seen, but it is reasonable to expect at least a solution with a periodic structure for this translationally invariant problem.

It is possible to build such a solution by taking a linear superposition of the above solutions, with the centers of the harmonic oscillator wave function regularly spaced on the x-axis, that is, by taking $x_0 = na$ with the integer n going from $-\infty$ to ∞, and a being the spacing between consecutive centers. This corresponds to taking $k_y = nb$ with $b = a\,e^*B/\hbar$. The simplest linear combination with equal coefficients is

$$\Delta(\mathbf{r}) = \sum_{n=-\infty}^{n=\infty} e^{inby} \exp\left[-\frac{(x-na)^2}{2\xi^2}\right] \tag{6.37}$$

Since e^{inby} has a period $2\pi/(nb)$ (or any multiple of it) along the y-axis, we see that Eq. (6.37) is periodic along the y direction with period $2\pi/b$, which is common to all the e^{inby}. Hence Eq. (6.37) has the symmetry of a rectangular periodic lattice. Since the x and y-axis are equivalent, we expect for symmetry reasons that the periods are the same for the two axes, which means that we have actually a square lattice. This implies $a = 2\pi/b$, which gives the value of the lattice period $ab = 2\pi = a^2 e^*B/\hbar$. As a result, the magnetic field flux through the unit square is $Ba^2 = 2\pi\,\hbar/e^* = h/2e = \Phi_0$, that is the flux quantum. From Eq. (6.35), this gives $a = \sqrt{2\pi}\xi$.

With these simplifications, Eq. (6.37) can be rewritten as[7]

$$\Delta(\mathbf{r}) = \sum_{n=-\infty}^{n=\infty} e^{2i\pi n\frac{y}{a}} \exp\left[-\pi\left(\frac{x}{a} - n\right)^2\right] \tag{6.38}$$

---

[7] The function appearing on the right-hand side is directly related to the mathematical special function ϑ_3, itself related to elliptic functions.

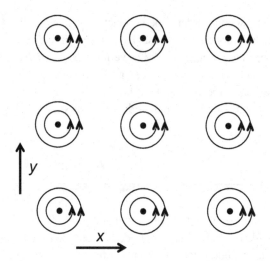

Fig. 6.3 Schematic representation of the Abrikosov square lattice. The lines are the electric current lines. Near the centers of the vortices they have the shape of circles, but farther away (not shown) they are progressively deformed.

From this expression, one can easily see a remarkable property. If we take $x/a = y/a = 1/2$, we have $\exp(2i\pi ny/a) = (-1)^n$ and the series $\sum_n \exp[-\pi(n-1/2)^2]$ for odd n cancels exactly the same series for even n in Eq. (6.38). Hence $\Delta(\mathbf{r}) = 0$ at this point, and the same is naturally true for all equivalent points in the square lattice $x_m/a = m + 1/2$ and $y_p/a = p + 1/2$ (with m and p being integers). This means that there are (point-like) normal regions appearing in the superconductor, and their positions form a square lattice.

In the vicinity of these points, we can expand the function $\Delta(\mathbf{r})$. The lowest-order terms are those proportional to $X = (x - x_m)/a$ and $Y = (y - y_p)/a$. The derivatives of $\Delta(\mathbf{r})$ are obtained from Eq. (6.38). The X derivative is $\sum(-1)^n 2\pi n \exp[-\pi(n-1/2)^2]$, and the Y derivative $\sum(-1)^n 2i\pi n \exp[-\pi(n-1/2)^2]$, so one sees that $\Delta(\mathbf{r})$ is proportional to $X + iY$. This dependence satisfies the physical equivalence of the x and y axes, and it results also from the expected rotational invariance in the close vicinity of the zeros of $\Delta(\mathbf{r})$, which implies that $\Delta(\mathbf{r})$ does not change under rotation up to a phase factor. We can indeed write $X + iY = \rho e^{i\phi}$, with $\rho = \sqrt{X^2 + Y^2}$ and $\phi = \arctan(Y/X)$.

In the expression for the corresponding currents resulting from the second Ginzburg–Landau equation Eq. (6.21), the last term is proportional to $|\Delta(\mathbf{r})|^2 \propto \rho^2$ and accordingly, close to the zeros of $\Delta(\mathbf{r})$, it is negligible compared to the first term. As a result, the electric currents are running counterclockwise along circles centered on the zeros of $\Delta(\mathbf{r})$. Hence in the vicinity of these zeros, the current distribution has the shape of a whirlpool, a vortex. Current lines are circles only when they are close to the zeros; further away, their shape is deformed so that the whole current structure appears as a packed array of vortices, called the Abrikosov[8] vortex lattice [40]. This square lattice is displayed schematically in Fig. (6.3). An important property of these vortices is that just as we have seen for the unit cell, the total magnetic flux through the whole surface of a single vortex is equal to the flux quantum Φ_0, so these vortices are often called quantized vortices.

[8] In 2003, A. A. Abrikosov received the Nobel Prize for his theoretical prediction of this vortex structure. The prize was shared with V. L. Ginzburg and A. J. Leggett.

This vortex structure can also be understood qualitatively from the negative surface energy we have found above for a normal-superconducting interface for type II superconductors. As we have mentioned, it is energetically favorable as a result to have a large number of normal-superconductor interfaces. On the other hand, it is not energetically favorable to have large normal regions; rather, it is better to have the normal region as small as possible. One way to obtain this result is, starting from the planar geometry shown in Fig. (1.6) and Fig. (6.2), to wrap the superconducting region around the normal one and to reach in this way a cylindrical geometry with a central cylinder containing the normal domain, surrounded by a superconducting domain. However, we have seen in Chapter 5 that the total magnetic flux through this whole structure is a multiple of the flux quantum ϕ_0. In order to have as much normal regions as possible, instead of having a single structure carrying a flux $N\Phi_0$, it seems better to have N structures each one carrying the minimum flux Φ_0. Packing these cylindrical regions together, and making the normal regions point-like to have them as small as possible, we arrive at a structure that is similar to the vortex lattice we have just considered.

On the other hand, the precise arrangement of the vortices is not a simple problem. The proper way to handle systematically all this matter is, after having recognized that the linearized Ginzburg–Landau equation has a large degeneracy for its solutions, to evaluate for any linear combination of these solutions the Ginzburg–Landau free energy Eq. (6.12), including the fourth-order term in $\Delta(\mathbf{r})$. It is this fourth-order term that lifts the degeneracy. More precisely, Abrikosov showed [40] that the lowest free energy is obtained when the ratio

$$\beta_A \equiv \frac{\overline{|\Delta(\mathbf{r})|^4}}{\left(\overline{|\Delta(\mathbf{r})|^2}\right)^2} \tag{6.39}$$

is minimal. Here the bar over the symbols is for the spatial average of the involved quantity $\overline{f(\mathbf{r})} \equiv \int d\mathbf{r} f(\mathbf{r})$. Abrikosov initial numerical calculations gave the square lattice, corresponding to our simple choice Eq. (6.37), as the lattice with the lowest free energy. But later more precise evaluations showed that it is actually the triangular lattice that has the lowest energy. The corresponding $\Delta(\mathbf{r})$ has essentially the same expression as Eq. (6.37), but with alternating coefficients instead of having them all equal to 1. However, the difference between these two lattices is very small since $\beta_A = 1.16$ for this triangular lattice while $\beta_A = 1.18$ for the square lattice. This shows that this vortex lattice is quite soft.

The existence of the Abrikosov lattice has been confirmed experimentally in the sixties, first by neutron diffraction, and then by visually quite attractive "decoration" experiments, where ferromagnetic particles are deposited on the plane surface of a superconductor perpendicular to the applied magnetic field and display by their arrangement the inhomogeneity due to the lattice. More recently scanning tunneling microscopy has allowed a quite direct observation of vortices and of the vortex lattice. These experiments have confirmed that the stablest lattice is triangular. But this lattice is indeed easily deformed. Often, it displays defects due to superconducting material imperfection, or the vortices may even show a fully disordered arrangement.

Since each vortex carries a flux Φ_0, for a field B the number N of vortices for a surface S of the superconductor is merely given by $BS = N\Phi_0$. At B_{c2}, we have from Eq. (6.35)

$N = S/(2\pi\xi^2(T))$. On the other hand, we have seen that starting from the normal part, which is right at the center of the vortex, the order parameter goes from zero to its bulk value over a length of order $\xi(T)$. The corresponding region is called the core of the vortex. Its surface is typically $\pi\xi^2(T)$. Hence the relation $2\pi\xi^2(T)N = S$ at B_{c2} means that the cores of the vortices are touching each other. This gives a simple physical understanding of the result Eq. (6.35) for B_{c2}. At this field the number of vortices is so high that they are at their maximum packing, with their cores being in contact so that there is no way to put additional vortices. In this way, one can understand that the field is at its maximum value and one has reached the upper critical field.

When the applied field H is lowered from H_{c2}, the induction B also decreases so the number of vortices goes down and they are more separated. When the field is low enough, vortices are well separated so that we can consider a single isolated vortex. It is made of its core, of typical size ξ, surrounded by a region of typical size λ where the order parameter has its bulk value and in which screening currents are circulating. This qualitative description is obtained by continuity from the situation we have found in the vicinity of H_{c2} where we could use quantitatively Ginzburg–Landau theory. However, far below H_{c2} the order parameter is not small and this Ginzburg–Landau theory is no longer justified, so a quantitative description of the isolated vortex is not simple.

However, in the limit of extreme type II superconductors, where $\kappa \gg 1$ so $\lambda \gg \xi$, we can have an essentially quantitative description of the vortex because the core region is very small, so that the vortex structure and its energy is dominated by the region where the order parameter has its bulk value and screening currents are present. These currents as well as the corresponding induction are obtained from a calculation analogous to the one we have performed in Chapter 1 when we have seen that London equation leads to screening currents which are responsible for the Meissner effect. The only difference is that we have now to deal with a cylindrical geometry, rather than a planar one. This is the physical situation we have already mentioned above when we have considered wrapping the superconducting region around the normal one.

Hence, taking the field and the vortex axis along the z-axis, the field $B_z(r)$ depends only on the distance r to this z-axis. In this case, Eq. (1.25) for the field becomes $\lambda^2\Delta B_z(\mathbf{r}) = B_z(\mathbf{r})$ when we take into account the definition Eq. (1.27) of the penetration depth. Making use of the expression of the Laplacian operator in cylindrical coordinates, we have $\Delta B_z(r) = (1/r)d(rB_z'(r))/dr$, where $B_z'(r) = dB_z(r)/dr$. Hence Eq. (1.25) becomes

$$\frac{d^2}{dr^2}B_z(r) + \frac{1}{r}\frac{d}{dr}B_z(r) - \frac{1}{\lambda^2}B_z(r) = 0 \tag{6.40}$$

This differential equation has for solutions Bessel functions, which very often play the same role as simple exponentials when one deals with problems in cylindrical coordinates instead of cartesian ones. More specifically two independent solutions of Eq. (6.40) are $I_0(r/\lambda)$ and $K_0(r/\lambda)$, and $K_0(r/\lambda)$ has to be chosen because it decreases exponentially[9]

---

[9] The properties of $K_0(x)$ can be quite conveniently obtained from its integral representation $K_0(x) = \int_0^\infty dt \exp(-x\cosh t)$. One can check, by making a by-parts integration, that this expression is indeed solution of $d^2K_0(x)/dx^2 + (1/x)dK(x)/dx - K(x) = 0$, corresponding to Eq. (6.40).

for large r, in the same way as we had to choose the exponentially decreasing solution Eq. (1.26). The precise solution for $B_z(r)$ is

$$B_z(r) = \frac{\Phi_0}{2\pi\lambda^2} K_0\left(\frac{r}{\lambda}\right) \tag{6.41}$$

The prefactor is obtained by requiring that the whole vortex carries a total flux Φ_0, that is $\int_0^\infty dr\,(2\pi r)B_z(r) = \Phi_0$, which is verified because $K_0(x)$ satisfies $\int_0^\infty dx\, xK_0(x) = 1$. Specifically for large r, one has $B_z(r) \simeq (\Phi_0/(2\pi\lambda^2))\sqrt{\pi\lambda/(2r)}\exp(-r/\lambda)$, which has the expected exponentially decreasing behavior. For $r \to 0$, this solution behaves as $B_z(r) \simeq (\Phi_0/(2\pi\lambda^2))\ln(2e^{-C}\lambda/r)$. The apparent singularity in $\ln(1/r)$ does not exist actually, since for $r \sim \xi$ one enters the vortex core in which our equation is no longer valid. Hence the field saturates to its $r \sim \xi$ value.

Even if we do not have a full description of the vortex, we can make in this regime a phenomenological calculation of the vortex energy. Indeed, compared to the situation without a vortex, the presence of the vortex brings no loss in condensation energy since the order parameter has its equilibrium value everywhere except in the core, which brings a negligible correction since its size, of order ξ, is very small compared to the vortex size, which is of order λ. Hence we are left with the magnetic contribution to the energy, given by the last term in Eq. (6.12). In addition, there is a contribution from the kinetic energy of the supercurrents forming the vortex. Within London theory we can write this kinetic energy density as $(1/2)n_s m v_s^2$. Taking into account $\mathbf{j}_s = n_s e \mathbf{v}_s$ and $\lambda_L^{-2} = n_s e^2 \mu_0/m$, this can be rewritten as $n_s m \mathbf{j}_s^2/[2n_s^2 e^2] = \mu_0\lambda_L^2 \mathbf{j}_s^2/2$. Note that this kinetic energy term corresponds to the gradient term in the free energy in Ginzburg–Landau theory Eq. (6.12), and we could show that it can be rewritten as we have just done by making use of the expression of the second Ginzburg–Landau equation Eq. (6.20) for $\mathbf{j}_s$ and the expression Eq. (6.23) for λ_L^2.

Taking into account Maxwell's equation curl $\mathbf{B} = \mu_0\mathbf{j}_s$ we can finally write the vortex energy as

$$E_V = \frac{1}{2\mu_0}\int d\mathbf{r}\left[\mathbf{B}^2(\mathbf{r}) + \lambda^2(\text{curl } \mathbf{B}(\mathbf{r}))^2\right] \tag{6.42}$$

Naturally, with our vortex being translationally invariant in the z direction, the vortex energy is proportional to its length in this direction. Taking a unit length in this z direction, the vortex energy becomes with our cylindrical geometry

$$E_V = \frac{1}{2\mu_0}\int_0^\infty dr\,(2\pi r)\left[B_z^2(r) + \lambda^2\left(\frac{dB_z(r)}{dr}\right)^2\right] \tag{6.43}$$

$$= \frac{\Phi_0^2}{4\pi\mu_0\lambda^2}\int_0^\infty dx\, x\left[K_0^2(x) + \left(\frac{dK_0(x)}{dx}\right)^2\right]$$

by making use in the last step of $B_z(r)$ given by Eq. (6.41), and making the change of variable $r = \lambda x$. The first term gives a finite contribution even if we let the core size ξ go to zero. However, this is not the case for the second term, which diverges in this case. Physically, this is due to the fact that the azimuthal currents have a $1/r$ behavior for small r, so that their contribution to the kinetic energy diverges if we do not take into account the existence of the core. From the small x behavior of $K_0(x)$, we have $dK_0/dx \simeq -1/x$

and the integral behaves as $\int_0 dx/x$, which diverges logarithmically. One has to replace in the r integral the lower boundary 0 by ξ, corresponding to ξ/λ in the x integration. Hence the dominant contribution is $\ln(\lambda/\xi)$. Actually, the integral in Eq. (6.43) can be performed exactly because $d(xKK')/dx = KK' + xK'^2 + xKK'' = xK'^2 + xK^2$ from the differential equation Eq. (6.40) (K' and K'' are the first and second derivatives of $K(x)$). Since $xK'(x) \simeq -1$ for small x, we find

$$E_V = \frac{\Phi_0^2}{4\pi\mu_0\lambda^2}K_0(\xi/\lambda) \simeq \frac{\Phi_0^2}{4\pi\mu_0\lambda^2}\ln\frac{\lambda}{\xi} \tag{6.44}$$

where we have only written the logarithmically dominant term, because the energy arising from the vortex core, which we have not evaluated, turns out to be comparable to the term we have omitted.

This result allows one to obtain the lower critical field H_{c1}. Physically, this is where, upon increasing the field from very low values, it is energetically favorable to let a first vortex enter the bulk of the superconductor. As we have seen the appropriate thermodynamic potential for this situation is $G = F - \mathbf{H} \cdot \int d\mathbf{r}\, \mathbf{B}(\mathbf{r})$, which we need to evaluate for the vortex. In the presence of the vortex, F is equal to the vortex energy E_V we have just obtained, and $\int d\mathbf{r}\, \mathbf{B}(\mathbf{r})$ is just the flux of the magnetic induction (for a unit length vortex), which is equal to Φ_0. Hence $G_V = E_V - H\Phi_0$. On the other hand, in the absence of vortex, F as well as the magnetic flux is zero, so that $G = 0$. Accordingly, the presence of the vortex is energetically favorable when $G_V \leq 0$, so that its first appearance is for the field

$$B_{c1} = \mu_0 H_{c1} = \mu_0\frac{E_V}{\Phi_0} = \frac{\Phi_0}{4\pi\lambda^2}\ln\frac{\lambda}{\xi} \tag{6.45}$$

It is worthwhile to compare this field to the upper critical field. From Eq. (6.35) we find

$$\frac{H_{c2}}{H_{c1}} = 2\frac{\lambda^2}{\xi^2}\frac{1}{\ln(\lambda/\xi)} = \frac{2\kappa^2}{\ln\kappa} \tag{6.46}$$

We see that in this limit of large κ superconductor, the mixed phase that corresponds to the whole domain between H_{c1} and H_{c2} occupies most of the $T - H$ phase diagram shown in Fig. (1.4). Let us finally mention that when the field is raised above H_{c1}, the growth in the number of vortices is limited by the repulsive interactions that can be shown to exist between vortices.

6.4 Application of Superconductivity to High Currents and Fields

We can now discuss more precisely how superconductors may be used to achieve very high currents and very high magnetic fields. The essential point is that vortices play a crucial role in all these applications. Indeed, as we have already mentioned in Chapter 1, type II superconductors are the ones in which superconductivity will exist in very high fields. This will occur in the mixed state, by going near the upper critical field H_{c2}. However, this means that many vortices are present in these conditions.

A simple typical situation is the one with the geometry represented in Fig. (6.3). Let us call z the direction perpendicular to the plane of the figure, corresponding to the direction of the magnetic field. We consider the situation where, in addition to the current distribution represented in the figure, we let an additional current run along the y-axis, going upward in this figure, with the x-axis going to the right-hand side of this figure. Naturally, since the starting situation in the absence of this current is inhomogeneous, the resulting modification of the current distribution in the presence of this additional current will also be inhomogeneous. Nevertheless, without entering into these details, let us call $\mathbf{J}$ the current density averaged over the x direction (naturally in the absence of our additional current this average is zero). $\mathbf{J}$ is along the y direction. Qualitatively, due to the presence of the magnetic field $\mathbf{B}$ along the z direction, the current structure will feel a Lorentz force, typically given by $\mathbf{F} = \mathbf{J} \times \mathbf{B}$. This force is oriented along the x direction. As a result of this force, the whole vortex structure represented in Fig. (6.3) will move in the x direction.

Hence, in contrast with what one might expect, letting an additional current run along the y-axis does not lead to a slightly modified equilibrium situation, with a net dissipationless current. It rather gives rise to a time-dependent situation. In particular, each vortex carrying a flux quantum Φ_0, the motion of the vortex lattice in the x direction implies that there is a flux flow in this direction. Hence, if we think of our additional current as going along a kind of wire along the y direction, this flux flow is equivalent on the average to an increase of the magnetic flux Φ on one side of the wire with respect to the other side. From Lenz law, this leads to an applied voltage $\mathcal{E} = -d\Phi/dt$, linked to the time-dependent vector potential $\mathbf{A}(t)$. The resulting average electric field $\mathbf{E}$ leads to a power generation $\mathbf{J} \cdot \mathbf{E}$ available for dissipation. Although the details of this dissipation process are quite complicated, the net result is that despite being in the superconducting phase, we cannot have a dissipationless permanent current in this simple situation.

The solution which has been found to this paradoxical situation is merely to lock the vortex lattice in some specific position, to avoid any possible motion. Indeed the above drift of the vortex lattice is basically due to the translational invariance of the simple situation we have considered. Since all the positions obtained by translation are energetically equivalent, the lattice is free to move. On the other hand, this equivalence will be broken if we create some inhomogeneity in the superconductor. More precisely since the core of a vortex is essentially in the normal state, it is energetically less costly if the vortex goes through some normal region since it saves condensation energy. More generally in the presence of any kind of inhomogeneity or defect, the vortex line will choose the path which is energetically more favorable. In particular, this will no longer correspond to a straight line for the vortex core, as was the case in Fig. (6.3), but may be a somewhat tortuous path. The important point is that there is a specific geometry corresponding to the energy minimum, so the vortex line will resist going away from this minimum and it will no longer move due to the Lorenz force, provided naturally that this force is not too strong. In this way, the vortex has been pinned in some position, and the same is naturally true for any vortex and so for the whole vortex lattice (which will no longer have a perfect geometry as in Fig. (6.3), but will rather adjust as a function of the defects' positions). Hence vortex pinning allows one to lock the vortex lattice and consequently to have a dissipationless current.

Starting from this principle, it is nevertheless quite a complicated matter to produce in a specific compound the optimal practical pinning structures. On the one hand, one has to provide a strong enough pinning to result in high critical currents before dissipation sets in. On the other hand, the whole process has to be reliable and practical enough to lead to an industrial production satisfying all the various reasonable constraints of this last stage for useful applications. As a result, a superconducting wire is not at all a simple object, but in many cases it is made of twisted fine superconducting filaments. Moreover, this whole structure is embedded in copper, which inter alia provides some buffer in case dissipation sets in and the heated superconductor goes normal, leading to a catastrophe due to the high energy stored when very high currents are in play.

In practice, progress in the performances of superconductors able to support very high currents and produce very high fields has been such that they are of current use in several high technology applications. Perhaps the closest to the wide public life is the production of the high magnetic fields required for magnetic resonance imaging (MRI) used for medical purposes. These MRI machines are by now widely used. It would not be possible to achieve the required performances with standard electromagnets. These superconducting magnets are mostly made of NbTi, which requires liquid helium cryogenics. The field can reach 20T. These magnets provide the very high homogeneity and stability necessary for this technique.

Naturally, further away from standard uses, superconducting magnets are also used worldwide in many research laboratories. The most well-known example is at the European Center for Nuclear Research (CERN) in Geneva, in the Large Hadron Collider (LHC), presently the largest high-energy particle accelerator working in the world. In its circular tunnel of 27 km perimeter length, ultra-relativistic proton beams collide head-on with a total collision energy of typically 10 TeV. To bend the proton trajectories into this circular path, and to fine-tune their details, a total of around 10,000 various superconducting magnets are necessary. The largest of them are 15 m long and weigh 27 tonnes, providing a field of approximately 8 T. The superconducting compound is again mostly NbTi, and the total length of the superconducting cable is around 7000 km. Cooling all these magnets with superfluid ^{4}He at the operating temperature of 1.9 K requires a huge cryogenic system, using approximately 100 tonnes of ^{4}He.

An even more fascinating use of superconducting magnets is about to take place in the future nuclear fusion experimental reactor ITER (International Thermonuclear Experimental Reactor). In this Tokamak under construction, a plasma will be heated to an extremely high temperature, around 150 million degrees. Such a plasma should naturally not touch the walls of the reactor, and it will be confined by large magnetic fields that will guide the trajectories of its charged ingredients. The central solenoid coils and the toroidal coils will be made of Nb_3Sn, and produce a field going up to 13 T, while other lower field magnets will use NbTi. The total length of the superconducting filaments is supposed to reach 100,000 km. This machine will provide a most fascinating proximity between the extremely high temperature of the plasma and the very low temperature at 4K above absolute zero of the superconducting magnets, a few meters away.

Finally, a more concrete application of the ability of superconductors to carry huge currents has been made recently in electric power transmission. This is to answer the need to provide large cities with the important electric power they use. While one uses big power lines going from pylon to pylon in the countryside to carry the required power from the production centers to the vicinity of the city, this is a very inconvenient solution in the last few kilometers for lack of available space. A buried superconducting line is obviously a much better solution. However, the need for helium cryogenics has made it in practice too inconvenient to use this possibility. The advent of high T_c cuprates superconductors has completely modified the situation, since the required nitrogen cryogenics is by far less demanding than helium cryogenics. Such a 600 m long superconducting supply line has been present since 2008 in Long Island, in New York state. The high T_c superconductor used is BSCCO, a bismuth compound. Another more recent superconducting line has connected two substations in Essen City Centre (Germany) since 2014. It is 1 km long. Other projects of this kind are expected to be realized in the near future, and it is clear that these superconducting supply lines should multiply fairly rapidly. It should be noted that such a large-scale use of high T_c superconductors has not occurred earlier to a large extent because these compounds are basically ceramics, so they are brittle and much more difficult to handle than standard conductors such as copper.

6.5 Validity of the Theory, the Ginzburg Criterium

As noted at the beginning of this chapter, Ginzburg–Landau theory is not exact. To see this in a more specific way let us start with Eq. (6.9), which we rewrite here for convenience, omitting for simplicity the unimportant $F_n(T)$ term, which is a kind of reference energy for the free energy. Hence

$$F_s(T, \{\Delta(\mathbf{r})\}) = \int d\mathbf{r} \left(\alpha(T) \, |\Delta(\mathbf{r})|^2 + \frac{\beta}{2} \, |\Delta(\mathbf{r})|^4 + \gamma \, |\nabla \Delta(\mathbf{r})|^2 \right) \qquad (6.47)$$

This expansion (that is, the integrand on the right-hand side of Eq. (6.47)), assumed by Ginzburg and Landau for the local microscopic contribution to the free energy from the order parameter $\Delta(\mathbf{r})$, is certainly quite reasonable. Let us take the simple case where no magnetic field is present, and the boundary conditions are such that the free energy is minimum for $\Delta(\mathbf{r}) = \Delta_0 = (|\alpha|/\beta)^{1/2}$, independent of $\mathbf{r}$, as we have found in Eq. (6.4).

The problem arises because one takes then systematically this value Δ_0 for $\Delta(\mathbf{r})$ whereas, from statistical physics, other configurations for $\Delta(\mathbf{r})$ are allowed, but their probability is merely proportional to $\exp(-F_s/k_B T)$. Naturally, $\Delta(\mathbf{r}) = \Delta_0$, which minimizes F_s, maximizes this probability. Nevertheless this does not mean that other configurations do not contribute significantly. Hence one has to worry about these departures $\delta\Delta(\mathbf{r}) = \Delta(\mathbf{r}) - \Delta_0$ of the order parameter from the value corresponding to the minimum of F_s, that is, the fluctuations of the order parameter around Δ_0.

If the fluctuations $\delta\Delta(\mathbf{r})$ stay small enough, which is the regime we are only interested in, $\Delta(\mathbf{r})$ will fluctuate symmetrically around Δ_0, which will be accordingly the average of $\Delta(\mathbf{r})$. Hence in Ginzburg–Landau theory, one replaces $\Delta(\mathbf{r})$ by its average. This is the procedure used systematically in mean-field theories, where this approximate replacement of a field by its average allows one to eliminate the difficult problem of handling the numerous fluctuating variables and their correlations. Hence Ginzburg–Landau theory is a mean-field theory.

Here, we merely want to find the range of validity of Ginzburg–Landau theory. For this purpose, we will evaluate the importance of these small fluctuations, in some volume $\mathbb{V}$ to be determined later on, and see under which conditions they are negligible. We cannot merely take, as a measure of the importance of fluctuations, the thermodynamic average of the sum of all fluctuations $\langle \int_{\mathbb{V}} d\mathbf{r}\, \delta\Delta(\mathbf{r}) \rangle$ over this volume. Indeed this quantity is equal to zero since, as we have just seen, Δ_0 is the average of $\Delta(\mathbf{r})$. In order to have a nonzero result, we rather have to square this quantity before taking its average. Then, in order to evaluate if the effect of fluctuations is small, we will compare this to the same average with the order parameter itself, that is, $\langle \int_{\mathbb{V}} d\mathbf{r}\, \Delta(\mathbf{r}) \rangle^2 = (\mathbb{V}\,\Delta_0)^2$. This leads us to introduce the ratio

$$R = \frac{\langle (\int_{\mathbb{V}} d\mathbf{r}\, \delta\Delta(\mathbf{r}))^2 \rangle}{\langle \int_{\mathbb{V}} d\mathbf{r}\, \Delta(\mathbf{r}) \rangle^2} \tag{6.48}$$

The averages have to be taken with a weight proportional to $\exp(-F_s(\{\Delta(\mathbf{r})\})/k_B T)$.

In order to simplify our presentation, we assume that $\Delta(\mathbf{r})$ and $\delta\Delta(\mathbf{r})$ are real quantities. It is actually not difficult at all to generalize to the case of complex quantities. We will also forget about the gradient term $\gamma\,|\nabla\Delta(\mathbf{r})|^2$ in the expression Eq. (6.47) of the free energy. It can be shown that this does not change the final result, and it will simplify our handling. Finally, in order to make reading easier, we will go from a continuously indexed variable $\Delta(\mathbf{r})$ to a discrete one. This can be done, for example, by retaining only discrete spatial positions $\mathbf{r}_i$, very close to each other, so that the continuum limit is obtained by letting the separation between these positions go to zero. This leads to the introduction of the discrete variables $x_i = \delta\Delta(\mathbf{r_i}) = \Delta(\mathbf{r_i}) - \Delta_0$.

Let us rewrite with these variables the partition function

$$Z = \int \{d\Delta(\mathbf{r})\}\, e^{-\frac{F_s(\{\Delta(\mathbf{r})\})}{k_B T}} = \int \{d\Delta(\mathbf{r})\} \exp\left(-\int d\mathbf{r}\,(\alpha'\Delta^2(\mathbf{r}) + \frac{\beta'}{2}\Delta^4(\mathbf{r}))\right) \tag{6.49}$$

with $\alpha' = \alpha/k_B T$ and $\beta' = \beta/k_B T$. When we expand $\alpha\Delta^2(\mathbf{r_i}) + (\beta/2)\Delta^4(\mathbf{r_i})$ around its minimum located at $\Delta^2(\mathbf{r_i}) = \Delta_0^2 = |\alpha|/\beta$, to second order in x_i (which is enough when the fluctuations x_i are small), we find $\alpha\Delta^2(\mathbf{r_i}) + (\beta/2)\Delta^4(\mathbf{r_i}) = -|\alpha|^2/2\beta + 2|\alpha|\,x_i^2$. In this way, Z becomes

$$Z = \int \left(\prod_i dx_i\right) \exp(-2|\alpha'| \sum_j x_j^2) = \left(\frac{\pi}{2|\alpha'|}\right)^{N/2} \tag{6.50}$$

where N is the number of discrete variables x_i, and in the last step we have performed all the Gaussian integrations. The integration boundaries have been extended to $-\infty$ and ∞, because the Gaussian e^{-x^2} converges very rapidly to zero for large $|x|$. Finally, we have not written in the expression Eq. (6.50) for Z a common factor $\exp\left[(\mathbb{V}|\alpha'|^2/(2\beta'))\right]$, which disappears when the average Eq. (6.51) is calculated.

Similarly the numerator in Eq. (6.48) leads to the evaluation of

$$\langle (\sum_m x_m)^2 \rangle = \sum_{m,n} \langle x_m x_n \rangle = Z^{-1} \sum_{m,n} \int (\prod_i dx_i)\, x_m x_n \, \exp(-2|\alpha'| \sum_j x_j^2) \qquad (6.51)$$

For terms with $m \neq n$ the result of the integration is zero because the integrand is an odd function of x_m and x_n. So we are left only with the contributions $m = n$. For these terms, the only modification with respect to the calculation of Z in Eq. (6.50) is that one has to calculate $\int dx_m\, x_m^2 \exp(-2|\alpha'|x_m^2)$, instead of $\int dx_m \exp(-2|\alpha'|x_m^2)$. The ratio between these two integrals is merely $1/(4|\alpha'|)$. Since all the other Gaussian integrals cancel out with the similar ones contained in the factor Z^{-1}, the result for the numerator is simply $N/(4|\alpha'|)$, since m takes N values, so there are N similar terms, all identical. Now, if we work with the continuous variables $\delta\Delta(\mathbf{r})$, instead of using discrete variables, we will obtain similarly a factor $\int_V d\mathbf{r} = \mathbb{V}$, instead of $\sum_m 1 = N$. Hence the actual result for our numerator in Eq. (6.48) is $\mathbb{V}/(4|\alpha'|)$. Dividing by the denominator $(\mathbb{V} \Delta_0)^2$, we obtain finally

$$R = \frac{\beta k_B T_c}{4|\alpha|^2 \mathbb{V}} \qquad (6.52)$$

where we have set $T = T_c$, since all these considerations apply in the immediate vicinity of the critical temperature. We recall that from Eq. (6.2) we have $|\alpha(T)| = \alpha_0(T_c - T)$.

The essential features of our result Eq. (6.52) for R are easy to understand physically. First, we have obtained that the numerator in Eq. (6.48), that is the amplitude of the order parameter fluctuations, diverges when $T \to T_c$. This is easy to understand because on the right side of Fig. (6.1), we see that the well around the minimum Δ_0 in which the order parameter fluctuates gets shallower and wider when $T \to T_c$, so there is less constraint for these fluctuations, which get naturally ever larger in this limit. An additional reason for a factor $|\alpha|$ in the denominator of Eq. (6.52) is merely that in Eq. (6.48) we have compared the fluctuations to the minimum value Δ_0 of the order parameter, which goes also to zero for $T \to T_c$.

The other noticeable feature is the $1/\mathbb{V}$ dependence of the ratio R. Actually, this is a well-known result from statistical physics, valid for the fluctuations of a quantity which is the sum of uncorrelated contributions. The fact that the relative importance of these fluctuations go to zero in the thermodynamic limit of infinite volume is the reason for which only the corresponding averages are relevant in thermodynamics, and, for example, for a system at temperature T one considers its energy without worrying about the fluctuations of this energy.

However, in the present case, this raises the question of the proper choice of the volume $\mathbb{V}$ for our estimate of the importance of the order parameter fluctuations. Naturally, it would be meaningless to take it infinite for our purpose. One has rather to argue in the following way. If one considers the region surrounding a specific point of the superconductor, say, for example, the origin of coordinates, the fluctuations of the order parameter in regions very far away from the origin are clearly completely irrelevant to what happens to the superconductor at the origin. Rather we have seen that the superconductor at the origin is sensitive to the values of the order parameter within a distance of the order of the coherence length. This is naturally within Ginzburg–Landau theory, but we are perfectly entitled to use it since we are aiming at a consistency check of this theory. Hence the volume $\mathbb{V}$ we have to consider is typically the one of a sphere of radius the coherence length $\xi(T)$, as given by Eq. (6.26). It is interesting to consider, for a moment, that we are living in a space which is D-dimensional, instead of taking immediately the standard case of our three-dimensional space $D = 3$. In this case, our volume $\mathbb{V}$ is of order $\xi^D(T)$ and our final result for R is

$$R = \frac{\beta k_B T_c}{4|\alpha(T)|^2 \xi^D(T)} \tag{6.53}$$

Since we have $|\alpha(T)| \sim (T_c - T)$, while from Eq. (6.26) $\xi(T) = (\gamma/|\alpha(T)|)^{1/2} \sim (T_c - T)^{-1/2}$, we see that the overall temperature dependence of R is $(T_c - T)^{D/2-2}$, that is the same as $\xi^{4-D}(T)$ where $\xi(T)$ diverges for $T \to T_c$. Hence if we had $D > 4$, R would go to zero for $T \to T_c$, and fluctuations would be completely negligible. This means that Ginzburg–Landau theory would turn out to be correct for such dimensions. This is not the case for $D = 3$, but this is not so far from $D = 4$, and this has led to the development of expansions in powers of $\epsilon = 4 - D$ in the general theory of phase transitions.

Coming back to the case of superconductors with $D = 3$, we have just to insert the specific values of α, β, and γ to obtain a specific value of R. Our evaluation is approximate enough, so that we do not have to worry about specific values of numerical coefficients. From Eq. (6.8) we have $|\alpha(T)| = \alpha_0(T_c - T) = N_0(1 - T/T_c)$ and $\beta \sim N_0/(k_B T_c)^2$, with $N_0 \sim mk_F/\hbar^2$. For the coherence length we have mentioned that, although not identical to the BCS result, it is a very similar quantity. Hence for its extrapolation at $T = 0$, we can take the BCS value Eq. (2.89) $\xi_0 = \hbar v_F/\pi \Delta$. Taking into account the divergent behavior in $(1 - T/T_c)^{-1/2}$ for $T \to T_c$, this leads to take $\xi(T) = \xi_0(1 - T/T_c)^{-1/2}$, in which we make $\Delta \sim k_B T_c$. In this way, we end up with

$$R \sim \left(\frac{k_B T_c}{E_F}\right)^2 \frac{1}{(1 - T/T_c)^{1/2}} \tag{6.54}$$

which shows indeed that Ginzburg–Landau theory is not valid for $T \to T_c$, since the right-hand side diverges in this case so that fluctuations can no longer be omitted.

However, if we look closely at this Ginzburg criterium, that is the condition of validity for Ginzburg–Landau theory, it requires that fluctuations are negligible $R \ll 1$. If we consider the opposite domain where it is not valid, which is called the critical domain, we obtain from Eq. (6.54) for its temperature extent

$$1 - T/T_c \lesssim \left(\frac{k_B T_c}{E_F}\right)^4 \tag{6.55}$$

For a standard superconductor where typically $k_B T_c/E_F \sim 10^{-3}$, the ratio on the right-hand side is exceedingly small, of order 10^{-12}. Clearly the uncertainty in the numerical prefactor does not change the situation. Accordingly, even though in principle Ginzburg–Landau theory is not valid in the immediate vicinity of the critical temperature, the temperature domain where it fails is so small that it is completely unobservable experimentally. This is not only because the temperature domain is so small, but mostly because the sample quality must be such that the critical temperature is extremely well defined.

The advent of high T_c superconductors, mostly cuprate superconductor, has modified to a large extent this situation. Indeed the critical temperature ~ 100 K is much higher, and Fermi energies are probably rather low, perhaps of order of 0.1 eV. So the ratio $k_B T_c/E_F$ might be of order 0.1. Taking into account a more precise prefactor, one could hope to reach the critical domain with very high precision thermometry. However, these materials are quite complex and solving the sample quality problem is difficult, which makes the overall situation unclear.

In conclusion, it must be emphasized that this remarkable situation with respect to the domain of validity of Ginzburg–Landau theory is quite specific of superconductors. It arises because, as we have seen, the physics of superconductors is characterized by several physical length scales. Not only do we have the atomic length scale (to which corresponds the Fermi energy for energy scale), but we have the coherence length scale (linked to the critical temperature with respect to the energy scale). This is the basic reason why such a small energy ratio can occur in Eq. (6.55). By contrast, in many other physical systems displaying similar phase transitions, a single length scale appears, usually of atomic size. As a result, no small parameter appears and the critical domain is of order unity. In these cases, there is basically no range of validity for mean-field theory.

6.6 Further Reading: The Little–Parks Experiment

In this experiment, the superconducting sample has the shape of a hollow cylinder. The inner radius and the outer radius of this cylinder are very close, so that this cylinder is very thin. A uniform magnetic field is applied along the axis of the cylinder. Since the direction along this axis plays no role, we rather consider for simplicity a thin annulus in a plane perpendicular to the magnetic field, as shown in Fig. (6.4).

Due to the small dimensions of the sample, the permanent currents in the superconductor are so weak that their effect can be completely omitted. The field is just the applied field B, and we choose the gauge $\mathbf{A}(\mathbf{r}) = (1/2)\mathbf{B} \times \mathbf{r}$ for the corresponding vector potential. Hence it has a simple azimuthal component that goes along the annulus in the counterclockwise direction. The annulus is so thin that this component has merely the constant value $A_\theta = BR/2$ on the annulus, with R being the radius of the annulus.

Regarding the order parameter, the annulus is assumed much thinner than the coherence length ξ, so that the modulus of the order parameter cannot vary in the radial direction across the annulus. This is consistent with the boundary condition in the Ginzburg–Landau theory (which we have not considered). In the present case of superconductor-vacuum interface, it states that the derivative of the order parameter perpendicular to the interface should be zero. Accordingly, the order parameter depends only on the azimuthal angle θ along the annulus.

More precisely, the modulus Δ_0 of the order parameter is constant around the annulus by rotational invariance, so only its phase $\varphi(\theta)$ varies. Moreover, since the superfluid velocity $\mathbf{v}_s = \hbar\nabla\varphi/2m$ has again a constant modulus along the annulus by rotational invariance, $\varphi(\theta)$ is merely proportional to θ. Since the order parameter is a single-valued function of

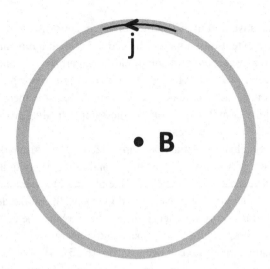

Fig. 6.4 Thin superconducting annulus with an applied magnetic field B perpendicular to the plane of the annulus and the resulting induced supercurrents **j**.

position, its phase must have changed by a multiple n of 2π after going once around the annulus. Taken together, these points show that the phase of the order parameter must be of the form $\varphi(\theta) = n\theta$, where n is any integer.

This leads, in the first Ginzburg–Landau equation Eq. (6.21), $-i\hbar\nabla\Delta(\mathbf{r})$ to be an azimuthal vector with this tangential component being equal to $n\hbar\Delta_0 \exp(in\theta)/R$. Repeating this derivative operation, and taking into account that A_θ is constant, we merely obtain for this first Ginzburg–Landau equation, after simplication by $\Delta_0 \exp(in\theta)$,

$$\alpha(T) + \beta \Delta_0^2 = -\frac{1}{2m^*}\left(\frac{n\hbar}{R} - e^*\frac{BR}{2}\right)^2 \tag{6.56}$$

Actually, we are here only interested in the variation of the critical temperature $T_c(B)$ as a function of the applied magnetic field B. Hence the modulus Δ_0 of the order parameter goes to zero, and we may neglect, as usual for this kind of situations, the $\beta \Delta_0^2$ term compared to $\alpha(T)$. Moreover, we see appearing on the right-hand side the ratio between the magnetic flux $\Phi = \pi R^2$ through the annulus and the flux quantum $\Phi_0 = h/e^* = h/2e$. In this way, Eq. (6.56) reduces to

$$T_c(0) - T_c(B) = \frac{\hbar^2}{2m^*\alpha_0 R^2}\left(n - \frac{\Phi}{\Phi_0}\right)^2 \tag{6.57}$$

where $T_c(0)$ is our standard critical temperature T_c in the absence of magnetic field.

To have a physically clearer expression it is better to go back to Eq. (6.26), which can be rewritten as $\xi_{GL0}^2 \equiv (T_c - T)\xi^2(T) = \hbar^2/(2m^*\alpha_0)$. From this definition ξ_{GL0} is some microscopic constant, independent of temperature, of order of a coherence length. In this way, our result becomes

$$T_c(0) - T_c(B) = \frac{\xi_{GL0}^2}{R^2}\left(n - \frac{\Phi}{\Phi_0}\right)^2 \tag{6.58}$$

Fig. 6.5 Dependence of the critical temperature $T_c(B)$ of the annulus on the flux Φ of the applied magnetic B through the annulus. The result is periodic, with a period equal to the flux quantum Φ_0.

Clearly since it involves the square of the ratio of microscopic length ξ_{GL0} to a macroscopic length R, the effect is fairly small. Nevertheless it can be seen [41], thanks to the brutal change of resistivity when one goes from the superconducting to the normal state.

The result Eq. (6.58) is not yet fully explicit since we have not indicated the value of the integer n. The answer can be obtained by finding the free energy and taking the integer n, which makes it minimal. However, it is simpler to find the result by understanding the physical origin of this effect. If we calculate the azimuthal current $\mathbf{j}$ from the second Ginzburg–Landau equation, we find

$$j_\theta = \frac{e^*\hbar}{m^*}\frac{n}{R}\Delta_0^2 - \frac{e^{*2}}{m^*}\frac{BR}{2}\Delta_0^2 = \frac{e^*\hbar}{m^*R}\Delta_0^2\left(n - \frac{\Phi}{\Phi_0}\right) \tag{6.59}$$

We see that the presence of the magnetic field imposes the existence of currents, because the single-valuedness of the order parameter in this geometry forces a specific angular dependence of the phase and consequently specific values for the superfluid velocity. There is naturally a kinetic energy associated with these currents, and they raise the energy of the superconducting phase, compared to the normal phase where these currents do not exist.

Hence this additional energy cost for the superconducting phase lowers the temperature at which it is favored, compared to the normal phase. This explains qualitatively the lowering of T_c given by Eq. (6.58). Now, the superconducting phase chooses naturally the currents with the lowest possible kinetic energy cost, to lower its total energy. This is done by taking the value of n, which makes $|n - \Phi/\Phi_0|$ as small as possible, or in other words, the proper value of n is the one that makes the $T_c(B)$ in Eq. (6.58) as high as possible. The resulting $T_c(B) - T_c(0)$ is displayed in Fig. (6.5). A quite remarkable feature of this result is that it is periodic in the applied magnetic flux Φ, with a period equal to the flux quantum Φ_0. This provides an additional example of macroscopic manifestations of quantum phenomena in superconductivity.

7

Bose–Einstein Condensation

The purpose of this chapter is first to provide somewhat more detail on Bose–Einstein condensation, a phenomenon we have frequently mentioned in the first chapters of this book, but without giving any quantitative description. Clearly more information makes it easier to appropriately grasp this basic phenomenon. This chapter is also a necessary introduction for a proper understanding of the next chapter on the BEC–BCS crossover, since Bose–Einstein condensation (BEC) is one of the two limiting cases for this crossover.

7.1 The Perfect Bose Gas

Einstein discovered this condensation phenomenon when he investigated the properties of massive non-interacting particles following Bose statistics, which Bose had introduced [42] shortly before for photons in order to rederive in a simple way the standard Planck's formula for black-body radiation.

Being non-interacting, the energy of these particles is merely their kinetic energy $\epsilon_{\mathbf{k}} = \hbar^2 k^2 / 2m$, where $\mathbf{k}$ is their wavevector and m is their mass. Following the Bose statistics, the average particle number $\bar{n}_{\mathbf{k}}$ in a state with wavevector $\mathbf{k}$ is given, at temperature T, by

$$\bar{n}_{\mathbf{k}} = \frac{1}{e^{\beta(\epsilon_{\mathbf{k}} - \mu)} - 1} \tag{7.1}$$

where, as usual, $\beta = 1/k_B T$ and μ is the chemical potential. We note that for the lowest kinetic energy, corresponding to $\mathbf{k} = \mathbf{0}$, this particle number is $\bar{n}_0 = 1/(e^{-\beta\mu} - 1)$. Since physically this number should be positive, this implies that the chemical potential is necessarily negative $\mu < 0$.

The particle number N is obtained by summing Eq. (7.1) over all possible eigenstates. We take our non-interacting Bose gas to be within a cube of size L, which has volume $\mathbb{V} = L^3$. We assume for convenience the periodic boundary conditions of Born–von Karman, so that the allowed values of the wavevector are $k_x = 2\pi n_x/L$, $k_y = 2\pi n_y/L$, and $k_z = 2\pi n_z/L$. Hence N is given by

$$N = \sum_{\mathbf{k}} \frac{1}{e^{\beta(\epsilon_{\mathbf{k}} - \mu)} - 1} = \sum_{n_x, n_y, n_z} \frac{1}{e^{\beta h^2 (n_x^2 + n_y^2 + n_z^2)/(2mL^2) - \beta\mu} - 1} \tag{7.2}$$

For large volume $\mathbb{V}$, it is natural to go from the discrete summation to an integral, as we have already systematically done, by $\sum_{\mathbf{k}} \rightarrow \mathbb{V}/(2\pi)^3 \int d\mathbf{k}$. This leads, after angular integration, to

$$N = \frac{4\pi \mathbb{V}}{(2\pi)^3} \int_0^\infty dk \, \frac{k^2}{e^{\beta(\hbar^2 k^2/2m - \mu)} - 1} \tag{7.3}$$

The crucial point is that the integral on the right-hand side is perfectly convergent and bounded, so that if the number N of bosons is too large, it is not possible to satisfy this Eq. (7.3). This can be seen by noticing that the integrand is largest when we take $-\mu = |\mu| \to 0$. This gives an upper bound for the integral, for which Eq. (7.3) can be written, after the change of variable $x = k^2(\hbar^2 \beta/2m)$, as

$$N = \frac{4\pi \mathbb{V}}{(2\pi)^3} \int_0^\infty dk \, \frac{k^2}{e^{\beta \hbar^2 k^2/2m} - 1} = \frac{\mathbb{V}}{\lambda_T^3} \frac{2}{\sqrt{\pi}} \int_0^\infty dx \, \frac{x^{1/2}}{e^x - 1} \tag{7.4}$$

where we have introduced the thermal de Broglie wavelength $\lambda_T = (2\pi \hbar^2/mk_B T)^{1/2}$, which is the standard de Broglie wavelength for a boson with typical energy $k_B T$. The coefficient of $\mathbb{V}/\lambda_T^3$ is a number, equal[1] to $\zeta(3/2) \simeq 2.612$. Hence for large enough number of these bosonic particles and low enough temperature so that $N\lambda_T^3/\mathbb{V} > 2.612$, Eq. (7.3) has no solution.

The physical solution of this puzzle comes from the consideration of the $T = 0$ situation. In this case, things are very simple. At equilibrium, the system must have its lowest possible energy. This is merely obtained by putting all the bosons in their ground state, which has wavevector $\mathbf{k} = \mathbf{0}$ and zero energy. There is no obstacle to do this since, in contrast with fermions, there is no exclusion principle for bosons, so one can put in any state as many bosons as one likes.

This solution is quite compatible with the general Bose distribution Eq. (7.1). Indeed, setting $\epsilon_{\mathbf{k}} = 0$ for this ground state, we have for the number of bosons N_0 in the ground state $N_0 = (e^{\beta|\mu|} - 1)^{-1}$ (since this number is large, the fluctuations are negligibly small and we can consider that the average number coincides with the actual number). This implies

$$|\mu| = k_B T \ln(1 + \frac{1}{N_0}) \simeq \frac{k_B T}{N_0} \tag{7.5}$$

where the last equality comes from the fact that we consider a situation where N_0 is large. Taking the limit $T \to 0$, we find that putting all the bosons in the ground state is a consistent solution provided the chemical potential goes to zero with temperature according to Eq. (7.5). Note that this result is actually obvious since the chemical potential is the energy required to put an additional particle in the system. In the present case, we will put this additional boson in the ground state, so that the energy cost is indeed zero. Coherently, for $T \to 0$, if we consider all the other states with nonzero energy $\epsilon_{\mathbf{k}}$, we have for them $\beta \epsilon_{\mathbf{k}} \to \infty$, while from Eq. (7.5) $\beta|\mu| = 1/N_0 \simeq 0$. Hence $\bar{n}_{\mathbf{k}} \to 0$ in this limit, and there are no bosons in excited states – all of them are in the ground state. This is the physical situation expected at $T = 0$.

---

[1] We can perform a series expansion of the integral by writing $\int_0^\infty dx \, x^{1/2}/(e^x - 1) = \int_0^\infty dx \, x^{1/2} e^{-x}/(1 - e^{-x}) = \sum_{n=0}^\infty \int_0^\infty dx \, x^{1/2} e^{-(n+1)x}$, with $\int_0^\infty dx \, x^{1/2} e^{-(n+1)x} = 2(n+1)^{-3/2} \int_0^\infty dy \, y^2 e^{-y^2} = \sqrt{\pi}(n+1)^{-3/2}/2$ by the change $y^2 = (n+1)x$ and by-parts integration. Hence $(2/\sqrt{\pi}) \int_0^\infty dx \, x^{1/2}/(e^x - 1) = \sum_{p=1}^\infty p^{-3/2} = \zeta(3/2)$.

If we now raise slightly the temperature above this $T = 0$ situation, we will naturally obtain a thermal population of these excited states. This produces a decrease in the number $N_0(T)$ of bosons staying in the ground state. However, as long as this number is large, the chemical potential will stay essentially equal to zero, according to Eq. (7.5). Accordingly, we will have a population of thermally excited bosons, their distribution being given by Eq. (7.1) with the chemical potential $\mu = 0$. All the other bosons are in the ground state. This solves our puzzle. In Eq. (7.3) we have only written the number of these thermally excited bosons. We have missed the bosons gathered in the ground state because they correspond to the term $n_x = n_y = n_z = 0$ in Eq. (7.2), and this term is very large. So it is somewhat singular among the other terms while, by going from the discrete summation to the integral, we have handled it uncarefully as if it was of the same order as the other terms. In conclusion, at low enough temperature and large enough particle number, there is from Eq. (7.4) no longer enough space to put all the bosons in low-energy excited states, so that the overflow of bosons condenses in the ground state.

Taking now into account the macroscopic number $N_0(T)$ of bosons in the ground state, together with the fact that their presence sets the chemical potential equal to zero, we can now rewrite properly Eq. (7.3) as

$$N = N_0(T) + \frac{4\pi \mathbb{V}}{(2\pi)^3} \int_0^\infty dk \, \frac{k^2}{e^{\beta \hbar^2 k^2/2m} - 1} \tag{7.6}$$

This is valid as long as $N_0(T) > 0$. Hence the critical temperature T_c, which is the upper limit for this Bose–Einstein condensed phase with a macroscopic occupancy of the ground state, is obtained by setting $N_0(T) = 0$ in Eq. (7.6). This is just the equation written in Eq. (7.4), so the critical temperature is given by

$$k_B T_c = \frac{2\pi \hbar^2}{m} \left(\frac{N}{2.612 \, \mathbb{V}} \right)^{2/3} = \frac{3.31 \, \hbar^2}{m} \left(\frac{N}{\mathbb{V}} \right)^{2/3} \tag{7.7}$$

We note that $\mathbb{V}/N$ is the typical volume occupied by each particle, so that $d = (\mathbb{V}/N)^{1/3}$ is the typical interparticle distance. Hence, at T_c, from Eq. (7.7) this interparticle distance d is of the order of the thermal de Broglie wavelength λ_T. This shows that Bose–Einstein condensation takes place when the degenerate statistical regime is reached. This is in contrast with the classical statistical regime, found at higher temperature, where the de Broglie wavelength is extremely small compared to the interparticle distance, so that it is physically irrelevant together with all quantum effects.

Since the second term on the right-hand side of Eq. (7.6) is proportional to $1/\lambda_T^3 \sim T^{3/2}$, and $N_0(T)$ is zero at T_c, we have at temperature T

$$\frac{N - N_0(T)}{N} = \left(\frac{T}{T_c} \right)^{3/2} \tag{7.8}$$

which gives explicitly the number $N_0(T)$ of bosons in the ground state. Since the bosons condensed in the ground state have zero energy, they do not contribute to the total energy which results only from the thermally excited bosons. As a result, the energy of the bosons is given by

$$E = \sum_{\mathbf{k}} \epsilon_{\mathbf{k}} \bar{n}_{\mathbf{k}} = \sum_{\mathbf{k}} \frac{\epsilon_{\mathbf{k}}}{e^{\beta \epsilon_{\mathbf{k}}} - 1} = \frac{\hbar^2 \mathbb{V}}{4\pi^2 m} \int_0^\infty dk \frac{k^4}{e^{\beta \hbar^2 k^2 / 2m} - 1} \tag{7.9}$$

$$= \frac{3 k_B T \mathbb{V}}{2 \lambda_T^3} \frac{4}{3\sqrt{\pi}} \int_0^\infty dx \frac{x^{3/2}}{e^x - 1}$$

where we have used $\mu = 0$ and made in the last step the same change of variable $x = (\lambda_T^2 / 4\pi) k^2$ as above. Just as in Eq. (7.4), the last coefficient in the final result is easily shown to be $\zeta(5/2) \simeq 1.341$.

Having the energy, with its temperature dependence in $T^{5/2}$, we easily obtain the specific heat $C_V = \partial E / \partial T$, the entropy from $C_V = T \partial S / \partial T$, and the free energy $F = E - TS$. One finds

$$C_V = \frac{5}{2} \frac{E}{T} \qquad S = \frac{5}{3} \frac{E}{T} \qquad F = -\frac{2}{3} E \tag{7.10}$$

We notice that with respect to extensive variables, these quantities are merely proportional to $\mathbb{V}$ and do not depend on particle number N. This is coherent with $\mu = \partial F / \partial N \big|_{T,\mathbb{V}} = 0$. From the thermodynamic identity $F - \mu N = -P\mathbb{V}$, with $\mu = 0$ in the present case, we conclude that this gas pressure P

$$P = \frac{2}{3} \frac{E}{\mathbb{V}} \tag{7.11}$$

is independent of volume, so that $\partial P / \partial \mathbb{V} = 0$. This implies that the compressibility $\kappa = -(1/\mathbb{V}) \partial \mathbb{V} / \partial P$ of this perfect Bose gas is infinite. This behavior is clearly not physical for a real system. One may accordingly worry about the physical relevance of this perfect Bose gas and of the Bose–Einstein condensation phenomenon. Clearly repulsive interactions between bosons will make the Bose gas able to stand against an increase in pressure; hence the compressibility should no longer be infinite and so these interactions should cure this pathology. However, one may wonder if the Bose–Einstein condensation will not disappear at the same time. This is what we consider in the next section.

7.2 The Effect of Interactions

Since we want mainly to see how Bose–Einstein condensation in the perfect gas is affected qualitatively by interactions, we may here restrict ourselves to the case of weak interactions, without bothering with the complications arising with strong interactions which we will consider later on. This allows one to investigate this question within a perturbative approach. Also, we restrict ourselves to the case of repulsive interactions. Indeed, since the Bose–Einstein condensate is at the border of instability, we expect that an attractive interaction between bosons would lead to an unstable system and that the gas would collapse. This has actually been seen experimentally to occur in dilute ultracold gases.

The systematic approach to the weakly interacting Bose gas will be considered in the next section. In the present section, we only want to show that repulsive interactions stabilize the Bose–Einstein condensate. This is a somewhat surprising result, since one could have expected that repulsion between bosons would push them out of the single-particle ground state corresponding to the condensate, leading perhaps to its complete disappearance. Actually, such an effect of interactions is present, but this is not physically the dominant effect.

For simplicity, we will display this effect by considering a very elementary situation. We will restrict ourselves to the case of two bosons, and will demonstrate that due to repulsive interactions, it is energetically more favorable to put them in the same eigenstate rather than in two different eigenstates. We present first the argument with the first quantization formalism. We will then repeat it in second quantization, which will bring us close to the formalism used in the following section.

We take as eigenstates two plane waves with wavevector $\mathbf{k}$ and $\mathbf{k}'$. Assuming the bosons to be in a box with volume unity, the corresponding normalized wave functions are $\varphi(\mathbf{r}) = e^{i\mathbf{k}\cdot\mathbf{r}}$ and $\varphi'(\mathbf{r}) = e^{i\mathbf{k}'\cdot\mathbf{r}}$. If we put the two bosons in the same eigenstate, say $\varphi(\mathbf{r})$, then the two bosons wave function $\psi(\mathbf{r}_1, \mathbf{r}_2)$ is merely

$$\psi(\mathbf{r}_1, \mathbf{r}_2) = \varphi(\mathbf{r}_1)\varphi(\mathbf{r}_2) = e^{i\mathbf{k}\cdot(\mathbf{r}_1+\mathbf{r}_2)} \tag{7.12}$$

where $\mathbf{r}_1$ and $\mathbf{r}_2$ are the position variables of each boson. Note that this wave function is properly normalized to unity $\int d\mathbf{r}_1 d\mathbf{r}_2 |\psi(\mathbf{r}_1, \mathbf{r}_2)|^2 = 1$ because each plane wave is normalized. This wave function satisfies also the essential requirement for bosons that it should be symmetric under the exchange $\mathbf{r}_1 \leftrightarrow \mathbf{r}_2$ of these two bosons variables $\psi(\mathbf{r}_1, \mathbf{r}_2) = \psi(\mathbf{r}_2, \mathbf{r}_1)$.

Let us now put the two bosons in the two different eigenstates corresponding to wavevectors $\mathbf{k}$ and $\mathbf{k}'$. In this case, the wave function is not $\varphi(\mathbf{r}_1)\varphi'(\mathbf{r}_2)$, nor $\varphi(\mathbf{r}_2)\varphi'(\mathbf{r}_1)$ because none of these two wave functions is symmetric under $\mathbf{r}_1 \leftrightarrow \mathbf{r}_2$. In order to obtain the properly symmetric wave function, we have to take the sum of these two trials:

$$\psi(\mathbf{r}_1, \mathbf{r}_2) = \varphi(\mathbf{r}_1)\varphi'(\mathbf{r}_2) + \varphi(\mathbf{r}_2)\varphi'(\mathbf{r}_1) = e^{i(\mathbf{k}\cdot\mathbf{r}_1+\mathbf{k}'\cdot\mathbf{r}_2)} + e^{i(\mathbf{k}\cdot\mathbf{r}_2+\mathbf{k}'\cdot\mathbf{r}_1)} \tag{7.13}$$

However, this wave function is not properly normalized since

$$\int d\mathbf{r}_1 d\mathbf{r}_2 |\psi(\mathbf{r}_1, \mathbf{r}_2)|^2 = \int d\mathbf{r}_1 d\mathbf{r}_2 \left[|\varphi(\mathbf{r}_1)|^2 |\varphi'(\mathbf{r}_2)|^2 + |\varphi(\mathbf{r}_2)|^2 |\varphi'(\mathbf{r}_1)|^2 \right. \tag{7.14}$$
$$\left. + (\varphi(\mathbf{r}_1)\varphi'(\mathbf{r}_2)\varphi^*(\mathbf{r}_2)\varphi'^*(\mathbf{r}_1) + c.c) \right] = 2$$

since the eigenstates are normalized $\int d\mathbf{r}_1 |\varphi(\mathbf{r}_1)|^2 = \int d\mathbf{r}_2 |\varphi'(\mathbf{r}_2)|^2 = 1$ and the different eigenstates are orthogonal $\int d\mathbf{r}_1 \varphi(\mathbf{r}_1)\varphi'^*(\mathbf{r}_1) = 0$. Hence for this case, the normalized wave function is

$$\psi(\mathbf{r}_1, \mathbf{r}_2) = \frac{1}{\sqrt{2}} \left(\varphi(\mathbf{r}_1)\varphi'(\mathbf{r}_2) + \varphi(\mathbf{r}_2)\varphi'(\mathbf{r}_1) \right) \tag{7.15}$$

We consider now a repulsive interaction $V(\mathbf{r}_1 - \mathbf{r}_2)$ between the two bosons, which is weak so that we can handle it as a perturbation. The corresponding change in energy δE is consequently obtained by taking the average of this interaction in the unperturbed wave function. When we deal first with the case of the two bosons in the same state, the wave function being given by Eq. (7.12), we have

$$\delta E = \int d\mathbf{r}_1 \, d\mathbf{r}_2 |\psi(\mathbf{r}_1, \mathbf{r}_2)|^2 \, V(\mathbf{r}_1 - \mathbf{r}_2) = \int d\mathbf{r}_1 \, d\mathbf{r}_2 \, V(\mathbf{r}_1 - \mathbf{r}_2) = \int d\mathbf{r} \, V(\mathbf{r}) \quad (7.16)$$

where the last step is obtained by going to the variables center of mass $\mathbf{R} = (\mathbf{r}_1 + \mathbf{r}_2)/2$ and relative position $\mathbf{r} = \mathbf{r}_1 - \mathbf{r}_2$, using $d\mathbf{r}_1 \, d\mathbf{r}_2 = d\mathbf{R} \, d\mathbf{r}$ and integrating over $\mathbf{R}$ in the whole unit volume. For a purely repulsive interaction, $\int d\mathbf{r} \, V(\mathbf{r})$ is naturally positive.

Switching to the case where the two bosons are in different states, we have from Eq. (7.15)

$$\delta E = \int d\mathbf{r}_1 \, d\mathbf{r}_2 |\psi(\mathbf{r}_1, \mathbf{r}_2)|^2 \, V(\mathbf{r}_1 - \mathbf{r}_2) = \frac{1}{2} \int d\mathbf{r}_1 \, d\mathbf{r}_2 \Big[|\varphi(\mathbf{r}_1)|^2 |\varphi'(\mathbf{r}_2)|^2 \quad (7.17)$$

$$+ \, |\varphi(\mathbf{r}_2)|^2 |\varphi'(\mathbf{r}_1)|^2 + \big(\varphi(\mathbf{r}_1)\varphi'(\mathbf{r}_2)\varphi^*(\mathbf{r}_2)\varphi'^*(\mathbf{r}_1) + c.c\big) \Big] \, V(\mathbf{r}_1 - \mathbf{r}_2)$$

Since $|\varphi(\mathbf{r}_1)|^2 = |\varphi'(\mathbf{r}_2)|^2 = 1$, the first two terms on the right-hand side of Eq. (7.17) lead together to the same result as Eq. (7.16). However, we have in addition the last term, which comes from the product of the two terms in Eq. (7.15), the last one coming from the first one by the exchange $\mathbf{r}_1 \leftrightarrow \mathbf{r}_2$. This contribution comes directly from the requirement that for identical bosons, the wave function must be symmetrical under the exchange of the position variables, which is a direct translation of the fact that these bosons are indiscernible particles. In contrast with what happened in normalizing the wave function, this contribution is nonzero and is called the "exchange energy." Since $\varphi(\mathbf{r}_1)\varphi'(\mathbf{r}_2)\varphi^*(\mathbf{r}_2)\varphi'^*(\mathbf{r}_1) = e^{i(\mathbf{k}-\mathbf{k}')\cdot(\mathbf{r}_1-\mathbf{r}_2)}$, this term gives to δE a contribution $V_{\mathbf{k}'-\mathbf{k}}/2$, if $V_\mathbf{k}$ is the Fourier transform of $V(\mathbf{r})$

$$V_\mathbf{k} = \int d\mathbf{r} \, e^{-i\mathbf{k}\cdot\mathbf{r}} \, V(\mathbf{r}) \quad (7.18)$$

If we assume for simplicity that the interaction potential is symmetric $V(\mathbf{r}) = V(-\mathbf{r})$, the Fourier transform is real $V_\mathbf{k}^* = V_\mathbf{k}$, so that in Eq. (7.17) the complex conjugate term gives also the same contribution $V_{\mathbf{k}'-\mathbf{k}}/2$. Gathering these results, we finally find for the two bosons in different states

$$\delta E = V_0 + V_{\mathbf{k}'-\mathbf{k}} \quad (7.19)$$

since $\int d\mathbf{r} \, V(\mathbf{r}) = V_0$.

Hence we see that compared to the case where the two bosons are in the same eigenstate, we have an additional contribution which is the exchange energy. Nevertheless the comparison is not completely satisfactory since we have two wavevectors $\mathbf{k}$ and $\mathbf{k}'$ coming in the present case, in contrast with the single wavevector $\mathbf{k}$ when the two bosons are in the same state. We can eliminate this asymmetry by letting $\mathbf{k}' \to \mathbf{k}$. In this case, we find from Eq. (7.19) that when the two bosons are in two different states with the same wavevector $\mathbf{k}$, they have an energy

$$\delta E = 2V_0 \quad (7.20)$$

while when they are in the same state with wavevector $\mathbf{k}$, their energy is merely

$$\delta E = V_0 \tag{7.21}$$

the difference being due to the exchange energy. In conclusion, we see that for reasons directly linked to the bosonic nature of the particles, it is energetically more advantageous (since $V_0 > 0$ for our repulsive potential) to "condense" our two bosons in a single eigenstate, rather than having them in two different ones.

We now take up the above argument with the second quantization formalism, which, in particular, makes it easier to generalize it to a higher number of bosons. We write the interaction term between bosons just as we have done in Eq. (2.23) for fermions, merely replacing fermionic operators c and $c^\dagger$ with the bosonic operators b and $b^\dagger$. Actually, the form Eq. (2.23) is extremely general and is valid even for a non-local interaction. If we assume, as above, a local interaction $V(\mathbf{r}_1 - \mathbf{r}_2)$, the factor $V_{\mathbf{k},\mathbf{k}',\mathbf{q}}$ does not depend on $\mathbf{k}$ and $\mathbf{k}'$, and the remaining factor $V_\mathbf{q}$ is just the Fourier transform of the interaction we have just considered. We will even further simplify our interaction by taking into account that the bosons we deal with have a low energy, because we will deal with low enough temperature. Accordingly, the involved wavevectors will be small enough that we can approximate any $V_\mathbf{q}$ by V_0. More precisely, if we have an interaction $V(\mathbf{r})$ with typical range r_0, which is the order of magnitude of $|\mathbf{r}|$ in the Fourier transform Eq. (7.18), we assume that $qr_0 \ll 1$, so that the value of $V_\mathbf{q}$ is essentially independent of $\mathbf{q}$ and can be replaced with V_0. Hence this implies an interaction with a short enough range r_0. In this case, the interaction reads merely

$$H_{int} = \frac{V_0}{2} \sum_{\mathbf{k},\mathbf{k}',\mathbf{q}} b^\dagger_{\mathbf{k}'+\mathbf{q}/2} b^\dagger_{-\mathbf{k}'+\mathbf{q}/2} b_{-\mathbf{k}+\mathbf{q}/2} b_{\mathbf{k}+\mathbf{q}/2} \tag{7.22}$$

In the very simple case we have considered above, we had only two plane wave states with wavevector $\mathbf{k}$ and $\mathbf{k}'$. However, since we ended up taking $\mathbf{k}' \to \mathbf{k}$, we have only to consider a single wavevector, and for simplicity we take it equal to zero. Nevertheless we want two plane wave states, and we merely call the corresponding bosonic operators b and b'. In the general expression Eq. (7.22), we only need to retain the terms involving these two operators and relevant to calculate the average of the interaction in the two states we consider. These are

$$H_{int} = \frac{V_0}{2} \left[b^\dagger b^\dagger bb + b^\dagger b'^\dagger bb' + b'^\dagger b^\dagger bb' + b^\dagger b'^\dagger b'b + b'^\dagger b^\dagger b'b \right] \tag{7.23}$$

(for example, we have not written the term $b'^\dagger b'^\dagger b'b'$ because we do not consider states with two b' bosons, so this term would play no role).

We call $|1, 1\rangle$ the state with one boson b and one boson b'. We have

$$|1, 1\rangle = b^\dagger b'^\dagger |0\rangle \tag{7.24}$$

where $|0\rangle$ is the normalized vacuum. This state is normalized since $\langle 1, 1 | 1, 1 \rangle = \langle 0 | b'bb^\dagger b'^\dagger |0\rangle = 1$, ($b$ and b' commute and $bb^\dagger |0\rangle = |0\rangle$). Calling $|2, 0\rangle$ the state with two b bosons and no b' boson, it is obtained by creating two b bosons from vacuum $|0\rangle$, so it is proportional to $b^\dagger b^\dagger |0\rangle = b^{\dagger 2} |0\rangle$. However,

$$\langle 0 | bbb^\dagger b^\dagger |0\rangle = \langle 0 | b(1 + b^\dagger b) b^\dagger |0\rangle = 2 \tag{7.25}$$

since $[b, b^\dagger] = 1$ and $bb^\dagger|0\rangle = |0\rangle$. Hence the properly normalized expression is

$$|2, 0\rangle = \frac{1}{\sqrt{2}} b^{\dagger 2} |0\rangle \tag{7.26}$$

We can now calculate the average of H_{int} for these two states. We have

$$\langle 0|bb(b^\dagger b^\dagger bb)b^\dagger b^\dagger|0\rangle = 2\langle 0|bb(b^\dagger b^\dagger)|0\rangle = 4 \tag{7.27}$$

from the calculation performed in Eq. (7.25). Taking into account the normalization of $|2, 0\rangle$, this gives

$$\langle 2, 0|H_{int}|2, 0\rangle = V_0 \tag{7.28}$$

in agreement with Eq. (7.21). On the other hand, in the calculation of $\langle 1, 1|H_{int}|1, 1\rangle$, the last four terms in Eq. (7.23) contribute in the same way, since b and b' commute. For the first one, we have

$$\langle 0|b'b(b^\dagger b'^\dagger bb')b^\dagger b'^\dagger|0\rangle = \langle 0|bb^\dagger bb^\dagger|0\rangle \langle 0|b'b'^\dagger b'b'^\dagger|0\rangle = 1 \tag{7.29}$$

Taking into account the four terms, this leads to

$$\langle 1, 1|H_{int}|1, 1\rangle = 2V_0 \tag{7.30}$$

again in agreement with our earlier calculation Eq. (7.20). Hence we recover the conclusion that the repulsive interaction makes it energetically more favorable to put the two bosons in the same state rather than in two different states. This is independent from kinetic energy considerations, which have been completely omitted in this argument.

It is easy and quite interesting to generalize this result to the case where we have n bosons b, and wonder if it is more favorable to put an additional boson in the same b state, or in b'. To do this, it is convenient to use the commutation relation

$$[b, b^{\dagger n}] = n b^{\dagger(n-1)} \tag{7.31}$$

This is proved by a recursive argument, starting from the basic bosonic commutator $[b, b^\dagger] = 1$. Assuming this relation Eq. (7.31) to be valid for $n - 1$, we have

$$[b, b^{\dagger n}] = [b, b^\dagger b^{\dagger(n-1)}] = [b, b^\dagger]b^{\dagger(n-1)} + b^\dagger[b, b^{\dagger(n-1)}] \tag{7.32}$$
$$= b^{\dagger(n-1)} + b^\dagger(n-1)b^{\dagger(n-2)} = n b^{\dagger(n-1)}$$

where we have used the standard algebraic identity for commutators $[A, BC] = [A, B]C + B[A, C]$. From this relation, one obtains the well-known result that the operator $\hat{n} = b^\dagger b$ gives the boson number n, when applied to a state with well-defined boson number

$$\hat{n} b^{\dagger n}|0\rangle = (b^\dagger b)b^{\dagger n}|0\rangle = b^\dagger(b^{\dagger n}b + n b^{\dagger(n-1)})|0\rangle = n b^{\dagger n}|0\rangle \tag{7.33}$$

making use of $b|0\rangle = 0$. Since, making use of Eq. (7.31), we have

$$\langle 0|b^n b^{\dagger n}|0\rangle = n\langle 0|b^{n-1}b^{\dagger(n-1)}|0\rangle = \cdots = n! \tag{7.34}$$

we find that the normalized state with n bosons b is

$$|n, 0\rangle = \frac{1}{\sqrt{n!}} b^{\dagger n}|0\rangle \tag{7.35}$$

in agreement with the well-known relation $b^\dagger|n-1,0\rangle = \sqrt{n}|n,0\rangle$ for the normalized eigenstates of the harmonic oscillator.

Coming back to our initial problem, if we put an additional boson in state b, we obtain the normalized state

$$|n+1,0\rangle = \frac{1}{\sqrt{(n+1)!}}\, b^{\dagger(n+1)}|0\rangle \tag{7.36}$$

Only the first term of Eq. (7.23) contributes to the average of H_{int} in this state. Hence we have to evaluate $\langle 0|b^{(n+1)}(b^\dagger b^\dagger bb)b^{\dagger(n+1)}|0\rangle$. This is easily done by, for example, noticing that $bb^{\dagger(n+1)}|0\rangle$ has n bosons so that $\hat{n}\, bb^{\dagger(n+1)}|0\rangle = n\, bb^{\dagger(n+1)}|0\rangle$. Repeating this argument for $\hat{n}\, b^{\dagger(n+1)}|0\rangle$, one finds in this way

$$\langle (n+1),0|H_{int}|(n+1),0\rangle = n(n+1)\frac{V_0}{2} \tag{7.37}$$

On the other hand, if we put the additional boson in state b', we have the normalized state

$$|n,1\rangle = \frac{1}{\sqrt{n!}}\, b'^\dagger b^{\dagger n}|0\rangle \tag{7.38}$$

Once again, the four last terms of Eq. (7.23) contribute equally. Since there is a single boson b', its contribution to the average of H_{int} is easily obtained, simply leading to a factor of 1. For the b boson, we have to deal with $\langle 0|b^n b^\dagger bb^{\dagger n}|0\rangle = n\langle 0|b^n b^{\dagger n}|0\rangle = n \times n!$. So the contribution from these four terms to $\langle n,1|H_{int}|n,1\rangle$ is $2n\,V_0$. On the other hand, there is also a contribution from the first term of Eq. (7.23), which is merely obtained from the result Eq. (7.37) by replacing $n+1$ by n, so that this contribution is $n(n-1)V_0/2$. Putting together these two contributions, we find

$$\langle n,1|H_{int}|n,1\rangle = n(n+3)\frac{V_0}{2} \tag{7.39}$$

In this way, we reach the conclusion that

$$\langle n,1|H_{int}|n,1\rangle - \langle (n+1),0|H_{int}|(n+1),0\rangle = n\,V_0 \tag{7.40}$$

It is again energetically more favorable to put the additional boson with all the other b bosons, rather than putting it in a different state b'. An important additional conclusion from Eq. (7.40) is that the energy difference is proportional to n, that is the number of particles involved. So it is an extensive quantity, relevant in the thermodynamic limit of large systems. Indeed one could have feared that this energy difference is an irrelevant microscopic quantity. Our analysis with two bosons was not able to eliminate this possibility.

7.3 Bogoliubov Theory

The problem of the weakly interacting Bose gas has been addressed systematically by Bogoliubov [43]. He found that even weak interactions lead to very important modification of the physical picture with respect to the perfect Bose gas. Nevertheless, from the physical situation found in the perfect Bose gas, he took the starting hypothesis that the condensation of a large number of bosons in the ground state $\mathbf{k} = \mathbf{0}$ still exists in the presence of weak interactions.

If we use for this ground state the standard language of first quantization, having a large number of bosons implies that for this bosonic state, we have to deal with very large quantum numbers. However, the regime of very large quantum numbers is the regime where quantum physics goes to classical physics, so that quantum fluctuations become negligible. In particular, operators can be replaced by standard numbers, without having to worry about the fact that operators may not commute. This is clear if we consider the creation and annihilation operators $b_\mathbf{0}^\dagger$ and $b_\mathbf{0}$ in the ground state. If, in the commutator $[b_\mathbf{0}, b_\mathbf{0}^\dagger] = 1$, these operators act on states with large quantum numbers, so that they lead themselves to large values, the value 1 of the commutator becomes negligible. In this case, everything happens as if $[b_\mathbf{0}, b_\mathbf{0}^\dagger] = 0$, so that these operators commute like standard classical quantities, and we can consider them as simple numbers instead of operators. Since the boson number in the ground state $N_0 = \langle b_\mathbf{0}^\dagger b_\mathbf{0} \rangle \simeq b_\mathbf{0}^\dagger b_\mathbf{0}$ is large, the numbers $b_\mathbf{0}$ and $b_\mathbf{0}^\dagger$ are themselves large. As they are complex conjugates, we can take them real and equal by eliminating a physically irrelevant phase factor. Hence $N_0 = b_\mathbf{0}^\dagger b_\mathbf{0} = b_\mathbf{0}^2$, so that we can take everywhere $b_\mathbf{0} = b_\mathbf{0}^\dagger = \sqrt{N_0}$.

We now investigate the various terms of the interaction, using the large value of $b_\mathbf{0}$ and performing an expansion in decreasing powers of $b_\mathbf{0}$. The dominant term in the interaction H_{int} given by Eq. (7.22) is clearly the single term obtained by taking all wavevectors equal to zero, so that we have four times $b_\mathbf{0}$ appearing in the interaction term. With $b_\mathbf{0}^2 = N_0$, this gives an interaction energy

$$E_{int} = \frac{1}{2} N_0^2 V_0 \tag{7.41}$$

which is just the result one would obtain by perturbation theory, starting with a state where all the bosons are in the ground state. The result Eq. (7.41) does not contain the gas volume $\mathbb{V}$ because we have assumed it to be equal to unity. It is easy to restore this dependence since this energy must be an extensive quantity. This leads us to write $E_{int} = (1/2\mathbb{V})N_0^2 V_0$ (for a general volume, H_{int} in Eq. (7.22) should have a prefactor $1/\mathbb{V}$).

At zero temperature, all the bosons are in the ground state $N_0 = N$, and there is no kinetic energy since the bosons have all $\mathbf{k} = \mathbf{0}$. Hence Eq. (7.41) is the total energy E

$$E = \frac{N^2}{2\mathbb{V}} V_0 \tag{7.42}$$

Accordingly, the pressure $P = -\partial E/\partial \mathbb{V} = N^2 V_0/2\mathbb{V}^2$ depends on volume, and the compressibility $\kappa = -(1/\mathbb{V})\partial \mathbb{V}/\partial P = \mathbb{V}^2/(N^2 V_0)$ is no longer infinite as in the perfect Bose gas. So we see that repulsive interactions cure, as expected, this pathological behavior. Actually, this compressibility is directly linked to sound velocity c_s which will be of interest later on. From hydrodynamics we have

$$mc_s^2 = \frac{\mathbb{V}}{N\kappa} = nV_0 \tag{7.43}$$

where $n = N/\mathbb{V}$ is the particle number density.

We go now, in H_{int}, to the next order term in powers of b_0. We cannot have three b_0 factors, because momentum conservation would force the last operator to have also momentum $\mathbf{0}$ and this last operator would necessarily be b_0 (or $b_0^\dagger$). So we would have four b_0 factors, but we have already considered this term. Hence the next terms we have to consider contain only two b_0 factors, which implies that they have two operators with nonzero wavevectors.

One possibility is that one of these operators is a creation operator, the other one being an annihilation operator. In order to satisfy momentum conservation, this means that we have in this case a factor $b_{\mathbf{k}}^\dagger b_{\mathbf{k}}$. There are four ways to obtain such a factor from Eq. (7.22), depending on which operator among the two creation operators we choose to be taken as $b_0^\dagger$, and which one among the two annihilation operators has to be taken as b_0 (one has naturally to make the appropriate change of variables on the wavevectors to end up with the product $b_{\mathbf{k}}^\dagger b_{\mathbf{k}}$).

The other possibility is that the two operators with nonzero wavevectors are either both creation operators or both annihilation operators. To ensure momentum conservation, this implies that we have either a factor $b_{\mathbf{k}}^\dagger b_{-\mathbf{k}}^\dagger$ or a factor $b_{\mathbf{k}} b_{-\mathbf{k}}$. In each case, there is only one way to obtain such a factor. As a result, we obtain from Eq. (7.22) at this order a contribution

$$H_{int}^{(1)} = \frac{V_0}{2} b_0^2 \sum_{\mathbf{k}} (4 b_{\mathbf{k}}^\dagger b_{\mathbf{k}} + b_{\mathbf{k}}^\dagger b_{-\mathbf{k}}^\dagger + b_{\mathbf{k}} b_{-\mathbf{k}}) \qquad (7.44)$$

However, when working at this order, we must be careful to handle the b_0^4 term with the same precision. Since $H_{int}^{(1)}$ contains a contribution $b_0^2 \sum_{\mathbf{k}} b_{\mathbf{k}}^\dagger b_{\mathbf{k}} = N_0 \sum_{\mathbf{k}} \hat{n}_{\mathbf{k}}$, where $\hat{n}_{\mathbf{k}}$ is the operator number of bosons in state $\mathbf{k}$, we have to take into account bosons that are not in the $\mathbf{k} = \mathbf{0}$ condensate and write the total particle number N as

$$N = N_0 + \sum_{\mathbf{k}} b_{\mathbf{k}}^\dagger b_{\mathbf{k}} \qquad (7.45)$$

as we have already done for the perfect Bose gas. Inserting the resulting expression of $b_0^2 = N_0$ in the $V_0 b_0^4/2$ term, we obtain

$$\frac{V_0 b_0^4}{2} = \frac{V_0}{2} \left(N - \sum_{\mathbf{k}} b_{\mathbf{k}}^\dagger b_{\mathbf{k}} \right)^2 \simeq \frac{V_0}{2} \left(N^2 - 2N \sum_{\mathbf{k}} b_{\mathbf{k}}^\dagger b_{\mathbf{k}} \right) \qquad (7.46)$$

The first term is just the result, Eq. (7.42), we have already found for the ground state energy, at the dominant order in our expansion. In Eq. (7.46) we have consistently omitted the last term in the square expansion because, in our b_0 expansion, it is negligible at the order we are considering. Similarly we can write, in the second term of Eq. (7.46), $N \simeq N_0 = b_0^2$, so that this term compensates half of the first term in Eq. (7.44).

As a result, in addition to the ground state energy, we find for the bosons the Hamiltonian

$$H = \sum_{\mathbf{k}} \epsilon_{\mathbf{k}} b_{\mathbf{k}}^\dagger b_{\mathbf{k}} + \frac{N V_0}{2} \sum_{\mathbf{k}} (2 b_{\mathbf{k}}^\dagger b_{\mathbf{k}} + b_{\mathbf{k}}^\dagger b_{-\mathbf{k}}^\dagger + b_{\mathbf{k}} b_{-\mathbf{k}}) \qquad (7.47)$$

where we have gathered the interaction terms coming from Eq. (7.44) and Eq. (7.46). In this interaction term, we have consistently made $b_0^2 = N_0 = N$. In addition, we have taken into account the kinetic energy. The last two terms of Eq. (7.47) look strange because they do not conserve the boson number. However, looking at the terms from which they originate, the physical significance of these terms is clear. The missing bosons originate from the condensate. In the term where two bosons are created, they come actually from the condensate, and when two bosons are destroyed, they actually go into the condensate.

Because the Hamiltonian Eq. (7.47) is quadratic with respect to the bosonic operators, it is easily diagonalized by a linear transformation, which is known as the Bogoliubov transformation. The situation is actually quite close to the one we have already encountered in Chapter 3 with Hamiltonian Eq. (3.19), and in the same way as in Eq. (3.11), we introduce a new operator $B_{\mathbf{k}}^{\dagger}$ that is a linear combination of $b_{\mathbf{k}}^{\dagger}$ and $b_{-\mathbf{k}}$. Together with its Hermitian conjugate, it reads

$$B_{\mathbf{k}}^{\dagger} = u_{\mathbf{k}} b_{\mathbf{k}}^{\dagger} + v_{\mathbf{k}} b_{-\mathbf{k}} \qquad B_{\mathbf{k}} = u_{\mathbf{k}} b_{\mathbf{k}} + v_{\mathbf{k}} b_{-\mathbf{k}}^{\dagger} \qquad (7.48)$$

We will try from the start to find a solution where the unknown coefficients $u_{\mathbf{k}}$ and $v_{\mathbf{k}}$ are real, and also even in the change $\mathbf{k} \to -\mathbf{k}$, since $\epsilon_{\mathbf{k}}$ in Eq. (7.47) has this property.

Similar to what we have seen for fermions, we want the new operators $B_{\mathbf{k}}$ and $B_{\mathbf{k}}^{\dagger}$, which allow the diagonalization of H, to be bosonic operators. This implies that they must satisfy the commutation relation

$$[B_{\mathbf{k}}, B_{\mathbf{k}}^{\dagger}] = 1 \qquad (7.49)$$

Replacing $B_{\mathbf{k}}$ and $B_{\mathbf{k}}^{\dagger}$ with their definitions in Eq. (7.48), and making use of the known commutators for $b_{\mathbf{k}}$ and $b_{\mathbf{k}}^{\dagger}$, we find immediately

$$u_{\mathbf{k}}^2 - v_{\mathbf{k}}^2 = 1 \qquad (7.50)$$

which is similar to the normalization Eq. (2.45) we had for fermions.

Changing $\mathbf{k}$ into $-\mathbf{k}$ in the second equation in Eq. (7.48) gives us

$$B_{-\mathbf{k}} = u_{\mathbf{k}} b_{-\mathbf{k}} + v_{\mathbf{k}} b_{\mathbf{k}}^{\dagger} \qquad (7.51)$$

Taken together with the first equation Eq. (7.48), this allows one to obtain the expression of the operators $b_{\mathbf{k}}^{\dagger}$ and $b_{-\mathbf{k}}$ in terms of $B_{\mathbf{k}}^{\dagger}$ and $B_{-\mathbf{k}}$. Making use of the relation Eq. (7.50), one finds

$$b_{\mathbf{k}}^{\dagger} = u_{\mathbf{k}} B_{\mathbf{k}}^{\dagger} - v_{\mathbf{k}} B_{-\mathbf{k}} \qquad b_{-\mathbf{k}} = u_{\mathbf{k}} B_{-\mathbf{k}} - v_{\mathbf{k}} B_{\mathbf{k}}^{\dagger} \qquad (7.52)$$

One obtains $b_{-\mathbf{k}}^{\dagger}$ and $b_{\mathbf{k}}$ by changing $\mathbf{k}$ into $-\mathbf{k}$, taking into account the even parity of $u_{\mathbf{k}}$ and $v_{\mathbf{k}}$.

We are now ready to find the values of $u_{\mathbf{k}}$ and $v_{\mathbf{k}}$ that lead to the diagonalization of the Hamiltonian Eq. (7.47). We substitute in Eq. (7.47) the $b_{\mathbf{k}}$'s in terms of the $B_{\mathbf{k}}$'s and request that the coefficient of $B_{\mathbf{k}}^{\dagger} B_{-\mathbf{k}}^{\dagger}$ is zero. The calculation is in practice easy because we do not have to worry about commutation relations of the operators, and we can handle them as simple algebraic numbers. Indeed any possible commutator, equal to 1, would give a simple number and produce as a result in H an additional constant, not an operator. Gathering all these constant contributions would give a number that should be added to the ground state energy. However, we will not be interested in the resulting modification of the ground state energy. As a result, we do not need to worry about commutators.

In Eq. (7.47) the term $b_{\mathbf{k}}^{\dagger} b_{\mathbf{k}}$ gives a $-u_{\mathbf{k}} v_{\mathbf{k}} B_{\mathbf{k}}^{\dagger} B_{-\mathbf{k}}^{\dagger}$ contribution; from $b_{\mathbf{k}}^{\dagger} b_{-\mathbf{k}}^{\dagger}$ we get $u_{\mathbf{k}}^2 B_{\mathbf{k}}^{\dagger} B_{-\mathbf{k}}^{\dagger}$ and from $b_{\mathbf{k}} b_{-\mathbf{k}}$ we have $v_{\mathbf{k}}^2 B_{\mathbf{k}}^{\dagger} B_{-\mathbf{k}}^{\dagger}$. Hence the condition that there is no overall $B_{\mathbf{k}}^{\dagger} B_{-\mathbf{k}}^{\dagger}$ term reads

$$\frac{NV_0}{2}(u_{\mathbf{k}}^2 + v_{\mathbf{k}}^2) = (\epsilon_{\mathbf{k}} + NV_0) u_{\mathbf{k}} v_{\mathbf{k}} \tag{7.53}$$

The condition that there is no $B_{\mathbf{k}} B_{-\mathbf{k}}$ term gives the same result, since it is the Hermitian conjugate term. From Eq. (7.53) and Eq. (7.50), we obtain

$$4u_{\mathbf{k}}^2 v_{\mathbf{k}}^2 = (u_{\mathbf{k}}^2 + v_{\mathbf{k}}^2)^2 - (u_{\mathbf{k}}^2 - v_{\mathbf{k}}^2)^2 = 4u_{\mathbf{k}}^2 v_{\mathbf{k}}^2 \left(\frac{\epsilon_{\mathbf{k}} + NV_0}{NV_0} \right)^2 - 1 \tag{7.54}$$

Introducing

$$E_{\mathbf{k}} = \sqrt{(\epsilon_{\mathbf{k}} + NV_0)^2 - (NV_0)^2} = \sqrt{\epsilon_{\mathbf{k}}(\epsilon_{\mathbf{k}} + 2NV_0)} \tag{7.55}$$

and, choosing the positive determination for the square root, we can rewrite Eq. (7.54) as

$$2u_{\mathbf{k}} v_{\mathbf{k}} = \frac{NV_0}{E_{\mathbf{k}}} \tag{7.56}$$

Introducing this result on the right-hand side of Eq. (7.53) and combining again with Eq. (7.50), we obtain the explicit expressions of $u_{\mathbf{k}}$ and $v_{\mathbf{k}}$

$$u_{\mathbf{k}} = \sqrt{\frac{1}{2}\left(\frac{\epsilon_{\mathbf{k}} + NV_0}{E_{\mathbf{k}}} + 1 \right)} \qquad v_{\mathbf{k}} = \sqrt{\frac{1}{2}\left(\frac{\epsilon_{\mathbf{k}} + NV_0}{E_{\mathbf{k}}} - 1 \right)} \tag{7.57}$$

Finally, we need to write the expression of the resulting diagonalized Hamiltonian. We collect now the $B_{\mathbf{k}}^{\dagger} B_{\mathbf{k}}$ and $B_{-\mathbf{k}}^{\dagger} B_{-\mathbf{k}}$ contributions resulting from the introduction of the Bogoliubov transformation in Eq. (7.47). The $b_{\mathbf{k}}^{\dagger} b_{\mathbf{k}}$ term gives a $u_{\mathbf{k}}^2 B_{\mathbf{k}}^{\dagger} B_{\mathbf{k}} + v_{\mathbf{k}}^2 B_{-\mathbf{k}} B_{-\mathbf{k}}^{\dagger}$ contribution, while $b_{\mathbf{k}}^{\dagger} b_{-\mathbf{k}}^{\dagger}$ gives $-u_{\mathbf{k}} v_{\mathbf{k}} (B_{\mathbf{k}}^{\dagger} B_{\mathbf{k}} + B_{-\mathbf{k}} B_{-\mathbf{k}}^{\dagger})$, whereas $b_{\mathbf{k}} b_{-\mathbf{k}}$ provides $-u_{\mathbf{k}} v_{\mathbf{k}} (B_{\mathbf{k}} B_{\mathbf{k}}^{\dagger} + B_{-\mathbf{k}}^{\dagger} B_{-\mathbf{k}})$. Just as we have done previously, we do not worry about non-zero commutators, since they lead to contributions to the ground state energy, which we do not consider. Hence we may replace $B_{\mathbf{k}} B_{\mathbf{k}}^{\dagger}$ with $B_{\mathbf{k}}^{\dagger} B_{\mathbf{k}}$, and similarly for $B_{-\mathbf{k}} B_{-\mathbf{k}}^{\dagger}$. Moreover, we can change $\mathbf{k}$ into $-\mathbf{k}$ in the summation over $\mathbf{k}$, so that we end up with the single operator $B_{\mathbf{k}}^{\dagger} B_{\mathbf{k}}$, with a factor found by collecting all the coefficients we have just found above, that is,

$$(u_{\mathbf{k}}^2 + v_{\mathbf{k}}^2)(\epsilon_{\mathbf{k}} + NV_0) - 2u_{\mathbf{k}} v_{\mathbf{k}} NV_0 = \frac{(\epsilon_{\mathbf{k}} + NV_0)^2 - (NV_0)^2}{E_{\mathbf{k}}} = E_{\mathbf{k}} \tag{7.58}$$

where we have used Eq. (7.53) and Eq. (7.56), and the definition Eq. (7.55) for $E_{\mathbf{k}}$. As a result, up to a constant which goes into the ground state energy, Hamiltonian Eq. (7.47) becomes merely

$$H = \sum_{\mathbf{k}} E_{\mathbf{k}} B_{\mathbf{k}}^{\dagger} B_{\mathbf{k}} \qquad (7.59)$$

which is just our desired result from the start.

From Eq. (7.59) we conclude that within this Bogoliubov approximation, the elementary excitations of this weakly interacting Bose gas are non-interacting bosons, with creation and annihilation operators $B_{\mathbf{k}}^{\dagger}$ and $B_{\mathbf{k}}$, and energy $E_{\mathbf{k}}$. For large wavevectors $\epsilon_{\mathbf{k}} \gg NV_0$, and it is clear from the start that in the Hamiltonian Eq. (7.47), the interaction term is negligible compared to the kinetic energy. Hence we should recover free boson physics. Indeed, in this case, Eq. (7.55) gives $E_{\mathbf{k}} \simeq \epsilon_{\mathbf{k}}$. In the same way, we have from Eq. (7.57) $u_{\mathbf{k}} \simeq 1$ and $v_{\mathbf{k}} \simeq 0$, so that from Eq. (7.48) $B_{\mathbf{k}}^{\dagger}$ is essentially identical to $b_{\mathbf{k}}^{\dagger}$. Hence in this regime the elementary excitations are identical to the bare bosons, creating an excitation is just the same as creating a boson.

On the other hand, the situation is completely different in the opposite limit of small wavevectors $\mathbf{k} \to \mathbf{0}$. In this case, $\epsilon_{\mathbf{k}} \ll NV_0$ and, from Eq. (7.55), the dispersion relation of the elementary excitations is given by $E_{\mathbf{k}} \simeq \hbar k \sqrt{NV_0/m} = \hbar k c_s$, where c_s is the sound velocity given by Eq. (7.43), since we have $N = n$ because we deal with a system having its volume equal to unity. Hence in this regime the elementary excitations coincide with the quanta of sound waves. Obviously it is natural to have sound waves propagating in the weakly interacting Bose gas, and this field of small oscillations should be quantized (in the same way as lattice vibrations are quantized into phonons in a solid). Hence it is physically clear that the quanta corresponding to sound waves should be present in the Bose gas at low energy. Accordingly, it is quite satisfactory that Bogoliubov theory gives precisely these quanta as low-energy elementary excitations.

Notice that in this case, from Eq. (7.57), $u_{\mathbf{k}}$ and $v_{\mathbf{k}}$ are essentially equal (and large). So $B_{\mathbf{k}}^{\dagger}$ is an equal combination of $b_{\mathbf{k}}^{\dagger}$ and $b_{-\mathbf{k}}$ (both corresponding to give a momentum $\hbar k$ to the system). This combination of adding and removing a boson is physically coherent with a sound wave, when one goes to the macroscopic limit of a large number of quanta. This result of Bogoliubov theory is also quite remarkable because the quadratic dispersion relation $\hbar^2 k^2 / 2m$ of the perfect Bose gas is transformed into the linear dispersion $\hbar k c_s$. This is a qualitative modification brought by a weak interaction, whereas one could have thought naively at first that a small interaction could only produce small modifications anyway.

Let us finally indicate that if we had explicitly calculated the ground state energy by including the terms we have left aside in the above calculation, we would have encountered a problem because, in summing over $\mathbf{k}$, we would have found an integral that diverges for large values of the wavevector. Ultimately this divergence is linked to the fact that we have immediately replaced in Eq. (7.22) the Fourier transform $V_{\mathbf{q}}$ of the interaction $V(\mathbf{r})$ by its low wavevector limit V_0, because all the involved wavevectors we had to consider were small. However, taking a constant interaction V_0 in Fourier transform is equivalent to take in direct space an interaction $V(\mathbf{r})$ proportional to a Dirac function $\delta(\mathbf{r})$, that is a contact interaction. This is obviously a quite singular potential, and it is known that its peculiar behavior for $\mathbf{r} \to \mathbf{0}$ leads correspondingly in Fourier transform to divergences for large wavevectors, as we have already mentioned for the Cooper problem. Hence this divergency problem can be cured by a more careful handling of the interaction at short distance.

This handling leads to the introduction of a physical quantity, namely the scattering length a, which is enough to wholly characterize the effect of the interaction for atoms (which are the bosons we have mostly in mind) with low enough energy. This scattering length will be introduced explicitly in the next chapter. One then finds a perfectly regular result for the ground state energy in terms of this scattering length. In a way this length a plays a role analogous to V_0 in our calculations. Actually, when the interaction is weak enough to be treated as a perturbation, the scattering length is directly linked to V_0 by $V_0 = 4\pi \hbar^2 a / m$. However, if we think of more practical applications of Bogoliubov theory, for example, for dilute atomic gases, the restriction to a weak interaction potential is a very strong limitation since, at short distances, the interatomic interaction is certainly not weak and cannot be treated as a perturbation. However, in these cases, the scattering length is perfectly well defined. This allows one to extend the domain of validity of Bogoliubov theory, since it relies essentially on the fact that the state $\mathbf{k} = \mathbf{0}$ is macroscopically occupied. The theory remains valid provided the result of interactions is still effectively weak, so that we are physically near the situation of the perfect Bose gas. For example, in Eq. (7.45) the difference $N - N_0$ should be small compared to N_0. In this way, we may stop the expansion at the b_0^2 term, the remaining terms in the interaction being negligible.

More specifically we can calculate at $T = 0$, where the weakly interacting Bose gas is in its ground state $|0\rangle$, the average value $\langle 0| \sum_{\mathbf{k}} b_{\mathbf{k}}^\dagger b_{\mathbf{k}} |0\rangle$. In this ground state we have no elementary excitation so that $B_{\mathbf{k}}|0\rangle = 0$ and $\langle 0|B_{\mathbf{k}}^\dagger = 0$. Hence, from the Bogoliubov transformation Eq. (7.52), we merely have

$$\langle 0|b_{\mathbf{k}}^\dagger b_{\mathbf{k}}|0\rangle = \langle 0|(u_{\mathbf{k}}B_{\mathbf{k}}^\dagger - v_{\mathbf{k}}B_{-\mathbf{k}})(u_{\mathbf{k}}B_{\mathbf{k}} - v_{\mathbf{k}}B_{-\mathbf{k}}^\dagger)|0\rangle = v_{\mathbf{k}}^2 \qquad (7.60)$$

Making use of Eq. (7.57) and going from the $\mathbf{k}$ summation to an integral, we find

$$\langle 0| \sum_{\mathbf{k}} b_{\mathbf{k}}^\dagger b_{\mathbf{k}} |0\rangle = \frac{1}{4\pi^2} \int_0^\infty dk\, k^2 \left(\frac{\epsilon_{\mathbf{k}} + NV_0}{E_{\mathbf{k}}} - 1 \right) = \frac{1}{3\pi^2} \left(\frac{mNV_0}{\hbar^2} \right)^{3/2} \qquad (7.61)$$

where in the last step we have made the change of variable $\epsilon_{\mathbf{k}} = NV_0 y$, that is, $k = (2mNV_0/\hbar^2)^{1/2}\sqrt{y}$. This leads to the appearance of the convergent integral $\int_0^\infty dy\,((y+2)^{1/2} - (y+2)^{-1/2} - y^{1/2})$, which is equal to $2^{3/2}/3$.

From Eq. (7.61) the validity condition $\langle 0|b_{\mathbf{k}}^\dagger b_{\mathbf{k}}|0\rangle \ll N$ for Bogoliubov theory becomes $mN^{1/3}V_0/\hbar^2 \ll 1$. When we substitute the above relation between V_0 and the scattering length a, and recall that $N = n$, the particle density, since the volume of our system is unity, this gives

$$n^{1/3}a \ll 1 \qquad (7.62)$$

that is the interparticle distance $\sim n^{-1/3}$ should be much larger than the scattering length. This implies that the dilution of the Bose gas should be large enough. When the interaction is not weak and cannot be treated as a perturbation, one can show that the validity condition for Bogoliubov theory is still given by Eq. (7.62) in terms of the scattering length. In most cases, this means in practice that the interparticle distance should be large compared to the range of the interatomic interaction.

7.4 The Case of ^{4}He

As we have seen, the discovery of superfluidity in liquid ^{4}He played a quite important role in the development of physical ideas leading to the understanding of standard super-conductivity. As we mentioned, it was suggested by London that this remarkable coherent behavior was in some way related to Bose–Einstein condensation. However, liquid ^{4}He is extremely far from the non-interacting Bose gas considered in Bose–Einstein conden-sation. Indeed, apart from the weak van der Waals forces responsible for the liquid stability, ^{4}He atoms have at shorter distance a very strong hard core repulsion, which makes them physically quite similar to billiard balls. Moreover, the liquid density is such that these balls are not so far from being, so to speak, in contact. Actually, applying some moder-ate pressure, specifically 25 atmospheres, is enough to turn ^{4}He into a solid. The liquid exists, at lower pressure and down to $T = 0$, only because the helium mass is so light that the quantum fluctuations implied by the Heisenberg uncertainty principle become quite important. The resulting zero-point motion of atoms is analogous to the thermal agitation brought by a temperature increase, and it produces on the solid a kind of quantum melting responsible for the existence of the liquid. Clearly, with ^{4}He, one has to deal with a very strongly interacting liquid, quite the opposite of a non-interacting gas.

This difficulty led Landau to develop [12] directly from quantum principles a hydro-dynamic theory, valid independently of the strength of the interactions and accordingly suited for a strongly interacting liquid, such as liquid ^{4}He. Based on this theory, Lan-dau could identify the elementary excitations in liquid ^{4}He. These elementary excitations form, following Landau, the basis of the physical understanding of quantum systems at low enough energy and temperature. All the low energy eigenstates of the system are made of a combination of these elementary excitations, which are so to speak the building blocks of the low energy states. We have already seen in Chapter 2 how this principle allows one to understand the low energy eigenstates for a (possibly strongly interacting) Fermi liquid. Here we have the same idea applied to the Bose liquid.

From his hydrodynamics, Landau found that the liquid supported sound waves, and as a result the quanta of this vibration field, namely the corresponding phonons, were among the elementary excitations that Landau identified. However, he pointed out the existence of another class of excitations. Indeed he found from his hydrodynamics that the liquid could not only support as expected flow fields without any vorticity, such as the one cor-responding to sound waves, but also flow fields carrying vorticity. However, in contrast to what happens for a classical fluid, the vorticity corresponding to these flows could not be arbitrarily small. This situation is analogous to the quantized vortices we have seen in Chapter 6. As a consequence, one cannot go continuously from an elementary excitation without any vorticity to an elementary excitation with a nonzero vorticity. Consequently,

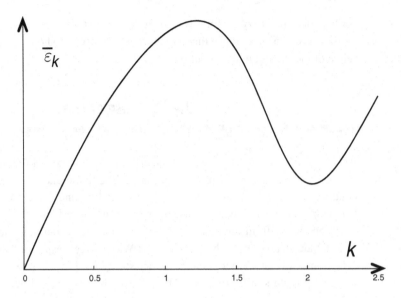

Fig. 7.1   Qualitative behavior of the dispersion relation $\bar{\epsilon}_k$ for elementary excitations in ^{4}He, showing in particular, the small k linear behavior corresponding to phonons and the roton minimum for k around 2 Å^{-1}. The k axis unit is Å^{-1}.

Landau introduced another class of elementary excitations, which he called "rotons." While at first, he took the (positive) minimum energy of these rotons to correspond to zero momentum, he then put it at a finite momentum. In this way, the dispersion relation $\bar{\epsilon}_k$ for elementary excitations could be a continuous function, with a local minimum energy corresponding to the rotons. Such a dispersion relation is displayed qualitatively in Fig. (7.1). It has been later on confirmed by experiments where neutrons were scattered inelastically by liquid ^{4}He.

A very interesting way to understand this peculiar dispersion relation has been found by Feynman [44]. Even if the resulting formula is approximate for ^{4}He, it is highly suggestive. It relates the dispersion relation $\bar{\epsilon}_k$ to the static structure factor $S(k)$. This structure factor itself is directly related to the two-atoms correlation function, that is, the probability $n_2(\mathbf{r}_1, \mathbf{r}_2)$ to find simultaneously one ^{4}He atom located at $\mathbf{r}_1$ and another at $\mathbf{r}_2$. When the system is uniform, as is the case for liquid ^{4}He, this probability depends only on the difference $\mathbf{r} = \mathbf{r}_1 - \mathbf{r}_2$, so that $n_2(\mathbf{r}_1, \mathbf{r}_2) = n_2(\mathbf{r})$. The static structure factor $S(k)$ is essentially the Fourier transform of $n_2(\mathbf{r})$. Feynman formula reads

$$\bar{\epsilon}_k = \frac{\hbar^2 k^2}{2m \, S(k)} \tag{7.63}$$

Details about the derivation of this formula are given in the Further Reading section at the end of this chapter.

While this relation is approximate for ^{4}He, it is exact for the case of the dilute Bose gas (see the Further Reading section). In this case we can apply Bogoliubov theory, and the dispersion relation is identical to the excitation energy given by Eq. (7.55), so that $\bar{\epsilon}_k = E_k$. This allows us to look directly from Eq. (7.63) at the behavior of $S(k)$ in limiting situations.

First, in the case $k \to \infty$, we merely have from Eq. (7.55) $E_k = \hbar^2 k^2 / 2m$, so that $S(k) = 1$ in this limit. This simple result is actually valid for any system. It results from the fact that in the density–density correlation function, there is always an autocorrelation contribution corresponding to the fact that a given atom is always correlated with itself. This gives rise to a contribution proportional to $\delta(\mathbf{r})$ in the correlation function $n_2(\mathbf{r})$. Corresponding to this very short distance behavior the Fourier transform $S(k)$ goes to 1 in the corresponding domain of large wavevectors.

Considering now the opposite limit of small wavevectors $k \to 0$, we have seen that $E_k = \hbar k c_s$ in this case. From Eq. (7.63) we obtain the linear behavior $S(k) = \hbar k / (2mc_s)$. Again this result is actually valid for any system, and in particular, for ^{4}He. In this case, we can understand it by the fact that the elementary excitations description becomes exact in the limit of zero energy. In this situation, the elementary excitations are the quanta of sound waves, and accordingly their dispersion relation is $\bar{\epsilon}_k = \hbar k c_s$. Moreover, in this zero energy limit, the Feynman formula becomes exact (see the Further Reading section). Hence the above behavior of $S(k)$ is also valid in the case of ^{4}He.

In the case of the dilute Bose gas, from Eq. (7.55) and Eq. (7.63), one sees readily that $S(k)$ grows monotonously from $S(k) = \hbar k / (2mc_s)$ to $S(k) = 1$ when k goes from zero to infinity. As we have just seen, for ^{4}He the two limiting behaviors are the same, but it is no longer true that one has a monotonous growth. Although a quantitative evaluation of $S(k)$ for ^{4}He is quite a complicated theoretical problem, it is fairly easy to qualitatively understand what happens.

Indeed, as we have indicated, due to the hard core repulsion, the ^{4}He atom behaves nearly as a hard ball of radius b. Hence if two atoms "touch" each other, the distance between their center is $2b$ and the probability to have them closer is essentially zero. So if we consider the correlation function $n_2(\mathbf{r})$, it is zero for $r < 2b$. Then we have mentioned that in liquid ^{4}He, these hard balls are not so far from touching each other. This implies that there is a high probability to find the nearest neighbor of a given atom at a distance that is higher than $2b$, but not so far from it. Correspondingly $n_2(\mathbf{r})$ has a marked maximum at some r_0, which is somewhat higher than $2b$.

Then the next-nearest neighbor is markedly farther from our given atom, since it is kept at some distance by the presence of the nearest neighbors. Hence $n_2(\mathbf{r})$ decreases beyond the nearest neighbor maximum and grows again to a second maximum corresponding to the next-nearest neighbors. If we had a one-dimensional solid, with the atoms regularly spaced, $n_2(\mathbf{r})$ would display in this way regular oscillations, up to infinity. In the present case, both because we deal with a three-dimensional system, and because it is a liquid, these oscillations are rapidly damped. For large distances, the atomic positions are essentially uncorrelated, so that $n_2(\mathbf{r})$ goes to a constant for $r \to \infty$. This situation is sketched in Fig. 7.2 for the reduced correlation function $\bar{n}_2(r)$, normalized to go to unity for large distance r. Also our qualitative description corresponds to a classical system, but as we have indicated ^{4}He is a quantum liquid: the light mass of the atom makes the zero-point motion quite important. This makes our hard balls effectively softer and larger than in the classical system. But this does not modify the qualitative picture.

We come now to $S(k)$, which is essentially the Fourier transform of $n_2(\mathbf{r})$. If $n_2(\mathbf{r})$ had regular oscillations, we would have in its Fourier transform a delta function at the

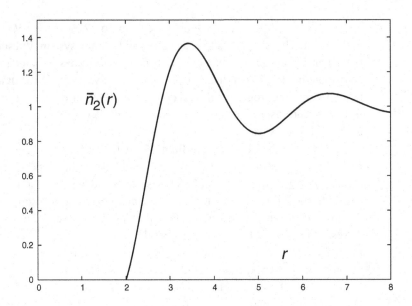

Fig. 7.2 Qualitative behavior of the reduced correlation function $\bar{n}_2(r)$ between two ^{4}He atoms as a function of their distance r in Å.

wavevector corresponding to these oscillations. Since the oscillations in $n_2(\mathbf{r})$ are actually strongly damped, the peak in the Fourier transform is broadened, but it is nevertheless present at the wavevector roughly equal to $2\pi/r_0$. Taking into account the known behavior for $k \to 0$ and $k \to \infty$, this leads for $S(k)$ to an overall dependence on k sketched in Fig. (7.3). When this $S(k)$ is inserted in Eq. (7.63), this leads for $\bar{\epsilon}_k$ to a qualitative behavior that is in agreement with Fig. (7.1), and in particular, to a minimum that is perfectly consistent with the experimental result for the roton wavevector, found near 2 Å^{-1}.

Feynman relation leads naturally to a more recent view of the roton minimum. If we imagine a sharper peak in $S(k)$ corresponding to a better correlation between the positions of the ^{4}He atoms, this would lead to a lower energy for the roton minimum. This could go down to the situation where this energy would reach zero. Such a situation is physically quite meaningful. This corresponds to the case where an elementary excitation with a nonzero wavevector, for example, a (very short wavelength) sound wave, is so to speak frozen, since zero frequency for this excitation means that there is no time evolution. However, the occurrence of a frozen sound wave, with a specific nonzero wavevector, merely signifies that the system has turned into a solid. This really makes sense for liquid ^{4}He, since we have mentioned that a moderate increase in pressure makes it turn into a solid. Hence a possible interpretation of the roton minimum is that it is a manifestation of the fact that liquid ^{4}He is not so far from becoming a solid. This is consistent with the experimental fact that the energy of the roton minimum decreases with increasing pressure. However, the transition toward the solid is not a continuous one, since it is a first-order transition.

In conclusion, all the above considerations on ^{4}He have no direct link with Bose–Einstein condensation, and one may wonder if there is some evidence of it in this strongly

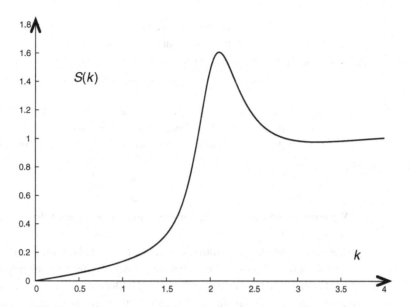

Fig. 7.3 Qualitative behavior of the static structure factor $S(k)$ in ^{4}He.

interacting liquid. Nevertheless it has been possible to experimentally access to the population of the $\mathbf{k} = \mathbf{0}$ state by neutron scattering experiments, and to show in this way that around 10% of the atoms are in this state at low temperature. Naturally, as we have seen in Eq. (7.61) for the weakly interacting Bose gas at $T = 0$, the condensate is also made of $\mathbf{k} \neq \mathbf{0}$ states coherently linked to the $\mathbf{k} = \mathbf{0}$ state.

7.5 Superfluidity

When superfluid properties are considered as the result of Bose–Einstein condensation, the coherent flow implied by these properties is naturally ascribed to the condensate. In the case of the perfect gas, the condensate is at $\mathbf{k} = \mathbf{0}$, which corresponds to an absence of flow. A uniform fluid flow at some velocity $\mathbf{v}_s$ is merely obtained by having the condensation occurring at the wavevector $\mathbf{k}_s = m\mathbf{v}_s/\hbar$, instead of $\mathbf{k} = \mathbf{0}$. As a result, the wave function of the N bosons in the condensate reads

$$\Psi_N(\mathbf{r}_1, \mathbf{r}_2, \cdots, \mathbf{r}_N) = \prod_i \exp\left(i\frac{m\mathbf{v}_s \cdot \mathbf{r}_i}{\hbar}\right) \tag{7.64}$$

One can also view it as the physical situation seen when one changes the reference frame, as the result of a Galilean transformation, with the new reference frame moving at velocity $-\mathbf{v}_s$ with respect to the $\mathbf{k} = \mathbf{0}$ condensate.

More formally, let us consider the action on the condensate of the boson annihilation operator $\psi(\mathbf{r}) = \sum_{\mathbf{k}} e^{i\mathbf{k}\cdot\mathbf{r}} b_{\mathbf{k}}$ (see Eq. (2.31)). If the condensate is at $\mathbf{k} = \mathbf{0}$, only the term $\mathbf{k} = \mathbf{0}$ is important in this sum and, with $\langle N - 1|b_0|N\rangle$

$= \sqrt{N}$, we find $\langle N - 1|\psi(\mathbf{r})|N\rangle = \sqrt{N}$, which is independent of $\mathbf{r}$ as expected for uniform condensate. On the other hand, for a uniform condensate moving with velocity $\mathbf{v}_s$, only the term $\mathbf{k} = \mathbf{k}_s$ is important in the sum, and we find accordingly

$$\langle N - 1|\psi(\mathbf{r})|N\rangle = \sqrt{N}\, e^{i\mathbf{k}_s \cdot \mathbf{r}} \tag{7.65}$$

Hence the average $\langle \psi(\mathbf{r})\rangle \equiv \langle N - 1|\psi(\mathbf{r})|N\rangle$ gets a space-dependent phase factor, with the corresponding phase $S(\mathbf{r}) = \mathbf{k}_s \cdot \mathbf{r}$ being directly related to the velocity $\mathbf{v}_s$. This phase factor is the same as the one we had for each particle in Eq. (7.64). This relation can be written as well:

$$\mathbf{v}_s = \frac{\hbar}{m} \nabla S(\mathbf{r}) \tag{7.66}$$

We have found this phase factor explicitly in the case of the perfect gas. However, we have noticed that this phase factor can also be obtained by a change of reference frame, and as such it results also from a Galilean transformation, although we do not go into the practical, technical details of performing this transformation. The important point is that the phase factor for $\langle \psi(\mathbf{r})\rangle$, resulting from this Galilean transformation, clearly comes from the characteristics of the transformation, namely the velocity $\mathbf{v}_s$, and is independent of whether or not there are interactions between the bosons. As a result, the relation Eq. (7.66) between the velocity $\mathbf{v}_s$ of the superfluid and the phase $S(\mathbf{r})$ is also valid for a Bose condensate when interactions between the bosons are present, so that the many-body wave function is naturally no longer given by Eq. (7.64). This could be checked in the case of a weakly interacting Bose gas, for which Bogoliubov theory provides an explicit solution. But the validity of this relation extends to the case of strongly interacting Bose liquid, such as ^{4}He, for which we do not have such a solution.

Another interest of Eq. (7.66) is that its validity can be extended to the case where the superfluid flow is no longer uniform, which arises naturally when the boundary conditions, due to the shape of the vessel containing the gas, are not compatible with a uniform flow. This occurs when we consider a flow which, at a macroscopic scale, is not uniform. Nevertheless, if we look at a small enough part of the fluid, it can be considered locally as uniform, so that the relation Eq. (7.66) between the local phase and the local value of the superfluid velocity is still valid. The conditions of validity are just the same as the ones for which hydrodynamic equations give a proper description of the fluid. Since in this case the superfluid velocity depends on position, we should make this explicit in the equation and write

$$\mathbf{v}_s(\mathbf{r}) = \frac{\hbar}{m} \nabla S(\mathbf{r}) \tag{7.67}$$

The validity of this relation also can be extended to the case where $\mathbf{v}_s(\mathbf{r})$ and $S(\mathbf{r})$ have slow enough time dependences.

An important physical consequence of Eq. (7.67) is that the superfluid velocity $\mathbf{v}_s(\mathbf{r})$ is a potential flow, so that it is irrotational

$$\operatorname{curl} \mathbf{v}_s(\mathbf{r}) = \mathbf{0} \tag{7.68}$$

This implies that $\mathbf{v}_s(\mathbf{r})$ cannot take the standard form $\Omega \times \mathbf{r}$ corresponding to a rotating fluid, so that it is not possible to set the superfluid in rotation. This is a remarkable and quite surprising property. This is at least what indeed happens for a slow enough motion of the walls through which one tries to make the superfluid rotate. However, when one tries to impart a rapid enough rotation to the superfluid and one goes beyond some critical velocity for these moving walls, a vortex appears in the superfluid. This vortex is quite similar to the ones we have seen appearing in the preceding chapter for type II superconductors beyond the lower critical field H_{c1}. For the flow corresponding to this vortex Eq. (7.68) is still valid, except at the center of the vortex, which is a singular point where curl $\mathbf{v}_s(\mathbf{r})$ has a contribution proportional to a delta function. Much as for a superconducting vortex, super-fluidity is destroyed in the vicinity of the center over a typical distance called the "healing length" ξ, which is precisely defined as $\xi = \hbar/(\sqrt{2}mc_s)$ in the case of the dilute Bose gas following Bogoliubov theory. If one keeps increasing the wall's velocity, additional vortices appear, and one can show that, when one reaches a large number of vortices, the resulting average velocity field gets close to the standard form $\Omega \times \mathbf{r}$ taken by an ordinary fluid under rotation.

7.6 Critical Velocity

At this stage, although we have described the velocity field corresponding to the condensate in some detail, we have not considered if the corresponding motion has the property of superfluidity. Actually, the opposite seems to be true for the case of the perfect gas. Indeed if one considers such a gas moving with velocity $\mathbf{v}_s$ with respect to a wall, there is apparently nothing to prevent bosons from this condensate to go, for example, by colliding with the wall, from the $\mathbf{k}_s = m\mathbf{v}_s/\hbar$ state to the $\mathbf{k} = \mathbf{0}$ state, which corresponds to the lowest energy, and, for example, to build up in this way a $\mathbf{k} = \mathbf{0}$ condensate corresponding to the equilibrium situation with respect to the wall. This would be a manifestation of a viscous behavior with respect to the wall, in contradiction with the expected superfluidity.

The required condition for a superfluid behavior has been examined by Landau [45] in general terms. Let us consider a condensate moving at velocity $\mathbf{v}_s$ with respect to a wall. The manifestation of a viscous, dissipative behavior for the condensate would be a momentum transfer to the wall, leading to a decrease in the condensate velocity. This would occur through the creation of excited states, carrying the lost momentum. For low superfluid velocity, these excited states are expected to have low energies and accordingly to be made of elementary excitations. Hence we are led to consider if the first step in this process, that is, the creation of a single elementary excitation, is energetically favorable.

We know that the energy of an elementary excitation of momentum $\mathbf{p} = \hbar\mathbf{k}$ is $\bar{\epsilon}_\mathbf{k}$. However, this is valid with respect to the condensate at rest. In the present case, the condensate is moving at velocity $\mathbf{v}_s$ with respect to the wall. So in order to be at rest with the wall and obtain the corresponding elementary excitation energy, we must perform a change to a reference frame going with velocity $-\mathbf{v}_s$ with respect to the condensate. In such a change, the interaction energy of the particles is not modified since it depends only on their relative positions. Only the kinetic energy is modified. When one goes to a new reference frame with velocity $-\mathbf{v}_s$, the momentum $\mathbf{p}$ of a particle becomes $\mathbf{p}' = \mathbf{p} + m\mathbf{v}_s$. Correspondingly its kinetic energy $\mathbf{p}^2/2m$ becomes $\mathbf{p}'^2/2m = \mathbf{p}^2/2m + \mathbf{v}_s \cdot \mathbf{p} + m\mathbf{v}_s^2/2$. Summing over all the N particles, and adding the interaction energy contribution, we find that the energy E' in the new reference frame is related to the energy E in the original frame by

$$E' = E + \mathbf{v}_s \cdot \mathbf{P} + \frac{1}{2}M\mathbf{v}_s^2 \qquad (7.69)$$

where $M = Nm$ is the total mass, and $\mathbf{P} = \sum_i \mathbf{p}_i$ is the total momentum of the system in the original frame. Since creating an elementary excitation at $\mathbf{k} = \mathbf{0}$ does not cost any energy, we can consider that the energy $\bar{\epsilon}_\mathbf{k}$ of an elementary excitation with momentum $\mathbf{p} = \hbar\mathbf{k}$ is the difference in the system energy $E_\mathbf{p} - E_\mathbf{0}$ between the system energy with elementary excitation $\mathbf{p}$ and the same system with elementary excitation $\mathbf{0}$, which is identical to the ground state energy. Now, if we look from Eq. (7.69) at the value of the corresponding energy difference $E'_\mathbf{p} - E'_\mathbf{0}$ in the new frame, we find

$$E'_\mathbf{p} - E'_\mathbf{0} = E_\mathbf{p} - E_\mathbf{0} + \mathbf{v}_s \cdot \mathbf{p} = \bar{\epsilon}_\mathbf{k} + \mathbf{v}_s \cdot \mathbf{p} \qquad (7.70)$$

Hence, in the new frame, the elementary excitation spectrum is shifted by $\mathbf{v}_s \cdot \mathbf{p}$.

Now, in order to result in the creation of this elementary excitation, the process must lead to a decrease in energy (the excess energy going into dissipative processes) or at least energy must be conserved. This leads to the condition

$$\bar{\epsilon}_\mathbf{k} + \mathbf{v}_s \cdot \mathbf{p} \leq 0 \qquad (7.71)$$

The left-hand side is minimal if $\mathbf{p} = \hbar\mathbf{k}$ is oriented in the direction opposite to $\mathbf{v}_s$, in which case its value is $\bar{\epsilon}_k - \hbar k v_s$. If this quantity is positive, there is no way in which condition Eq. (7.71) can be satisfied. Hence no elementary excitation with wavevector k can be created if the superfluid velocity is less than $\bar{\epsilon}_k/(\hbar k)$. This quantity itself depends on k, and the creation of an elementary excitation, whatever its wavevector, is strictly impossible only if v_s is less than its minimum with respect to k. In this way, we arrive at Landau's criterion for superfluidity: for a condensate with velocity v_s smaller than the critical velocity v_c defined by

$$v_c = \min_k \left| \left(\frac{\bar{\epsilon}_k}{\hbar k} \right) \right. \qquad (7.72)$$

no dissipative processes due to the creation of elementary excitations can occur. In such a case no viscous processes will arise with this condensate, so it will indeed behave as a superfluid.

Let us now evaluate this critical velocity for the various condensates we have considered. In the graph of $\bar{\epsilon}_k$ versus k, $\bar{\epsilon}_k/k$ is geometrically the slope of the straight line going from the origin to the point with coordinates $(k, \bar{\epsilon}_k)$. So $\bar{\epsilon}_k/k$ is minimal when this straight line is just touching the line representing $\bar{\epsilon}_k$. This means that it is tangent to $\bar{\epsilon}_k$, and the critical velocity is directly related to the slope of this tangent.

In the case of the perfect Bose gas, $\bar{\epsilon}_k = \hbar^2 k^2 / 2m$, and the tangent to this parabola starting from the origin is just the k-axis. The corresponding slope is zero, so that $v_c = 0$. The minimum of $\bar{\epsilon}_k / k$ is indeed zero, obtained for $k = 0$. Hence we conclude that the perfect Bose gas does not behave as a superfluid, since at any nonzero velocity, viscous behavior will occur. This corresponds to the physical suspicion we have mentioned above.

On the other hand, for the weakly interacting Bose gas described by Bogoliubov theory, from Eq. (7.55) the dispersion relation E_k starts from the origin with a slope equal to $\hbar c_s$ and bends upward toward the $\hbar^2 k^2 / 2m$ behavior valid at large k. So the tangent we are looking for is just the tangent at the origin, and its slope is $\hbar c_s$. This means from Eq. (7.72) that the critical velocity is equal to the sound velocity c_s, which is nonzero. Therefore, we conclude that this weakly interacting Bose gas is superfluid. We see that starting with the non-superfluid perfect gas, interactions play an essential role for converting it into a superfluid. This is quite remarkable since this does not depend on the strength of these interactions. Hence the perfect Bose gas appears as a kind of singular situation. Nevertheless, if the interactions are very weak, the sound velocity will be small, as will the critical velocity. Hence even if this gas behaves as a superfluid, this is for such small velocities that this will hardly be observable. So in this way, one goes continuously to the perfect gas limit, for which no superfluid behavior is observable.

In the case of superfluid ^{4}He we have also the phonon branch for the elementary excitations, and we might think that the critical velocity is equal to the sound velocity, just as for the weakly interacting Bose gas. However, due to the roton minimum, we have also to consider the possibility that the tangent from the origin touches the dispersion relation in the vicinity of this minimum, as can be seen from Fig. (7.4). This occurs if the energy of

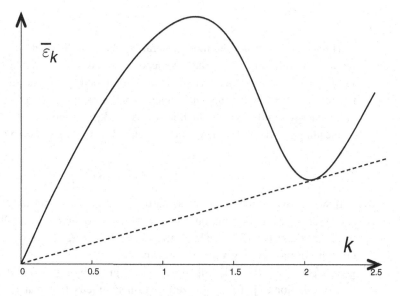

Fig. 7.4   Dashed line: Tangent, through the origin, to the ^{4}He elementary excitation dispersion relation Fig. (7.1), the slope giving the superfluid ^{4}He critical velocity, according to Landau's criterion Eq. (7.72).

this minimum is low enough, and this is indeed what happens with the experimental values. In this way, the critical velocity corresponding to Landau's criterion is due to the creation of rotons, and since the tangent touches the dispersion relation very near the roton minimum, with energy Δ_r and wavevector k_r, this critical velocity is roughly equal to $\Delta_r/\hbar k_r$. However, experimentally the critical velocities observed in superfluid ^{4}He are much lower than this critical velocity given by Landau's criterion. This has been attributed to the creation of excitations linked to vortices. This might be the creation of vortex loops from the superflow, or preexisting vortices in the fluid might help with the generation of such excitations.

Up to now we have only considered the condensate, which is appropriate at zero temperature. However, elementary excitations may also be present, as will be the case at nonzero temperature, which will produce them by thermal excitation. Collectively these elementary excitations form what is called the "normal part" of the fluid, in contrast with the superfluid part corresponding physically to the condensate. This normal fluid behaves physically as an ordinary gas, and in particular, it may display viscosity. This two-fluid picture has been introduced initially by Tisza [46]. However, this picture does not mean at all that some atoms belong to the superfluid and others to the normal fluid. Such a distinction is not possible and is physically incorrect. Rather this two-fluid picture is just a way to describe the whole complicated statistical wave function corresponding to the fluid.

In standard situations, the normal part comes in thermal equilibrium, due to small residual interactions between elementary excitations, which allow them to exchange energy and momentum. In such a case, since these excitations behave as essentially non-interacting bosons, as we have seen in the case of the weakly interacting Bose gas, the standard distribution they follow is

$$\bar{n}_{\mathbf{k}} = \frac{1}{e^{\beta \bar{\epsilon}_{\mathbf{k}}} - 1} \tag{7.73}$$

This is, however, not the most general equilibrium distribution, since momentum conservation in interactions between elementary excitations is compatible with a general drift velocity of these excitations. If $\mathbf{v}_n$ is this drift velocity, the distribution Eq. (7.73) is valid in a reference frame moving with velocity $\mathbf{v}_n$. But we have now to use the elementary excitation energy in this frame, which is $\bar{\epsilon}_{\mathbf{k}} - \mathbf{v}_n \cdot \hbar \mathbf{k}$, by following the same argument as the one leading to Eq. (7.70). Hence in this general case, the excitation distribution is

$$\bar{N}_{\mathbf{k}} = \frac{1}{e^{\beta(\bar{\epsilon}_{\mathbf{k}} - \mathbf{v}_n \cdot \hbar \mathbf{k})} - 1} \tag{7.74}$$

If we are interested in the mass current $\mathbf{j}_n$ due to this normal fluid, each elementary excitation gives a contribution $\hbar \mathbf{k}$, and we have to sum it over all the excitations with the distribution Eq. (7.74), which gives $\mathbf{j}_n = \sum_{\mathbf{k}} \hbar \mathbf{k} \bar{N}_{\mathbf{k}}$. If we restrict ourselves to small enough normal velocity $\mathbf{v}_n$, we may expand the distribution $\bar{N}_{\mathbf{k}}$ to first order in $\mathbf{v}_n$. This gives $\bar{N}_{\mathbf{k}} = \bar{n}_{\mathbf{k}} - (\mathbf{v}_n \cdot \hbar \mathbf{k}) \partial \bar{n}_{\mathbf{k}}/\partial \bar{\epsilon}_{\mathbf{k}}$, with $\bar{n}_{\mathbf{k}}$ given by Eq. (7.73). For symmetry reasons, the contribution to $\mathbf{j}_n$ from $\bar{n}_{\mathbf{k}}$ is zero. Again symmetry tells that $\mathbf{j}_n$ will be along $\mathbf{v}_n$. If we call x the direction of $\mathbf{v}_n$, we get a factor k_x^2 in the integral giving $\mathbf{j}_n$. As usual, the angular average of this term gives a $1/3$ factor. This leads for this normal current to the result

$$\mathbf{j}_n = \rho_n \mathbf{v}_n \qquad\qquad \rho_n = \frac{\hbar^2}{6\pi^2} \int_0^\infty dk\, k^4 \left(-\frac{\partial \bar{n}_k}{\partial \bar{\epsilon}_k} \right) \qquad (7.75)$$

where ρ_n is called the "normal density." One finds easily that this normal density is zero at $T = 0$, as is physically expected.

In the above considerations, we have assumed implicitly that the condensate was at rest and $\mathbf{v}_n$ was the drift velocity of the normal part with respect to this condensate. On the other hand, let us consider, as we have done above, a wall with the condensate moving at velocity $\mathbf{v}_s$ with respect to this wall. In this case, the normal part velocity relative to the condensate is $\mathbf{v}_n - \mathbf{v}_s$, and it is this difference which should now enter the normal current Eq. (7.75) with respect to the condensate. If we want the fluid mass current $\mathbf{j}$ in the wall reference frame, we have to go from the condensate frame to the wall frame, and performing this Galilean transformation adds a contribution $\rho \mathbf{v}_s$ to the mass current, where ρ is the mass density of the whole fluid. As a result, adding this contribution to the normal fluid contribution Eq. (7.75) (with $\mathbf{v}_n$ replaced by $\mathbf{v}_n - \mathbf{v}_s$), the total mass current reads

$$\mathbf{j} = \rho_s \mathbf{v}_s + \rho_n \mathbf{v}_n \qquad\qquad \rho_s \equiv \rho - \rho_n \qquad (7.76)$$

where ρ_s is called the "superfluid density," and $\mathbf{v}_s$ as well as $\mathbf{v}_n$ are velocities with respect to the wall. It is clear from its very definition Eq. (7.76) that ρ_s is different from the fraction of bosons present in the Bose condensate.

We have gone up to Eq. (7.76) to show quantitatively how the normal fluid enters the physical description of the system. This is a first step in setting up the general equations of the hydrodynamic theory formulated by Landau for such a fluid. We will not go further in this elaboration. It is however important to note that this quantum hydrodynamics does not require explicitly the existence of a Bose–Einstein condensate, even if ultimately the existence of the superfluid component of the fluid is physically related to Bose–Einstein condensation. In particular, this hydrodynamic theory would apply as well to a BCS type superfluid, made of fermions, where the superfluid part is due to the condensation of Cooper pairs.

7.7 Further Reading: Feynman Relation

Feynman relation can be conveniently obtained from an exact equality known as the f-sum rule. This sum rule is a property of the dynamic structure factor $S(\mathbf{q}, \omega)$, which generalizes the static structure factor $S(\mathbf{q})$ to take into account a frequency dependence. These structure factors are directly related to the density–density correlation function. Considering first the static factor, it is related to the equal time correlation function

$$n_2(\mathbf{r}) = \langle 0 | \rho(\mathbf{r}) \rho(\mathbf{0}) | 0 \rangle \qquad (7.77)$$

We restrict ourselves from the start to a uniform system, so that the density correlation depends only on the relative position $\mathbf{r}$. Here we consider only the $T = 0$ case, so that the average is taken in the ground state $|0\rangle$. In Eq. (7.77), $\rho(\mathbf{r})$ is the density operator

$$\rho(\mathbf{r}) = \sum_i \delta(\mathbf{r} - \mathbf{r}_i) \tag{7.78}$$

where $\mathbf{r}_i$ represents the particles' positions. Its Fourier transform is

$$\rho_{\mathbf{q}} = \int d\mathbf{r}\, \rho(\mathbf{r}) e^{-i\mathbf{q}\cdot\mathbf{r}} = \sum_i e^{-i\mathbf{q}\cdot\mathbf{r}_i} \qquad \rho(\mathbf{r}) = \sum_{\mathbf{q}} e^{i\mathbf{q}\cdot\mathbf{r}} \rho_{\mathbf{q}} \tag{7.79}$$

The second quantized versions of these expressions are (see Eq. (4.8))

$$\rho(\mathbf{r}) = \psi^\dagger(\mathbf{r})\psi(\mathbf{r}) \qquad \rho_{\mathbf{q}} = \sum_{\mathbf{k}} c^\dagger_{\mathbf{k}-\mathbf{q}/2} c_{\mathbf{k}+\mathbf{q}/2} \tag{7.80}$$

The static structure factor $S(\mathbf{q})$ is essentially the Fourier transform[2] of $n_2(\mathbf{r})$. Substituting in Eq. (7.77) the expression Eq. (7.79) for $\rho(\mathbf{r})$ and $\rho(\mathbf{0})$, we have

$$n_2(\mathbf{r}) = \sum_{\mathbf{q},\mathbf{q}'} \langle 0|\rho_{\mathbf{q}}\rho_{\mathbf{q}'}|0\rangle\, e^{i\mathbf{q}\cdot\mathbf{r}} = \sum_{\mathbf{q}} \langle 0|\rho_{\mathbf{q}}\rho_{-\mathbf{q}}|0\rangle\, e^{i\mathbf{q}\cdot\mathbf{r}} \tag{7.81}$$

The last step takes into account that $\langle 0|\rho_{\mathbf{q}}\rho_{\mathbf{q}'}|0\rangle$ is zero unless $\mathbf{q}' = -\mathbf{q}$. Indeed, since from Eq. (7.80) $\rho_{\mathbf{q}}$ increases the momentum of each particle by $-\mathbf{q}$, in order to obtain a nonzero average in the ground state, one must have a global operator $\rho_{\mathbf{q}}\rho_{-\mathbf{q}}$ creating zero total momentum, because the ground state has a well-defined momentum equal to zero. Eq. (7.81) implies that $S(\mathbf{q})$ is given by

$$NS(\mathbf{q}) = \langle 0|\rho_{\mathbf{q}}\rho_{-\mathbf{q}}|0\rangle = \langle 0|\rho_{\mathbf{q}}\rho^\dagger_{\mathbf{q}}|0\rangle \tag{7.82}$$

since $\rho^\dagger_{\mathbf{q}} = \rho_{-\mathbf{q}}$. The factor N (particle number) is the standard convention.

The dynamic structure factor $S(\mathbf{q}, \omega)$ is defined as

$$S(\mathbf{q}, \omega) = \sum_n |\langle n|\rho^\dagger_{\mathbf{q}}|0\rangle|^2\, \delta\left(\hbar\omega - (E_n - E_0)\right) \tag{7.83}$$

where the sum is over all the eigenstates $|n\rangle$ of the Hamiltonian, with energy E_n. Integrating over frequency, one recovers the static structure factor Eq. (7.82)

$$\hbar \int_{-\infty}^{\infty} d\omega\, S(\mathbf{q}, \omega) = \sum_n |\langle n|\rho^\dagger_{\mathbf{q}}|0\rangle|^2 = \sum_n \langle 0|\rho_{\mathbf{q}}|n\rangle\langle n|\rho^\dagger_{\mathbf{q}}|0\rangle = NS(\mathbf{q}) \tag{7.84}$$

where we have used the closure relation $\sum_n |n\rangle\langle n| = 1$. The Fourier transform of $S(\mathbf{q}, \omega)$ with respect to frequency is

$$\hbar \int_{-\infty}^{\infty} d\omega\, e^{-i\omega t} S(\mathbf{q}, \omega) = \sum_n e^{iE_0 t/\hbar} \langle 0|\rho_{\mathbf{q}}|n\rangle e^{-iE_n t/\hbar} \langle n|\rho^\dagger_{\mathbf{q}}|0\rangle \tag{7.85}$$

$$= \sum_n \langle 0|e^{iHt/\hbar}\rho_{\mathbf{q}} e^{-iHt/\hbar}|n\rangle\langle n|\rho^\dagger_{\mathbf{q}}|0\rangle = \langle 0|\rho_{\mathbf{q}}(t)\rho^\dagger_{\mathbf{q}}|0\rangle$$

----

[2] The precise definition is often slightly modified to avoid the peculiar behavior resulting from $n_2(\mathbf{r})$ going at large distances to a nonzero constant, equal to the square of the density. This gives in Fourier transform a singular contribution for $\mathbf{q} = \mathbf{0}$.

where we have introduced the time-dependent operator $\rho_{\mathbf{q}}(t) = e^{iHt/\hbar}\rho_{\mathbf{q}}e^{-iHt/\hbar}$. This is the Fourier transform of the time-dependent density operator $\rho(\mathbf{r}, t) = e^{iHt/\hbar}\rho(\mathbf{r})e^{-iHt/\hbar}$. Hence we see that the dynamic structure factor $S(\mathbf{q}, \omega)$ is directly related to the Fourier transform, with respect to space and time, of the time-dependent density–density correlation function $n_2(\mathbf{r}, t) = \langle 0|\rho(\mathbf{r}, t)\rho(\mathbf{0}, 0)|0\rangle$.

The f-sum rule states that

$$\hbar^2 \int_{-\infty}^{\infty} d\omega\, \omega\, S(\mathbf{q}, \omega) = N \frac{\hbar^2 q^2}{2m} \tag{7.86}$$

It can be proved in the following way. Inserting the definition Eq. (7.83), the left-hand side becomes

$$\hbar^2 \int_{-\infty}^{\infty} d\omega\, \omega\, S(\mathbf{q}, \omega) = \sum_n (E_n - E_0)\langle 0|\rho_{\mathbf{q}}|n\rangle\langle n|\rho_{\mathbf{q}}^\dagger|0\rangle \tag{7.87}$$

$$= \sum_n \langle 0|\rho_{\mathbf{q}}H - H\rho_{\mathbf{q}}|n\rangle\langle n|\rho_{\mathbf{q}}^\dagger|0\rangle = \langle 0|[\rho_{\mathbf{q}}, H]\rho_{\mathbf{q}}^\dagger|0\rangle$$

However, our system is invariant by parity, so $S(\mathbf{q}, \omega)$ is unchanged if $\mathbf{q}$ is changed into $-\mathbf{q}$. Making use of $\rho_{\mathbf{q}}^\dagger = \rho_{-\mathbf{q}}$, we can write

$$\hbar^2 \int_{-\infty}^{\infty} d\omega\, \omega\, S(-\mathbf{q}, \omega) = \sum_n (E_n - E_0)\langle 0|\rho_{\mathbf{q}}^\dagger|n\rangle\langle n|\rho_{\mathbf{q}}|0\rangle \tag{7.88}$$

$$= \sum_n \langle 0|\rho_{\mathbf{q}}^\dagger|n\rangle\langle n|H\rho_{\mathbf{q}} - \rho_{\mathbf{q}}H|0\rangle = \langle 0|\rho_{\mathbf{q}}^\dagger[H, \rho_{\mathbf{q}}]|0\rangle$$

Combining Eq. (7.87) and Eq. (7.88) with parity invariance, we obtain

$$\hbar^2 \int_{-\infty}^{\infty} d\omega\, \omega\, S(\mathbf{q}, \omega) = \frac{1}{2}\langle 0|[[\rho_{\mathbf{q}}, H], \rho_{\mathbf{q}}^\dagger]|0\rangle \tag{7.89}$$

On the other hand, the resulting double commutator on the right-hand side can be directly evaluated. Since our interaction depends only on the particle coordinates $\mathbf{r}_i$, the only part of the Hamiltonian that does not commute with $\rho_{\mathbf{q}} = \sum_i e^{-i\mathbf{q}\cdot\mathbf{r}_i}$ is the kinetic energy $\sum_i \mathbf{p}_i^2/2m$. Hence, to obtain $[\rho_{\mathbf{q}}, H]$, we have to calculate $[e^{-i\mathbf{q}\cdot\mathbf{r}}, \mathbf{p}^2] = \mathbf{p} \cdot [e^{-i\mathbf{q}\cdot\mathbf{r}}, \mathbf{p}] + [e^{-i\mathbf{q}\cdot\mathbf{r}}, \mathbf{p}] \cdot \mathbf{p}$ for each particle, and sum over the particles. For any wave function $\psi(\mathbf{r})$, we have

$$(e^{-i\mathbf{q}\cdot\mathbf{r}}\mathbf{p} - \mathbf{p}\,e^{-i\mathbf{q}\cdot\mathbf{r}})\psi = \frac{\hbar}{i}\left(e^{-i\mathbf{q}\cdot\mathbf{r}}\nabla\psi - \nabla(e^{-i\mathbf{q}\cdot\mathbf{r}}\psi)\right) = \hbar\mathbf{q}\,e^{-i\mathbf{q}\cdot\mathbf{r}}\psi \tag{7.90}$$

which gives for our commutator

$$[e^{-i\mathbf{q}\cdot\mathbf{r}}, \mathbf{p}] = \hbar\mathbf{q}\,e^{-i\mathbf{q}\cdot\mathbf{r}} \tag{7.91}$$

leading to $[e^{-i\mathbf{q}\cdot\mathbf{r}}, \mathbf{p}^2] = \hbar\mathbf{q} \cdot \mathbf{p}\,e^{-i\mathbf{q}\cdot\mathbf{r}} + e^{-i\mathbf{q}\cdot\mathbf{r}}\hbar\mathbf{q} \cdot \mathbf{p}$. We then take the commutator of this result with $\rho_{\mathbf{q}}^\dagger = \sum_i e^{i\mathbf{q}\cdot\mathbf{r}_i}$, proceeding as above. We have to use $[\mathbf{p}, e^{i\mathbf{q}\cdot\mathbf{r}}] = \hbar\mathbf{q}\,e^{i\mathbf{q}\cdot\mathbf{r}}$, which is the Hermitian conjugate of Eq. (7.91). The two terms of $[e^{-i\mathbf{q}\cdot\mathbf{r}}, \mathbf{p}^2]$ give the same contribution $\hbar^2 q^2$, and summing the resulting scalar over all the particles gives merely $2N\hbar^2\mathbf{q}^2$. Carrying this result into Eq. (7.89) leads to the f-sum rule Eq. (7.86).

In the definition Eq. (7.83) for $S(\mathbf{q}, \omega)$, all the possible eigenstates $|n\rangle$ contribute provided they have the appropriate energy $E_n = E_0 + \hbar\omega$ and have a nonzero matrix element

$\langle n|\rho_{\mathbf{q}}^{\dagger}|0\rangle$, which implies that they have a momentum $\hbar\mathbf{q}$. Feynman relation is obtained if one makes the hypothesis that the only contributing eigenstates are the elementary excitations. For a given $\mathbf{q}$, there is only one elementary excitation with energy $E_n - E_0 = \bar{\epsilon}_{\mathbf{q}} = \hbar\omega$, where $\bar{\epsilon}_{\mathbf{q}}$ is the dispersion relation of the elementary excitations. Since in this case we must have $\hbar\omega = \bar{\epsilon}_{\mathbf{q}}$, this implies that $S(\mathbf{q}, \omega)$ is proportional to $\delta(\hbar\omega - \bar{\epsilon}_{\mathbf{q}})$. In order to satisfy the relation Eq. (7.84) with $S(\mathbf{q})$, the coefficient must be $NS(\mathbf{q})$, so that with this hypothesis $S(\mathbf{q}, \omega)$ is given by

$$S(\mathbf{q}, \omega) = NS(\mathbf{q})\,\delta(\hbar\omega - \bar{\epsilon}_{\mathbf{q}}) \tag{7.92}$$

Inserting this explicit expression in the f-sum rule gives

$$\bar{\epsilon}_{\mathbf{q}}\, S(\mathbf{q}) = \frac{\hbar^2 q^2}{2m} \tag{7.93}$$

which is Feynman relation.

In the case of Bogoliubov theory, the excited states are given by Eq. (7.59), and one can see easily that only the elementary excitations with dispersion relation E_k contribute to $S(\mathbf{q}, \omega)$. Hence the above hypothesis is correct in this case, and Feynman relation is exact for the dilute Bose gas described by Bogoliubov theory. In contrast, in the case of liquid ^{4}He, the hypothesis is correct only for vanishing energy. Indeed it is only in this limit that elementary excitations are exact eigenstates. When the energy is raised they have a finite lifetime, which gets shorter at higher energy, so that they are only approximate eigenstates. Moreover, there is also the possibility that states involving multiple excitations contribute. One can nevertheless show that Feynman relation provides an upper bound for the energy of elementary excitations.

The BEC–BCS Crossover

In this chapter, we describe the BEC–BCS crossover that has been recently realized experimentally in ultracold fermionic gases. Together with displaying beautiful experiments, these recent developments in atomic physics have allowed physicists to create and study a number of new superfluid systems. However, the essential motivation for its presence in this book is that it provides explicitly a conceptually very important link between the Bose–Einstein condensation and the BCS condensate. Indeed, in the course of our study of superconductivity, we have often relied on physical ideas coming from Bose–Einstein condensation. Nevertheless the whole development of the physical understanding of superconductivity can be made without any specific reference to Bose–Einstein condensation, and seeing, roughly speaking, superconductivity as a Bose-condensation of Cooper pairs has often been a disputed physical idea. The experimental realization of the BEC–BCS crossover, going continuously from a system displaying a BCS condensation to another one presenting Bose-condensation, has provided a concrete relation between these two condensates that clarifies their relations, allowing one to explicitly see their differences and their similarities.

8.1 Cold Atoms

Obtaining cold atoms has always been a very important aim in atomic physics. The basic reason is a very practical one. Spectral lines, observed in light emission or absorption processes, provide essential information on atomic physical properties. In principle these lines have an intrinsic natural width, due to the intrinsic lifetime of the atomic energy levels corresponding to the initial and final states involved in the process. However, these natural widths are often extremely small and do not correspond to the widths experimentally observed in standard situations. There are indeed extrinsic physical processes that contribute dominantly to the actual spectral width. These are mostly interatomic collisions and the Doppler effect. The resulting spectral line broadening is a nuisance since it not only leads to a limitation in the precision of the determination of the line frequency, but it also quite frequently makes several narrow lines merge into a broader single line, so that rich detailed structural information is completely lost.

Atomic spectral lines are usually observed from a fairly large number of atoms, forming a gas contained in some vessel. This is required in order to have a conveniently strong signal. These atoms are moving around, and so there are occasionally collisions that occur between these atoms. The time between two consecutive collisions produces a limit to the lifetime τ of a free moving atom and so, from Heisenberg uncertainty principle, a corresponding uncertainty $\Delta E \sim \hbar/\tau$ on the energy of a given energy level. This translates naturally into an uncertainty in the frequency of specific spectral lines, which means the existence of a corresponding linewidth. Similarly, collisions against the walls of the container lead to lifetime limitations and corresponding linewidths. Clearly, if one succeeds in having the atoms moving more slowly, this will increase the time between collisions and contribute to reduce this extraneous source of linewidth.

Another source of spectral line broadening is the Doppler effect. If ω_0 is the frequency of a photon emitted by an atom at rest, the observed photon frequency is shifted when the atom is moving with velocity $\mathbf{v}$ with respect to the detector. For a small-enough velocity, the frequency seen by the detector is $\omega_0(1+\mathbf{v}\cdot\mathbf{u}/c)$, where c is the light velocity and $\mathbf{u}$ is the unit vector in the direction linking the emitting atom to the photon detector. Since the atom may move in any direction, the observed frequency can be anywhere between $\omega_0(1 - v/c)$ and $\omega_0(1 + v/c)$. Moreover, the atom velocity itself is distributed, in the classical regime, according to the Maxwell-Boltzmann distribution, so that the probability for an atom of mass m to have velocity v is proportional to $\exp(-mv^2/(2k_BT))$. The typical corresponding velocity $v_m \sim \sqrt{2k_BT/m}$ for, say, a ^{6}Li atom at ordinary temperature $T \simeq 300$ K is of the order of 10^3 m/s, which is quite large. While the corresponding relative linewidth $v_m/c \sim 3\times 10^{-6}$ may seem satisfactory in absolute terms, it is nevertheless huge compared to the typical precision of order 10^{-16}, which can be reached nowadays in optical frequency measurements. Going to cryogenic temperature of the order of a fraction of 1 K improves the matter, but not very much since the width is proportional to $\sqrt{T}$. Hence it is of high practical interest to slow down atoms as much as possible by any means.

However, cooling down atoms also has a more fundamental interest. Indeed, whereas a proper account of the physical properties of electrons inside an atom requires quantum mechanics, the whole atom in an atomic gas can be treated within classical mechanics. This can be seen from the value of its thermal de Broglie wavelength $\lambda_T = (2\pi\hbar^2/mk_BT)^{1/2}$. If we evaluate it for a ^{6}Li atom at $T = 1$ K, we obtain a result of order 1 nm. This is not so different from the de Broglie wavelength of an electron in a standard metal, the large mass ratio between the ^{6}Li atom and the electron being essentially compensated by the fact that we consider an energy much smaller than a typical Fermi energy. However, this de Broglie wavelength is very small compared to the typical distance between atoms in a gas under the standard experimental conditions, which is in the range of 10^2–10^3 nm. Hence we are indeed in a regime where quantum effects for the atoms are negligible so they can be treated as classical objects. We note, in contrast, that in the case of liquid ^{4}He considered in the preceding chapter (which has a mass similar to ^{6}Li and a similar temperature), the interatomic distance is also of the order of 1 nm, so that liquid ^{4}He is indeed a quantum liquid.

It has been a fascinating goal to try and reach in atomic gases the quantum regime where atoms, instead of behaving as classical objects, display properties showing that they are ruled by quantum mechanics. In particular, important differences should appear depending on whether these atoms are fermions or bosons. For fermions, one should see the formation of a Fermi sea, while for bosons one should observe the appearance of a Bose–Einstein condensate. This aim of having the de Broglie wavelength of the order of the interatomic distance is often formulated, through the gas density n, in terms of the phase-space density $\rho_{ps} = n\lambda_T^3$, which should be of order 1 in the quantum regime, whereas it is much smaller than unity in the classical regime.

An obvious way to increase this phase-space density is to increase the gas density n, and one evidently tries to do the best on this side. However, there is a fundamental limit that arises in this direction. Indeed, when two atoms collide in a gas, energy and momentum conservation prohibit that they recombine into a possible lower energy molecule. This is obvious in their center of mass frame since momentum conservation implies that the resulting molecule does not move and there is no way to evacuate the excess energy resulting from the recombination. On the other hand, when a third atom is present, there is no limitation of this kind since the kinetic energy of this additional atom allows the atomic system to carry away this excess energy.

These inelastic three-body processes are a major problem since not only an unwanted molecule is formed, but the third atom, which gets in practice very energetic, is also lost from the relevant gas region, after having in general contributed to an important heating of the gas through two-body collisions. More generally the cold gases we consider in this chapter are highly metastable, since at finite density and very low temperature the thermodynamical stable state of matter for such an assembly of atoms is a solid, not a gas. However, if the gas is kept at a low-enough density, it is very unlikely to have three atoms within short distance of each other, as it is required to have such processes occurring. In this case, these three-body processes are unfrequent enough, and these metastable gases may last a long enough time to allow the full performance of experiments. The typical time for the stability of these gaseous samples is of the order of 1 second. When the sample is lost, another one is fabricated and the experiment is repeated. This is the standard way in which experiments on cold gases proceed. The corresponding workable range of gas densities is of the order of 10^{12} cm^{-3}, corresponding to a typical interatomic distance of order 10^3 nm.

Since the increase in gas density meets the above limitations, the only other way to increase the phase space density is to increase the de Broglie wavelength by decreasing the temperature. As we have already mentioned, standard cryogenics is not enough to reach the quantum regime for gases. Indeed the lowest temperatures reached in this way have been obtained in cooling liquid ^{3}He and they are in the mK and submK range, the technical cooling process being mostly adiabatic demagnetization. At $T = 0.1$ mK the above Li gas, with density 10^{12} cm^{-3}, has a phase space density ρ_{ps} about 10^{-3}, whereas a temperature of order 1μK is necessary to achieve $\rho_{ps} \sim 1$. Accordingly, in order to reach the quantum regime, the cooling method has to act on the gas directly, rather than cooling the surrounding environment.

The main first step in this direction has been the laser cooling method. Its principle is simple. If an atom in its ground state, with mass m and velocity $\mathbf{v}$, absorbs a photon with momentum $\hbar\mathbf{k}$ opposite to $\mathbf{v}$, from momentum conservation its velocity becomes $\mathbf{v}'$, with $mv' = mv - \hbar k$, so that the atom is slowed in this process. The photon energy $\hbar\omega = \hbar kc$ has to be such that the atom can absorb it by going from its ground state to an excited state. Naturally the atom does not stay for long in this excited state, and it goes rapidly back to its ground state by emitting a photon with the same energy $\hbar\omega$. However, the direction of this emitted photon is at random, so that in this re-emission process the component of the atom velocity in the direction of **v** is either increased or decreased, so that on average its velocity in the direction of interest is not modified. So, by repetition of this process, the atom is slowed on the whole. Indeed it is easily seen that compared to the atom momentum at $T \sim 300$ K, the photon momentum is quite small, so that the decrease in atom velocity in a single process is also small. But the high photon flux produced by a laser allows this process to be repeated rapidly a large number of times, so that these multiple absorptions-emissions lead indeed, on average, to an efficient slowing down of the atoms.

A major problem met with this method is due to the Doppler effect. Since, as we have seen, the frequency of the photon corresponding to a specific transition is shifted due to the atom velocity, the photons which are absorbed, say, by high-velocity atoms will no longer be absorbed by low-velocity atoms. This means that in order to keep the absorption process always possible during the atomic slowing down, one should continuously change the laser frequency. This is very inconvenient. In practice, one rather uses a laser with a fixed frequency and one modifies appropriately the energy difference between the atomic ground state and the considered excited state, so that the laser light still has the proper frequency to be absorbed. This is done by making use of the Zeeman effect, which corresponds to the fact that the energy levels of an atom depend in general on the magnetic field seen by this atom. Since during the slow down process the atom velocity varies continuously, one must have a magnetic field strength that also varies continuously. In practice, one sends the collimated beam of atoms, coming out of an oven, along the axis of an elongated coil, which has the number of its wire windings varying progressively with position, so that the magnetic field strength varies also with position. The laser beam is also propagating along this axis. The overall device is a Zeeman slower. At the end of the coil, the atoms are collected in a trap that keeps them in an appropriate region by the combined action of laser beams and/or a static magnetic field.[1]

Nevertheless, laser cooling has not been enough to reach the quantum regime, and an additional procedure has been used to obtain the necessary temperatures. This is the evaporative cooling. Again, its principle is simple. As suggested by its name, it works in a way analogous to the cooling effect of water evaporation. The molecules escaping from liquid water into the water vapor carry a high energy, and so the remaining liquid water loses energy and gets cooled in this way. Similarly, one can successfully cool atoms by removing the atoms in the high energy tail of their distribution. The leftover atoms have a lower energy, which leads to a lower temperature after thermalization. Naturally all the operations are performed in an ultra-high vacuum.

In practice this is achieved by making use of the magnetic properties of the involved atoms maintained in a magnetic trap. This trap corresponds to the spatial region in the vicinity of a local minimum of an applied static magnetic field **B** (one can show from Maxwell equations that the magnetic field can not have a maximum in vacuum). Corresponding to the Zeeman effect, the energy degeneracy of atoms having their magnetic

---

[1] The 1997 Nobel Prize in physics was shared by S. Chu, C. Cohen-Tannoudji, and W. D. Phillips for their work on the manipulation of atoms with photons, with the aim of atom cooling.

momentum **M** oriented in various directions with respect to the magnetic field is lifted, and there is a splitting into energy levels having different magnetic quantum number m_z with respect to the field. Roughly speaking it is possible, through irradiation by radio-frequency fields having appropriate frequency and polarization, to select the m_z value in such a way that the atomic magnetic moment is oriented in the direction antiparallel to the magnetic field. In this way, the magnetic energy $-\mathbf{M} \cdot \mathbf{B} = MB$ is minimum where B is minimum. Atoms with the lowest energy stay in the close vicinity of this minimum. On the other hand, atoms with higher energy are able to go farther away from this minimum, and as a result they may experience higher magnetic field values. One can take advantage of this to remove these higher energy atoms. This is done for example by irradiating with radio-frequency field pulses having the proper frequency and time duration to flip the atom magnetization **M** into $-\mathbf{M}$. The resulting atom has in this way an energy $-MB$, which is opposite to the case where the magnetization is **M**. Hence this energy decreases when the atom goes away from the point where B is minimum. Accordingly, the corresponding magnetic force will push this atom out of the trap region. This magnetization flip is due to a resonant process, corresponding to a specific value of the magnetic field. Lower energy atoms, closer to the field minimum, will not see this specific field value and therefore they will not flip, so they will stay in the trap.

This evaporative cooling is in practice carried out in several steps. One removes the atoms having an energy higher than a specific energy by sweeping the radio-frequency through the corresponding frequency values, and one then waits until the remaining atoms thermalize to a lower temperature. This thermalization relies on inter-atomic collisions, and so it is important that interactions between atoms are strong enough for these processes to be quite efficient. When thermalization is reached, one evaporates another slice of atomic energy, slightly lower than the preceding one, and so on. Naturally a main problem of this method is that at each step one loses from the trap the atoms with higher energy. Hence one might reach quite low temperatures, but with so few atoms left that they are of no practical experimental interest. So one has naturally to start this evaporative cooling process with a large number of atoms and to optimize the successive evaporations in such a way that one gets, at the aimed low temperature, an atom number that is as large as possible.

This whole cooling process presents a number of possible pitfalls that must be avoided, and as a result one has to use plenty of tricks belonging to the atomic physics arsenal. Even so, it has not been possible to cool any atom to low-enough temperatures. As a consequence, the effort has been mainly focused on alkali atoms because, belonging to the first column of Mendeleev classification, they have a single electron in their outermost shell. This endows them with comparatively simple spectroscopic properties, which make all the necessary optical manipulations more convenient, or even merely possible. An indication of the difficulties met in carrying out the complete path to the quantum regime is given by the atom on which the first successful cooling has been achieved. An obvious candidate is the hydrogen atom since, being the lightest of all these atoms, it has for a given temperature the largest de Broglie wavelength λ_T. Hence it looks easier to make this wavelength comparable to the interatomic distance. Nevertheless, despite many efforts, hydrogen has not been the atom with which Bose–Einstein condensation has first been obtained. It is rather with ^{87}Rb that E. Cornell and collaborators succeeded in 1995 to reach a temperature low enough to produce the Bose–Einstein condensation in the gas, and with ^{23}Na that, the same year, W. Ketterle and collaborators obtained this same transition.[2] Afterward, all the alkali atoms have been progressively brought to low-enough temperature to undergo Bose–Einstein condensation.

To be specific, we should mention that these Bose–Einstein condensations have naturally been obtained for trapped atoms, which makes the quantitative situation somewhat different from the one we considered in the preceding chapter. Indeed, near its local minimum, the trapping potential can be approximated by a harmonic potential, this approximation being very good for most cold atoms experiments. This leads to a modification of the atoms' density of states which plays, as we have seen, a crucial role in the phenomenon of Bose–Einstein condensation. For a one-dimensional harmonic oscillator, with characteristic frequency ω_0, the energy levels are given by $\epsilon_n = (n + 1/2)\hbar\omega_0$. The corresponding density of states, that is the number of states per unit energy, is then obviously $\rho_{1h}(\epsilon) = 1/\hbar\omega_0$. A two-dimensional harmonic oscillator is the tensorial product of two one-dimensional harmonic oscillators, its energy being the sum of the energy of each one-dimensional oscillator $\epsilon = \epsilon_1 + \epsilon_2$. The corresponding density of states $\rho_{2h}(\epsilon)$ is just the convolution of the density of states of these one-dimensional oscillators

$$\rho_{2h}(\epsilon) = \int_0^\infty d\epsilon_1 \int_0^\infty d\epsilon_2 \, \rho_{1h}(\epsilon_1)\rho_{1h}(\epsilon_2)\,\delta(\epsilon_1 + \epsilon_2 - \epsilon) = \frac{\epsilon}{\hbar^2\omega_0^2} \tag{8.1}$$

Similarly, the density of states of a three-dimensional harmonic oscillator is obtained by taking the convolution of the density of the states of a two-dimensional oscillator with that of a one-dimensional oscillator

$$\rho_{3h}(\epsilon) = \int_0^\infty d\epsilon_1 \int_0^\infty d\epsilon_2 \, \rho_{2h}(\epsilon_1)\rho_{1h}(\epsilon_2)\,\delta(\epsilon_1 + \epsilon_2 - \epsilon) = \frac{1}{\hbar\omega_0}\int_0^\epsilon d\epsilon_1\rho_{2h}(\epsilon_1) = \frac{\epsilon^2}{2\hbar^3\omega_0^3} \tag{8.2}$$

When we compare this three-dimensional density of states to the corresponding one for free atoms, which is proportional to $\epsilon^{1/2}$, we see that the density of states for trapped atoms is quite reduced at low energies. This is easily understood physically since in the case of the trapped atoms, the low energy eigenstates correspond to atoms that are forced to be located in the vicinity of the trapping potential minimum, instead of being extended as for free atoms. Hence their number, as well as their density of states, is reduced. As a result, Bose–Einstein condensation is favored since, as we have seen, it is due to the lack of low energy states to accommodate all the atoms at low temperature.

Eq. (7.6) gives the critical temperature for $N_0(T) = 0$, and in terms of the density of states $\rho(\epsilon)$, it merely reads $N = \int_0^\infty d\epsilon \, \rho(\epsilon)/(e^{\epsilon/k_B T_c} - 1)$. Substituting Eq. (8.2) for the density of states gives

$$N = \frac{1}{2\hbar^3\omega_0^3}\int_0^\infty d\epsilon \, \frac{\epsilon^2}{e^{\epsilon/k_B T_c} - 1} = \frac{1}{2}\left(\frac{k_B T_c}{\hbar\omega_0}\right)^3 \int_0^\infty dx \, \frac{x^2}{e^x - 1} \tag{8.3}$$

---

[2] The 2001 Nobel Prize in physics was awarded to E. Cornell, W. Ketterle, and C. Wieman for the achievement of Bose–Einstein condensation in dilute gases of alkali atoms.

The integral on the right-hand side can be shown to be equal to $2\zeta(3)$, with $\zeta(3) \simeq 1.202$, in the same way (see footnote) as the integral in Eq. (7.4) was $\sqrt{\pi}\,\zeta(3/2)/2$. This leads for the critical temperature to

$$k_B T_c = 0.94\,\hbar\omega_0 N^{1/3} \tag{8.4}$$

A typical experimental value for ω_0 is of the order of a few kHz. On the other hand, this critical temperature is only weakly sensitive to the atom number N, due to the 1/3 exponent. With a typical number of atoms of order 10^4, one finds a critical temperature of the order of 100 nK. Note that when the trap is anisotropic, as is most often the case, ω_0 in Eq. (8.4) is merely replaced with $(\omega_x\omega_y\omega_x)^{1/3}$, where ω_x, ω_y and ω_z are the frequencies corresponding to the principal directions of the trapping potential.

Another interesting point about Bose–Einstein condensation in a trap is that it occurs also in a two-dimensional space (which can be experimentally realized in practice with very anisotropic pancake-like traps), since the corresponding density of states Eq. (8.1) also gives rise to a convergent integral. This is in contrast with the situation in free space (in practice a rectangular box) where Bose–Einstein condensation does not occur, strictly speaking, because one gets a divergent integral with the corresponding density of states, which is energy independent.

Let us finally indicate that in most cases, the atomic cloud corresponding to these cold gases is too small to be seen directly by optical means. Indeed the typical size of these clouds is a few microns. The standard procedure to observe them is to open the trap by removing the trapping field. Since the confining potential is no longer present, the atomic cloud expands freely. When it is large enough to allow a convenient observation, the atoms are seen by detecting their absorption of a light with appropriate resonant frequency. In this way, one obtains the expanded cloud density as a function of the position with respect to the small original trapped cloud. Since the distance traveled by atoms during their free flight is directly linked to their velocity, one obtains in this way direct information on the velocity distribution inside the original cloud.

In the case where interatomic interactions are small enough to be negligible, the relation is quite simple since one can show that the spatial density of the expanded cloud is directly linked to the atomic momentum distribution inside the original trapped cloud. In particular, the existence of the Bose–Einstein condensate is strikingly seen in this way. The thermally excited atoms with momentum of order $\sqrt{mk_B T}$ give rise, after expansion, to a large cloud. On the other hand, the condensate reflects the properties of the ground-state wave function in the trap, with a typical scale $\sqrt{m\hbar\omega_0}$ for the momentum distribution that is much smaller than the thermal scale. As a result, the condensate appears as a narrow and very strong peak in the center of the expanded cloud, a quite spectacular experimental result. The strength of this peak grows as the temperature is lowered below T_c, until at low-enough temperature it is the only remaining part of the expanded cloud, the thermal part having fully disappeared. Moreover, in the standard case of an anisotropic trap, the condensate peak displays a corresponding anisotropy, whereas the thermal part is always isotropic. All these features provide unambiguous proof that Bose–Einstein condensation has indeed been realized in these ultracold gases.

8.2 Scattering Length

In the simple situations we deal with, the interatomic interaction $V(r)$ is isotropic and so it depends only on the modulus $r = |\mathbf{r}|$ of the position vector. To be specific, we consider for this interaction, as a very simple model, a square well potential, with $V(r) = U_0$ for $r < R$, and $V(r) = 0$ for $r > R$. Here the constants U_0 and R characterize respectively the strength and the spatial extension of the potential. In most cases, we will consider an attractive interaction, so that we will have $U_0 < 0$. Note that real interatomic interactions decrease quite rapidly with distance r, so that taking $V(r) = 0$ for $r > R$ is in this respect quite a good description of a real interaction. On the other hand, taking a constant $V(r) = U_0$ for $r < R$ is just a convenient oversimplification, made to have the calculations to be performed as easy as possible.

In order to investigate the scattering between two atoms, we have to solve the Schrödinger equation for their relative motion, described by the corresponding wave function $\psi(\mathbf{r})$. This equation reads

$$\left(\frac{\mathbf{p}^2}{2m_r} + V(r) \right) \psi(\mathbf{r}) = E\,\psi(\mathbf{r}) \tag{8.5}$$

where m_r is the reduced mass of the atoms (with $m_r = m/2$ if we have identical atoms with mass m). We set as usual for the energy $E = \hbar^2 k^2 / 2m_r$. Because of the rotational invariance of the interaction, it is convenient to separate the longitudinal and the transverse parts of the $\mathbf{p}^2 = -\hbar^2 \Delta$ operator. As is well known, the action of the longitudinal part p_r^2 on the wave function gives $p_r^2\,\psi(\mathbf{r}) = -\hbar^2 (1/r)\partial^2(r\psi(\mathbf{r}))/\partial r^2$. On the other hand, the transverse part $p_\perp^2$ is readily expressed, in terms of the angular momentum operator $\mathbf{L} = \mathbf{r} \times \mathbf{p}$, as $p_\perp^2 = \mathbf{L}^2/r^2$. Hence the Schrödinger equation reads

$$-\frac{\hbar^2}{2m_r} \frac{1}{r} \frac{\partial^2(r\,\psi(\mathbf{r}))}{\partial r^2} + \left(\frac{\mathbf{L}^2}{2m_r r^2} + V(r) \right) \psi(\mathbf{r}) = \frac{\hbar^2 k^2}{2m_r}\,\psi(\mathbf{r}) \tag{8.6}$$

For the scattering problem, we are interested in what happens to two atoms, moving freely toward each other. For the relative motion of the two atoms, this free motion is described by a wave function behaving as a plane wave $e^{i\mathbf{k}\cdot\mathbf{r}}$. Since this is a free motion, in a region where $V(r) = 0$, the wave vector $\mathbf{k}$ is related to the energy by $E = \hbar^2 k^2 / 2m_r$. We take the standard convention that this incident motion is along the z-axis, so that this plane wave becomes e^{ikz}. On the other hand, after having interacted through $V(r)$, the two atoms will move outward in any direction. In the relative motion, this outward propagation is naturally described, at large distance r, by an outgoing spherical wave $f(\theta)\,e^{ikr}/r$, with an amplitude $f(\theta)$ which, for symmetry reasons, depends only on the angle θ with the incident z direction. This spherical wave is again checked to be a solution of the Schrödinger equation Eq. (8.6) for the free motion $V(r) = 0$, since at large distances the $\mathbf{L}^2/2m_r r^2$ becomes negligible and only the first term on the left-hand side of Eq. (8.6) remains.

In order to describe this scattering process, we may think of a stationary state, where there is a permanent flux of incoming atoms and another permanent flux of outgoing atoms. Looking at the outgoing flux in some direction will allow us to determine the probability of

having the atoms scattered in this direction. This stationary state is naturally a solution of
the Schrödinger equation Eq. (8.6). However, the boundary conditions are specific to this
scattering process. Accordingly, for the relative motion, the wave function must, at large
distance, display the e^{ikz} contribution corresponding to the incoming flux and the $f(\theta)\,e^{ikr}/r$
contribution corresponding to the outgoing spherical wave. Hence we must have

$$\psi(\mathbf{r}) \simeq e^{ikz} + f(\theta)\,\frac{e^{ikr}}{r} \qquad r \to \infty \tag{8.7}$$

Because of the rotational invariance of the interaction $V(r)$, the Hamiltonian H com-
mutes with the angular momentum $\mathbf{L}$, and it is convenient to use as a basis wave functions
that are common eigenstates of H and $(\mathbf{L}^2, L_z)$. Their angular dependence is given by the
spherical harmonics $Y_{\ell m}(\theta, \varphi)$, and the corresponding eigenvalues are $\ell(\ell + 1)\hbar^2$ for $\mathbf{L}^2$
and $m\hbar$ for L_z. Naturally the solution $\psi(\mathbf{r})$ we are looking for does not correspond to
specific values of ℓ and m, as is clear from the presence of the term e^{ikz} in the boundary condition Eq. (8.7), which does not have the appropriate angular dependence. It is
rather a linear combination of these basis wave functions with different values of ℓ and m.
Actually, by rotational invariance around the z-axis, it is clear that our solution does not
have any dependence on the azimuthal angle φ. Since $Y_{\ell m}(\theta, \varphi)$ contains a $e^{im\varphi}$ factor, this
means that only the eigenstates with $m = 0$ will appear in this linear combination. Corre-
spondingly the scattering amplitude $f(\theta)$, which we want to find from the solution of the
Schrödinger equation Eq. (8.6), will be a combination of various $Y_{\ell, m=0}(\theta)$. This can also
be written as

$$f(\theta) = \sum_{\ell=0}^{\infty} (2\ell + 1) f_\ell \, P_\ell(\cos\theta) \tag{8.8}$$

in terms of the Legendre polynomials $P_\ell(\cos\theta)$, which are related by $Y_{\ell, m=0}(\theta) = \sqrt{(2\ell + 1)/4\pi}\, P_\ell(\cos\theta)$ to the spherical harmonics. They have the explicit expression
$P_\ell(u) = \left[(-1)^\ell/(2^\ell\,\ell!)\right] d^\ell(1 - u^2)^\ell/du^\ell$. In particular, $P_0(u) = 1$.

We will now make use of a very simplifying and interesting feature arising because we
are interested in cold atoms. These atoms have naturally a very low kinetic energy. More
specifically, in the quantum regime, we have seen that their de Broglie wavelength λ_T is
of the order of the interatomic distance at the critical temperature, and even larger below
T_c. As we have seen, the interatomic distance in these dilute gases is of the order of 10^2
to 10^3 nm. In contrast, the range of the interatomic interaction $V(r)$, specifically R in our
simplified square well model, is typically of the order of 1 nm. Since the wavevectors k we
have to deal with are of order $1/\lambda_T$, this means that we are in a regime where $kR \ll 1$.
In this very low energy regime, the scattering problem is strongly simplified. Indeed it can
be shown that the s-wave $\ell = 0$ component becomes completely dominant, because the
contributions from the $\ell \geq 1$ components get in comparison vanishingly small in the limiting situation $k \to 0$. We refer to standard textbooks on scattering theory for the detailed
proof of this low energy feature, but we can give qualitative arguments allowing one to
physically understand the origin of this result.

We note first that for a specific ℓ, the $\mathbf{L}^2/(2m_r r^2)$ term in Eq. (8.6) becomes $\hbar^2\,\ell(\ell+1)/$
$(2m_r r^2)$. For $\ell \geq 1$, this gives effectively to the interaction $V(r)$ an additional repulsive

contribution, called the centrifugal barrier. This barrier gets high for small r, and so it keeps the relative motion away from the small r region. Hence the interaction $V(r)$ is hardly felt in this case. This explains why the $\ell = 0$ component, for which the barrier does not exist, plays the dominant role at low-enough energy. One finds the same conclusion by looking at the classical limit, valid for high angular momentum. In this case, the angular momentum $\mathbf{L} = \mathbf{r} \times \mathbf{p}$ is given by $L = r_\perp p$, where $r_\perp$ is the component of $\mathbf{r}$ perpendicular to $\mathbf{p}$, that is, the direction of the incoming motion. For very low energy $p^2/2m_r$, p is very small, and so for a given L, the nearest distance $r_\perp$ at which the incoming particle passes from the center gets very large, so the interaction potential $V(r)$ is indeed not felt. This classical view overlaps actually with the preceding one, since at the quantum level the classical turning point, that is the value r_0 of the distance r for which the height of the centrifugal barrier is equal to the particle energy, is given by $\hbar^2 \ell(\ell + 1)/(2m_r r_0^2) = \hbar^2 k^2/2m_r = p^2/2m_r$, that is, for large ℓ, $L^2 = (\hbar\ell)^2 = (r_0 p)^2$, which is exactly the relation we have seen in the classical regime.

An immediate and important consequence of the dominance of the s-wave component is that scattering becomes isotropic in this low energy limit, since in Eq. (8.8) we can neglect f_ℓ for $\ell \geq 1$ so that we are left with $f(\theta) = f_0$, which does not depend on θ. This result can also be seen as a natural consequence of the fact that in the low energy limit $\mathbf{k} \to \mathbf{0}$, there is no longer any reference direction left, so that scattering is necessarily isotropic. Conversely, if we take this argument for granted, this is another way to understand that at low energy s-wave scattering is dominant and the other angular momenta are negligible.

Considering only $\ell = 0$ simplifies our analysis of the Schrödinger equation Eq. (8.6) since there is no angular dependence anymore. We are only left with the radial dependence of the wave function. Setting $\chi(r) = r\psi(r)$ this equation becomes for $\ell = 0$

$$\chi''(r) + \left[k^2 - \frac{2m_r}{\hbar^2} V(r)\right] \chi(r) = 0 \tag{8.9}$$

where $\chi''(r)$ is the second derivative of $\chi(r)$ with respect to r. We consider first the region far enough from the scattering potential to have $V(r) = 0$. In our square well model, this is merely $r > R$. In this case, the equation becomes $\chi'' + k^2\chi = 0$, and the most general solution is any combination of $\sin(kr)$ and $\cos(kr)$. Introducing the phase shift $\delta_0(k)$ for s-wave scattering, this can also be written as

$$\chi(r) = \sin(kr + \delta_0(k)) \tag{8.10}$$

where we have taken the simplest normalization constant. For a free particle, with $V(r) = 0$ everywhere, we would have to satisfy the boundary condition $\chi(r) = 0$ for $r = 0$, so that $\psi(r)$ does not diverge for $r = 0$. This would give $\delta_0(k) = 0$. In the presence of a potential, this is no longer the case, and $\delta_0(k)$ has to be determined from the solution of the Schrödinger equation for $r < R$. For our square well potential, we merely have $V(r) = U_0$ for $r < R$, so that Eq. (8.9) becomes

$$\chi''(r) + k'^2\chi(r) = 0 \qquad\qquad k'^2 = k^2 - \frac{2m_r}{\hbar^2} U_0 \tag{8.11}$$

The general solution is again any combination of $\sin(k'r)$ and $\cos(k'r)$, but we must now satisfy the boundary condition $\chi(r) = 0$ for $r = 0$, so we must take the $\sin(k'r)$ solution. The final step is to match, at $r = R$ this solution and its derivative to our solution Eq. (8.10) for $r > R$. It is more convenient to match only the logarithmic derivatives $\chi'(r)/\chi(r)$, which eliminates the question of the normalization of the wave function. This gives the condition

$$k' \cot(k'R) = k \cot(kR + \delta_0(k)) \tag{8.12}$$

leading for $\delta_0(k)$ to the solution

$$\delta_0(k) = -kR + \arctan\left[\frac{k}{k'}\tan(k'R)\right] \tag{8.13}$$

Coming back to our scattering problem, we have to demand that the boundary condition Eq. (8.7) is satisfied. Since in our low energy regime scattering for $\ell \geq 1$ is negligible, we have only to deal with its s-wave part in order to obtain the corresponding part f_0 of the scattering amplitude. This is done by expanding the wave function given by Eq. (8.7) in Legendre polynomials, just as in Eq. (8.8). We have to use the fact that, just as the spherical harmonics, Legendre polynomials are orthogonal and satisfy $\int_{-1}^{1} du\, P_\ell(u)P_{\ell'}(u) = (\ell + 1/2)^{-1}\delta_{\ell,\ell'}$. Since $P_0(u) = 1$ the s-wave component is merely obtained, in Eq. (8.7), by integrating the right-hand side over $u = \cos\theta$ and dividing by 2. The scattered part $f_0\, e^{ikr}/r$ is naturally unchanged since it corresponds already to the s-wave part of the scattered wave function. The plane wave part $e^{ikz} = e^{ikr\cos\theta} = e^{ikru}$ gives the contribution $(1/2)\int_{-1}^{1} du\, e^{ikru} = \sin(kr)/kr = (e^{ikr} - e^{-ikr})/(2ikr)$. Hence the whole s-wave part of the wave function Eq. (8.7) is $\sin(kr)/kr + f_0\, e^{ikr}/r$, corresponding to $\chi(r) = r\psi(r) = \sin(kr)/k + f_0\, e^{ikr}$.

This s-wave part must coincide, within a normalization constant, with the solution Eq. (8.10) of the Schrödinger equation. The normalization is eliminated by merely requesting that the ratio between the factor of e^{ikr} and the factor of e^{-ikr} is the same in these two solutions. This leads to the condition

$$1 + 2ikf_0 = e^{2i\delta_0(k)} \tag{8.14}$$

that is

$$f_0 = \frac{e^{2i\delta_0(k)} - 1}{2ik} \tag{8.15}$$

Actually, if we had considered the angular momenta $\ell \geq 1$, we would have found exactly the same relation between f_ℓ, the component of the scattering amplitude coming in Eq. (8.8), and the corresponding phase shift $\delta_\ell(k)$, defined in the same way as in Eq. (8.10) for the solution of the Schrödinger equation for the angular momentum ℓ.

The above analysis is actually valid for any energy. Let us now come back to our specific energy range of interest, namely the low energy regime. From the solution Eq. (8.13) for our specific square well model, we see that when $k \to 0$, the phase shift $\delta_0(k)$ is proportional to k, and goes to zero in this limit. From Eq. (8.15) we have correspondingly $f_0 \simeq \delta_0(k)/k$, so introducing the proportionality constant a we have

$$\delta_0(k) = -ka \qquad f_0 = -a \tag{8.16}$$

This behavior is actually valid for a general potential $V(r)$, and the length a is called the scattering length. For a general angular momentum ℓ, one finds generically that $\delta_\ell(k)$ is proportional to $k^{2\ell+1}$, and similarly $f_\ell \simeq \delta_\ell(k)/k \sim k^{2\ell}$. This shows explicitly that at low energy, the angular momentum $\ell = 0$ is dominant since in this regime all the f_ℓ are negligible compared to f_0.

Specifically for the case of our square well model

$$a = R\left(1 - \frac{\tan(k'_0 R)}{k'_0 R}\right) \qquad k'^2_0 = -\frac{2m_r}{\hbar^2} U_0 \tag{8.17}$$

Here the constant k'_0 is real only when the potential is attractive $U_0 < 0$. In the opposite case of a repulsive potential $U_0 > 0$, it is purely imaginary $k'_0 = i\kappa'_0$ and from $\tan(ix) = i\tanh x$, Eq. (8.17) becomes explicitly

$$a = R\left(1 - \frac{\tanh(\kappa'_0 R)}{\kappa'_0 R}\right) \qquad \kappa'_0 = \left(\frac{2m_r}{\hbar^2} U_0\right)^{1/2} \tag{8.18}$$

The physical meaning of the scattering length can now be seen most directly from the expression Eq. (8.10) of the wave function far away from the potential region. Making use of Eq. (8.16), it becomes

$$\chi(r) = \sin(k(r - a)) \tag{8.19}$$

When in this low energy regime k is small enough, this expression can even be more simply written $\chi(r) = k(r - a)$. Hence, from these expressions, we see that, with respect to the free particle wave function $\chi(r) = \sin(kr)$, the effect of the potential on the wave function is merely to push it by a length simply given by the scattering length a. In the case where the potential is repulsive, one has $a > 0$ from Eq. (8.18) and the wave function is pushed away from the potential, which sounds physically quite reasonable. In particular, when the repulsive potential U_0 gets infinite, which corresponds physically to an infinitely hard ball, κ'_0 is infinite and one merely has $a = R$, so that the wave function is indeed pushed away by a length that is just the radius of the hard ball. This is the physically expected result, and it is quite obvious from the start since in this case the boundary condition for the wave function is that it should be zero for $r = R$, in agreement with its expression $\chi(r) = \sin(k(r - R))$.

On the other hand, when the potential is attractive $U_0 < 0$ and is assumed to be weak enough so that $\tan(k'_0 R) \simeq k'_0 R + (k'_0 R)^3/3$, we have from Eq. (8.17) $a = -k'^2_0 R^3/3 < 0$. Hence in this case the wave function is pulled toward the attractive potential, which sounds again quite reasonable physically. However, we see from the exact expression Eq. (8.17) for a that this behavior is not valid in the general case, where the attractive potential is strong enough. We examine in the next section the dependence of the scattering length on the specific features of the potential in this more complex situation.

8.3 Feshbach Resonance

8.3.1 Resonance

The behavior of the scattering length for an attractive potential is not a simple one, as is clear from the explicit result Eq. (8.17) for the specific example of our square well model. This is seen in Fig. (8.1), where the ratio a/R is displayed as a function of $k_0'R$. The typical value of a is in general of the order of the spatial extension R of the potential, as we might expect physically by analogy with the repulsive potential case. However, it may have much larger values, linked to the divergences of $\tan(k_0'R)$. These occur for $k_0'R = \pi(n - 1/2)$, where n is any positive integer. Let us find a physical interpretation for these conditions by looking at the possible bound states in this attractive potential. Since we are only interested in a s-wave dependence of the wave function, we restrict also ourselves coherently to the case $\ell = 0$ for these bound states.

Bound states correspond to negative values of the energy E in the Schrödinger equation Eq. (8.5), so setting $E = \hbar^2 k^2/2m_r$ is inappropriate. We should rather write $E = -\hbar^2\kappa^2/2m_r$, but this is equivalent to set $k = i\kappa$. Regarding the large distance behavior of the wave function, where $V(r) = 0$, the solutions of the Schrödinger equation $\chi'' - \kappa^2\chi = 0$ are any combination of $e^{\kappa r}$ and $e^{-\kappa r}$, but only the non-divergent solution $e^{-\kappa r}$ is physically acceptable in this range $r \geq R$, which serves as a boundary condition for the bound state. On the other hand, in the range $r \leq R$, it is physically clear (and easy to check) that a solution exists only for high enough energy $E > U_0$. Then the Schrödinger equation reads

$$\chi''(r) + \kappa'^2\chi(r) = 0 \qquad \kappa'^2 = -\frac{2m_r}{\hbar^2}U_0 - \kappa^2 = k_0^2 - \kappa^2 > 0 \qquad (8.20)$$

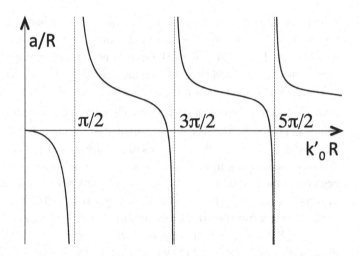

Fig. 8.1   Full line: Reduced scattering length a/R, given by Eq. (8.17), for an attractive square well potential of size R and strength $|U_0|$, as a function of $k_0'R = R\sqrt{2m_r|U_0|}/\hbar$. The dashed lines show the values $k_0'R = \pi(n - 1/2)$, for which this scattering length diverges.

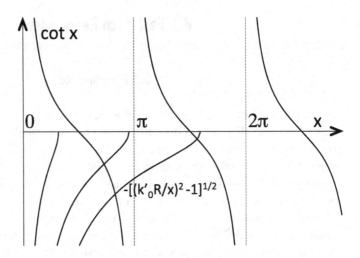

Fig. 8.2 Graphical solution of Eq. (8.21) for $x = \kappa'R$, obtained from the intersections of $\cot x$ by $-\sqrt{(k_0'R/x)^2 - 1}$. This gives the energy E of bound states by $E = -\hbar^2\kappa^2/2m_r = -\hbar^2[(k_0'R)^2 - x^2]/2m_rR^2 = -|U_0| + \hbar^2x^2/2m_rR^2$. Three curves representing $-\sqrt{(k_0'R/x)^2 - 1}$ are shown for the three cases $k_0'R = 1, 3$, and 5.

for which we must choose the solution $\chi(r) = \sin(\kappa'r)$ in order to satisfy the boundary condition $\chi(r = 0) = 0$. Matching the logarithmic derivatives of the two solutions at $r = R$ yields the condition

$$\cot(\kappa'R) = -\frac{\kappa}{\kappa'} = -\sqrt{\left(\frac{k_0'R}{\kappa'R}\right)^2 - 1} \qquad (8.21)$$

Note that $k_0'R = (|U_0|/(\hbar^2/2m_rR^2))^{1/2}$ compares the strength $|U_0|$ of the attractive potential to the kinetic energy $\hbar^2/(2m_rR^2)$ resulting from the localization of the particle in a well of limited size R, following Heisenberg uncertainty principle. Because of this positive kinetic energy, no bound state exists if the attractive potential is too weak.

The graphical solution of this equation for $x = \kappa'R$ is obtained by finding the intersections of the curve representing $\cot x$ by the curve corresponding to $-\sqrt{(k_0'R/x)^2 - 1}$, as shown in Fig. (8.2). This last curve is limited to the range $0 \leq x \leq k_0'R$, corresponding to the condition $U_0 < E < 0$, and it goes to $-\infty$ for $x \to 0$. On the other hand, $\cot x$ displays divergences, jumping from $-\infty$ to $+\infty$ for $x = n\pi$, and for our domain of interest it is negative in the ranges $(n - 1/2)\pi < x < n\pi$, with n a positive integer. Hence the number of solutions is equal to the number of the negative branches of $\cot x$ crossed by the square root. As is clear from Fig. (8.2), this number is directly obtained from the value of $k_0'R$ since for $x = k_0'R$, the square root is equal to zero. For $0 < k_0'R < \pi/2$ there is no bound state, the attractive potential being too weak to create one. For $(n - 1/2)\pi \leq k_0'R < (n + 1/2)\pi$, there are n solutions, n being a positive integer. In particular, considering $k_0'R$ increasing from zero, a new bound state appears each time we have $k_0'R = (n - 1/2)\pi$. When this bound state appears, it has zero energy $E = 0$, since in this case we have for Eq. (8.21) the solution $x = \kappa' R = k'_0 R = (n - 1/2)\pi$ and from Eq. (8.20) $\kappa' = k'_0$ implies $\kappa = 0$. When $k'_0 R$ is increased from this limiting value, the bound state goes from zero to increasingly negative energy.

We now see that the condition $k'_0 R = (n - 1/2)\pi$ we have found for the existence of a bound state with zero energy coincides with the values for which the scattering length diverges. This can be understood physically as a resonance phenomenon. Indeed the bound states in this attractive potential can be seen as eigenmodes of the particle in this potential, that is the frequencies at which the resulting system vibrates naturally. On the other hand, when we consider the scattering length, we deal with a particle arriving on the potential with an energy $E \to 0$. When there is a bound state with energy $E = 0$, this incoming particle has just the energy corresponding to an eigenfrequency in the attractive potential. Hence it is natural that the scattering amplitude diverges in this case, in the same way as a classical system excited at its resonance frequency has an infinitely strong response in the absence of any damping.

Conversely, this physical argument shows that this result is not specific of our square well model, and that it is valid for a general potential. When this potential is such that it has a bound state with zero energy, this resonance phenomenon occurs and accordingly the scattering length diverges. The mathematical point is that the scattering length characterizes the solution of the Schrödinger equation for $E = 0$, and that its calculation can be done by approaching this value from above $E \to 0_+$ as we have done in the preceding section, or from below $E \to 0_-$ as it arises from a bound state calculation. And when this binding energy goes to zero, the extension of the corresponding wave function, which is over a distance of order $1/\kappa$ beyond the range of action of the potential, goes to infinity in this zero-energy limit. This implies that the scattering length also goes to infinity.

More specifically and quantitatively we have noticed that in the region beyond the range of the potential, the form Eq. (8.19) of the wave function can be written $\chi(r) = k(r - a) \equiv c_1 r + c_2$, when k is small enough. Hence in this region the wave function is just a linear function of r, and the scattering length is merely obtained from the ratio of the coefficients by $a = -c_2/c_1$. On the other hand, for a weakly bound state, we have seen that the wave function reads $\chi(r) = Ce^{-\kappa r}$ where we introduce a normalization constant C. When κ is small enough, this can be expanded as $\chi(r) = C(1 - \kappa r) \equiv c_1 r + c_2$, which is exactly of the same form as the above wave function for small enough positive energy. This is natural since both correspond to the solution of the Schrödinger equation for zero energy $E = 0$, which reads in this case $\chi''(r) = 0$ and has indeed as solutions linear functions of r. Accordingly, we have $a = -c_2/c_1 = 1/\kappa$, so the weakly bound state energy is directly related to the scattering length by

$$E = -\frac{\hbar^2 \kappa^2}{2m_r} = -\frac{\hbar^2}{2m_r a^2} \qquad (8.22)$$

We note that in this case, where a weakly bound state is present, the scattering length $a = 1/\kappa$ is large and positive. On the other hand, if we think of decreasing slightly the attractive strength $|U_0|$ of the potential, we reach the situation where the bound state energy is zero $E = 0$, corresponding to a scattering length that has grown to $+\infty$. If we keep decreasing $|U_0|$, corresponding to decrease $k'_0 R$ in our square well model, the scattering length jumps from $+\infty$ to $-\infty$ and so it becomes negative. This is just what happens in Eq. (8.17) when $k_0'R$ decreases and goes slightly below $(n - 1/2)\pi$. In this case, there is strictly speaking no longer any bound state, but one refers often to this situation as having a "virtual bound state" since it is quite close to being present and the corresponding scattering amplitude is quite high, in a way similar to the case where a real bound state is present.

The existence of these resonances allows one to understand the complex behavior displayed in Fig. (8.1) for the dependence of the scattering length as a function of the strength of the potential. Since their existence is a general feature of any potential as we have just stressed, this allows one to understand that, qualitatively, this complex behavior is present for a general potential. Nevertheless specific atoms have naturally a specific interaction potential, leading to some specific value of the scattering length. The whole point about the complex behavior discussed above is that it is not possible to guess easily what is the value of the scattering length, which might well be large or small. This determination requires actually complex and delicate calculations together with experimental work. This lack of control is one of the reasons which may make one specific alkali atom, with its resulting specific scattering properties, more convenient than others to reach low temperature and Bose–Einstein condensation.

8.3.2 Feshbach Resonance

If the situation for the scattering length was just fully as we have described, the crossover we want to investigate in the present chapter could not be realized experimentally. Fortunately, there are most often additional degrees of freedom we have not mentioned yet and which come into play. These are the spin degrees of freedom, both the electronic spin of the single electron in the atomic s-wave outer shell (for alkali) and the nuclear spin. The hyperfine coupling between these two spins gives rise to the hyperfine structure of the lowest energy levels of the electron. For example, the ^{6}Li atom has nuclear spin $I = 1$, while the isotope ^{7}Li has nuclear spin $I = 3/2$. Notice that ^{6}Li, having, in addition to its three electrons, a nucleus made of three protons and three neutrons, is made of nine fermions and is accordingly a fermion. Since ^{7}Li has an additional neutron it is a boson.

In the following, to be specific, we will have essentially in mind the case of ^{6}Li, although a number of analogous situations have been investigated both theoretically and experimentally. For ^{6}Li, taking into account the electronic (degeneracy 2) and nuclear (degeneracy $2I + 1 = 3$) degrees of freedom, the ground state is $2 \times 3 = 6$-fold degenerate if we forget about the hyperfine coupling. However, adding the electronic and the nuclear spin, we find a total angular momentum that is either $F = I + 1/2 = 3/2$ or $F = I - 1/2 = 1/2$. Since electronic and nuclear spins are not oriented in the same way in these two cases, the value of the hyperfine coupling is different, which gives rise to the hyperfine structure. In the present case, the $F = 1/2$ manifold, with degeneracy $2F + 1 = 2$, has the lower energy while for $F = 3/2$, with degeneracy $2F + 1 = 4$, the energy is slightly higher. Moreover, in the presence of a magnetic field, all these energy levels are split due to the Zeeman effect, and no degeneracy remains. At large enough field, which will be the case most relevant for us in the following, the energy is dominated by the coupling of the electronic magnetic moment with the field (the nuclear magnetic moment is much smaller and plays no role in this respect) and the lowest energy is obtained with the electronic magnetic moment parallel to the field. In order to minimize the hyperfine coupling, the nuclear spin is essentially antiparallel to the electronic spin. The corresponding eigenstate belongs to the $F = 1/2$ manifold.

When we consider the scattering properties of two atoms, we have to specify in which eigenstates they are with respect to these spin degrees of freedom. First of all, if these atoms are in the same eigenstate, we have to deal with the scattering of identical atoms. The above considerations about scattering did not consider this point and are actually valid for non-identical atoms. When they are identical, we additionally need to take care of the restrictions imposed by statistical properties. We assume that the spin degrees of freedom are unchanged during scattering, which corresponds to the assumption of elastic scattering. Accordingly, the symmetry properties are affecting the orbital part of the wave function describing the relative motion. If we deal with bosons, the wave function must be unchanged under the exchange of the two atoms. This implies that it must be even in the change $\mathbf{r} \rightarrow -\mathbf{r}$. In our angular momentum expansion, only components with even ℓ satisfy this condition. Therefore in our above analysis of scattering properties, only even ℓ contributions should be retained. Since at low energy we have kept only the $\ell = 0$ contribution, which satisfies this condition, this additional symmetry requirement does not affect our analysis at all.

On the other hand, if we deal with fermions, as is the case for ^{6}Li atoms, the wave function must be odd when the two atoms are exchanged. With unchanged spin degrees of freedom, we end up with a wave function for the relative motion, which must be odd in $\mathbf{r} \rightarrow -\mathbf{r}$, meaning that only the odd ℓ contributions in the angular momentum expansion have to be kept. This implies that the $\ell = 0$ component is irrelevant in this case, which completely modifies our conclusions on the scattering properties in the low energy regime. Since $\ell = 0$ is forbidden, the dominant component in this case is the $\ell = 1$ one. However, as we have seen, it goes to zero for $k \rightarrow 0$. Hence in this limit all the components of the scattering amplitude go to zero, so the scattering amplitude itself is zero. This means physically that there is no scattering at all. One has the very surprising result that in the zero-energy limit, identical fermions go through each other without, so to speak, seeing each other. The scattering process is forbidden by Pauli principle. This is as if the two atoms did not interact. As a result, at $T = 0$, an assembly of identical fermionic atoms forms a perfect non-interacting Fermi sea.

Although the preceding situation is quite neat, it is not from a physical point of view so interesting since we have quite a good understanding of the behavior of a non-interacting Fermi gas. Actually we are rather more interested in the much more complex physics of interacting fermions in the degenerate regime. However, from the above discussion, this means that we have to deal with (at least) two fermionic species. These could correspond for example to two different atomic species. However, it is enough to have two identical atoms in different states with respect to the spin degrees of freedom, since in this case there is no restriction from the Pauli principle, in the same way as there is no similar constraint on electrons in a metal having different spins $\uparrow$ and $\downarrow$. From an experimental point of view, this use of a same atomic species is very convenient. Indeed, if one wants to work with different atomic species, one has to cool each of them down to very low temperature.

However, the corresponding precise experimental setup for each atomic species is different, so cooling two atomic species is markedly more complicated than cooling a single one.

The convenient choice for these two fermionic states is to take the two hyperfine states with the lowest energy in the presence of the magnetic field. In the case of ^{6}Li, these are the two states of the $F = 1/2$ manifold. Actually there is not so much freedom in choosing these states. Indeed we want to have elastic scattering between these states, which implies that the chosen hyperfine states should not be changed by scattering. However, this is not the general situation and, for higher energy hyperfine states, the result of scattering may well imply a modification of the hyperfine states. On the other hand, one can easily see that as a result of conservation laws, the hyperfine states are unchanged by scattering when one works with the two lowest energy ones. This is the situation we will restrict ourselves to in the following. This case is completely analogous to the one encountered in metals, when one has to consider electrons with ↑ and ↓ spins. The similarity is such that in the literature, these hyperfine states are most often referred to as the ↑ and ↓ states, although this is not a proper description of their actual state with respect to their spin degrees of freedom. We will follow this standard convenient convention. In practice, these states are experimentally obtained by a number of manipulations, implying in particular radiofrequency pulses with appropriate frequencies and lengths resulting in the transfer of atoms from one hyperfine state to another.

Once these two fermionic states we want to deal with have been chosen, their behavior at low energy is fully described by their corresponding scattering length a. Naturally the scattering length depends in general on the specific hyperfine states that are considered. More generally, if we want to calculate the scattering length, the electronic and nuclear spins are necessarily involved because, in contrast with our above simple assumption, the interaction between two atoms depends on these degrees of freedom, since for example it contains the dipole-dipole interaction between the two electronic magnetic moments. Since this interaction depends on the distance r between the two atoms, this implies that the full scattering problem must be solved within the six-dimensional space for each atom involving these degrees of freedom, rather than being the simple one-dimensional problem considered above. Nevertheless this strong complication in the actual calculation of the scattering length seems at first sight unimportant, since our interest is merely in the result, which we might for example think of as being determined experimentally. This is indeed the standard situation. However, proceeding in this way would make us miss the understanding of the extremely interesting phenomenon of Feshbach resonance.

This Feshbach resonance is quite analogous to the resonances we have discussed above, and is linked in the same way to the existence of a bound state. However, in contrast with this simple case, this bound state does not occur for the spin degrees of freedom corresponding to our two hyperfine states of interest (our ↑ and ↓ spins) as they are when the atoms are far away from each other. Rather, it occurs for a different arrangement of these spin degrees of freedom. For this arrangement, the two atoms form a bound state, and accordingly cannot be far away; hence the name of "closed channel" used to specify the corresponding spin degrees of freedom. This is in contrast with the "open channel," where the spin degrees of freedom are just as in our two hyperfine states ↑ and ↓ of interest, and for which, at essentially the same energy as the bound state in the closed channel, the two atoms can be far away from each other. The important point is that these two channels are coupled. This is because in solving the scattering problem for our ↑ and ↓ states, we have to solve a Schrödinger equation in a six-dimensional space, which implies that the spin degrees of freedom evolve with the space separation r. Hence, at close distance r, they can have an overlap with the closed channel arrangement for the spin degrees of freedom. In this way, the open and the closed channels feel each other.

The essential point is now that since the spin degrees of freedom are different for the closed and the open channel, their energy depends in a different way on the applied static magnetic field. In a simplified version, we might think that in the open channel the electronic spins of the two atoms are parallel, so that their energy is fully sensitive to the magnetic field, whereas in the closed channel these electronic spins would be antiparallel so that the energy of the bound state would not depend on the magnetic field. The net result is that the difference between the energy of the bound state in the closed channel and the energy of the two far separated atoms (with nearly zero kinetic energy) in the open channel depends on the applied magnetic field. This is as if, in our discussion at the beginning of this section, we had a knob to change at will the bound state energy (by changing $k'_0 R$, for example). Since, when the bound state energy goes through zero, we have the resonance phenomenon where the scattering length a diverges and in this way goes from large negative values to large positive values, we have a way to control a by the magnetic field. Hence cold atoms provide a truly remarkable physical system where, at low energy, not only are interatomic interactions fully characterized by a single parameter, the scattering length, but also this parameter can be modified at will by turning the knob that sets the value of the magnetic field.

Feshbach resonances are actually fairly frequent, but most of the time the coupling to the closed channel is quite weak. This makes the range of magnetic fields where the resonance is felt very small, resulting in narrow Feshbach resonances. Clearly these are not so convenient experimentally. On the other hand, ^{6}Li, for example, displays a quite wide Feshbach resonance centered at $B_0 = 834$ G and a width of the order of a few hundred G. This is experimentally an agreeable situation. As a function of the magnetic field B, the scattering length behaves, in the vicinity of the resonance, as

$$a = a_b \left(1 + \frac{w_B}{B - B_0} \right) \tag{8.23}$$

where w_B gives the width of the resonance and a_b is a background value for the scattering length.

8.4 The BEC–BCS Crossover, the Unitary Limit

In the vicinity of a Feshbach resonance, we have a remarkable physical situation. The scattering length can go from a large negative value to a large positive value. In principle, if on the negative side, we let the magnetic field go far away from the resonance value B_0, we can go down to a fairly small negative value of the scattering length, as results from the dominant dependence $a \sim 1/(B - B_0)$. Similarly, on the positive side, we can go to small positive values of the scattering length by going far enough from the resonance. Pushing this behavior to the extreme, we might ideally think that we could go from a situation with almost zero scattering length where $a = 0_-$ to a very large negative a, then jump to a large positive a and go to a nearly zero value on the positive side $a = 0_+$. This is more easily described with the variable $1/a$, which can ideally go from $-\infty$ to $+\infty$ when one goes through a Feshbach resonance. Naturally this is an idealized picture, and experimentally the scattering length meets bounds when one goes away from the resonance, as it results for example from the background contribution in Eq. (8.23). More generally experimental problems could make smaller values of a inaccessible when one goes away from the resonance. Nevertheless, in the following discussion we will forget about these practical limitations, and we will consider the idealized situation in its full scope.

Let us consider which effective Hamiltonian can describe our Fermi gas, made up of our ↑ and ↓ fermionic states. As we have already mentioned the actual interaction between two atoms is fairly complicated. However, at low energy the only relevant information from the result of this interaction is contained in the value of the scattering length, since in this dilute gas most of the physical situations correspond to the atoms being far apart from each other and not feeling the interatomic interaction. Hence, in order to describe our interacting Fermi gas, we may as well take a simpler Hamiltonian giving exactly the same value of the scattering length. Since the interatomic interaction has effectively a very short range, we may as well take a very short range interaction, which can be essentially represented by an expression proportional to a Dirac function $\delta(\mathbf{r})$. So we can take $V(\mathbf{r}) = g\,\delta(\mathbf{r})$, which introduces the simple coupling constant g, instead of a complex structure.

We have still to find which g has to be taken in order to reproduce a given scattering length a. This can be obtained from our square well model, since the function $\delta(\mathbf{r})$ is just the limit of such a square well when the size R goes to zero. We just need to make sure that the overall strength $\int d\mathbf{r}\, V(\mathbf{r})$ of the potential is the same. For $g\,\delta(\mathbf{r})$, this integral is merely equal to g, while for the square well potential it is equal to the strength U_0 of the potential multiplied by the volume $4\pi R^3/3$ occupied by this square well. Hence we merely have

$$g = \frac{4\pi}{3} U_0 R^3 \tag{8.24}$$

On the other hand, in the limit $R \to 0$, the result Eq. (8.17–8.18) for the scattering length becomes

$$\frac{a}{R} = -\frac{1}{3}(k_0' R)^2 = \frac{2m_r}{3\hbar^2} U_0 R^2 \tag{8.25}$$

which is valid regardless of the sign of U_0. Comparing Eq. (8.24) and Eq. (8.25) gives

$$g = \frac{2\pi \hbar^2 a}{m_r} \tag{8.26}$$

This relation can actually be obtained for a general interaction potential from a lowest-order perturbative calculation, which is called the Born approximation for scattering theory. From Eq. (8.26) we can write our simplified interaction as

$$V(\mathbf{r}) \simeq \frac{2\pi \hbar^2 a}{m_r} \delta(\mathbf{r}) \tag{8.27}$$

Actually going to this limit of a contact interaction is a procedure involving some singular behavior (hence our $\simeq$), a problem we have already encountered. This can be seen if we notice that our derivation (as well as the Born approximation) implies that a is small, which is a problem if we want to consider situations with large a. A natural first step is to go to the next order in our expansion of Eq. (8.17) in powers of $G \equiv (k_0'R)^2$. This gives $a/R \simeq -G/3 - 2G^2/15$. Making use of Eq. (8.24), we have $G = -(3m_r/2\pi \hbar^2) g/R$, so we see that the G^2 term gives to the scattering length a a contribution proportional to g^2/R, which diverges in the limit $R \to 0$ (at fixed g). Higher-order terms would give even stronger divergences. Hence one cannot stop at some order in this expansion; one rather must consider the whole series expansion. If, instead of our square well model, we had the simpler series expansion $a/R = -(G/3)(1 + G + G^2 + \cdots) = -(G/3)/(1 - G)$, we would obtain the simple result $1/a = (2\pi \hbar^2)/(m_r g) + 3/R$. This is a satisfactory relation between a and g, provided we stop in our small R limit at some small but nonzero cut-off R_c, rather than taking the full limit $R \to 0$. This means that we can take our potential with a very short range R_c, but not strictly a zero range.

Actually the standard handling of this problem is made in $\mathbf{k}$ space, rather than in $\mathbf{r}$ space. So it introduces in a similar way a cut-off k_c, which plays the same role as $1/R_c$, and which can be taken as very large but not infinite. In this case, the relation between a and g is very similar to the preceding one, and it reads

$$\frac{m_r}{2\pi \hbar^2 a} = \frac{1}{g} + \frac{m_r k_c}{\pi^2 \hbar^2} \tag{8.28}$$

The last term is just the result of the divergent sum $(2m_r/\hbar^2) \sum 1/\mathbf{k}^2$, with a cut-off at k_c. This relation reduces to Eq. (8.26) in the case of small a and g. On the other hand, it allows one to have any value for $1/a$ in the range $]-\infty, +\infty[$ by an appropriate choice of finite values for g and k_c.

We may now consider a Fermi gas made up of $\uparrow$ and $\downarrow$ fermions with mass m and write the corresponding Hamiltonian. Here we will assume that the number of $\uparrow$ and $\downarrow$ fermions are the same, just as for electrons in a metal. Although this is not so obvious, this can be realized experimentally with cold atoms with very good precision. Since the number n of $\uparrow$ and $\downarrow$ fermions per unit volume is the same, they have also the same Fermi momentum k_F with $n = k_F^3/6\pi^2$ and the same Fermi energy $E_F = \hbar^2 k_F^2/2m$. Because we take an interaction that is essentially a contact potential $V(\mathbf{r}) = g\,\delta(\mathbf{r})$, its Fourier transform is the constant g and the interaction is quite similar to the one we had in Eq. (7.22) for bosons in Chapter 7 with similar conditions. As a result, the Hamiltonian of this Fermi gas is

$$H = \sum_{\mathbf{k}\alpha} \epsilon_{\mathbf{k}} c_{\mathbf{k}\alpha}^\dagger c_{\mathbf{k}\alpha} + g \sum_{\mathbf{k},\mathbf{k}',\mathbf{q}}^{k_c} c_{\mathbf{k}'+\mathbf{q}/2\uparrow}^\dagger c_{-\mathbf{k}'+\mathbf{q}/2\downarrow}^\dagger c_{-\mathbf{k}+\mathbf{q}/2\downarrow} c_{\mathbf{k}+\mathbf{q}/2\uparrow} \tag{8.29}$$

where $\alpha = \uparrow, \downarrow$ is the spin index, and $\epsilon_{\mathbf{k}} = \hbar^2 \mathbf{k}^2/2m$ is the standard kinetic energy. As we have discussed, the interaction has a cut-off at k_c. We assume that this gas is contained in a cubic box, the case where it is in a harmonic trap being most often obtained from this ideal situation via a local density approximation.

We note that this Hamiltonian is extremely similar to the one Eq. (2.24) we had in Chapter 2 for BCS theory, the main difference being that in this reduced BCS Hamiltonian we have, in the interaction, taken $\mathbf{q} = \mathbf{0}$ for the momentum transfer, since we have restricted ourselves to Cooper pairs with zero total momentum. The interaction term had also a cut-off, but the reason was the disappearance of the mechanism for attractive interaction beyond this cut-off.

Let us start by considering the physical situation when the scattering length a is small and negative, corresponding to the dimensionless parameter $1/k_F a$ going to $-\infty$. Following BCS theory, the weak attractive coupling g, given by Eq. (8.26), leads to the formation of Cooper pairs resulting in a corresponding superfluid ground state completely analogous to the BCS wave function Eq. (2.44). Naturally, since the coupling is small, we expect the critical temperature for this superfluid phase to be exponentially small, just as in BCS theory. Then if we go toward the Feshbach resonance by having the scattering length take larger and larger negative values, so that $1/k_F a$ is increasing while still being negative, the coupling constant also takes larger negative values. Hence the binding energy of the Cooper pairs gets larger, and we naturally expect the critical temperature to grow accordingly.

When we reach the Feshbach resonance, where the scattering length jumps from $-\infty$ to $+\infty$, and $1/k_F a = 0$, we arrive at a threshold where two isolated $\uparrow$ and $\downarrow$ fermions form a bound state. As soon as the scattering length becomes positive, the binding energy becomes nonzero and is related to the scattering length by Eq. (8.22), as we have seen above. This is in contrast with the situation where the scattering length is negative, for which the presence of the Fermi sea is necessary in order to obtain a bound state corresponding to the Cooper pair. Naturally the fact that beyond the Feshbach resonance, a bound state arises even for isolated fermions, makes the formation of Cooper pairs even easier, and similarly for the BCS condensate. Accordingly, one expects its critical temperature to keep growing as a function of $1/k_F a$.

However, if one keeps increasing $1/k_F a$ and going away from the Feshbach resonance on the $a > 0$ side, the physical nature of the system changes progressively. Indeed, as we have seen, in the vicinity of the resonance, the bound state wave function behaves as $e^{-\kappa r}$ with $\kappa = 1/a$. So its size, of the order of $1/\kappa$, is equal to a. When one goes away from the Feshbach resonance, a decreases and becomes smaller than $1/k_F$ when $1/k_F a \gtrsim 1$. Since $1/k_F \sim \lambda_F$ is of the order of the distance between fermions, this means that we progressively enter a regime where the size of the bound state gets smaller than the interparticle separation. This implies that $\uparrow$ and $\downarrow$ fermions form a bound state, and that the separation between the resulting bound states gets larger and larger. In other words, we form $\uparrow - \downarrow$ dimers, and we have a gas of these dimers that gets more and more dilute when a keeps decreasing and $1/k_F a$ goes to $+\infty$. Since these dimers contain two fermions, they behave as bosons. So we go progressively toward a dilute Bose gas of these dimers. At this stage, the $\uparrow$ and $\downarrow$ Fermi seas have completely disappeared, with this transformation occurring roughly in the domain $k_F a \sim 1$.

Now we know that a dilute Bose gas forms at low temperature a Bose–Einstein condensate, often abbreviated as BEC. This BEC is, as we have seen, superfluid. Hence, despite the transformation of the physical nature of the system, its superfluid nature has not been lost. In this way, by having $1/k_F a$ go from $-\infty$ to $+\infty$, we have progressively transformed a BCS-like superfluid into a Bose–Einstein condensate. This is called the BEC–BCS cross-over. There is no sudden jump in this transformation, no sharp transition, only a smooth evolution. This is indeed the natural result of our reasoning, and this is actually what experiments show in these cold-atoms systems. But this is by no means obvious. It is quite conceivable that a sharp transition would have occurred. One could even think that these systems would collapse, as a natural consequence of strong attractive interactions between these particles, as happens for Bose gases with a sufficiently strong attractive interaction. Hence it is a very nice feature of these cold atoms systems that they allow to display experimentally this whole BEC–BCS crossover.

In this crossover, a most remarkable situation is right at the Feshbach resonance, where the scattering length diverges. Nevertheless let us remark that if a diverges, the whole scattering amplitude f_0 remains finite provided we keep the energy small, but nonzero. Indeed, going back to Eq. (8.15), it can also be written $f_0 = e^{i\delta_0(k)} \sin \delta_0(k)/k$, so that

$$\frac{1}{f_0} = k \cot \delta_0(k) - ik \qquad (8.30)$$

In the small energy limit, we have $k \cot \delta_0(k) \simeq k/\delta_0(k) = -1/a$, which goes indeed to zero for $a \to \pm\infty$. However, the imaginary part of $1/f_0$, equal to $-ik$, is still present.

Actually it can be shown that the result $\mathrm{Im}(1/f_0) = -k$ is completely general, as it results from particle conservation in the scattering process. This conservation implies that a general operator, called the S matrix, which characterizes the scattering, must be unitary $SS^\dagger = 1$. In our case the s-wave component S_0 of S turns out to be given by $S_0 = e^{2i\delta_0(k)} = 1 + 2ikf_0$, so that the unitarity condition $|S_0|^2 = 1$ leads indeed to $\mathrm{Im}(1/f_0) = -k$. One can show that the total cross section σ, which gives the total probability of scattering for an incident particle flux equal to unity, is merely obtained from the scattering amplitude by $\sigma = 4\pi \sum_\ell (2\ell + 1)|f_\ell|^2$, that is $\sigma = 4\pi |f_0|^2$ in our low energy case where s-wave is dominant. At fixed k, $|f_0|^2$ is maximal and equal to $1/k^2$ when the real part of $1/f_0$ in Eq. (8.30) is zero, that is for $1/a = 0$ in our small k limit. In this case, the total cross section reaches its maximal value, the existence of this upper bound being due to the unitarity condition. This unitarity limit is merely given by $\sigma = 4\pi/k^2$. Hence the $1/k_F a = 0$ situation is called the unitarity (or unitary) limit.

Unitarity is indeed a remarkable situation in the BEC–BCS crossover. The reason is that since $1/k_F a = 0$ at unitarity, we have the disappearance of any parameter characteristic of the scattering processes, and so of interparticle interaction. The only parameter left to characterize the Fermi gas is its density, or equivalently k_F. The only characteristic length is $1/k_F$, and the only characteristic energy is $E_F = \hbar^2 k_F^2/2m$. One often refers to results at unitarity as universal results, because they are independent of any other parameter, and any physical system at unitarity follows exactly the same laws. In a way, this is quite similar to the extreme situations $a \to 0_\pm$, that is $1/k_F a \to \pm\infty$. In the case $1/k_F a \to -\infty$, one finds the non-interacting Fermi gas, with also its density as only characteristic parameter. Similarly, in the limit $1/k_F a \to +\infty$, one has the Bose gas of dimers, which is also non-interacting, since in this very dilute situation interaction between dimers is negligible. Again the only parameter left is the dimer density, which is half the underlying fermion density.

However, in both these limiting situations, the physics is the one of a non-interacting gas, so all the properties are perfectly known. In contrast, at unitarity, the physics is the one of a very strongly interacting gas. This is an extremely complicated problem, for which to date no full solution has been found. The major part of the knowledge is from approximate methods or numerical work, in particular Monte-Carlo calculations, and naturally from experimental studies. Nevertheless, somewhat surprisingly, in contrast with its peculiar situation, the physical properties of the Fermi gas are, from all the studies, perfectly continuous when the parameter $1/k_F a$ goes across the zero value. The singularity linked to the appearance of a bound state for two isolated fermions is washed out due to the presence of the Fermi sea.

8.5 The BCS Approximation

In this section, we consider some of the results obtained from the BCS modeling when it is applied to the BEC–BCS crossover. Naturally we do not expect the BCS approach to give a proper quantitative description of this fermionic system. A first basic reason is that the BCS theory starts from a reduced Hamiltonian in order to focus on the effect of pairing with Cooper pairs having zero total momentum. This implies that in the Hamiltonian Eq. (8.29), all the $\mathbf{q} \neq \mathbf{0}$ terms are omitted in the reduced Hamiltonian Eq. (2.24–2.26). In particular, BCS theory gives above the critical temperature a perfect non-interacting Fermi gas, while our cold atoms system is above T_c a Fermi gas, which is particularly strongly interacting around unitarity. Clearly we should not rely quantitatively on BCS theory for an accurate description.

On the other hand, it sounds natural to expect on the BCS side $1/k_F a \to -\infty$ that BCS theory gives reasonable results. This is indeed a regime with a Fermi gas in the presence of a weakly attractive interaction, which is exactly the kind of physical situation that BCS theory is supposed to describe properly. There are nevertheless some physical differences, since the small energy scale provided by the Debye frequency ω_D in standard superconductors is not present in cold gases. So there is clearly a need for adjustment, but we expect semi-quantitatively a correct description.

A more surprising range where we expect BCS theory to provide reasonable results is the opposite BEC limit $1/k_F a \to +\infty$. This property has been noted fairly early after the publication of BCS theory. This can be understood easily from our starting expression Eq. (2.27) for the BCS wave function $\Phi(\mathbf{r}_1 - \mathbf{r}_2)\, \Phi(\mathbf{r}_3 - \mathbf{r}_4)\cdots$. As we have noted at that stage, this expression is inappropriate as a wave function for fermions since it is not properly antisymmetrized. In order to obtain a correct fermionic wave function, we should add (with appropriate signs) to Eq. (2.27) all the similar expressions obtained by exchanging identical fermion variables. However, let us take a configuration for $\mathbf{r}_1, \mathbf{r}_2, \mathbf{r}_3, \mathbf{r}_4, \cdots$ corresponding with a large probability to a dilute dimer gas, that is $\mathbf{r}_1$ and $\mathbf{r}_2$ close together (within typically the scattering length a, which is the range of the bound state wave function $\Phi(\mathbf{r})$), similarly for $\mathbf{r}_3$ and $\mathbf{r}_4$, but with $\mathbf{r}_3$ and $\mathbf{r}_4$ far away from $\mathbf{r}_1$ and $\mathbf{r}_2$, and so on for all the fermionic particle positions. If we consider, for example, the term obtained by exchanging $\mathbf{r}_1$ and $\mathbf{r}_3$, the resulting factors in the wave function are $\Phi(\mathbf{r}_3 - \mathbf{r}_2)\,\Phi(\mathbf{r}_1 - \mathbf{r}_4)$. Since $\mathbf{r}_3$ and $\mathbf{r}_2$, as well as $\mathbf{r}_1$ and $\mathbf{r}_4$, are far away from each other, and the bound state wave function $\Phi(\mathbf{r})$ decreases exponentially with r, this contribution obtained by the anti-symmetrization process is vanishingly small. The same conclusion applies to all the other terms obtained from antisymmetrization. Hence the only important term in the BCS wave function is the one written explicitly in Eq. (2.27), and it indeed properly describes a Bose condensate of dimers.

Since the BCS model provides a description that is expected to be reasonable for the two limiting cases of the BEC–BCS crossover, it is naturally interesting to see the results it provides in the whole crossover. This is an obviously tempting interpolation scheme to explore. We restrict ourselves for the moment to the $T = 0$ case. To write the starting equations, we have just two adaptations to make. The first one regards the gap equation Eq. (2.78), which reads $1/V = \sum_{\mathbf{k}}^{\omega_D} 1/(2E_{\mathbf{k}})$, with $E_{\mathbf{k}} = \sqrt{\xi_{\mathbf{k}}^2 + \Delta^2}$ and $\xi_{\mathbf{k}} = \epsilon_{\mathbf{k}} - \mu$, $\epsilon_{\mathbf{k}} = \hbar^2 k^2/2m$. Instead of our notation V for the attractive potential, we should write $-g$, as is clear by comparing the Hamiltonian expressions Eq. (2.26) and Eq. (8.29). Making use of Eq. (8.28), of $m_r = m/2$, and $(mk_c)/(2\pi^2\hbar^2) = (m/\hbar^2)\int^{k_c} d\mathbf{k}/\mathbf{k}^2$, this leads to

$$\frac{m}{4\pi\hbar^2 a} = \sum_{\mathbf{k}}\left(\frac{1}{2\epsilon_{\mathbf{k}}} - \frac{1}{2E_{\mathbf{k}}}\right) = \frac{m}{2\pi^2\hbar^2}\int_0^\infty dk\left(1 - \frac{\epsilon_{\mathbf{k}}}{E_{\mathbf{k}}}\right) \qquad (8.31)$$

This equation has the advantage of making use, instead of V, of a directly accessible physical quantity, namely the scattering length a, to characterize the attractive interaction. Moreover we see that the sum on the right-hand side is convergent, so that there is no need to write the cut-off k_c. This is because our introduction of k_c in Eq. (8.28) is fully consistent with the way we should cut off the attractive interaction in BCS theory, if we did not have the physical cut-off provided by the phonon Debye frequency ω_D. This had been mentioned when we have handled the Cooper problem in Section 2.1.

The second adaptation we have to make is to write explicitly the equation giving the particle number, in order to have a relation giving the value of the chemical potential μ. In BCS theory, we did not make use of such an equation, first of all because we could in practice take the chemical potential $\mu = E_F$ fixed at its zero temperature value, since the characteristic energies of superconductivity are so small that we could consider that the chemical potential was unaffected by superconductivity and temperature. We had also remarked that due to Coulomb interaction, the electronic density was locked to its value ensuring total charge neutrality in the metal. These two arguments do not hold anymore for our neutral atoms in the crossover BEC–BCS. The number equation we have to write is merely provided by Eq. (2.76), so that

$$n \equiv \frac{k_F^3}{6\pi^2} = \sum_{\mathbf{k}}\frac{1}{2}\left(1 - \frac{\xi_{\mathbf{k}}}{E_{\mathbf{k}}}\right) = \frac{1}{4\pi^2}\int_0^\infty dk\, k^2\left(1 - \frac{\xi_{\mathbf{k}}}{E_{\mathbf{k}}}\right) \qquad (8.32)$$

where n is the single spin (i.e., hyperfine state) atomic density, and the first equality serves as our definition for k_F.

It is fairly easy to see from Eq. (8.32) how the chemical potential μ evolves qualitatively, as a function of Δ. Let us start from the $\Delta = 0$ situation where $(1/2)(1 - \xi_{\mathbf{k}}/E_{\mathbf{k}})$ is just a step function, equal to 1 for $\epsilon_{\mathbf{k}} < \mu$ and to 0 for $\epsilon_{\mathbf{k}} > \mu$. Hence the right-hand side is just the integral $(1/2\pi^2) \int_0^{k_\mu} dk\, k^2$, with $k_\mu = (2m\mu)^{1/2}/\hbar$. The result $k_\mu^3/6\pi^2$ for this integral leads to the usual free fermion result $k_\mu = k_F$, so that $\mu = \hbar^2 k_F^2/2m \equiv E_F$.

If we now consider the case where Δ is nonzero, but small, the step function we had for $\Delta = 0$ is smeared, over a small typical range Δ for $\xi_{\mathbf{k}}$. More precisely, if we write the difference between the $\Delta \neq 0$ and the $\Delta = 0$ function, we find $(1/2)(1 - \xi_{\mathbf{k}}/E_{\mathbf{k}})$ for $\xi_{\mathbf{k}} > 0$, that is $k > k_\mu$, and $-(1/2)(1 - |\xi_{\mathbf{k}}|/E_{\mathbf{k}})$ for $\xi_{\mathbf{k}} < 0$. This is an odd function of $\xi_{\mathbf{k}}$. If the integration variable was merely $\xi_{\mathbf{k}}$, the integral of this difference would be zero (for small enough Δ), so the result would be unchanged with respect to the $\Delta = 0$ case, and we would still have $k_\mu = k_F$, that is $\mu = E_F$. However, the integration is over k, with the weight k^2, corresponding to the fact that the density of states increases with $\epsilon_{\mathbf{k}}$ as $\epsilon_{\mathbf{k}}^{1/2}$. Hence, in the preceding integration, the weight of the $\xi_{\mathbf{k}} > 0$ part is increased with respect to the $\xi_{\mathbf{k}} < 0$ part, so that the actual result for this integration is positive, rather than zero. If we keep $k_\mu = k_F$ unchanged, the total integral on the right-hand side of Eq. (8.32), including the contribution from the step, is increased, so the result gets larger than the left-hand side n. So, in order to satisfy the equation when Δ is increased, we have to decrease the contribution from the step, which means to decrease k_μ, that is to decrease μ. This argument is reinforced, for larger Δ, by the fact that the $\xi_{\mathbf{k}} > 0$ part of the integral is unbounded, whereas the $\xi_{\mathbf{k}} < 0$ part is bounded by $-\mu < \xi_{\mathbf{k}} < 0$, corresponding to the lower boundary $k = 0$ in the k integration.

Actually this situation is quite similar to the well-known case for the evolution of the chemical potential μ for free fermions, as a function of temperature T. Here Eq. (8.32) is replaced with $n = \sum_{\mathbf{k}} 1/(\exp[\beta(\epsilon_{\mathbf{k}} - \mu)] + 1)$, with $\beta = 1/k_B T$. At $T = 0$ the Fermi distribution gives again a step function at $\epsilon_{\mathbf{k}} = \mu$, which gets smeared when temperature is increased. As a result, the chemical potential decreases, as is well known, being pushed away from the domain of high density of states to the one with lower density of states, and finally zero density of states. So it goes from E_F, at $T = 0$, to large negative values, when T gets large and one reaches the classical regime. The situation is the same with Eq. (8.32), where the chemical potential μ decreases and finally goes to large negative values when Δ increases.

To consider now more quantitative results, let us first address the question of the expression of the gap on the BCS side. We expect naturally the equivalent of the exponential dependence $\exp(-1/N_0 V)$ we have found in Eq. (2.81). Since $N_0 = mk_F/(2\pi^2\hbar^2)$ and V should be replaced with $-g \simeq 4\pi\hbar^2 |a|/m$ from Eq. (8.28), taking into account $a < 0$, we should find for Δ a dominant behavior proportional to $\exp[-\pi/(2k_F|a|)]$. On the other hand, since the cut-off is different (there are no phonons coming in), the prefactor should be different. So we have to find it starting from Eq. (8.31). Writing in this equation $\epsilon_{\mathbf{k}} = \xi_{\mathbf{k}} + \mu$, we can evaluate the contribution from $\int_0^\infty dk\,(1 - \xi_{\mathbf{k}}/E_{\mathbf{k}})$ when $\Delta \to 0$. It is merely equal to $\int_0^\infty dk\,(1 - \xi_{\mathbf{k}}/|\xi_{\mathbf{k}}|) = 2k_\mu$, so we are left with the calculation of the convergent integral

$$\int_0^\infty dk\, \frac{1}{E_{\mathbf{k}}} = \left(\frac{2m}{\hbar^2\mu}\right)^{1/2} \int_0^\infty dx\, \frac{1}{\left[(x^2 - 1)^2 + \delta^2\right]^{1/2}} \tag{8.33}$$

where we have set $\delta = \Delta/\mu$ and made the change of variable $k = k_\mu x$. In the limit of small δ, the integral on the right-hand side is found[3] to be equal to $3\ln 2 - \ln \delta$, which leads to

$$\Delta = \frac{8}{e^2} \mu \exp\left(-\frac{\pi}{2k_\mu |a|}\right) \simeq 1.08 \, \mu \exp\left(-\frac{\pi}{2k_\mu |a|}\right) \tag{8.34}$$

Consistent with our above discussion on the chemical potential, we have in this BCS limit $a \to 0_-$ to replace μ with E_F and k_μ with k_F in this formula. We note that in this way Δ is expressed in terms of the single energy scale available in our system, namely the Fermi energy.

This above calculation within BCS theory may give the feeling that the prefactor in Eq. (8.34) is the correct one. Actually this is not true, and only the exponential dependence $\exp[-\pi/(2k_F|a|)]$ is correct. The basic reason is that, as we have mentioned, BCS theory omits all interaction effects in the normal state. In order to obtain the correct prefactor, one has to go to the next order in the expansion of Δ in powers of a, or more precisely in the expansion of $\ln(\Delta/E_F)$ in powers of $k_F a$. At this order, one has to take into account modifications in the normal state of the effective interaction leading to pairing. This correct calculation, which is a kind of perturbative calculation involving the proper handling of the whole problem in powers of $k_F a$, has been done early after the publication of BCS theory. It gives for Δ the following result

$$\Delta = \left(\frac{2}{e}\right)^{7/3} E_F \exp\left(-\frac{\pi}{2k_F|a|}\right) \simeq 0.49 \, E_F \exp\left(-\frac{\pi}{2k_F|a|}\right) \tag{8.35}$$

Note that the simplicity of this result is linked to the remarkable simplicity of the physical system, characterized by a single energy scale E_F and a single parameter a to describe interactions.

Similar considerations are also relevant for the chemical potential. From BCS theory we have seen in Eq. (2.88) that the condensation energy is proportional to Δ^2; hence from Eq. (8.35) this is an exponentially small correction to the total gas energy. Since the chemical potential is obtained from the derivative of the energy, the same conclusion applies to the modification of this chemical potential we have discussed above from pure BCS theory. Once again this omits interaction effects in the normal state. For the chemical potential, these effects are much stronger than the ones coming from BCS theory. Indeed they are linear in $k_F a$. One finds to first order in $k_F a$

---

[3] Writing the numerator as $(1-x) + x$, this integral becomes the sum of two divergent integrals, but each one can be integrated analytically when a cut-off x_c is introduced. The first one has a finite limit when $\delta \to 0$, so we can set immediately in it $\delta = 0$. One obtains for this integral $\int_0^1 dx \, 1/(x+1) - \int_1^{x_c} dx \, 1/(x+1) = 2\ln 2 - \ln x_c$. Introducing the variable $y = x^2$, the second integral becomes $(1/2)\int_0^{x_c^2} dy \, [(y-1)^2 + \delta^2]^{-1/2} = (1/2)\left[\ln(\sqrt{(y-1)^2 + \delta^2} + y - 1)\right]_0^{x_c^2}$. In evaluating the upper bound contribution, we can set $\delta = 0$, which gives for large x_c a result $\ln x_c + (1/2)\ln 2$. For small δ, the lower bound contribution can be written as $(1/2)\ln(\sqrt{1 + \delta^2} - 1) \simeq \ln \delta - (1/2)\ln 2$. This gives for the second integral $\ln x_c - \ln \delta + \ln 2$, which together with the first integral leads to the result given in the text.

$$\mu \simeq E_F(1 + \frac{4}{3\pi}k_Fa) = E_F + \frac{4\pi\hbar^2 na}{m} \tag{8.36}$$

where the last term can be obtained perturbatively from the interaction Eq. (8.27).

Let us now consider the situation away from this BCS limit. For the quantitative analysis, it is slightly more convenient to perform in Eq. (8.32) a by-part integration, making use of $d(1 - \xi_\mathbf{k}/E_\mathbf{k})/dk = -d(\xi_\mathbf{k}/E_\mathbf{k}))/d\xi_\mathbf{k} \times d\xi_\mathbf{k}/dk = -\hbar^2 k\Delta^2/(mE_\mathbf{k}^3)$. This leads to

$$k_F^3 = \frac{\hbar^2 \Delta^2}{2m} J_4 \qquad J_4 = \int_0^\infty dk \frac{k^4}{E_\mathbf{k}^3} \tag{8.37}$$

Similarly, a by-part integration in Eq. (8.31), with $1 - \epsilon_\mathbf{k}/E_\mathbf{k} = (1 - \xi_\mathbf{k}/E_\mathbf{k}) - \mu/E_\mathbf{k}$ and $d(1/E_\mathbf{k})/dk = -\hbar^2 k(\epsilon_\mathbf{k} - \mu)/(mE_\mathbf{k}^3)$, gives

$$\frac{\pi}{2a} = \frac{\hbar^2(\Delta^2 + \mu^2)}{m} J_2 - \frac{\hbar^4 \mu}{2m^2} J_4 \qquad J_2 = \int_0^\infty dk \frac{k^2}{E_\mathbf{k}^3} \tag{8.38}$$

Just as in Eq. (8.33), we go to adimensional integrals by the change of variable $k = k_\mu x$, with now $k_\mu = (2m|\mu|)^{1/2}/\hbar$ to take care of the fact that μ can be negative. In this way, $J_2 = (2m/\hbar^2|\mu|)^{3/2}I_2(\delta)$ and $J_4 = (2m/\hbar^2)^{5/2}/|\mu|^{1/2}I_4(\delta)$, with $\delta = \Delta/|\mu|$. Finally we write μ in terms of E_F, defining $\beta = |\mu|/E_F$. We obtain

$$\beta^{-3/2} = \delta^2 I_4(\delta) \qquad I_4(\delta) = \int_0^\infty dx \frac{x^4}{[(x^2 - s)^2 + \delta^2]^{3/2}} \tag{8.39}$$

and

$$\frac{\pi}{4k_Fa} = \beta^{1/2}\left[(1 + \delta^2)I_2(\delta) - sI_4(\delta)\right] \qquad I_2(\delta) = \int_0^\infty dx \frac{x^2}{[(x^2 - s)^2 + \delta^2]^{3/2}} \tag{8.40}$$

where $s = \mu/|\mu|$ is the sign of μ. In Eq. (8.39-8.40) δ works as a running parameter. The integrals $I_2(\delta)$ and $I_4(\delta)$ are calculated numerically. Then Eq. (8.39) gives β, which provides μ, and Eq. (8.40) gives $1/k_Fa$. Finally Δ is obtained from δ and μ.

We look first at the BEC limit, where we have physically a dilute gas of dimers. As we have seen in Eq. (8.22) the binding energy of these dimers is $\hbar^2/ma^2$. So, in order to create a single dimer, one must provide to each of the ↑ and ↓ fermion making up the dimer an energy $-\hbar^2/2ma^2$, which is therefore their common chemical potential μ. In the BEC limit $a \to 0_+$, this chemical potential goes to $-\infty$, in agreement with our above qualitative discussion. Hence this limit should correspond to $\delta = \Delta/|\mu| \to 0$. Indeed, in this case where $s = -1$, one has the analytical[4] results $I_2(0) = \pi/16$ and $I_4(0) = 3\pi/16$. Then Eq. (8.40) gives $\beta = 1/(k_Fa)^2 = |\mu|/E_F$, in perfect agreement with $\mu = -\hbar^2/2ma^2$. Eq. (8.39) yields $\delta^2 = 16(k_Fa)^3/3\pi$, that is $\Delta/|\mu| = (32\pi na^3)^{1/2}$, taking into account $n = k_F^3/6\pi^2$ for the single spin atomic density n. This ratio goes indeed to zero in the dilute limit $na^3 \to 0$.

It is interesting and not difficult to go to the next order for μ in powers of k_Fa. In Eqs. (8.40), we just need to expand $I_2(\delta)$ and $I_4(\delta)$ to first order in δ^2, and use for δ^2 the expression given by Eq. (8.39), with β given by its lowest-order expression found

---

[4] There are general formulas for this kind of integrals.

above, that is $\beta^{-3/2} = (k_F a)^3$, corresponding to $\delta^2 = 16(k_F a)^3/3\pi$. One finds $I_2(\delta) \simeq (\pi/16)(1 - 15\delta^2/32)$ and $I_4(\delta) \simeq (3\pi/16)(1 - 3\delta^2/32)$. This leads to

$$\mu = -\frac{\hbar^2}{2ma^2} + \frac{2\pi\hbar^2 na}{m} \tag{8.41}$$

On the other hand, we may think of directly calculating this last term in the expression of μ, because we have a dilute Bose gas of dimers and we have already considered this situation in Chapter 7. We have found the interaction energy in Eq. (7.42), as $E/\mathbb{V} = n_d^2 V_0/2$, where $n_d = N/\mathbb{V}$ is the dimer density, which is equal to our single fermion density n, since each dimer contains a $\uparrow$ fermion and a $\downarrow$ fermion. On the other hand, V_0 is the equivalent, for bosons, to the interaction g between fermions we have considered in Section 8.4, and it is related in the same way as Eq. (8.26) to the dimer–dimer scattering length a_d by $V_0 = 4\pi\hbar^2 a_d/m_d$, as we have also indicated in Chapter 7. Here we have to take into account that the dimer mass m_d is twice the fermion mass $m_d = 2m$. The resulting dimer chemical potential is $\mu_d = \partial(E/\mathbb{V})/\partial n_d = V_0 n_d = 4\pi\hbar^2 n_d a_d/m_d = 2\pi\hbar^2 n a_d/m$. On the other hand, as we have seen, the fermion chemical potential is half the dimer chemical potential. Taking into account the dimer binding energy $-\hbar^2/ma^2$ in the dimer chemical potential, we obtain in this way

$$\mu = -\frac{\hbar^2}{2ma^2} + \frac{\pi\hbar^2 na_d}{m} \tag{8.42}$$

Comparing with Eq. (8.41), we see that the two results agree provided $a_d = 2a$. Naturally this result is not the correct one. Finding the exact a_d, corresponding to dimer–dimer scattering, implies dealing with a four-body problem that is already somewhat complicated. Clearly the simple BCS framework is not handling such a problem. The exact calculation has been done and yields $a_d \simeq 0.6\,a$. It can be seen that the BCS result $a_d = 2a$ is actually the outcome of a simple perturbative calculation, that is, it corresponds to the Born approximation for this scattering length. Hence we find on the BEC side a situation analogous to the one we have met on the BCS side. The BCS approximation gives correctly the dominant term, but the next order term is not the proper one. Nevertheless we note that on the BEC side, the situation is somewhat better since the $\hbar^2 na/m$ dependence for the next order term is the correct one. Only the coefficient is not the good one, which is not unexpected.

If we turn to the situation between the BCS and the BEC limits, we have to handle the full numerical problem found in Eq. (8.39-8.40), which is quite simple. The result for μ/E_F is displayed in Fig. (8.3), and we see that it interpolates regularly between the two limiting BCS and BEC behaviors we have considered above. A specific point of interest is unitarity $1/k_F a = 0$. From Eq. (8.40), δ satisfies $I_4(\delta)/I_2(\delta) = 1 + \delta^2$, which gives numerically $\delta = 1.16$, and correspondingly one obtains $\beta = \mu/E_F = 0.59$. Just as in the BCS and BEC limits, the result of this BCS approximation is naturally not the correct one. A proper evaluation of this ratio, often called the Bertsch parameter, requires difficult numerical calculations, mostly Monte-Carlo calculations. Similarly it is difficult to obtain experimentally a precise value for this parameter. Both experiment and theory agree to find β in the 0.35–0.4 range.

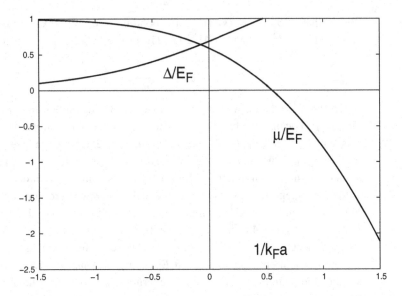

BEC–BCS crossover: BCS approximation for the chemical potential μ/E_F and the gap Δ/E_F, in units of E_F, as a function of $1/k_Fa$.

Another point of interest is the one where the chemical potential goes through zero to become negative. In the BCS approximation, $\mu = 0$ is found numerically to occur for $1/k_Fa \simeq 0.55$. This means that with respect to unitarity, it is on the BEC side. This is also the case in the real system since at unitarity one finds a positive chemical potential. This point is of physical interest because, so to speak, this is the point where one goes from a BCS-like physics to a BEC-like physics (although here again, for all physical quantities, there is a progressive transition from one regime to another). This is seen in the expression $E_{\mathbf{k}} = \sqrt{(\epsilon_{\mathbf{k}} - \mu)^2 + \Delta^2}$ for the excitation energy. For $\mu > 0$, the minimum occurs at $k_\mu = (2m\mu/\hbar^2)^{1/2}$, and at this minimum it is equal to the standard gap Δ. This is the usual BCS physical picture. By contrast, for $\mu < 0$, the minimum occurs at $k = 0$, and it is equal to $\sqrt{\mu^2 + \Delta^2}$. This becomes rapidly equal, on the BEC side, to $|\mu|$. Physically this is the energy required to break a dimer. The physical situation is no longer ruled by the presence of a Fermi sea. The physics is rather the one of a dimer gas.

To conclude on this BCS approximation at $T = 0$, we see that it gives, for the description of the BEC–BCS crossover, results which are qualitatively, and even semi-quantitatively, correct. This may be expected since we have seen that the two extreme cases of the BCS and the BEC limit are correctly described. Nevertheless, one might have feared that this approximation gives, for the interpolation between these two limits, somewhat wild results. This is not the case, and for example the chemical potential given in Fig. (8.3) is not so far from the actual results, to the best of the present knowledge. Naturally one may be somewhat disappointed by the quite significant discrepancies, we have seen above, between this approximation and the correct results. However, one could hardly expect better results, taking into account on one hand the complexity of the system, which is around unitarity a strongly interacting fermionic system requiring a full many-body treatment, and on the other hand, the simplicity of the BCS approximation for such a problem. On balance, it is the ability of this very simple approximation to deliver reasonable results that is more striking, rather than its expected deficiencies.

Finally, let us consider the finite temperature situation by looking briefly at the critical temperature within the BCS approximation. On the BCS side, the situation is the same as the one we have found above for the gap value. This is because we can actually relate directly T_c to Δ by equating the value of $1/N_0 V$ from Eq. (2.79) (giving Δ) to the one from Eq. (3.52) (yielding T_c). The result is a convergent integral, so $\hbar\omega_D$ can be replaced by ∞, and phonons disappear completely from the picture, together with the mechanism for attraction between fermions. The resulting integral relates implicitly Δ to T_c. Clearly the result is the ratio $\Delta/k_B T_c \simeq 1.76$ we have found in Eq. (3.55). Actually, in order to obtain the proper prefactor to the exponential dependence, we have to make for T_c the same correction as in Eq. (8.35) for Δ, for the same physical reasons. This leads to

$$k_B T_c = 0.28\, E_F \exp\left(-\frac{\pi}{2k_F|a|}\right) \tag{8.43}$$

Unfortunately, this result is not in practice very useful since in this BCS regime $1/k_F|a| \to \infty$ this critical temperature is exponentially small, and so it is not possible to observe the corresponding transition experimentally. If we try to go to smaller values of $1/k_F|a|$, the above result is no longer valid and the problem of finding the critical temperature is a very complex one. Nevertheless, in order to have a very rough idea of the result at unitarity, let us just extrapolate the above result Eq. (8.43) to $1/k_F a = 0$. This leads to

$$k_B T_c \simeq 0.28\, E_F \tag{8.44}$$

Compared to the best calculations and experimental determinations of the critical temperature at unitarity, which give $k_B T_c \simeq 0.16 E_F$, this estimate is not unreasonable.

On the other hand, if we try to go on the BEC side with the BCS approximation for T_c, we obtain a meaningless result. This is a good example showing that this approximation can go very wrong. What happens is that in BCS theory, the critical temperature is due to the full breaking of Cooper pairs. When one goes on the BEC side, this corresponds to the thermal breaking of dimers. This requires an energy of the order of the dimer binding energy. This would lead to a critical temperature $k_B T_c \sim \hbar^2/ma^2$, which diverges in the BEC limit $1/a \to \infty$. However, such a thermal dimer breaking has nothing to do with the critical temperature corresponding to the Bose–Einstein condensation into the superfluid phase. We have seen in Chapter 7 that this condensate appears when the excited-states phase space for the involved bosons is no longer large enough, when temperature is reduced, to accommodate all the bosons. However, this excited-states phase space is relative to the motion of the boson as a whole, that is the motion of its mass center. This has nothing to do with breaking the boson into its fermionic constituents. Hence the BCS physical picture relative to the critical temperature is completely wrong on this BEC side, and we cannot expect any meaningful results out of it. On the other hand, we know quite well the critical temperature in this BEC limit; this is the perfect gas T_c given by Eq. (7.7). We can express this result in terms of our single energy scale E_F. Since the dimer density is equal to the single spin fermion density $N/\mathbb{V} = n = k_F^3/6\pi^2$, Eq. (7.7) is equivalent to

$$k_B T_c = \frac{3.31}{(6\pi^2)^{2/3}} E_F \simeq 0.218 \, E_F \qquad (8.45)$$

taking into account that the dimer mass is twice the fermion mass. We see that this critical temperature is quite similar to the T_c at unitarity discussed above. Actually improved theoretical treatments, allowing to generalize the BCS theory so as to recover the proper physical behavior on the BEC side, give a critical temperature increasing slightly when one goes on the BEC side of unitarity, and then decreasing gently toward the above T_c of the BEC limit Eq. (8.45). Hence the maximal critical temperature is reached on the BEC side, slightly beyond unitarity.

8.6 The Vortex Lattice

In the preceding sections, we have mostly described what was expected to occur in the BEC–BCS crossover. For our purposes, the most important point is the existence, at low-enough temperature, of a superfluid phase over the whole crossover. In the $1/k_F a \to -\infty$ limit, this phase is perfectly described as a BCS superfluid, with the existence of Cooper pairs due to the attractive interaction between $\uparrow$ and $\downarrow$ fermions in the presence of their respective Fermi seas. On the other hand, in the $1/k_F a \to +\infty$ limit, this phase is a Bose–Einstein condensate of dimers, which are the result of a bound state formed by $\uparrow$ and $\downarrow$ fermions, due to the same attractive interaction. However, it is quite essential that this phase does exist experimentally, and is not merely a very interesting intellectual construction, as was the case before the development of cold atom physics, since this crossover has been indeed investigated theoretically before it could be realized experimentally. In particular, it is quite conceivable that an instability could occur between the BCS and the BEC limits, due to a collapse of the system produced by the attractive interaction. Such a collapse is known to occur in Bose–Einstein condensates in the presence of an attractive interaction between bosons. We review here the main evidences for the existence of this superfluid phase throughout the crossover.

Surprisingly the first evidence of superfluidity in this system has been seen on the BEC side of the crossover. It has been possible to produce, at low-enough temperature, a Bose condensate of dimers. The existence of this condensate has been proved in much the same way as we have seen above in Section 8.1 for the case of atomic bosons. After expansion a clear central peak in the density has been observed, surrounded by a much broader cloud corresponding to the thermally excited part of the dimer system. Actually merely producing these dimers, and getting rid of the excess of lone fermions, is experimentally a very convenient way to obtain equal populations in the two fermionic hyperfine states.

Then a type of experiment that has been developed very successfully with standard Bose condensates has also been performed in these fermionic systems. This is the exploration of the eigenmodes of the quantum fluid in the harmonic trap in which it is confined. Just as for a classical fluid, if one considers the small departures from equilibrium of our quantum fluid in its confined geometry, one finds that their natural oscillations occur at specific frequencies, with corresponding specific displacements. These frequencies are quite conveniently observed and can be measured with fairly good precision. On the other hand, at low temperature the superfluid is ruled by the perfect fluid hydrodynamics that we have already considered in Chapter 7. The results depend on the equation of states (which in our case is directly obtained from the dependence of the chemical potential on density), as well as on the specific shape of the trap. In the simplest situations, for example when the trap has a spherical symmetry, these eigenmodes can be classified according to their symmetries, and one has for example monopole, dipole or quadrupole oscillations, and so on.

Since the eigenfrequencies of these modes are quite specific, one can check that the experimental results are in agreement with the expected hydrodynamics and equation of state. One can then, starting from the BEC side, go on exploring the BEC–BCS crossover and check that one gets the expected evolution of these eigenfrequencies. Hence these experiments provide a fairly good evidence that the system stays superfluid, as it is on the BEC side. In principle one would like to explore the whole BEC–BCS crossover. However, experiments meet practical limitations. Most notably, on the BCS side, since the critical temperature decreases rapidly when $1/k_F a \to -\infty$, one meets the limitation that at some stage the achieved low temperature (assuming for example that it stays constant while a is modified by changing the magnetic field) becomes nevertheless equal to the critical temperature. In this way, the quantum fluid is no longer superfluid. Even before the critical temperature is reached, the fact that temperature is comparable to the critical temperature leads immediately to complications. Similarly it is also quite frequent that on the BEC side, one meets limitations due to instabilities, or merely to the fact that the scattering length cannot go to very small values when one goes away from the Feshbach resonance. Nevertheless it has been possible, starting from values reasonably far on the BEC side, to go across unitarity and then reasonably far on the BCS side. This coverage of the central part of the $1/k_F a$ domain provides a fairly convincing evidence that the quantum gas stays superfluid all along the BEC–BCS crossover, in the $T = 0$ limit.

Although the above experiments give a clear feeling that the BEC–BCS crossover is physically realized in these fermionic cold gases, a most direct and striking evidence comes from the observation of vortices and vortex lattices in these systems. As we have seen in Section 7.5, these vortices appear in a Bose condensed system when a strong enough rotation is imparted to the fluid. Since these cold gases are not in contact with a vessel, one may perhaps wonder how to proceed to produce a rotation. This can be done by giving to a trapping potential with cylindrical symmetry an additional contribution that gives it an ellipsoidal shape, and having this contribution rotate. Another even simpler way is to stir, with an additional potential force, the inside of the gas, just as one would stir with a stick the inside of a cup of coffee. In all cases, laser beams are very convenient ways to produce these additional potentials since, just as with laser tweezers, they can repel or attract atoms depending on the choice of their light frequency. The position of these laser beams can be easily modified by optical means, which makes them very convenient tools.

In the case of BEC condensates, these vortices can be observed fairly conveniently. Indeed in the case of a weakly interacting Bose gas, all the bosons are in the same wave function $\Phi(\mathbf{r})$, so the atomic boson density $n(\mathbf{r})$ is proportional to $|\Phi(\mathbf{r})|^2$. Now, as we have already discussed in Section 7.5, the center of the vortex presents some singularity. Indeed if we consider the simplest geometry of a vortex with cylindrical symmetry, the velocity

$v_s(\mathbf{r})$ of the currents circulating around the vortex is oriented perpendicular to the radial direction, tangential to the circles centered on the vortex. By symmetry, this velocity is constant along a circle with radius r. Hence its circulation around the vortex $\oint \mathbf{v}_s \cdot d\boldsymbol{\ell}$ is merely equal to $2\pi r v_s$. On the other hand, from Eq. (7.67) we have $\mathbf{v}_s(\mathbf{r}) = (\hbar/m) \nabla S(\mathbf{r})$, where $S(\mathbf{r})$ is the phase of the wave function $\Phi(\mathbf{r})$. Accordingly, $\oint \mathbf{v}_s \cdot d\boldsymbol{\ell} = (\hbar/m)\delta S$, where δS is the variation of the phase of the wave function when a single tour on the circle has been done. Since the wave function itself is single-valued we have necessarily $\delta S = 2\pi n$. Now, just as we have seen in Section 6.3, it can be shown that it is energetically more favorable to have $n = 1$ for the integer n. In other words, it is more favorable, instead of having a single vortex with $n = 2$, to have two separated vortices with $n = 1$. This leads us finally to

$$\oint \mathbf{v}_s \cdot d\boldsymbol{\ell} = 2\pi r v_s = \frac{h}{m} \qquad (8.46)$$

Naturally this expression for the velocity field satisfies the irrotational condition Eq. (7.68) on the velocity field, characteristic of a superfluid behavior.

Here we see appearing explicitly the singular behavior already mentioned in Section 7.5, since $v_s = (\hbar/m)(1/r)$ diverges for $r \to 0$, which would entail a divergent kinetic energy. Similarly, at $r = 0$, we have a problem with the phase of the wave function, which for symmetry reasons is equal to $S = 2\pi\theta$ on any circle (with θ being the azimuthal angle), and so becomes undefined for $r = 0$. The solution is the same as the one we have seen in Section 6.3 with the order parameter for the Abrikosov lattice. The modulus of the wave function $\Phi(\mathbf{r})$ goes to zero when $r \to 0$, so that $\Phi(\mathbf{r} = \mathbf{0}) = 0$ and the problem with its phase disappears. Similarly this leads to a convergent integral for the kinetic energy. This decrease of the wave function to zero occurs over a typical distance corresponding to the coherence length for the superconducting case, which is rather called the healing length for Bose gases, as we have mentioned in Section 7.5. Since this healing length $\xi \sim 1/c_s$ gets large in the case of a weakly interacting Bose gas, this implies that there is a fairly large region of size ξ around the center of the vortex where the wave function is small, so that the gas density is also small. Naturally this low-density region marking the vortex center cannot be seen directly since the whole gas cloud in its trap is quite small, but it appears very clearly after a long enough expansion of this cloud obtained by opening the trap.

If the stirring of the gas is strong enough one obtains a number of vortices, instead of single one. Indeed, as we have already mentioned in Section 7.5, for a large number of vortices, the resulting velocity field mimics, on average, the velocity field of a standard fluid under uniform rotation. Since each vortex carries only the quantum of velocity circulation given by Eq. (8.46), one needs a number of vortices to imitate the large velocity circulation corresponding to a strong rotation. This is completely analogous to the superconducting case where a vortex carries a single flux quantum Φ_0, and where in the presence of a strong field, a large number of vortices are necessary in order to obtain the required total magnetic flux. Actually the Hamiltonians corresponding to these two situations can be shown to be formally essentially identical, with the magnetic field $\mathbf{B}$ corresponding to the angular velocity Ω, but we do not need to enter this detailed analysis. Finally if, after a strong stirring, one waits long enough to have the resulting vortices relax to their equilibrium position (in the rotating referential), one finds that they form a triangular lattice, just as we have seen in Section 6.3 for the Abrikosov lattice. This vortex lattice, marked by the low-density regions corresponding to the vortex centers, has again been seen quite clearly in Bose condensates, after letting the gas cloud expand.

These experiments on Bose condensates can also be performed on the BEC side of the BEC–BCS crossover, with similar results. It is then natural to go toward unitarity and the BCS side to see if the same vortex lattice can be observed. However, one rapidly meets problems upon going in this direction. At first, one gets a more strongly interacting Bose gas of dimers, so the healing length is shorter and as a result it is harder to see the depleted region corresponding to the vortex centers. However, a much more important problem is that this density depletion becomes weaker and disappears progressively. The reason is that the relation between particle density and wave function does not hold anymore, because we are no longer in a situation where the many-particle wave function is given by the simple form Eq. (2.27) corresponding to a very dilute Bose gas. The terms corresponding to the necessary antisymmetrization become progressively more important, and it is no longer possible to omit them.

Nevertheless, our above analysis with the ground-state wave function $\Phi(\mathbf{r})$ remains correct if we replace this wave function with the order parameter $\Delta(\mathbf{r})$. It is then essentially the same situation as the one we discussed in detail in Section 6.3 for the Abrikosov lattice. What happens is that the particle density is no longer simply related to the order parameter, in contrast to what happens on the BEC side. This is clear in the BCS limit. In BCS theory, we had an electronic density that we have taken as a constant, independent of the position inside the superconducting sample, and in particular independent of the value of the order parameter, be it zero or not. Actually our argument for this was that the energy cost for changing electronic density, thereby going away from electric neutrality, is of the order of the plasma frequency, which is much higher by order of magnitudes than the characteristic energies of the superconducting state. This argument does not hold in the present case since the constituents of our atomic gases are not charged objects. However, this does not change the situation much. Indeed the energy cost for a change in density is directly related to the cost in energy to add a particle, which is the chemical potential, equal to E_F on the BCS side. This is again much larger than the typical condensation energies on this side, such as the critical temperature or the gap. Hence the conclusion remains that there is essentially no density change at the center of a vortex in the BCS limit.

However, experimentalists have found a very nice way to go around this problem. The experiment is performed at some location in the BEC–BCS crossover by setting appropriately the magnetic field with respect to the Feshbach resonance. One proceeds to the stirring of the gas and its relaxation, as described previously. Then, instead of performing a simple expansion, the magnetic field is modified very rapidly[5] to bring the system on

---

[5] One obvious problem one might fear with this procedure is that the sudden change of magnetic field, which produces a strong change of the scattering length and accordingly of the gas properties, would disorganize or destroy the vortex structure in such a way that it would quite be different from the original one. On the other hand, there is some topological rigidity with the organization of the vortex lattice and the corresponding order parameter, which makes it resistant to changes. Anyway, the observation of the triangular lattice is a clear proof that this destruction does not occur in these experiments (with their appropriate experimental settings).

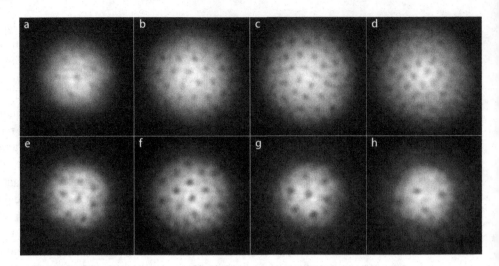

Fig. 8.4   Vortex lattices observed in the BEC–BCS crossover in ^{6}Li. The magnetic fields are (a) 740 G, (b) 766 G, (c) 792 G, (d) 812 G, (e) 833 G, (f) 843 G, (g) 853 G, and (h) 863 G. The corresponding Feshbach resonance is located at $B_0 = 834$ G. The BEC side of the crossover corresponds to fields below B_0. The other parameters in Eq. (8.23) are $a_b = -1405\, a_0$ (a_0 being the Bohr radius) and $w_B = 300$ G. Reproduced with the kind permission of M. Zwierlein, from *Nature* **435**, 1047–1051 (2005) by M. W. Zwierlein, J. R. Abo-Shaeer, A. Schirotzek, C. H. Schunck, and W. Ketterle.

the BEC side. This is done almost immediately after the beginning of the expansion. The vortices are then seen as is done usually with a BEC condensate. Results of such an experiment with a ^{6}Li gas are shown in Fig. (8.4). They are very striking by the almost-perfect triangular lattice present in all the figures.

The implications of this perfect order are quite strong. Such an ordering means that all the vortices are identical objects. Indeed, if these vortices were different from each other, some would be smaller, others would be bigger, and these different sizes would induce corresponding deformations in the vortex arrangement. This identity of all the vortices in the lattice is basically equivalent to stating that these vortices are quantized, with their identity resulting from satisfying a same quantification rule, such as the circulation quantification Eq. (8.46). And this quantification rule is itself directly related to the fact that the circulating currents in a vortex are supercurrents, resulting directly from the link Eq. (7.67) between the velocity and the phase of the order parameter. Hence these beautiful experimental results provide a most direct proof that our fermionic gas is superfluid throughout the whole BEC–BCS crossover.

The existence of this experimental continuity between a system displaying superfluidity in a Bose–Einstein condensate, and a system showing the same superfluid behavior within a physical regime fully described by a BCS-like theory, is of major conceptual importance. Indeed it shows explicitly that the same basic physical principle, namely the existence of a condensate, is present in these two limiting situations. Extending the BEC regime to the case of superfluid ^{4}He and the BCS limit to the case of superconductors, it proves the deep link that exists between superconductivity and superfluidity in liquid ^{4}He. Although we have mentioned from the start that the similarities between the two phenomena have provided a source of inspiration in the elaboration of a full understanding of superconductivity, at the time of the elaboration of BCS theory, for historical reasons, the focus was rather on the differences between Bose-like and Fermi-like superfluidities. Indeed we have seen that the presence of the Fermi sea is an essential ingredient in our understanding of the physics of superconductors. However, there is no contradiction, and the existence of the BEC–BCS crossover shows that the bound state corresponding to the dimers on the BEC side survives surprisingly to the development of a Fermi sea, giving progressively rise in this way to Cooper pairs. Correspondingly it is still an active debate to determine whether or not, in some superconductors such as the cuprates, the electronic pairs responsible for superconductivity are actually on the BEC side of the crossover, instead of far on the BCS side, as is the case for standard superconductors.

8.7 Spin-Polarized Fermi Gases

A very interesting feature of cold gases is the possibility they offer to play on parameters and display physical situations that are difficult or impossible to reach in other real systems. In this Section we briefly address a specific example that is of interest for superconductivity, the ability to strongly polarize at will a Fermi gas, while this is much more difficult with an electronic gas. Indeed, as we have seen, there is no specific reason to have in a cold gas equal populations in the hyperfine states that play the role of the electronic spin populations in a metal. Actually the practical situation is rather the opposite and one needs to operate specifically to reach equal hyperfine states populations, in order to obtain a physical situation analogous to the one found for electrons in a metal.

In a standard metal, the situation is the opposite, and in the absence of magnetic field, the two electronic spin populations are equal, with equal chemical potentials. Applying a magnetic field in a normal state metal naturally produces an imbalance between the spin populations, but in most cases it is rather weak because of the experimental limitation in the strength of the magnetic fields that are currently possible to achieve. In the superconducting phase, this is even worse since the field is expelled from the bulk and can be present only in the penetration depth.

For superconductivity, we previously disregarded any spin polarization, because it has in standard situations a negligible effect. Indeed, as we have seen, when the strength of the magnetic field is increased, this is the energy cost related to setting up the screening currents, which produces the transition to the normal state. Hence these are the electronic orbital degrees of freedom that are involved, not the spins. This occurs much before spin polarization has any sizeable effect.

There are, however, cases where this spin polarization may have an observable effect. One of them arises when we consider a strongly two-dimensional superconductor. This is, for example, what happens when it is made of a stack of conducting planes, with a weak

electronic coupling between the different planes, as is the case for cuprate superconductors. As a result, currents flowing perpendicular to the planes are much weaker than those
flowing in-plane. If the magnetic field is oriented parallel to the planes, the geometry may
be such that the screening currents have to flow perpendicular to the planes. Since these
currents are strongly hindered, the orbital suppression of superconductivity may become
negligible compared to the effect of spin polarization. This opens the possibility to reach
very high critical fields, which is of extreme interest for applications of superconductivity.
This is this kind of situation that we assume in the following, where we consider the effect
of spin polarization alone. Naturally, since they are not charged, superfluid cold gases on
the BCS side are a perfect system to display this physical situation.

Clearly the Fermi gas polarization is unfavorable to pairing, since singlet Cooper pairs
are made by associating one ↑ and one ↓ fermions. Hence the ↑ and ↓ fermions come
naturally in equal numbers in the condensate, and any field that polarizes the gas has a
pair-breaking effect. Near the critical temperature, the effect of the gas polarization is not
so strong, since the excited states, the bogolons, are heavily populated by thermal excitation
and they can carry the gas polarization.

More precisely, let us start with equal spin populations $n_\uparrow = n_\downarrow$, corresponding to equal
chemical potentials $\mu_\uparrow = \mu_\downarrow = \mu$. A bogolon energy is as usual given by $E_\mathbf{k} = \sqrt{\xi_\mathbf{k}^2 + \Delta^2}$
with $\xi_\mathbf{k} = \epsilon_\mathbf{k} - \mu$. If we assume that the polarization is due to an applied field B, this gives
to ↑ and ↓ fermions an additional energy $\pm \mu_B B \equiv \pm \mu^*$, where μ_B is the Bohr magneton.
Correspondingly the ↑↓ chemical potentials are shifted by μ^*, so that $\mu_\uparrow = \mu + \mu^*$ and
$\mu_\downarrow = \mu - \mu^*$. It is more convenient to keep working with the average chemical potential
$\mu = (\mu_\uparrow + \mu_\downarrow)/2$ and the "effective" polarizing field $\mu^* = (\mu_\uparrow - \mu_\downarrow)/2$, rather than
with the spin populations, which are thermodynamically their conjugate variables. When
necessary we can use thermodynamics to relate particle number to chemical potential.

Now since the bogolons are physically just single ↑ or ↓ fermions, their energy is also
merely shifted by $\pm \mu^*$, so they are given by $E_\mathbf{k} \pm \mu^*$. This leads to a corresponding modification of the gap equation Eq. (3.45), since we have seen at the level of Eq. (3.39) that
in the factor $(1 - 2f_k)$, the term $2f_k$ corresponds to the probability that bogolons are present. The energy E_k in f_k now needs to be replaced with $E_\mathbf{k} \pm \mu^*$ since the energy of these
bogolons is modified. Hence the gap equation Eq. (3.51) becomes

$$\frac{1}{N_0 V} = \int_0^{\hbar\omega_D} d\xi_\mathbf{k} \, \frac{1}{2E_\mathbf{k}} \left(\tanh(\frac{\beta(E_\mathbf{k} + \mu^*)}{2}) + \tanh(\frac{\beta(E_\mathbf{k} - \mu^*)}{2}) \right) \tag{8.47}$$

In particular, the critical temperature $k_B T_c = 1/\beta_c$, as a function of μ^*, is obtained by
setting $\Delta = 0$, which gives

$$\frac{1}{N_0 V} = \int_0^{\hbar\omega_D} d\xi_\mathbf{k} \, \frac{1}{2\xi_\mathbf{k}} \left(\tanh(\frac{\beta_c(\xi_\mathbf{k} + \mu^*)}{2}) + \tanh(\frac{\beta_c(\xi_\mathbf{k} - \mu^*)}{2}) \right) \tag{8.48}$$

The result for $T_c = 0$ is particularly simple. The integrand is nonzero only for $\mu^* < \xi_k <
\hbar\omega_D$ and is equal to $1/\xi_k$. Hence the integral is $\ln(\hbar\omega_D/\mu^*)$. When this is compared to
Eq. (2.81), which gives for the $T = 0$ BCS gap Δ_0, $1/N_0 V = \ln(2\hbar\omega_D/\Delta_0)$, one sees
that the critical field μ_c^* beyond which the superconducting phase disappears is equal to

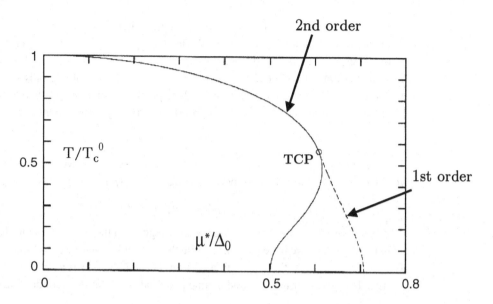

Fig. 8.5 Full line: Critical temperature T as a function of the pair-breaking polarizing effective field μ^*, as it results from the BCS formula Eq. (8.48). The critical temperature is expressed in units of the critical temperature T_c^0 in absence of the field, and the field is expressed in units of the $T = 0$ BCS gap Δ_0. The line goes to $\mu^*/\Delta_0 = 0.5$ at zero temperature. Dashed line: Clogston–Chandrasekhar first-order transition (for which the full line represents the spinodal instability), which goes to $\mu^*/\Delta_0 = 0.707$ at zero temperature. The two lines meet at the tricritical point (TCP).

$\mu_c^* = \Delta_0/2$ at $T = 0$. This is the explicit manifestation of the pair-breaking effect of the polarizing field.

The whole result for $T_c(\mu*)$ is obtained by a simple numerical calculation and is displayed in Fig. (8.5). The full line shows the location of the BCS second-order phase transition toward the superconducting phase. It has a fairly strange shape, with a reentrant behavior: at a fixed field μ^*, slightly beyond its $T = 0$ value, one enters the superconducting phase by increasing temperature, and then goes back to the normal state when temperature is further increased.

However, this natural extension of BCS theory to take into account the effect of the polarizing field is not the whole story. This is quite clear if we consider the $T = 0$ situation. In this case, we have in the superconducting phase the standard condensate made of Cooper pairs. This implies automatically the equality of the spin populations $n_\uparrow = n_\downarrow$. Nevertheless this leads to an energy cost with respect to the normal state, where the spin populations are different due to the polarizing field and adjust in order to satisfy thermodynamic equilibrium $\mu_\uparrow = \mu_\downarrow$. This energy cost is partially compensated by the condensation energy of the superconducting phase, but clearly, if the polarizing field is too strong, the energy balance gets unfavorable to the superconducting phase, and the stable state becomes the normal state. The existence of this phenomenon has been realized by Clogston and Chandrasekhar [47].

If we want to consider this situation more quantitatively, we have to write the proper thermodynamic potential. Since we deal with a system in the presence of a fixed polarizing field, corresponding to a fixed chemical potential, the appropriate potential is not the energy E of the system, but rather $E - \mu n$ (since we consider for simplicity only the $T = 0$ case, we do not have to care about entropy). More precisely, since we deal with two different populations with $\uparrow$ and $\downarrow$ spins, the appropriate thermodynamic potential is

$$\mathcal{G} = E - \mu_\uparrow n_\uparrow - \mu_\downarrow n_\downarrow \tag{8.49}$$

It is more convenient to rewrite this potential in terms of our average chemical potential $\mu = (\mu_\uparrow + \mu_\downarrow)/2$ and effective polarizing field $\mu^* = (\mu_\uparrow - \mu_\downarrow)/2$. This gives

$$\mathcal{G} = E - \mu(n_\uparrow + n_\downarrow) - \mu^*(n_\uparrow - n_\downarrow) \tag{8.50}$$

The term $\mu(n_\uparrow + n_\downarrow)$ does not change when we go from the normal state to the BCS state.

In the BCS state we have $n_\uparrow = n_\downarrow$, while the energy is given by the condensation energy Eq. (2.88), if we take the convention that the normal state energy with $\mu^* = 0$ has zero energy. This gives for the thermodynamic potential $\mathcal{G}_s$ in the superconducting state

$$\mathcal{G}_s + \mu(n_\uparrow + n_\downarrow) = -\frac{1}{2} N_0 \Delta_0^2 \tag{8.51}$$

On the other hand, in the normal state, the spin populations adjust to the polarizing field as we have seen in Section 3.8, when we have considered Pauli paramagnetism. The difference in spin population is given by $n_\uparrow - n_\downarrow = 2N_0\mu^*$. The differential of the thermodynamic potential Eq. (8.49) is $d\mathcal{G} = -n_\uparrow d\mu_\uparrow - n_\downarrow d\mu_\downarrow$. Taking $d\mu_\uparrow = -d\mu_\downarrow = d\mu^*$ into account, this gives $d\mathcal{G} = -(n_\uparrow - n_\downarrow)d\mu^* = -2N_0\mu^* d\mu^*$. When this is integrated from zero to the actual value of μ^*, this yields for the thermodynamic potential $\mathcal{G}_n$ in the normal state

$$\mathcal{G}_n + \mu(n_\uparrow + n_\downarrow) = -N_0\mu^{*2} \tag{8.52}$$

Physically this contribution merely results from the normal state susceptibility.

We see that for a zero or small polarization field, the thermodynamic potential $\mathcal{G}_s$ Eq. (8.51), which is independent of μ^*, is lower, so that the equilibrium state of the metal is the BCS state. On the other hand, if one keeps increasing μ^*, the normal state $\mathcal{G}_n$ Eq. (8.52) gets lower so that the equilibrium state switches to the normal state. The transition from the superconducting to the normal state occurs when $\mathcal{G}_s = \mathcal{G}_n$, that is

$$\mu_c^* = \frac{\Delta_0}{\sqrt{2}} \tag{8.53}$$

This result is higher than the result $\Delta_0/2$ we have found from the BCS formula Eq. (8.48) at the zero temperature. The physical relation between these two transitions is the following. The transition we have just found at $\Delta_0/\sqrt{2}$ is a first-order transition, since there is a discontinuous change of the order at the transition. Indeed, when the field μ^* is lowered so that the metal goes from the normal to the superconducting state, the order parameter jumps from zero in the normal state to its value Δ_0 in the superconducting state. In contrast, the transition at $\Delta_0/2$ is a second-order transition, as we have seen in detail in Chapter 6 on the Ginzburg–Landau theory.

If we start at high μ^* in the normal state and lower progressively μ^*, a first-order transition may occur at $\mu_c^* = \Delta_0/\sqrt{2}$. However, a phenomenon analogous to supercooling is possible, because there is, for the order parameter, a free energy barrier to go from zero to Δ_0. Hence, for $\Delta_0/2 < \mu^* < \Delta_0/\sqrt{2}$, the metal may stay in the normal state, although this does not correspond to the lowest potential $\mathcal{G}$ and is a metastable situation. On the other hand, when μ^* reaches $\Delta_0/2$, there is no longer any barrier in the free energy since the situation is just as in Fig. (6.1) in Ginzburg–Landau theory. The normal state is absolutely unstable with respect to the superconducting state, and nothing like supercooling is possible any longer. This is often called the spinodal instability, corresponding to the first-order transition.

The calculation for the position of this Clogston–Chandrasekhar instability (also called the Pauli limiting transition) can be extended to nonzero temperature, and the result is displayed in Fig. (8.5). One finds that the corresponding first-order transition line meets the spinodal BCS line at a tricritical point (TCP) located at a reduced temperature $T_{TCP}/T_c^0 = 0.56$ and a reduced field $\mu_{TCP}/\Delta_0 = 0.61$. Above T_{TCP} the transition from the normal to the superconducting phase is second-order; below T_{TCP}, it is first order. The analysis of the transition in the vicinity of the TCP can be done within Ginzburg–Landau theory, but we will not go into these details. The interesting point is that, as usual for a second-order phase transition, the coefficient $\alpha(T)$ of the second-order term is zero at the TCP. However, because the transition becomes first order at the TCP, the coefficient of the fourth-order term β is also zero at this point, and so one has naturally to take into account its temperature dependence. Accordingly, in contrast with the standard Ginzburg–Landau theory, one has to include the sixth-order term in order to obtain a consistent Ginzburg–Landau analysis.

8.7.1 Fulde–Ferrell–Larkin–Ovchinnikov Phases

Although the above structure of the phase diagram in the presence of the polarizing field is of interest, the actual physical situation, as far as theory is concerned, is even much more involved and surprising. Indeed it has been proposed by Fulde and Ferrell [48], and Larkin and Ovchinnikov [49], that a polarizing field in a superconductor may result in the formation of Cooper pairs with nonzero center of mass momentum $\mathbf{q}$. Accordingly, this gives rise to a superconducting order parameter $\Delta(\mathbf{r})$, which is space-dependent. This means that there is a spontaneous breaking of translational invariance, since a uniform polarizing field μ^* gives rise to a non-uniform superconducting order. This is quite a remarkable phenomenon, which is fairly analogous to what happens in the mixed phase of a type II superconductor. Indeed, in this case, the application of a uniform magnetic field $\mathbf{B}$ gives rise to a non-uniform superconducting order, with the appearance of the vortex lattice.

The qualitative reason for this phenomenon is the following. As we have seen, when we make a Cooper pair, we gain energy from the attractive interaction, but we do not want to waste kinetic energy by making use of plane waves $\mathbf{k}$ that are too far from the Fermi surface. From Fig. (2.6) we have seen that these plane waves have, with respect to the Fermi energy, an energy difference that is typically of the order of the gap Δ. When the $\uparrow$ and $\downarrow$ Fermi surfaces have the same size, this is done simultaneously for both components $\mathbf{k} \uparrow$

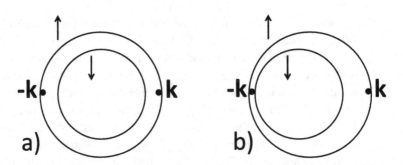

Fig. 8.6 (a) Fermi surfaces for ↑ and ↓ spins, when the corresponding chemical potentials $\mu_\uparrow$ and $\mu_\downarrow$ are different. In the case where the total momentum of the pair is zero, when the $\mathbf{k}$ ↑ fermion is near the ↑ Fermi surface, the $-\mathbf{k}$ ↓ fermion in the pair is far from the ↓ Fermi surface. (b) Same situation when the pair has a nonzero total momentum $\mathbf{q}$, which is equivalent to shifting the ↓ Fermi surface by $-\mathbf{q}$. In this case, the $\mathbf{k}$ ↑ fermion can be near the ↑ Fermi surface and the $-\mathbf{k} + \mathbf{q}$ ↓ fermion near the ↓ Fermi surface.

and $-\mathbf{k}$ ↓ of the Cooper pair. However, as seen in Fig. (8.6a), when the ↑ and ↓ Fermi surfaces have different radii because the chemical potential $\mu_\uparrow$ and $\mu_\downarrow$ are different, if we take $\mathbf{k}$ near the ↑ Fermi surface, the wavevector $-\mathbf{k}$ is far from the ↓ Fermi surface, so that we are wasting kinetic energy.

A possible way to remedy partially this problem is to shift the wavevector of the ↓ fermion by some wavevector $\mathbf{q}$, so that we are pairing $\mathbf{k}$ ↑ with $-\mathbf{k} + \mathbf{q}$ ↓, with both $\mathbf{k}$ and $-\mathbf{k} + \mathbf{q}$ being in the vicinity of their respective Fermi surface. With respect to the preceding $(-\mathbf{k}, \mathbf{k})$ pairing, this is equivalent to shift the ↓ Fermi surface by $-\mathbf{q}$, as shown in Fig. (8.6b). In this way, we obtain a partial matching of the ↑ and ↓ Fermi surfaces. This is somewhat analogous to the nesting of the Fermi surface we have seen in Section 3.8, which is favorable to antiferromagnetic ordering.

However, the shifting wavevector $\mathbf{q}$ is naturally the same for all wavevectors $\mathbf{k}$ involved in the formation of the Cooper pair. Indeed the interaction conserves momentum which makes for example the fermion pair $(\mathbf{k}$ ↑, $-\mathbf{k} + \mathbf{q}$ ↓$)$ to be scattered into $(\mathbf{k}'$ ↑, $-\mathbf{k}' + \mathbf{q}$ ↓$)$, with the same total momentum $\mathbf{q}$. Naturally, as a result, the matching of the two Fermi surfaces does not work for all wavevectors, as is clear from Fig. (8.6b). Hence it is a matter of quantitative analysis to show that the gain with kinetic energy from some parts of the Fermi surfaces is larger than the loss from some other parts.

We will not enter the corresponding detailed calculations within BCS theory, which become rapidly complicated, and we will merely give the results. However, it is interesting and easy to see that already at the simple level of the Cooper problem, it is favorable for a Cooper pair to have a nonzero total momentum when the ↑ and ↓ Fermi surfaces have different chemical potential. Indeed let us take the Cooper problem, as we have seen it in Section 2.1 and, instead of taking a pair with zero momentum, make a pair $(\mathbf{k}_1 = \mathbf{k} + \mathbf{q}/2$ ↑, $\mathbf{k}_2 = -\mathbf{k} + \mathbf{q}/2$ ↓$)$ that has total momentum $\mathbf{q}$, following the more convenient notations of Section 2.1. The exclusion principle forces $|\mathbf{k}_1| > k_\uparrow$ and $|\mathbf{k}_2| > k_\downarrow$, with $\hbar^2 k_\uparrow^2 / 2m - \mu^* = \hbar^2 k_\downarrow^2 / 2m + \mu^* = \mu$.

When we write the Schrödinger equation for the Fourier transform $a_{\mathbf{k},\mathbf{q}}$ of the pair wave function, we find naturally a kinetic energy term $\hbar^2(\mathbf{k}_1^2 + \mathbf{k}_2^2)/2m = \hbar^2(\mathbf{k}^2 + \mathbf{q}^2/4)/m$. We will restrict ourselves to first-order terms in $\mathbf{q}$. This is justified in standard situations where the optimal value of $\mathbf{q}$ is small compared to the Fermi wavevector. Hence we neglect the $\mathbf{q}^2/4$ term, and the kinetic energy term is $\hbar^2\mathbf{k}^2/m$, unchanged with respect to the $\mathbf{q} = 0$ case. The interaction energy is also naturally unchanged, so by following the same steps as in Section 2.1, we end up basically with the same equation Eq. (2.11), except that we have to keep the angular integration $d\Omega_{\mathbf{k}'}/4\pi$ over $\mathbf{k}'$ because the lower boundary for the $\xi_{k'}$ integration, which has to be modified, depends on the angle between $\mathbf{k}'$ and $\mathbf{q}$.

Indeed $|\mathbf{k}_1|^2 > k_\uparrow^2$ implies $|\mathbf{k}_1|^2 = \mathbf{k}^2 + \mathbf{k}\cdot\mathbf{q} + \mathbf{q}^2/4 \simeq k^2 + kq\cos\theta > 2m(\mu+\mu^*)/\hbar^2$, where θ is the angle between $\mathbf{k}$ and $\mathbf{q}$. This gives $\xi_k = \hbar^2k^2/2m - \mu > \mu^* - Q\cos\theta$, where we have set $Q = \hbar^2 k_F q/2m$, using the fact that k is anyway close to k_F. Similarly $|\mathbf{k}_2|^2 > k_\downarrow^2$ implies $\xi_k > -\mu^* + Q\cos\theta$. We must choose for ξ_k the largest of these two above bounds, which is $|\mu^* - Q\cos\theta|$. Accordingly, Eq. (2.11) is modified into

$$(2\xi_k - \epsilon)a_{\mathbf{k},\mathbf{q}} = N_0 V \int \frac{d\Omega_{\mathbf{k}'}}{4\pi} \int_{|\mu^*-Q\cos\theta|}^{\hbar\omega_D} d\xi_{k'}\, a_{\mathbf{k}',\mathbf{q}} \tag{8.54}$$

where θ is now the angle between $\mathbf{k}'$ and $\mathbf{q}$. Just as in Section 2.1, this shows that $a_{\mathbf{k},\mathbf{q}}$ is proportional to $1/(2\xi_k - \epsilon)$. When this is inserted on the right-hand side, this leads for the binding energy $|\epsilon|$ to the equation

$$\frac{1}{N_0 V} = \frac{1}{2}\int_{-1}^{1} du \int_{|\mu^*-Qu|}^{\hbar\omega_D} \frac{d\xi_{k'}}{2\xi_{k'} + |\epsilon|} \tag{8.55}$$

where, in the angular integration $\int d\Omega_{\mathbf{k}'} = \int d\varphi \sin\theta d\theta$, the azimuthal integration over φ gives a factor 2π, and we have made the change of variable $u = \cos\theta$.

It is interesting to consider first the simple situation $Q = 0$. In this case, the only difference with Section 2.1 is that the lower boundary in the $\xi_{k'}$ integration is μ^*, instead of zero. This leads, in the weak coupling limit $|\epsilon| \ll \hbar\omega_D$, to the following result for the binding energy

$$|\epsilon| = 2\hbar\omega_D\, e^{-\frac{2}{N_0 V}} - 2\mu^* = |\epsilon_C| - 2\mu^* \tag{8.56}$$

where $|\epsilon_C|$ is the Cooper result Eq. (2.15) for the binding energy. The interpretation is simple. Since we are forced to use only wavevectors $\mathbf{k}$ with kinetic energy larger than $\mu_\uparrow$, all the states with kinetic energy between $\mu_\downarrow$ and $\mu_\uparrow$ are lost, which pushes up the energy of the pair by $\mu_\uparrow - \mu_\downarrow = 2\mu^*$. Hence the binding energy is decreased by this amount.

It is also interesting to look at the case $\mu^* = 0$. After integration over $\xi_{k'}$ in Eq. (8.55), one finds $\ln|\epsilon_C| = \int_0^1 du\, \ln(|\epsilon| + 2Qu)$, which can be integrated by parts. This leads to an implicit equation for $|\epsilon|$. The result for small Q is directly obtained by expansion of the logarithm. One obtains for the binding energy

$$|\epsilon| = |\epsilon_C| - Q = |\epsilon_C| - \frac{\hbar^2 k_F}{2m}q \tag{8.57}$$

This shows that in the Cooper problem, it is indeed not favorable to have a nonzero momentum $\mathbf{q}$ for the pair. But the dominant effect is not the center of mass kinetic energy of order q^2. It is rather of order q, and is due to a decrease, due to Pauli exclusion, of the **k** domain available for making up the pair. One finds also that one reaches a zero binding energy for a wavevector q_0 satisfying $(\hbar^2 k_F/m)q_0 = e |\epsilon_C|$. This shows that the relevant range of wavevector q is indeed small compared to k_F, as stated above.

Finally, coming back to our problem of interest with both Q and μ^* nonzero, one has from Eq. (8.55) after $\xi_{\mathbf{k}'}$ integration

$$\ln |\epsilon_C| = \frac{1}{2} \int_{-1}^{1} du \, \ln(|\epsilon| + 2\mu^* - 2Qu) \tag{8.58}$$

where we have restricted ourselves to the case $Q < \mu^*$ for simplicity. This can again be integrated to give an implicit equation for $|\epsilon|$. However, it is enough for our purposes to consider only the situation where Q is small. We expand again the logarithm, but we need to go to second order in Q in order to obtain a significant result. This yields

$$|\epsilon| = |\epsilon_C| - 2\mu^* + \frac{2}{3}\frac{Q^2}{|\epsilon_C|} = |\epsilon_C| - 2\mu^* + \frac{1}{6|\epsilon_C|}\left(\frac{\hbar^2 k_F}{m}\right)^2 q^2 \tag{8.59}$$

This shows indeed that taking a nonzero momentum **q** increases the Cooper pair binding energy with respect to the result Eq. (8.56) we have found with $\mathbf{q} = \mathbf{0}$.

Let us come back to the Fulde–Ferrell–Larkin–Ovchinnikov phases (FFLO or LOFF phases) and indicate some results found within BCS theory. One finds indeed that a nonzero center of mass momentum **q** lowers the energy of the superconducting phase. At $T = 0$ the transition occurs for an effective field $\mu^* = 0.754 \, \Delta_0$, which is slightly higher than the Clogston–Chandrasekhar limit, and the optimal wavevector is $q = 2.4 \, m\mu^*/(\hbar^2 k_F)$. Fulde and Ferrell investigated the physical properties of an order parameter with a simple plane wave dependence $\Delta(\mathbf{r}) \sim e^{i\mathbf{q}\cdot\mathbf{r}}$. This leaves an important degeneracy since the wavevector **q** can be in any direction.

Larkin and Ovchinnikov noticed that this degeneracy can be at least partially lifted by a coupling between different plane waves. They have explored this possibility at zero temperature within a Ginzburg–Landau theory. If we introduce the Fourier transform $\Delta_{\mathbf{q}}$ of the order parameter $\Delta(\mathbf{r}) = \sum_{\mathbf{q}} \Delta_{\mathbf{q}} e^{i\mathbf{q}\cdot\mathbf{r}}$, the second-order term in Ginzburg–Landau theory is proportional to $\sum_{\mathbf{q}} |\Delta_{\mathbf{q}}|^2$, where the modulus of **q** is fixed at its optimal value. So this term does not couple plane waves with different **q**. This leaves the possibility to make any combination of these plane waves. On the other hand, the fourth-order term is proportional to

$$\sum_{\mathbf{q}_1+\mathbf{q}_2=\mathbf{q}_3+\mathbf{q}_4} J(\mathbf{q}_1, \mathbf{q}_2, \mathbf{q}_3, \mathbf{q}_4) \, \Delta_{\mathbf{q}_1} \Delta_{\mathbf{q}_2}^* \Delta_{\mathbf{q}_3} \Delta_{\mathbf{q}_4}^* \tag{8.60}$$

which couples indeed different values of wavevector **q**. The coefficient $J(\mathbf{q}_1, \mathbf{q}_2, \mathbf{q}_3, \mathbf{q}_4)$ can be calculated within BCS theory. However, finding the combination which minimizes this free energy is in general a very complicated problem.

Larkin and Ovchinnikov simplified this problem by assuming that the solution should correspond to a lattice. This is analogous to the appearance of the vortex lattice we have seen in Section 6.3. The translational invariance of the system is broken by the solution, but this solution is at least invariant under the discrete translations corresponding to the lattice. With this restriction, Larkin and Ovchinnikov solved completely the problem and found that the solution with the lowest free energy is an equal superposition of two plane waves with opposite wavevectors $\mathbf{q}$ and $-\mathbf{q}$. In other words, they found that

$$\Delta(\mathbf{r}) \sim \cos(\mathbf{q} \cdot \mathbf{r}) \tag{8.61}$$

However, Larkin and Ovchinnikov investigated only the possibility of a second-order transition. It turns out that a first-order transition is possible, with a critical field even higher than the one corresponding to the FFLO transition line. This can be seen in particular, by a Ginzburg–Landau theory going up to sixth-order terms, in the vicinity of the TCP, from which the FFLO transition line starts in BCS theory.

Hence the problem of the FFLO phases is still quite open and very interesting. The same is valid on the experimental side. In superconductors, many experiments have shown anomalous behaviors in the critical field at low temperature, which have been interpreted as possible evidence for a transition to FFLO phases. However, it is difficult to eliminate the possibility of other explanations. Nevertheless there are good indirect evidence of the validity of the basic idea, that is the fact that a polarizing field induces order parameter oscillations. In cold gases, the Clogston–Chandrasekhar transition has been clearly observed. There are also good indications in low-dimensional systems of FFLO-like oscillations of the order parameter. However, one never reaches in cold gases the regime where BCS theory is valid. In superconductors, one may also wonder if BCS is a strong enough theory to handle the whole question.

9 Strong Coupling Superconductivity

9.1 Introduction

In this final chapter, we will deal with an extension of BCS theory. As we have seen, BCS theory makes important use of the fact that the typical energies characterizing the superconducting state (critical temperature, gap, and so on) are much smaller than all other physical energy scales entering the problem. These small values result from the weakness of the effective coupling entering the interaction between electrons. As a result, BCS theory is a weak coupling theory. In our extension, this coupling will no longer be small so that some of these physical simplifications disappear. So, by contrast, this theoretical development is known as the strong coupling theory of superconductivity.

Unfortunately, the words "strong coupling" are nowadays quite often used in many other contexts, most notably when one tries to handle beyond any perturbative treatment the effect of the strong electronic interactions on the properties of these electrons, as happens in particular in many model Hamiltonians. Naturally, although there are a number of common features to all these various strong coupling theoretical treatments, this is somewhat confusing. Hence the extension we deal with is also often more recently called Eliashberg [50] theory or Migdal-Eliashberg (ME) theory of superconductivity.

To be more specific, we have seen that to describe the attractive interaction between electrons responsible for the Cooper pair formation, the BCS theory makes use of an instantaneous inter-electronic interaction $V(\mathbf{r})$. However, this attractive interaction is physically due, for the classical superconductors considered by BCS theory, to lattice vibrations and can be seen as the result of phonon exchange between the two involved electrons. In the rest of the chapter, we will keep considering this specific mechanism for the attractive interaction responsible for the formation of the Cooper pairs.

Nevertheless let us stress immediately that the extension to other possible pairing mechanisms, as might arise in non-standard superconductors such as some high T_c superconductors, would in many cases lead to an analysis similar to the one to which we will proceed. These cases are all the theories that attribute the attractive interaction to the exchange of bosons, which are different from the phonons involved in the standard mechanism. These mechanisms may be the exchange of magnetic excitations ("spin-fluctuation mechanism"), which has quite often been considered and which we have discussed briefly at the end of Chapter 3. But they could also be electronic excitations such as plasmons or excitons. In all cases, the characteristic frequency of these bosons will come into play, instead of the standard phonon frequency, but this will nevertheless lead to similar considerations.

Finally, let us mention that this strong coupling theory of superconductivity is necessary for the analysis of the physical properties of the very recently discovered high T_c hydride superconductors. However, in this case the presently favored pairing interaction is through the standard electron–phonon interaction, the very light hydrogen mass leading to high phonon frequencies and accordingly high critical temperature.

As we have seen in Chapter 2, the interaction resulting from phonon exchange implies that this interaction is not instantaneous. Instead the presence of an electron is felt by another one after some time delay, which is of the order of the inverse of the phonon typical frequency, which we keep calling ω_D as we have done in Chapter 2. In order to be able to neglect this time delay, so that the interaction can be nevertheless treated as instantaneous, it must be small enough compared to the characteristic times of the superconducting state. As we have seen in Section 2.2.3, this implies for the corresponding superconducting energies ϵ the condition $\epsilon \ll \hbar\omega_D$. These superconducting energies may be for example the gap or the critical temperature, so that the weak coupling condition, required for the validity of BCS theory, reads

$$\Delta\,, k_B T_c \ll \hbar\omega_D \tag{9.1}$$

In practice this inequality means, for example, that the ratio $k_B T_c / \hbar\omega_D$ should be at most 0.1.

In the strong coupling theory we consider now, the weak coupling condition Eq. (9.1) is no longer required. This implies that we will have to consider the time dependence of the interaction. Equivalently we will have to consider that quantities directly related to the interaction, such as the gap, acquire a frequency dependence. Since breaking the condition Eq. (9.1) implies that temperature can be of the order of phonon frequencies, we will necessarily have to take into account that real thermally excited phonons can be present, whereas satisfying the condition Eq. (9.1) means that in BCS theory we could consider that with respect to phonon degrees of freedom, we were at $T = 0$. Finally we will have to take into account that electrons have a finite lifetime, since we will find that this lifetime can be as short as $1/\omega_D$ for electrons having an energy ξ of order $\hbar\omega_D$, so that it can clearly not be omitted. In contrast with BCS theory, we take into account that the electron–phonon interaction brings already some of these effects in the normal state. Hence we will begin our investigation by the simpler and necessary step of looking at the normal state properties.

9.2 Normal State

An immediate problem we meet with strong coupling, when addressing the question of the effect of the electron–phonon interaction on an electronic Fermi sea, is that there is no simple systematic approach to get results, such as, for example, standard perturbation theory which can be used when the coupling is weak. We are facing what is called a many-body problem, which has no easy solution. The standard tools which have been developed to address this kind of problem are coming from quantum field theory, they are fairly complex and technical, making use of Green's functions and Feynman diagrams. It is naturally beyond our scope in this chapter to present and make use of all this machinery. We refer interested readers to the standard textbooks on the matter. As a result, we will not be able in this chapter to work with the level of precision found in the preceding chapters. Hence we will rather in many instances give the relevant results. However, we will provide physical justifications allowing us to understand the meaning of these results, so as to make them as natural as possible. In particular, we will make use of the quantum field theoretic vocabulary, which is the standard one in this development of superconductivity.

There is an apparent contradiction between our above statements, emphasizing that we face a very difficult problem, in the normal state and even more in the superconducting one, and the fact that we will be able to have specific results, that is, to solve this many-body problem. The reason for this success is that we will make use of the existence (in the superconductors we deal with) of a small parameter we have not yet mentioned. It comes directly from the fact that we will assume, as is the case in standard metals, that the Fermi energy E_F is very high. In particular, it is much higher than a typical phonon energy. Similarly it is also much higher than the characteristic energies of the superconducting state, which are anyway expected in strong coupling to be of the same order as a phonon energy. Hence in dealing with the strong coupling problem we will make use of the following assumptions

$$\Delta, \, k_B T_c, \, \hbar \omega_D \ll E_F \tag{9.2}$$

Hence in weak-coupling BCS theory we had actually two small parameters, namely $k_B T_c / \hbar \omega_D$ and $\hbar \omega_D / E_F$. In strong-coupling we keep only the last one.

The above small parameter is indeed essential, because it is the basic validity condition for what is known as "Migdal's theorem" [51]. It is the existence of this theorem that makes possible the treatment of the normal and the superconducting states in the strong coupling regime. Since this theorem is not such an easy matter, we will first assume its validity and look for the consequences. This will at the same time give us more familiarity with the relevant questions. Then the justification of this theorem will be considered in the Further Reading section at the end of this chapter.

9.2.1 Self-Energy

We will consider the effect of the electron–phonon interaction only on the electrons. Actually phonons are also affected by this interaction, but the effect turns out to be weak, so we may as well forget it. The effect of the interaction on the electrons is characterized by the electronic "self-energy." To grasp what this quantity means, let us start with a free electron with wavevector $\mathbf{k}$. Its wave function contains accordingly a factor $e^{i\mathbf{k}\cdot\mathbf{r}}$. But it has also a time dependence $e^{-i\omega t}$. For free electrons, the frequency ω is obtained from $\mathbf{k}$ by the corresponding dispersion relation, which gives the energy $\hbar^2 \mathbf{k}^2 / 2m \equiv \epsilon_{\mathbf{k}}$. In the following it will be more convenient to take the Fermi energy E_F as reference energy so, instead of

$\hbar\omega = \epsilon_{\mathbf{k}}$, we rather write $\hbar\omega = \epsilon_{\mathbf{k}} - E_F \equiv \xi_{\mathbf{k}}$. Let us now consider the effect of an instantaneous interaction on this electron, and take more specifically the case where it interacts with a periodic potential, which gives rise to a band structure. The space dependence of the wave function is modified into Bloch waves, but there is still in it a factor $e^{i\mathbf{k}\cdot\mathbf{r}}$, and a wavevector $\mathbf{k}$ to characterize the electronic state. Correspondingly we have still a time dependence $e^{-i\omega t}$, but the frequency corresponding to the wavevector $\mathbf{k}$ is modified into $\hbar\omega = \xi_{\mathbf{k}} + v(\mathbf{k})$, where $v(\mathbf{k})$ is the modification to the free electron dispersion relation brought by the interaction, the overall result being equal to the band dispersion relation.

If we consider now our case of interest, namely an interaction which has not only a space dependence but also a time dependence, it is natural to write that the modification to the dispersion relation depends not only on the wavevector $\mathbf{k}$ (linked to the space dependence of the wave function) but also on the frequency ω (linked to its time dependence). Hence, in this general case, we are led to write

$$\hbar\omega = \xi_{\mathbf{k}} + \Sigma(\mathbf{k}, \omega) \tag{9.3}$$

where $\Sigma(\mathbf{k}, \omega)$ is called the self-energy of the electron (due to the electron–phonon interaction). Naturally, for the actual determination of the frequency, Eq. (9.3) turns out to be rather an implicit equation for finding ω in terms of $\mathbf{k}$.

Actually, $\Sigma(\mathbf{k}, \omega)$ is a complex quantity, which is fairly obvious if we look at the physical interpretation of its imaginary part Im Σ. From Eq. (9.3) such an imaginary part implies that ω has also an imaginary part. Writing $\omega = \omega' + i\omega''$ in terms of its real and imaginary parts, ω' and ω'', leads the time dependence of the electronic wave function to be $e^{-i\omega t} = e^{-i\omega' t}e^{\omega'' t}$. This expression contains an exponentially decaying factor (physically we must have $\omega'' < 0$), corresponding to the fact that the electronic state has a finite lifetime. However, this is physically obvious since the existence of an electron–phonon interaction implies that an electron can emit a phonon, and in this way change its wavevector by momentum conservation. This means that the initial electronic state has decayed into another state, and so it has indeed a finite lifetime. This finite lifetime is due to real processes of phonon emission by the electron.

By contrast the real part Re Σ will rather be linked physically to the modification of the dispersion relation (if, because the lifetime is long enough, we can neglect the imaginary parts Im Σ and ω'', Eq. (9.3) merely reads $\hbar\omega' = \xi_{\mathbf{k}} + \Sigma(\mathbf{k}, \omega')$). This implies in particular that the effective mass of the electronic excitation is modified. This is often referred to as mass renormalization by the electron–phonon interaction. The corresponding processes are virtual processes.

These general considerations enlighten the physical meaning of the self-energy, but they are of no help to calculate it effectively. Here comes Migdal's theorem, which states that this self-energy can essentially be calculated by second-order perturbation theory, a quite astonishing result. Indeed this is by no means true because second-order perturbation is correct, since we want the result to be valid for strong coupling where the electron–phonon interaction cannot be considered as small. This is rather a consequence of the smallness of the parameter $\hbar\omega_D/E_F$. To be complete, the calculation must be done in a self-consistent way. However, this turns out to be unimportant in the normal state. Here we will take Migdal's theorem for granted, and refer to the Further Reading section for justifications of this result.

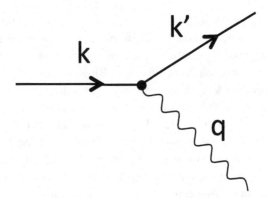

Diagrammatic representation of the process by which a **k** electron becomes **k'** by emission of a phonon with wavevector **q**. The probability amplitude for this process is $g_{\mathbf{k},\mathbf{k}',\mathbf{q}}$, see Eq. (9.4)

9.2.2 Calculation of Σ

We proceed now to the calculation of the self-energy, making use of Migdal's theorem. We have first to write explicitly the term H_{ep} in the Hamiltonian corresponding to the electron–phonon interaction. It reads

$$H_{ep} = \sum_{\mathbf{k},\mathbf{k}',\mathbf{q}} g_{\mathbf{k},\mathbf{k}',\mathbf{q}} \, c^{\dagger}_{\mathbf{k}'} c_{\mathbf{k}} \left(a^{\dagger}_{\mathbf{q}} + a_{-\mathbf{q}} \right) \qquad (9.4)$$

corresponding to the process displayed in Fig. (9.1). Operators $a^{\dagger}_{\mathbf{q}}$ and $a_{-\mathbf{q}}$ are creation and annihilation phonon operators. They come together as written due to the fact that the corresponding physical field, namely the atomic displacement field, is real. This is quite similar to the displacement in a standard harmonic oscillator, expressed in terms of its creation and annihilation operators. In writing Eq. (9.4), we have made the unimportant simplifying assumption that we have only one kind of phonons coming in, instead of the usual transverse and longitudinal ones.

-Imaginary Part of Σ

We begin by the calculation of the imaginary part Im $\Sigma(\mathbf{k}, \omega)$ of the self-energy. As we have seen, it produces in the electronic wave function $\psi(t)$ an exponential decay. (For simplicity we do not write any spatial dependence for this wave function since we concentrate on the time dependence.) From Eq. (9.3), we have $\hbar\omega'' = \text{Im } \Sigma$, so the wave function has an exponential factor $e^{\text{Im }\Sigma t/\hbar}$. Naturally we should have Im $\Sigma < 0$ to have this factor corresponding actually to a decay. The probability to have an electron is proportional to the squared modulus of its wave function $|\psi(t)|^2 \sim e^{-2|\text{Im }\Sigma||t|/\hbar}$. This implies a rate of decay for this presence probability given by

$$\frac{d|\psi(t)|^2}{dt} = -\frac{2}{\hbar} |\text{Im } \Sigma| \, |\psi(t)|^2 = -P_{\mathbf{k}} \, |\psi(t)|^2 \qquad (9.5)$$

In the last equality in Eq. (9.5), we have indicated that the decay rate $d|\psi(t)|^2/dt$ is also obviously equal to the presence probability $|\psi(t)|^2$, multiplied by the transition probability per unit time $P_\mathbf{k}$ for this electron with wavevector $\mathbf{k}$ to go to any other wavevector $\mathbf{k}'$ under the influence of the electron–phonon interaction H_{ep} given by Eq. (9.4).

On the other hand, since we can use second-order perturbation theory according to Migdal's theorem, this transition probability $P_\mathbf{k}$ is merely given by Fermi's golden rule, treating H_{ep} as a perturbation. Hence we can obtain in this way, from $2|\mathrm{Im}\,\Sigma| = \hbar P_\mathbf{k}$, the expression of $\mathrm{Im}\,\Sigma$. Putting for $\mathrm{Im}\,\Sigma(\mathbf{k},\omega)$ the appropriate negative sign, we obtain in this way

$$\mathrm{Im}\,\Sigma(\mathbf{k},\omega) = -\pi \sum_{\mathbf{k}',\mathbf{q}} |g_{\mathbf{k},\mathbf{k}',\mathbf{q}}|^2 \,\delta(\hbar\omega - \xi_{\mathbf{k}'} - \hbar\Omega_\mathbf{q})\,\theta(k' - k_F) \tag{9.6}$$

In the usual Dirac delta function for energy conservation, $\hbar\Omega_\mathbf{q}$ is the energy of a phonon with wavevector $\mathbf{q}$. Hence the part, not yet written, of the Hamiltonian describing the phonons without interaction is $H_p = \sum_\mathbf{q} \hbar\Omega_\mathbf{q} a_\mathbf{q}^\dagger a_\mathbf{q}$. In addition, we have put in Eq. (9.6) a factor $\theta(k' - k_F)$, where θ is the Heaviside step function, to take into account that according to the exclusion principle the final electronic state $\mathbf{k}'$ should be empty for this decay process to occur, which implies that it should be above the Fermi surface $k' > k_F$. We finally remark that in a full perturbative calculation, we should consistently replace $\hbar\omega$ with its lowest-order value $\xi_\mathbf{k}$. However, this deprives our result of any dependence on ω. The fact that we must keep ω results from the self-consistent character of Migdal's theorem and is ascertained by going into more technical detail.

The above result is actually only valid at $T = 0$, as is clear from the presence of the θ function. It is however not difficult to generalize it to nonzero temperature. This is done in much the same way as we have justified the statistical factors in Chapter 4, at the level of Eq. (4.30). First, instead of $\theta(k' - k_F)$, the probability of having the final electronic state $\mathbf{k}'$ empty is in general $1 - f(\xi_{\mathbf{k}'})$, where $f(x) = 1/(e^{\beta x} + 1)$ is the Fermi distribution. Moreover, we have to take into account the phenomenon of induced emission, well known in the optical domain. In the presence of $N(\Omega_\mathbf{q})$ thermally excited phonons, where $N(\Omega_\mathbf{q}) = 1/(e^{\beta\hbar\Omega_\mathbf{q}} - 1)$ is the Bose distribution, the probability for emission of an additional phonon is multiplied by $1 + N(\Omega_\mathbf{q})$. Hence $\theta(k' - k_F)$ is replaced with $(1 - f(\xi_{\mathbf{k}'}))$ $(1 + N(\Omega_\mathbf{q}))$.

In addition, we have to take into account that at nonzero temperature, there is a possibility that the reverse process occurs, where an electron with wavevector $\mathbf{k}'$ absorbs a thermally excited phonon $\mathbf{q}$ to give a $\mathbf{k}$ electron. The statistical factor for this process is $f(\xi_{\mathbf{k}'})N(\Omega_\mathbf{q})$, taking into account the probability $f(\xi_{\mathbf{k}'})$ to have a $\mathbf{k}'$ electron present and the probability $N(\Omega_\mathbf{q})$ to have a $\mathbf{q}$ phonon present. If the $\mathbf{k}$ electronic state were empty, this process would contribute to replenish it. However, we are in the opposite situation where the $\mathbf{k}$ electronic state is instead full. Hence these $f(\xi_{\mathbf{k}'})N(\Omega_\mathbf{q})$ processes are actually blocked by the presence of the $\mathbf{k}$ electron. So we have a lack of replenishment, which is equivalent to an additional source of decay for the $\mathbf{k}$ state. Hence the corresponding contribution should be added to the preceding one and the total statistical factor is

$$\left(1 - f(\xi_{\mathbf{k}'})\right)\left(1 + N(\Omega_\mathbf{q})\right) + f(\xi_{\mathbf{k}'})N(\Omega_\mathbf{q}) = 1 - f(\xi_{\mathbf{k}'}) + N(\Omega_\mathbf{q}) \tag{9.7}$$

Finally, to write the complete expression for the imaginary part of Σ, we have to take into account that the electron–phonon interaction Eq. (9.4) allows also the absorption of a phonon by the $\mathbf{k}$ electron, instead of its emission. This did not come in our above evaluation

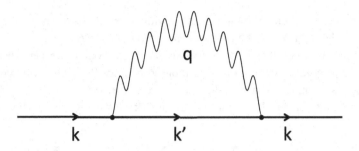

Fig. 9.2 Diagrammatic representation of the second-order process involved in the calculation of the real part of Σ.

at $T = 0$ since no thermally excited phonon to be absorbed was present. The matrix element is the same as in Eq. (9.6), namely $|g_{\mathbf{k},\mathbf{k}',\mathbf{q}}|^2$. But the energy conservation gives now rise to a Dirac function $\delta(\hbar\omega - \xi_{\mathbf{k}'} + \hbar\Omega_{\mathbf{q}})$. The statistical factor is calculated as above, with a contribution $(1 - f(\xi_{\mathbf{k}'}))\, N(\Omega_{\mathbf{q}})$ from the direct process, the factor $N(\Omega_{\mathbf{q}})$ corresponding to the probability to have the absorbed phonon present. On the other hand, the blocked reverse process has similarly a statistical factor $f(\xi_{\mathbf{k}'})\,(1 + N(\Omega_{\mathbf{q}}))$, since it involves a phonon emission with an induced emission factor. The overall result is merely $f(\xi_{\mathbf{k}'}) + N(\Omega_{\mathbf{q}})$, so that at nonzero temperature Eq. (9.6) is replaced with

$$\text{Im } \Sigma(\mathbf{k}, \omega) = -\pi \sum_{\mathbf{k}',\mathbf{q}} |g_{\mathbf{k},\mathbf{k}',\mathbf{q}}|^2 \left\{ (1 - f(\xi_{\mathbf{k}'}) + N(\Omega_{\mathbf{q}}))\, \delta(\hbar\omega - \xi_{\mathbf{k}'} - \hbar\Omega_{\mathbf{q}}) \right. \tag{9.8}$$

$$\left. + (f(\xi_{\mathbf{k}'}) + N(\Omega_{\mathbf{q}}))\, \delta(\hbar\omega - \xi_{\mathbf{k}'} + \hbar\Omega_{\mathbf{q}}) \right\}$$

-Real Part of Σ

We consider next the real part of the self-energy, which we calculate only at $T = 0$. As we mentioned previously, it corresponds physically to a modification of the energy of the $\mathbf{k}$ electron, and we can rely on Migdal's theorem to obtain it from Eq. (9.4) for H_{ep} by standard second-order perturbation theory (at first order the result is zero since the phonon creation and annihilation operators in Eq. (9.4) have zero diagonal elements). The corresponding process is displayed in Fig. (9.2). Second-order perturbation theory gives for the energy change δE for state $|0\rangle$.

$$\delta E = \sum_i \frac{|\langle i|H_{ep}|0\rangle|^2}{E_0 - E_i} \tag{9.9}$$

The initial state $|0\rangle$ is electron $\mathbf{k}$ (in the presence of the Fermi sea), as pictured in Fig. (9.3)b). It goes through H_{ep} to an intermediate state $|i\rangle$, where a phonon $\mathbf{q}$ has been emitted, the electron ending up in state $\mathbf{k}'$. According to Eq. (9.4), the matrix element of the perturbation H_{ep} between this initial state and this intermediate state is $g_{\mathbf{k},\mathbf{k}',\mathbf{q}}$. On the other hand, the energy difference between the initial state and the intermediate state has to be written in the same way as we have done in Eq. (9.6) for the imaginary part of the

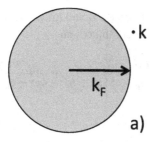 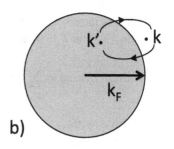

a)   b)

Fig. 9.3 (a) Initial state $|0\rangle$ in the second-order calculation Eq. (9.9) of Real part of Σ, with electron $\mathbf{k}$ in the presence of the Fermi sea. This gives rise to the first term on the right-side of Eq. (9.11). (b) Processes blocked by the presence of the $\mathbf{k}$ electron in the calculation of the ground-state energy of the Fermi sea. This gives rise to the second term on the right-side of Eq. (9.11). We have drawn for simplicity spherical Fermi surfaces, but the actual shape is unimportant.

self-energy, namely as $\hbar\omega - \xi_{\mathbf{k}'} - \hbar\Omega_{\mathbf{q}}$. According to second-order perturbation theory, this leads to

$$\mathrm{Re}\,\Sigma(\mathbf{k}, \omega) = \sum_{\mathbf{k}',\mathbf{q}} \frac{|g_{\mathbf{k},\mathbf{k}',\mathbf{q}}|^2}{\hbar\omega - \xi_{\mathbf{k}'} - \hbar\Omega_{\mathbf{q}}}\, \theta(k' - k_F) \tag{9.10}$$

Just as in Eq. (9.6), we have inserted a factor $\theta(k' - k_F)$ to take into account that this process is only possible if the $\mathbf{k}'$ state is empty in the initial state, which implies that $\mathbf{k}'$ must be above the Fermi surface. (If the $\mathbf{k}'$ state is filled in the initial state, the matrix element of $c_{\mathbf{k}'}^{\dagger}$ in Eq. (9.4) between initial and intermediate state is zero.)

This result Eq. (9.10) is actually incomplete because we have forgotten that the Fermi sea energy is also modified, due to H_{ep}, when the $\mathbf{k}$ electronic state is filled. The argument is analogous to what we have seen above for the imaginary part of the self-energy. Due to the presence of an electron $\mathbf{k}$, processes due to H_{ep} that contribute to the energy of the Fermi sea are blocked, whereas they are present when electron $\mathbf{k}$ is absent. This is pictured in Fig. (9.3b). Since our aim is to calculate what is the energy cost of adding an electron $\mathbf{k}$, we have also to take into account this modification of the Fermi sea energy. The involved processes are those where the intermediate state $|i\rangle$, in second-order perturbation theory, corresponds to any electron $\mathbf{k}'$ in the Fermi sea (hence the factor $\theta(k_F - k')$ in Eq. (9.11) below) being transferred in $\mathbf{k}$ while a phonon $\mathbf{q}$ is emitted. Writing as before the $\mathbf{k}$ electron energy as $\hbar\omega$, the energy difference between the initial and the intermediate state is $\xi_{\mathbf{k}'} - (\hbar\omega + \hbar\Omega_{\mathbf{q}})$. On the other hand, one can see that the corresponding matrix element leads to a factor $|g_{\mathbf{k},\mathbf{k}',\mathbf{q}}|^2$ when Eq. (9.9) is used to calculate the energy change of the Fermi sea. Since these processes are suppressed, and the Fermi sea energy correspondingly diminished, this energy change must be subtracted from Eq. (9.10) to obtain the global energy cost of electron $\mathbf{k}$. This leads to modifying Eq. (9.10) as

$$\mathrm{Re}\,\Sigma(\mathbf{k}, \omega) = \sum_{\mathbf{k}',\mathbf{q}} |g_{\mathbf{k},\mathbf{k}',\mathbf{q}}|^2 \left(\frac{\theta(k' - k_F)}{\hbar\omega - \xi_{\mathbf{k}'} - \hbar\Omega_{\mathbf{q}}} + \frac{\theta(k_F - k')}{\hbar\omega - \xi_{\mathbf{k}'} + \hbar\Omega_{\mathbf{q}}} \right) \tag{9.11}$$

It is of interest to gather the $T = 0$ results for Re Σ and Im Σ into a single formula. Indeed Eq. (9.8) (at $T = 0$) and Eq. (9.11) can be combined into

$$\Sigma(\mathbf{k}, \omega) = \sum_{\mathbf{k}',\mathbf{q}} |g_{\mathbf{k},\mathbf{k}',\mathbf{q}}|^2 \left(\frac{\theta(k' - k_F)}{\hbar\omega - \xi_{\mathbf{k}'} - \hbar\Omega_{\mathbf{q}} + i\epsilon} + \frac{\theta(k_F - k')}{\hbar\omega - \xi_{\mathbf{k}'} + \hbar\Omega_{\mathbf{q}} + i\epsilon} \right) \quad (9.12)$$

with $\epsilon \to 0_+$.

9.2.3 Analysis of Σ

We now make use of this formula to analyze this electronic self-energy. We will be interested in energies $\hbar\omega$ typically of the order of temperature T and phonon frequencies $\Omega_{\mathbf{q}}$. As usual, we replace the summation over $\mathbf{k}'$ with an integral over energy together with an angular average $\sum_{\mathbf{k}'} \to \int (d\bar{\Omega}_{\mathbf{k}'}/4\pi) \, d\xi_{\mathbf{k}'} N(\xi_{\mathbf{k}'})$. For $\xi_{\mathbf{k}'}$ large compared to $\hbar\omega$ and $\Omega_{\mathbf{q}}$, the integral over $\xi_{\mathbf{k}'}$ in Eq. (9.12) converges rapidly, due to the compensation between the parts with $\xi_{\mathbf{k}'} > 0$ and with $\xi_{\mathbf{k}'} < 0$. Hence $\xi_{\mathbf{k}'}$ stays effectively small and we can replace the density of states $N(\xi_{\mathbf{k}'})$ with its value N_0 at the Fermi energy.

Then, regarding the dependence of the result on $\mathbf{k}$, and more specifically on $\xi_{\mathbf{k}}$, when we consider values of $|\xi_{\mathbf{k}}|$ of order T or $\Omega_{\mathbf{q}}$, the variation of $g_{\mathbf{k},\mathbf{k}',\mathbf{q}}$ is small because its typical range of variation is of the order of k_F or the size of the Brillouin zone, the Fermi surface being nothing particular for $g_{\mathbf{k},\mathbf{k}',\mathbf{q}}$. Hence the dependence of $\Sigma(\mathbf{k}, \omega)$ on $\xi_{\mathbf{k}}$ is weak. One could still have a sizeable variation of $\Sigma(\mathbf{k}, \omega)$ when $\mathbf{k}$ moves on the Fermi surface, corresponding to some strong anisotropy. For simplicity we will restrict ourselves to an isotropic situation where such a variation does not arise. In conclusion, the dependence of $\Sigma(\mathbf{k}, \omega)$ on $\mathbf{k}$ is weak, in contrast with the strong variation we will find as a function of ω. Hence we may as well forget about this $\mathbf{k}$ dependence of the self-energy, which we denote simply as $\Sigma(\omega)$ in the following discussion.

We next take into account that the matrix element $g_{\mathbf{k},\mathbf{k}',\mathbf{q}}$ contains necessarily a factor $\delta_{\mathbf{q},\mathbf{k}-\mathbf{k}'}$ because quasi-momentum must be conserved in the process described by the electron–phonon interaction shown in Eq. (9.4) (One could add to this conservation law a wavevector belonging to the reciprocal lattice, taking in this way into account Umklapp processes, but this complication does not bring any qualitative difference, so we omit it.) It is just for simplicity that we have not introduced explicitly this factor up to now. This conservation law implies that having $\mathbf{k}$ fixed, the summation over $\mathbf{k}'$ and over the phonon wavevector $\mathbf{q}$ are linked, so that there is only a single summation over $\mathbf{k}'$ in Eq. (9.12). However, we are interested in collecting the contributions of phonons having a given frequency. For this purpose, we introduce formally a $\delta(\Omega - \Omega_{\mathbf{q}})$ function, to single out the phonons having a specific frequency Ω. When this is summed over frequency, one obtains naturally $\int d\Omega \, \delta(\Omega - \Omega_{\mathbf{q}}) = 1$. Introducing this identity on the right-hand side of Eq. (9.12), and taking into account the above simplifications, we obtain, with integration over the solid angle $\bar{\Omega}_{\mathbf{k}'}$,

$$\Sigma(\omega) = N_0 \int d\Omega \int \frac{d\bar{\Omega}_{\mathbf{k}'}}{4\pi} |g_{\mathbf{k},\mathbf{k}',\mathbf{k}-\mathbf{k}'}|^2 \delta(\Omega - \Omega_{\mathbf{k}-\mathbf{k}'}) \tag{9.13}$$

$$\times \int d\xi_{\mathbf{k}'} \left(\frac{\theta(k' - k_F)}{\hbar\omega - \xi_{\mathbf{k}'} - \hbar\Omega + i\epsilon} + \frac{\theta(k_F - k')}{\hbar\omega - \xi_{\mathbf{k}'} + \hbar\Omega + i\epsilon} \right)$$

We see that the phonon frequency Ω comes with the weight

$$\alpha^2 F(\Omega) = N_0 \int \frac{d\bar{\Omega}_{\mathbf{k}'}}{4\pi} |g_{\mathbf{k},\mathbf{k}',\mathbf{k}-\mathbf{k}'}|^2 \delta(\Omega - \Omega_{\mathbf{k}-\mathbf{k}'}) \tag{9.14}$$

This $\alpha^2 F(\Omega)$ function is called the Eliashberg function. The notation $\alpha^2 F$, instead of a single letter for example, looks fairly involved. Nevertheless it is the traditional notation, and we keep it. It is a reminder that this function contains implicitly a coupling strength (the α^2 part) as well as a phonon density of states (the F part). With this notation, and switching to the variable $\omega' = \xi_{\mathbf{k}'}/\hbar$ for the integration, Eq. (9.13) for $\Sigma(\omega)$ now reads

$$\Sigma(\omega) = \int d\Omega \, \alpha^2 F(\Omega) \int d\omega' \left(\frac{\theta(\omega')}{\omega - \omega' - \Omega + i\epsilon} + \frac{\theta(-\omega')}{\omega - \omega' + \Omega + i\epsilon} \right) \tag{9.15}$$

The integration over ω' is easily performed by putting as boundaries, instead of $\pm\infty$, large finite values that one then lets go to infinity. The result is expressed in terms of logarithms, properly taking care of the determination to get the imaginary part, which can also be directly obtained from $\int_0^\infty d\omega' \, \delta(\omega - \omega' - \Omega) + \int_{-\infty}^0 d\omega' \, \delta(\omega - \omega' + \Omega)$. One finds

$$\Sigma(\omega) = - \int d\Omega \, \alpha^2 F(\Omega) \left[\ln \left| \frac{\Omega + \omega}{\Omega - \omega} \right| + i\pi\theta(|\omega| - \Omega) \right] \tag{9.16}$$

The physical meaning of the imaginary part is clear: for $\omega > 0$ for example, it is nonzero only if the electron energy ω is larger than the phonon frequency Ω, in order to allow for the emission of a real phonon that leads to the decay of the electronic state.

- Results

Let us first consider the situation when the frequency $\omega \to 0$. Defining the positive dimensionless coupling constant λ as

$$\hbar\lambda = \int d\Omega \, \frac{2\alpha^2 F(\Omega)}{\Omega} \tag{9.17}$$

we find that

$$\text{Re} \, \Sigma(\omega) \simeq -\lambda \, \hbar\omega \tag{9.18}$$

When we carry this result into the definition Eq. (9.3) of the self-energy, we have $\hbar\omega = \xi_{\mathbf{k}} - \lambda \, \hbar\omega$, which gives the dispersion relation

$$\hbar\omega = \frac{\xi_{\mathbf{k}}}{1 + \lambda} \tag{9.19}$$

This means that instead of having in the vicinity of the Fermi surface the free electron dispersion relation $\omega = \hbar(k - k_F)k_F/m$, we obtain $\omega = \hbar(k - k_F)k_F/m(1 + \lambda)$. In other words, the electronic mass has been renormalized by the electron–phonon interaction into

$$m^* = m(1 + \lambda) \tag{9.20}$$

Physically this increase in the effective mass of the electron results from the fact that in its motion, the electron has to drag a cloud of virtual phonons, corresponding in practice to the lattice deformation induced by the electron–phonon interaction. The motion of this deformation implies an additional inertia, leading to an increase in the electron mass.

Turning now to the imaginary part of the self-energy in this same limit $\omega \to 0$, from Eq. (9.16) it is due to phonons having a frequency smaller than ω. This means that only the very-low-frequency acoustic phonons are involved, and it is well known that the density of states of these phonons is proportional to Ω^3. This low density of states implies similarly a dependence of the imaginary part of the self-energy proportional to ω^3. This goes to zero much more rapidly than the electron energy $\hbar\omega$, and so the decay of the electronic state becomes negligible in this limit, corresponding to a lifetime τ going to infinity. Hence this very long lifetime for low energy electrons is due to a phase space restriction for the phonons available for decay.

Let us consider the opposite limit of high electronic energy, which means in practice $\omega \gg \Omega_D$, where Ω_D is again a typical phonon frequency. The logarithm in Eq. (9.16) becomes $2\Omega/\omega$, and from the definition in Eq. (9.17) for the coupling constant, we see that Re $\Sigma(\omega)$ behaves in this regime as $\lambda\Omega_D^2/\omega$. Hence it goes to zero in this large frequency limit, which implies that the energy of the electron is unaffected by the electron–phonon interaction. This is physically easy to understand because when an electron has a very fast motion, the lattice does not have time to react to the presence of the electron, which is already gone before the lattice has moved.

On the other hand, for the imaginary part of the self-energy, we have in this high-frequency regime $\theta(|\omega| - \Omega) = 1$ in Eq. (9.16) so that all the phonons can contribute to the decay of the electronic state. Im $\Sigma(\omega)$ becomes a constant independent of ω, with a value of order $-\lambda\Omega_D$. Nevertheless for large ω this imaginary part is negligible compared to the electronic energy $\hbar\omega$, so that we have again a physical situation with an electronic state having a well-defined energy and comparatively negligible decay.

In contrast with these two limiting regimes, the physical situation is quite different in the case where the frequency is of the order of a typical phonon frequency $\omega \sim \Omega_D$. Assuming a coupling constant λ of order unity, we see that from Eq. (9.16), with respect to orders of magnitude, Re $\Sigma(\omega) \sim$ Im $\Sigma(\omega) \sim \Omega_D \sim \omega$. This implies that the inverse lifetime $1/\tau$ of the electron is of the same order as the frequency ω. In other words, the wave function barely has the time to have a few oscillations at frequency ω that the decay arises. This implies that there is a large uncertainty on the electronic energy, which is accordingly an ill-defined quantity. In such a situation the electronic state becomes physically meaningless. There is no way anymore to use it for physical arguments. This means that the quasiparticle concept, on which the Fermi liquid picture is built, breaks down completely in this energy range once a strong electron–phonon interaction is taken into account. The only thing we are left with is the possibility of performing calculations following the methods of quantum statistical physics, and, in particular, to make use of Migdal's theorem.

Let us conclude this discussion of the self-energy by looking for an estimate of the coupling constant λ in standard physical situations. We take the example of a standard monovalent isotropic metal, and consider the interaction of an electron with long-wavelength longitudinal phonons. These correspond to compression-dilatation waves of the ionic lattice, resulting in the build-up of an ionic charge. Due to Coulomb interaction, this charge is felt by the electron. In the long wavelength regime, the ionic lattice can be considered as a continuum, which makes the calculation fairly easy. Actually, in this limit we are back to the hypotheses we made in Section 3.7 in handling the jellium model. Note that in this jellium model we have taken into account the screening of the Coulomb interaction by the electron gas, as should be done to obtain a physically meaningful result. Hence we have already written in Eq. (3.108) the result for $g_{\mathbf{k}\,\mathbf{k}'\mathbf{q}}$ we are looking for. Since we are in the long wavelength regime, we have to consistently neglect the wavevector q compared to the screening wavevector k_s. Making use of the expression Eq. (3.94) for this screening wavevector, we find accordingly

$$|g_{\mathbf{k}\mathbf{k}'\mathbf{q}}| = \left(\frac{\hbar}{2nM\Omega_{\mathbf{q}}}\right)^{1/2} \frac{qn}{2N_0} \delta_{\mathbf{k},\mathbf{k}'+\mathbf{q}} \tag{9.21}$$

Note that we use, consistently with the notations of the present chapter, $\Omega_{\mathbf{q}}$ instead of $\omega_{\mathbf{q}}$ for the phonon dispersion relation.

In order to estimate the coupling constant, we extrapolate this result from small wavevectors q to typical wavevector $|\mathbf{k} - \mathbf{k}'| \sim k_F$ occurring in Eq. (9.14). It is reasonable to believe that this extrapolation gives a correct order of magnitude. From Eq. (9.14) and the definition given in Eq. (9.17), we obtain

$$\lambda \sim \frac{N_0|g|^2}{\hbar\Omega_D} \sim \left(\frac{E_F}{\hbar\Omega_D}\right)^2 \frac{m}{M} \sim 1 \tag{9.22}$$

In the second step, we have made use of Eq. (9.21), with $\Omega_{\mathbf{q}} \sim \Omega_D, q \sim k_F$ and $N_0 \sim n/E_F$. The last step results from a standard order of magnitude evaluation in an ordinary metal of the ratio $\hbar\Omega_D/E_F$ between a typical phonon energy $\hbar\Omega_D$ and a typical electronic energy E_F. This ratio is of order $(m/M)^{1/2}$, that is, typically 10^{-2} in a standard metal, and we have seen it appear explicitly in our treatment of the jellium model, in the paragraph below Eq. (3.91). Naturally our evaluation of λ is quite rough, and clearly it does not exclude at all, for example, that λ turns out to be 0.1, or 10 in the opposite direction. The result Eq. (9.22) means only that there appear, in the evaluation of this coupling constant, no physical parameters ratio, which would make it obviously small or large from the start. It implies that in the general situation, the electron–phonon interaction is not a priori small and that a simple lowest-order perturbative treatment is not in general valid.

9.3 Eliashberg Equations

We will not provide the derivation of Eliashberg equations, which are the basic equations of strong coupling theory. This would require us to go into the lengthy and complicated details of many-body theory, which is far beyond our scope. Instead we will give these equations below, as Eq. (9.26) and Eq. (9.27), and try to justify them physically as much as possible. However, we will first give at an intuitive level the principles of the derivation. A first interest is that this allows a better physical understanding of the various involved quantities appearing in the equations. Moreover, provided one stays at the intuitive and qualitative picture provided by drawing Feynman's diagrams, it is fairly easy to understand the basic ideas of this derivation. In order to make an easier reading, we switch now to the standard theoretical habit of taking $\hbar = 1$. It is always easy to restore the missing $\hbar$ from dimensional considerations.

-Normal State

For this purpose, it is useful to consider first the normal state and formulate our definition of the self-energy in a more precise and more technical way. A basic object in the many-body treatment is the electronic propagator or Green's function. This is essentially[1] the average $\langle c_{\mathbf{k}}(t)c_{\mathbf{k}}^{\dagger}(0)\rangle$: a $\mathbf{k}$ electron is created at time $t = 0$ and destroyed at time t, so this quantity tells how the electron propagates over this time interval. Precisely $c_{\mathbf{k}}(t) = e^{iHt}c_{\mathbf{k}}e^{-iHt}$ is the destruction operator in Heisenberg representation, and the average $\langle \cdots \rangle$ is over the ground state at $T = 0$, and implies a statistical average at nonzero temperature. If we deal with non-interacting electrons, we can ignore all the electrons except our $\mathbf{k}$ electron, and the time evolution is given by the corresponding wave function. This means that the propagator is merely $e^{-i\xi_{\mathbf{k}}t}$ (taking zero energy at the Fermi level). If we want to go to the Fourier transform of this propagator with respect to time, we will meet singularities as a function of the frequency ω. So in order to specify them properly, we have to slightly modify our propagator definition. If we consider, for example, the so-called retarded propagator $-i\langle c_{\mathbf{k}}(t)c_{\mathbf{k}}^{\dagger}(0)\rangle\,\theta(t)$, which is nonzero only for $t \geq 0$, its Fourier transform is $-i\int_0^\infty dt\, e^{i\omega t}e^{-i\xi_{\mathbf{k}}t}e^{-\epsilon t} = 1/(\omega - \xi_{\mathbf{k}} + i\epsilon)$, where we have to let $\epsilon \to 0_+$ in our convergence factor $e^{-\epsilon t}$. Hence we have for the Fourier transform $G_0(\mathbf{k}, \omega)$ of our propagator for non-interacting electrons

$$G_0(\mathbf{k}, \omega) = \frac{1}{\omega - \xi_{\mathbf{k}} + i\epsilon} \tag{9.23}$$

We note that by looking at the frequency location of the pole of $G_0(\mathbf{k}, \omega)$, we obtain the electronic dispersion relation $\omega = \xi_{\mathbf{k}}$.

We consider now the case where electrons interact through the electron–phonon interaction. The definition of the propagator is unchanged, but its calculation is naturally much more difficult since the Hamiltonian H contains now the electron–phonon interaction term. When one goes to Fourier transform one finds for the propagator a new result $G(\mathbf{k}, \omega)$. The principle of the treatment of this problem by the many-body approach is to perform a systematic perturbative expansion to all orders in powers of the electron–phonon interaction. Naturally this cannot be done in a general and explicit way. Nevertheless a number of results, either exact or approximate, can be reached by this approach. The exact result for $G(\mathbf{k}, \omega)$ is naturally different from the non-interacting Eq. (9.23). We can define the self-energy by writing it as

---

[1] More precisely, the propagator used in standard quantum field theory for perturbative expansions is the time-ordered Green's function, defined precisely as $G(\mathbf{k}, t) = -i\langle T\{c_{\mathbf{k}}(t)c_{\mathbf{k}}^{\dagger}(0)\}\rangle$, where the time-ordering operator T means $T\{c_{\mathbf{k}}(t)c_{\mathbf{k}}^{\dagger}(0)\} = c_{\mathbf{k}}(t)c_{\mathbf{k}}^{\dagger}(0)$ for $t > 0$, and $T\{c_{\mathbf{k}}(t)c_{\mathbf{k}}^{\dagger}(0)\} = -c_{\mathbf{k}}^{\dagger}(0)c_{\mathbf{k}}(t)$ for $t < 0$.

$$[G(\mathbf{k}, \omega)]^{-1} = [G_0(\mathbf{k}, \omega)]^{-1} - \Sigma(\mathbf{k}, \omega) = \omega - \xi_\mathbf{k} - \Sigma(\mathbf{k}, \omega) \qquad (9.24)$$

If we look for a dispersion relation modified by the effect of interactions, we will naturally, by analogy with Eq. (9.23), consider the existence of a pole for $G(\mathbf{k}, \omega)$, which is a zero for $[G(\mathbf{k}, \omega)]^{-1}$. This leads us to Eq. (9.3), and so we recover our earlier definition. However, this is physically meaningful only if the solution of this equation for ω is real or at least has a small imaginary part, so that we can consider that it corresponds physically to an eigenstate with a very long lifetime, as we have found in Section 9.2.3. If this is not the case, as we have found at the end of Section 9.2.3, Eq. (9.24) remains nevertheless a proper definition of $\Sigma(\mathbf{k}, \omega)$.

We may now write Eq. (9.24) in a slightly different way to see how $\Sigma(\mathbf{k}, \omega)$ can arise from a perturbative expansion. Multiplying it by GG_0, it also reads

$$G(\mathbf{k}, \omega) = G_0(\mathbf{k}, \omega) + G_0(\mathbf{k}, \omega)\Sigma(\mathbf{k}, \omega)G(\mathbf{k}, \omega) \qquad (9.25)$$
$$= G_0 + G_0\Sigma G_0 + G_0\Sigma G_0\Sigma G_0 + G_0\Sigma G_0\Sigma G_0\Sigma G_0 + \cdots$$

where in the last equality we have dropped all the arguments $\mathbf{k}, \omega$ for clarity. This last equality is obtained from the first one by repeatedly inserting $G = G_0 + G_0\Sigma G$ for G in the last term, or merely by performing the series expansion of $G = G_0/(1 - \Sigma G_0)$. In this series expansion for the propagator G, the first term G_0 in the series corresponds to a free propagation of the electron, without any electron–phonon interaction occurring. In the next term $G_0\Sigma G_0$, after an initial free propagation described by G_0, the factor Σ contains all the possible contributions, at all order in perturbations, from processes due to electron–phonon interaction. The last factor G_0 describes the final free propagation of the electron. In the following term $G_0\Sigma G_0\Sigma G_0$, this free propagator G_0 is not the final step since it is followed by another factor Σ describing again all possible processes from electron–phonon interaction. And so on for the following terms in this series expansion. The existence of all these terms makes it intuitively clear that Σ itself contains only "irreducible" processes. This means that in all the processes occurring within Σ, there is no intermediate stage where one has only a single lonely $\mathbf{k}$ electron. This is because this would produce an additional factor G_0, whereas these factors are already written explicitly in the series expansion Eq. (9.25). Hence this would correspond to a double-counting of the processes taken into account in the perturbative expansion, which would naturally be erroneous.

In order to have a good understanding and visualization of all these processes arising in a perturbative expansion, it is very convenient to represent them graphically. These graphical representations are called Feynman's diagrams. Actually, these Feynman's diagrams are much more than simple graphical representations, because for any diagram there are precise rules that allow one write the specific algebraic expression for the corresponding term occurring in the perturbative expansion. However, we will not enter into this, and we will rather stay at the intuitive level provided by these diagrams. In these diagrams, we will represent the propagator G_0 of a free electron by a simple line with arrows $\rightarrow\!\!\!\rightarrow$. On the other hand, we represent the full electronic propagator G by a double line with arrows $\Rightarrow\!\!\!\Rightarrow$. Finally we represent the self-energy by a box, with Σ written inside. In this way, Eq. (9.25) reads graphically as in Fig. (9.4)

Naturally the above considerations would be of a somewhat formal interest without Migdal's theorem. This theorem gives us a practical mean to calculate Σ, since it states

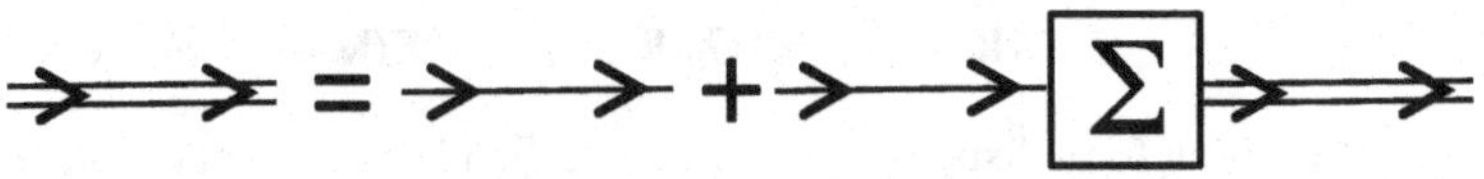

Fig. 9.4
Diagrammatic representation of Eq. (9.25).

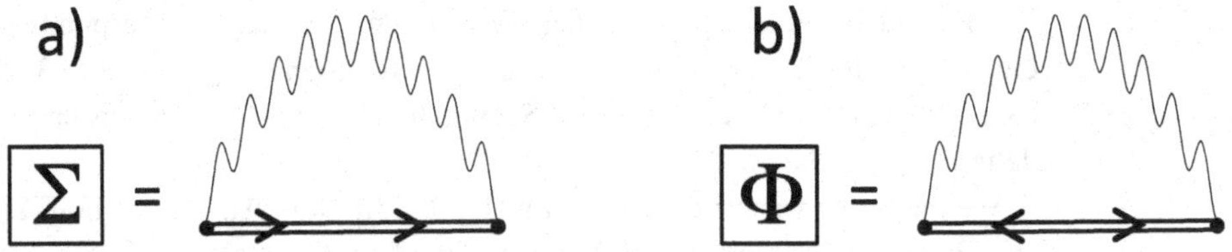

Fig. 9.5
(a) Normal self-energy Σ and (b) anomalous self-energy Φ, according to Migdal's theorem.

precisely that as shown in Fig. (9.5a), the only processes we have to keep, at dominant order in $(m/M)^{1/2}$, are those where a phonon is emitted, with the electron evolving according to the full propagator G, before the phonon is finally reabsorbed by the electron. As we indicate in the Further Reading section, in the normal state, this self-energy reduces to the explicit diagram given in Fig. (9.2).

-Superconducting State

After these considerations on the normal state, let us come to the situation in the superconducting state. The basic modification to the above picture results from a general consequence of the BCS theory we have seen appearing in Chapter 3, in Section 3.1, namely the appearance of the anomalous averages $\langle c_{-\mathbf{k}} c_{\mathbf{k}} \rangle$, and correspondingly $\langle c_{\mathbf{k}}^{\dagger} c_{-\mathbf{k}}^{\dagger} \rangle$ (we do not write explicitly the electron spin). These averages apparently violate electron number conservation, since by destroying, for example, two electrons, the final state has two electrons less than the initial state. However, we have encountered these kinds of things at length in dealing with BCS theory. Physically the understanding of this situation is due to the existence of the condensate, which provides a reservoir of particles. Hence when two electrons are destroyed, they do not disappear; they actually go into the condensate. Similarly, when two electrons are created, they are coming out from the condensate.

The extension of these features to time-dependent situations leads to the consideration of the anomalous propagators $\langle c_{-\mathbf{k}}(t) c_{\mathbf{k}}(0) \rangle$ and $\langle c_{\mathbf{k}}^{\dagger}(t) c_{-\mathbf{k}}^{\dagger}(0) \rangle$, which are naturally related to each other by complex conjugation. We represent these full propagators in a way that is analogous to the one we have taken for the full normal propagator, namely by a double line with arrows pointing in opposite directions, either outward or inward, which is respectively $\Longleftrightarrow$ and $\Rightarrow\!\!\!=\!\!\!\Leftarrow$. Following for these anomalous propagators the same reasonings as above for normal propagators, one can introduce anomalous self-energies (which merely reduce to the BCS gap in the weak coupling limit). Just as for normal state, we represent these self-energies by a box with Φ or Φ' written inside. Making use of Migdal's theorem Φ has the diagrammatic expression shown in Fig. (9.5b); Φ' is merely obtained by changing

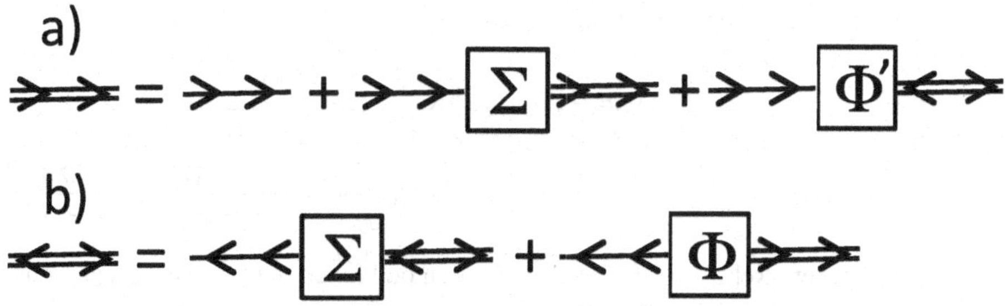

a)

b)

Fig. 9.6 (a) Generalization to the superconducting state of the diagrammatic representation of the normal propagator. (b) Corresponding diagrammatic representation for the anomalous propagator.

the sense of the arrows on the anomalous propagator from outward to inward going and is simply related to Φ.

Now, if we want to generalize Eq. (9.25) to the superconducting state, we have to take into account that in addition to simply propagating with any kind of electron–phonon interaction, an electron can also go into the condensate or come out of the condensate. This is done by taking into account the anomalous processes represented by the anomalous propagators and self-energies. This leads to an additional term when we want to extend to the superconducting state the diagrammatic expansion Fig. (9.4) of the full propagator. This term is the last one in Fig. (9.6a), which represents this generalization. Correspondingly we need such an expansion for the anomalous propagator appearing in this last term. It is obtained by a similar reasoning and is represented in Fig. (9.6b). Note that in this last expansion there is no term analogous to the first term G_0 appearing in the G expansion.

The equations corresponding to the diagrams in Fig. (9.6) are simple algebraic equations. Denoting F the anomalous propagator and skipping details, they are in rough approximate terms $G = G_0 + G_0 \Sigma G + G_0 \Phi' F$ and $F = G_0 \Sigma F + G_0 \Phi G$. They are easily solved for G and F, in terms of Σ, Φ, and Φ'. When the results are inserted in the self-energy expressions corresponding to the diagrams in Fig. (9.5), one obtains basically Eliashberg equations. Rather than Σ and Φ, these equations are usually written in terms of $Z(\omega) = 1 - \Sigma(\omega)/\omega$ and $\Delta(\omega) = \Phi(\omega)/Z(\omega)$.

The resulting Eliashberg equations are

$$\omega \left[1 - Z(\omega) \right] = \Sigma(\omega) = \left\langle \int_{-\infty}^{\infty} d\omega' \, \mathrm{Re} \left(\frac{\omega'}{\sqrt{\omega'^2 - \Delta^2(\omega')}} \right) \right. \tag{9.26}$$

$$\left. \times \left[\frac{1 - f(\omega') + N(\Omega)}{\omega - \omega' - \Omega} + \frac{f(\omega') + N(\Omega)}{\omega - \omega' + \Omega} \right] \right\rangle$$

where the brackets $\langle \cdots \rangle$ are for an average over phonon frequency, with a weight given by Eliashberg function: $\langle h(\Omega) \rangle \equiv \int d\Omega \, \alpha^2 F(\Omega) \, h(\Omega)$. The frequency ω is understood as having a very small imaginary part, the details of which we do not need to enter. The function $\mathrm{Re}[\omega/\sqrt{\omega^2 - \Delta^2(\omega)}]$ is an even function of ω.

The second Eliashberg equation is

$$\Delta(\omega)Z(\omega) = -\langle \int_{-\infty}^{\infty} d\omega' \, \mathrm{Re} \left(\frac{\Delta(\omega')}{\sqrt{\omega'^2 - \Delta^2(\omega')}} \right) \tag{9.27}$$
$$\times \left[\frac{1 - f(\omega') + N(\Omega)}{\omega - \omega' - \Omega} + \frac{f(\omega') + N(\Omega)}{\omega - \omega' + \Omega} \right] \rangle$$

where $\mathrm{Re}[\Delta(\omega)/\sqrt{\omega^2 - \Delta^2(\omega)}]$ is an odd function of ω. These equations form a set of two non-linear integral equations, which is not so easy to handle.

- Discussion

For $\Delta(\omega) = 0$, which corresponds to the normal state, $\Sigma(\omega)$ from Eq. (9.26) reduces, at $T = 0$, to the normal state result Eq. (9.15) we have found in the preceding section. Indeed, in this case, we have no thermally excited phonons $N(\Omega) = 0$, and the Fermi distribution is a step function $f(\omega) = 1 - \theta(\omega) = \theta(-\omega)$. At nonzero temperature we have to take into account occupation factors for electrons and phonons, but for the imaginary part of $\Sigma(\omega)$ these coincide exactly (with slightly different notations) with those we have directly obtained in Eq. (9.8). Hence, except for the fact that we have not derived these statistical factors for the real part, we recover in the normal state the result of the preceding section, so the factor in large brackets in Eq. (9.26) looks familiar. Now, when we go to the superconducting state, the only change is the presence of the additional factor

$$n(\omega) = \mathrm{Re} \left(\frac{\omega}{\sqrt{\omega^2 - \Delta^2(\omega)}} \right) \tag{9.28}$$

In the BCS weak coupling limit, where $\Delta(\omega)$ reduces to a constant equal to the gap Δ, we recognize the standard BCS density of states Eq. (3.23) $n(E) = E/\sqrt{E^2 - \Delta^2}$. Hence this factor Eq. (9.28) is clearly the appropriate generalization in the present case of the density of states for fermionic excitations, which should reasonably be present in the expression of the self-energy corresponding to Migdal's theorem, as represented in Fig. (9.5).

We turn now to the second Eliashberg equation Eq. (9.27), which is essentially a self-consistent equation for $\Delta(\omega)$ and which plays the same role as the gap equation Eq. (3.50) in BCS theory. The large bracket on the right-hand side is the same as in the first equation Eq. (9.26). The only major difference is that the factor in front is modified. However, we see that with respect to the density of states Eq. (9.28), we have only a multiplication by $\Delta(\omega)/\omega$ inside the real part. But this factor is quite analogous to the factor Δ/E_k appearing in the gap equation Eq. (3.50). Hence all the ingredients of this second equation look understandable. We go now further in this direction by considering in more detail the weak coupling limit of Eq. (9.27), and showing that one recovers indeed the BCS gap equation.

9.3.1 Weak Coupling Limit

The weak coupling limit corresponds to the case where the electron–phonon interaction is small. Accordingly, the Eliashberg function $\alpha^2 F(\Omega)$ is small in this case, and we have from

Eq. (9.17) for the coupling constant $\lambda \to 0$ in the weak coupling limit. As a result, the self-energy $\Sigma(\omega)$ goes to zero, and from its definition in Eq. (9.26) $Z(\omega) \to 1$. Hence we are left with analyzing the second Eliashberg equation. We expect from BCS theory (and can check at the end of the calculation) that Δ and T_c will be small with respect to phonon frequency Ω in this limit. Hence we have no thermally excited phonons $N(\Omega) = 1/e^{\Omega/k_BT} \simeq 0$. So the second Eliashberg equation Eq. (9.27) reduces to

$$\Delta(\omega) = -\langle \int_{-\infty}^{\infty} d\omega' \, \mathrm{Re}\left(\frac{\Delta(\omega')}{\sqrt{\omega'^2 - \Delta^2(\omega')}}\right)\left[\frac{1-f(\omega')}{\omega - \omega' - \Omega} + \frac{f(\omega')}{\omega - \omega' + \Omega}\right]\rangle \quad (9.29)$$

As long as the frequency ω is small compared to a typical phonon frequency $\omega \ll \Omega$, we can neglect it in the denominators in Eq. (9.29) and set $\omega = 0$. This implies, from Eq. (9.29) itself, that the characteristic energy scale for the variation of $\Delta(\omega)$ is the phonon frequency Ω. Considering this specific case $\omega = 0$, we can make use of the odd parity of $\mathrm{Re}(\Delta(\omega)/\sqrt{\omega^2 - \Delta^2(\omega)})$, together with $f(-\omega) = 1 - f(\omega)$, to reduce the integration interval to $[0, \infty]$ (the two terms in the bracket in Eq. (9.29) are exchanged in $\omega' \to -\omega'$). This leads to

$$\Delta(0) = 2\langle \int_{0}^{\infty} d\omega' \, \mathrm{Re}\left(\frac{\Delta(\omega')}{\sqrt{\omega'^2 - \Delta^2(\omega')}}\right)\left[\frac{1-f(\omega')}{\omega' + \Omega} + \frac{f(\omega')}{\omega' - \Omega}\right]\rangle \quad (9.30)$$

As long as ω' stays reasonably small compared to Ω, that is $\omega' \lesssim \Omega$, we may replace $\Delta(\omega')$ with $\Delta(0)$. Similarly $\omega' \pm \Omega \simeq \pm\Omega$. In the wide domain k_BT, $\Delta(0) \ll \omega' \ll \Omega$, we have $f(\omega') \simeq 0$ and $\sqrt{\omega'^2 - \Delta^2(\omega')} \simeq \omega'$. Hence the integrand behaves as $1/\omega'$, a fairly slow decrease, giving rise to the dominant contribution to the integral. Indeed, when on the other hand ω' goes beyond Ω, that is $\omega' \gtrsim \Omega$, we have rather $1/(\omega' \pm \Omega) \simeq 1/\omega'$. Moreover, from Eq. (9.29), $\Delta(\omega')$ itself decreases in this regime. Hence the integrand decreases much more rapidly as a function of ω', leading to a rapidly convergent integral. So, roughly speaking, the integration is essentially cut off beyond $\omega' \simeq \Omega$. If we replace the upper bound with a cut-off at Ω, and to be specific take an Einstein spectrum at frequency Ω_0, which implies $\alpha^2 F(\Omega) = (\lambda\Omega_0/2)\delta(\Omega - \Omega_0)$, we find from Eq. (9.30)

$$\Delta = \lambda \int_{\Delta}^{\Omega_0} d\omega' \, \frac{\Delta}{\sqrt{\omega'^2 - \Delta^2}} \tanh\left(\frac{\omega'}{2k_BT}\right) \quad (9.31)$$

Here we have taken into account that as the result of our above analysis of Eq. (9.30), $\Delta(0)$ is real, and we set $\Delta(0) = \Delta$. As a consequence, we have used the fact that $\mathrm{Re}(\Delta/\sqrt{\omega'^2 - \Delta^2})$ is zero for $\omega' < \Delta$, which fixes the lower boundary. Finally we used $1 - 2f(\omega') = \tanh(\omega'/2k_BT)$. Going back to the variable $\xi = \sqrt{\omega'^2 - \Delta^2}$, and using the standard notation E instead of ω', this equation can be rewritten as

$$\Delta = \lambda \int_{0}^{\Omega_0} d\xi \, \frac{\Delta}{E} \tanh\left(\frac{E}{2k_BT}\right) \quad (9.32)$$

which is exactly the gap equation Eq. (3.51), provided we identify N_0V with the coupling constant λ, and the Debye frequency ω_D with Ω_0. Clearly, due to our approximate handling, only the dominant logarithmic term of the integral, of order $\ln \Omega_0/\Delta$, is recovered by our treatment. A more precise calculation is required to obtain the proper next order term. In the following investigation of the critical temperature, we will obtain this term properly. Nevertheless our calculation has shown clearly how the Eliashberg theory directly recovers the BCS theory result in the weak coupling limit.

9.4 Critical Temperature

We look now at the answers provided by Eliashberg theory to the essential question of the critical temperature. Just as in BCS theory or Ginzburg–Landau theory, its value is found by letting $\Delta(\omega) \to 0$ in the general equations. In this case, Eq. (9.26) for the self-energy reduces to its normal state expression

$$\omega\,[1 - Z(\omega)] = \langle \int_{-\infty}^{\infty} d\omega' \left[\frac{1 - f(\omega') + N(\Omega)}{\omega - \omega' - \Omega} + \frac{f(\omega') + N(\Omega)}{\omega - \omega' + \Omega} \right] \rangle \qquad (9.33)$$

For this equation, as well as for the equation for $\Delta(\omega)$, it turns out to be much more convenient to work technically with frequencies located on the imaginary axis, instead of using real frequencies ω as in Eq. (9.33). The resulting imaginary axis Eliashberg equations are easier to handle than the real axis Eliashberg equations Eq. (9.26) and Eq. (9.27), mainly because they do not have the denominators $\omega - \omega' \pm \Omega$, which give rise to poles difficult to handle analytically as well as numerically. Moreover, they contain only a discrete frequency variable, instead of the continuous ω. The resulting discrete equations are much more convenient to deal with, in particular, numerically. Indeed the imaginary frequencies we will consider are given by

$$\omega = i\,\omega_n \qquad\qquad \omega_n = (2n + 1)\pi k_B T \qquad\qquad (9.34)$$

where n is an integer. These ω_n are called Matsubara frequencies.

These imaginary axis Eliashberg equations can be obtained from the real axis ones by continuing analytically the various involved physical quantities in the complex frequency plane, leading to consider $Z(i\omega_n)$ and similarly $\Delta(i\omega_n)$, which can be shown to be real quantities. Let us show in the simple specific case of Eq. (9.33) how this transformation can be done. We first make the change $\omega' \to -\omega'$ in the last term in the bracket, so we can factorize the statistical factor $f(-\omega') + N(\Omega) = (1/2)[\tanh(\omega'/2k_BT) + \coth(\Omega/2k_BT)]$. Hence, after the substitution $\omega \to i\omega_n$, where we assume $\omega_n > 0$, Eq. (9.33) becomes

$$i\omega_n\,[1 - Z(i\omega_n)] = \frac{1}{2}\langle \int_{-\infty}^{\infty} d\omega' \left(\tanh\frac{\omega'}{2k_BT} + \coth\frac{\Omega}{2k_BT} \right) \qquad (9.35)$$
$$\times \left[\frac{1}{i\omega_n - \omega' - \Omega} + \frac{1}{i\omega_n + \omega' + \Omega} \right] \rangle$$

The integral is calculated by closing the integration contour, so that it encloses the whole upper complex plane. The last term in the bracket has its pole at $\omega' = -(i\omega_n + \Omega)$, which is in the lower complex plane. The apparent pole from the first term in the upper complex plane at $\omega' = i\omega_n - \Omega$ does not actually exist because, for this value of ω', $\tanh(\omega'/2k_BT) = \tanh[i(n + 1/2)\pi - \Omega/2k_BT] = -\coth(\Omega/2k_BT)$, so that the statistical factor is zero and the pole is canceled. As a result, the only singularities in the upper complex plane come from the poles of $\tanh(\omega'/2k_BT)$. Making use of its expansion over these poles: $\tanh x = \sum_n 1/(x - i\pi(n + 1/2))$, the result of the integral comes from the residues at all the poles $\omega_{n'} = i\pi(2n' + 1)k_BT$ located in the upper complex plane. This gives

$$i\omega_n\left[1 - Z(i\omega_n)\right] = 2i\pi k_BT\langle\sum_{n'\geq 0}\left[\frac{1}{2i\pi k_BT(n'+n+1)+\Omega} - \frac{1}{2i\pi k_BT(n'-n)+\Omega}\right]\rangle \qquad (9.36)$$

Starting from $n' = 2n + 1$, the terms from the last sum in the bracket cancel exactly all those coming from the first sum. So we are merely left with the terms $n' = 0, \cdots, 2n$ at the beginning the last sum, that is

$$\omega_n\left[1 - Z(i\omega_n)\right] = -2\pi k_BT\langle\sum_{p=-n}^{p=n}\frac{1}{2i\pi pk_BT + \Omega}\rangle \qquad (9.37)$$

Taking into account $\langle(1/\Omega)\rangle = \lambda/2$, from the definition Eq. (9.17), this leads to

$$\omega_n Z(i\omega_n) = \omega_n + \lambda\pi k_BT + 2\pi k_BT\sum_{p=1}^{p=n}\langle\frac{2\Omega}{\Omega^2 + (2\pi pk_BT)^2}\rangle \qquad (9.38)$$

An analogous treatment can be used for the general Eliashberg equations,[2] leading to the following set of equations:

$$\omega_n Z_n = \omega_n + \pi k_BT\langle\sum_{m=-\infty}^{m=\infty}\frac{\omega_m}{(\omega_m^2 + \Delta_m^2)^{1/2}}\frac{2\Omega}{\Omega^2 + (\omega_n - \omega_m)^2}\rangle \qquad (9.39)$$

and

$$\Delta_n Z_n = \pi k_BT\langle\sum_{m=-\infty}^{m=\infty}\frac{\Delta_m}{(\omega_m^2 + \Delta_m^2)^{1/2}}\frac{2\Omega}{\Omega^2 + (\omega_n - \omega_m)^2}\rangle \qquad (9.40)$$

where we have switched to the simpler notations $Z_n \equiv Z(i\omega_n)$ and $\Delta_n \equiv \Delta(i\omega_n)$. Actually these equations can be derived much more directly from the standard many-body formalism at nonzero temperature, which works from the start with imaginary frequencies and Matsubara frequencies.

- General Handling of the T_c Calculation

Coming back to our critical temperature problem, Eq. (9.40) reduces in the limit of small Δ_n to

$$\Delta_n Z_n = \pi k_BT\langle\sum_{m=-\infty}^{m=\infty}\frac{\Delta_m}{|\omega_m|}\frac{2\Omega}{\Omega^2 + (\omega_n - \omega_m)^2}\rangle \qquad (9.41)$$

We further simplify the writing by introducing the notation

$$\lambda_p \equiv \langle\frac{2\Omega}{\Omega^2 + (2\pi pk_BT)^2}\rangle = \int d\Omega\, \alpha^2F(\Omega)\frac{2\Omega}{\Omega^2 + (2\pi pk_BT)^2} \qquad (9.42)$$

[2] The real parts arising in these equations are handled by writing $\text{Re}\, G(\omega) = (1/2)(G(\omega) + G^*(\omega))$ for $G(\omega) = \omega/\sqrt{\omega^2 - \Delta^2(\omega)}$ or $G(\omega) = \Delta(\omega)/\sqrt{\omega^2 - \Delta^2(\omega)}$.

The set of linearized Eliashberg equations Eq. (9.38) for Z_n and Eq. (9.41) for Δ_n reduces in this way, for $\omega_n > 0$, to

$$\omega_n Z_n = \omega_n + \pi k_B T \left(\lambda + 2 \sum_{p=1}^{n} \lambda_p \right) \tag{9.43}$$

$$\Delta_n Z_n = \pi k_B T \sum_m \frac{\Delta_m}{|\omega_m|} \lambda_{n-m} \tag{9.44}$$

or, by introducing $\overline{\Delta}_m = \Delta_m/|\omega_m|$,

$$\overline{\Delta}_n = \sum_m K_{nm} \overline{\Delta}_m \qquad K_{nm} = \frac{\lambda_{n-m}}{f_n} \qquad f_n = 2n + 1 + \lambda + 2 \sum_{p=1}^{n} \lambda_p \tag{9.45}$$

where the definition of f_n holds for $n \geq 0$ (for $n = 0$ it means $f_0 = 1 + \lambda$), that is, for positive Matsubara frequency $\omega_n > 0$. For negative Matsubara frequency one uses the definition $f(-\omega_n) = f(\omega_n)$, since one can check from Eq. (9.39) that this same symmetry property holds for Z_n and also for Δ_n. Explicitly this means $f_{-n} = f_{n-1}$.

From Eq. (9.45), one sees that finding T_c reduces to an eigenvalue problem: K_{nm} has at T_c an eigenvalue equal to 1. We note that all the eigenvalues of K_{nm} are real, because by setting $\overline{\Delta}_n = \overline{\Delta}'_n/\sqrt{f_n}$ one finds that these eigenvalues are also those of the matrix $K'_{nm} = \lambda_{n-m}/\sqrt{f_n f_m}$, which is real symmetric and has accordingly all its eigenvalues real.

If we look at the situation for very large temperature $T \to \infty$, we see that from their definition Eq. (9.42) all the λ_p are getting very small, of order $1/T^2$, except for $p = 0$, for which $\lambda_0 = \lambda$. Hence K_{nm} is essentially a diagonal matrix $K_{nm} = \delta_{nm}\lambda/f_n$. Since in the definition Eq. (9.45) of f_n, the sum $\sum_p \lambda_p \sim 1/T^2$ is negligible, we have merely $f_n = 2n + 1 + \lambda$. Hence the eigenvalues of K_{nm} are its diagonal elements $\lambda/(2n + 1 + \lambda)$. All these eigenvalues are positive, and they are all less than 1. Hence no critical temperature exists when we look in the very-high-temperature domain, which is naturally an expected result physically.

When we lower T, the eigenvalues are growing. Indeed, if we look at the extreme situation $T = 0$, we see that $\lambda_p = \lambda$, independent of p. Accordingly, $f_n = (1 + \lambda)|2n + 1|$ (including cases where $2n + 1 < 0$). Hence $K_{nm} = \lambda/(1 + \lambda)/|2n + 1|$. Clearly $\overline{\Delta}_m = 1/|2m + 1|$ is an eigenvector, but the eigenvalue is $\lambda/(1 + \lambda) \sum_m 1/|2m + 1| = \infty$, since this sum is divergent. So in this limit we have eigenvalues much larger than 1. In summary, when we start from high temperature and lower it, we have growing eigenvalues. When the largest of these eigenvalues reaches 1, we are at the critical temperature.

- Very Strong Coupling Limit

We first look at the very strong coupling limit $k_B T_c \gg \Omega$, because it is easy to solve and the result is physically quite interesting. We naturally expect this very high T_c to occur for quite a large coupling constant λ. This is clearly a somewhat formal investigation, since one does not expect in practice to have a real compound satisfying the condition for this limit.

We write the eigenvalue equation Eq. (9.45) as

$$\sum_m \lambda_{n-m} \overline{\Delta}_m = f_n \overline{\Delta}_n \tag{9.46}$$

that is explicitly (for $n \geq 0$)

$$\lambda \overline{\Delta}_n + \sum_{m \neq n} \lambda_{n-m} \overline{\Delta}_m = \left(2n + 1 + \lambda + 2 \sum_{p=1}^{n} \lambda_p \right) \overline{\Delta}_n \tag{9.47}$$

so that the contribution $\lambda \overline{\Delta}_n$, dominant at large λ, disappears from both sides of the equation. On the other hand, as we have seen, for this large temperature situation, in the definition Eq. (9.42) of λ_p we have $\Omega^2 \ll (2\pi p k_B T)^2$, so that

$$\lambda_p = \frac{1}{(2\pi p k_B T)^2} \int d\Omega \, \frac{2\alpha^2 F(\Omega)}{\Omega} \Omega^2 = \frac{R}{p^2} \qquad R \equiv \frac{\lambda \langle\langle \Omega^2 \rangle\rangle}{(2\pi k_B T)^2} \tag{9.48}$$

where we have defined the normalized average over phonon frequency as

$$\langle\langle h(\Omega) \rangle\rangle \equiv \frac{1}{\lambda} \int d\Omega \, \frac{2\alpha^2 F(\Omega)}{\Omega} h(\Omega) \tag{9.49}$$

so that $\langle\langle 1 \rangle\rangle = 1$. In this way, Eq. (9.47) becomes for $n \geq 0$

$$R \sum_{m \neq n} \frac{\overline{\Delta}_m}{(n-m)^2} = \left(2n + 1 + 2R \sum_{p=1}^{n} \frac{1}{p^2} \right) \overline{\Delta}_n \tag{9.50}$$

the equations for $n < 0$ being obtained as indicated above.

In this equation, all the physical quantities are gathered in R, so that we are just left with the purely numerical problem to find for which value of R this equation is satisfied. The numerical solution is quite easy because the denominator $(n-m)^2$ makes it a rapidly convergent problem. Let us take the simplest approximation by retaining only the $n = 0$ and $n = -1$ components, that is, $\overline{\Delta}_0$ and $\overline{\Delta}_{-1}$, corresponding to the lowest positive and negative Matsubara frequencies $\omega_0 = \pi k_B T$ and $\omega_{-1} = -\pi k_B T$. The $n = 0$ equation merely reads $R \overline{\Delta}_{-1} = \overline{\Delta}_0$ while the $n = -1$ equation is $R \overline{\Delta}_0 = \overline{\Delta}_{-1}$. This shows that $R = 1$ (the solution $R = -1$ leading for T_c to an unphysical result), and one finds also the symmetry property $\overline{\Delta}_{-1} = \overline{\Delta}_0$ mentioned previously. This gives for the critical temperature $k_B T_c = (1/2\pi) \lambda^{1/2} \langle\langle \Omega^2 \rangle\rangle^{1/2} \simeq 0.16 \lambda^{1/2} \langle\langle \Omega^2 \rangle\rangle^{1/2}$. One can improve on this result by retaining the $n = -2, -1, 0, 1$ components. Making use of the symmetry for $\overline{\Delta}_n$, this leads to a system of two linear homogeneous equations. Progressively enlarging the number of retained components, one goes to a system of three linear homogeneous equations, and so on. This converges very rapidly to the following result, barely different from our above simplest approximation,

$$k_B T_c = 0.1827 \, \lambda^{1/2} \langle\langle \Omega^2 \rangle\rangle^{1/2} \tag{9.51}$$

This result is of interest in several respects. First of all it is quite different from the standard BCS formula for T_c. In particular, in contrast with the BCS result, which grows as $e^{-1/\lambda}$, it does not saturate for large λ. It rather gives a critical temperature that grows indefinitely as $\sqrt{\lambda}$ with increasing coupling constant. Naturally this specific dependence is only valid in our asymptotic regime of very large λ, but it proves clearly that there is no saturation of the critical temperature. Finally the result depends only on a specific characteristic frequency $\langle\langle\Omega^2\rangle\rangle^{1/2}$, which is obtained from a weighted average over the whole phonon spectrum. This is a quite simple result, holding even in the case of a complicated phonon spectrum.

- Weak Coupling Limit

We consider now the opposite limiting regime where the coupling constant $\lambda \to 0$, and where we expect similarly the critical temperature to go to zero. For simplicity, we first take an Einstein spectrum for the phonons, which means that their frequencies take a single value, which we call Ω_0. Since we have $\lambda \to 0$, it is natural to neglect in the expression Eq. (9.45) of f_n all the λ terms and write $f_n \simeq |2n+1|$. We note that since our temperature goes to zero, in the definition Eq. (9.42) of λ_p, we have in the denominator $(2\pi p k_B T)^2 \ll \Omega_0^2$, so that all the λ_p's are equal to $\lambda_0 = \lambda$. However, this holds only as long as $2\pi p k_B T \ll \Omega_0$, that is, $p \lesssim \Omega_0/(2\pi k_B T) = p_c$. This upper bound is a large number in our limit $T \to 0$. With this simplification, we have merely from its definition Eq. (9.45) $K_{nm} \simeq \lambda/|2n+1|$, which is independent of m. However, this holds only provided that $|n-m| \lesssim p_c$. When m goes beyond this bound, K_{nm} decreases rapidly with m, according to its definition Eq. (9.45) and the general expression Eq. (9.42) for λ_p. Hence, as a first approximation, we can evaluate $\sum_m K_{nm}\overline{\Delta}_m$ by putting a cut-off at $|n-m| = p_c$. With this simplification, the eigenvalue equation Eq. (9.45) reads

$$\overline{\Delta}_n = \sum_m K_{nm}\overline{\Delta}_m = \frac{\lambda}{|2n+1|} S_n \qquad S_n = \sum_{m=n-p_c-1}^{m=n+p_c} \overline{\Delta}_m \qquad (9.52)$$

Assuming S_n weakly dependent on n, this means that the dominant dependence of $\overline{\Delta}_n$ is $1/|2n+1|$. Setting $\overline{\Delta}_n = A/|2n+1|$, one can check that S_n is indeed slowly varying since, for example, $S_1 - S_0 = A/(2p_c+3) - A/(2p_c+1) \simeq -A/2p_c^2$, and so on. Hence inserting this expression for $\overline{\Delta}_n$ in Eq. (9.52) for $n = 0$, we find to dominant order

$$1 = \lambda \sum_{m=-p_c-1}^{m=p_c} \frac{1}{|2m+1|} = 2\lambda(1 + \frac{1}{3} + \frac{1}{5} + \cdots + \frac{1}{2p_c+1}) \simeq \lambda \ln p_c \qquad (9.53)$$

This leads finally for the critical temperature to

$$k_B T_c = \frac{\Omega_0}{2\pi} e^{-\frac{1}{\lambda}} \qquad (9.54)$$

which coincides for the dominant factor $e^{-1/\lambda}$ with the BCS result.

If we consider now a general spectrum instead of an Einstein spectrum, it sounds reasonable to average the above relation over the spectrum and write $1 = \lambda\langle\langle\ln p_c\rangle\rangle$. This leads to the appearance of the new [52] characteristic frequency $\Omega_{\log}$

$$\ln \Omega_{\log} = \langle\langle\ln \Omega\rangle\rangle = \frac{1}{\lambda} \int d\Omega \frac{2\alpha^2 F(\Omega)}{\Omega} \ln \Omega \qquad (9.55)$$

This $\Omega_{\log}$ frequency is quite analogous to the characteristic frequency $\langle\langle\Omega^2\rangle\rangle^{1/2}$ entering the critical temperature in the strong coupling limit Eq. (9.51). It is reasonable to expect this $\Omega_{\log}$ to enter the weak coupling result for T_c in the case of a general spectrum, instead of Ω_0 appearing in Eq. (9.54).

Naturally, just as in Section 9.3.1, the above approximate treatment gives only properly the dominant exponential factor, but the prefactor in Eq. (9.54) is not the correct one. Nevertheless the calculation of this prefactor can be done analytically, but the derivation is much more complex [53]. Hence we give only the result. Considering first an Einstein spectrum with frequency Ω_0, one finds

$$k_B T_c = \frac{2}{\pi} e^{C-3/2} \Omega_0 e^{-\frac{1}{\lambda}} \simeq 0.253 \, \Omega_0 \, e^{-\frac{1}{\lambda}} \tag{9.56}$$

where C is the Euler constant. This is to be compared with the BCS result Eq. (3.54), where $N_0 V$ is to be replaced with λ, and one would naturally put $\hbar\Omega_0$ instead of $\hbar\omega_D$. The coefficient in the prefactor in Eq. (3.54) is 1.13, whereas the correct one is 0.253 in Eq. (9.56). Hence the BCS result overestimates this coefficient by a factor $1.13/0.253 \simeq 4.47$. This is quite a large ratio, which can, for example, easily transform a fairly low critical temperature result into a high T_c one. Accordingly, the correct coefficient in Eq. (9.56) is quite a relevant result.

Turning now to the case of a general phonon spectrum, one obtains not only that Ω_0 is indeed replaced with $\Omega_{\log}$ but that there is an additional dependence on the spectral shape, so that the critical temperature reads

$$k_B T_c = 0.253 \, \Omega_{\log} \, e^{-\mathcal{R}} e^{-\frac{1}{\lambda}} \tag{9.57}$$

where $\mathcal{R}$ is a positive quantity, which is a kind of measure of the intrinsic width of the spectrum. If the spectrum is not extremely wide, an appropriate expression for $\mathcal{R}$ is

$$\mathcal{R} \simeq \frac{1}{3} \langle\langle \left(\ln \frac{\Omega}{\Omega_{\log}} \right)^2 \rangle\rangle \tag{9.58}$$

The effect of $\mathcal{R}$ is not very strong. For example, if we take a spectrum (for $2\alpha^2 F(\Omega)/\Omega$) made of two delta peaks at $\Omega_1 = \Omega_{\log}/e$ and $\Omega_2 = e\,\Omega_{\log}$, with equal weight (one checks that the frequency calculated from Eq. (9.55) is indeed $\Omega_{\log}$), one finds merely $\mathcal{R} = 1/3$, and $e^{-1/3} \simeq 0.71$. One would need even wider spectra to have a stronger effect of $\mathcal{R}$. On the other hand, the critical temperature depends only on the large-scale structure of the spectrum. The fine-scale structure is essentially irrelevant. Since this last property is also true in the very strong coupling limit, it is quite likely to be valid regardless of the strength of the coupling.

- McMillan Formula

Let us start from our above rough estimate of the critical temperature in the weak coupling limit, leading to the result Eq. (9.54), and try to improve it for larger values of the coupling constant λ. We might notice that because we have $\lambda_p = \lambda$ for $p \lesssim p_c$, an obviously better approximation for f_n is $f_n \simeq (1+\lambda)|2n+1|$. This leads to $K_{nm} \simeq \lambda/|2n+1|/(1+\lambda)$. The rest of the argument is unchanged provided λ is replaced by $\lambda/(1 + \lambda)$. This gives for the critical temperature the result $T_c = \Omega_0/(2\pi) \exp(-(1+\lambda)/\lambda)$. Actually, this is not really different from Eq. (9.54), since the new result has just a further factor e^{-1}, and we have anyway seen that the prefactor of Eq. (9.54) is not the proper one, so that its modification is of no real significance. However, this modification is of physical interest since it results from taking into account properly the electronic mass renormalization by the electron–phonon interaction, that is, the replacement of m with m^* as we have seen in Eq. (9.20), which gives a factor $1 + \lambda$. This mass renormalization results from the modification of the normal state self-energy, and it is the same factor that is coming in the expression for f_n in Eq. (9.45).

This physical reason has led McMillan [54] to look for this functional dependence when he fit his numerical results for larger values of the coupling constant λ, such that the weak coupling limit is no longer valid. He investigated the specific case of Niobium. We give here the complete formula he found for this case:

$$k_B T_c = \frac{k_B \Theta_D}{1.45} \exp\left[-\frac{1.04(1+\lambda)}{\lambda - \mu^*(1 + 0.62\lambda)}\right] \tag{9.59}$$

where Θ_D is niobium Debye temperature. We see appearing in this formula the coupling constant μ^*, which describes the effect of Coulomb repulsion. We have not yet investigated its effect, and we come to this in the next section.

9.5 Coulomb Repulsion

As we have already seen, including Coulomb repulsion in the theoretical description is difficult. There are basically two steps in taking it into account. The first one has already been described in Section 2.2. It leads to the concept of quasi-particle, or in the present case quasi-electron, which corresponds qualitatively to the bare electron surrounded by its polarization cloud, which is the deformation it induces in the whole electronic medium. This is the basic picture of Fermi liquid theory. Actually, the Eliashberg theory we have seen previously is built on this Fermi liquid picture, that is, the basic objects we started with are not actually bare electrons, but quasi-electrons with, in particular, a mass which is their effective mass, renormalized by the existence of the polarization cloud. In practice, this does not make much of a difference since we have not specified the mass of the electron we started with in Eliashberg theory. Similarly, all the structure of this theory is unaffected by the replacement of bare electrons with quasi-electrons, because the typical energy entering Eliashberg theory is a phonon energy. This is much lower than the characteristic energy entering the formation of the quasi-electron, which is of the order of the Fermi energy. Hence we do not have to care about the quasi-electron structure, in the same way as we do not care about the structure of the proton when we deal with atomic physics, which has typical energies much lower than the ones entering the structure of the proton.

In the second step, we have to take into account the residual effect of Coulomb interaction in the pair formation. This interaction is expected to be repulsive, and has accordingly an unfavorable effect on pairing. However, in contrast with the description in Eliashberg theory of the effect of the electron–phonon interaction, since the characteristic energy for the Coulomb interaction is the Fermi energy, we have here no small parameter allowing us to set up a clean approximate theoretical approach and here we do not have the equivalent of Migdal's theorem. Accordingly, our approach will be fully phenomenological. We try to modify in a physically reasonable way Eliashberg equations. We will immediately restrict ourselves to considering only the critical temperature, so the equations we deal with are Eq. (9.43)–Eq. (9.44). Since as we have seen the quasi-electron mass renormalization is included from the start in Σ, we have only to modify the equation for Δ.

We will take Coulomb repulsion into account by writing a term analogous to the one produced by phonons, but corresponding to a repulsion rather than an attractive term. Moreover we put, in place of a phonon energy, a characteristic energy that is of the order of a Fermi energy and that we call Ω_F. This leads us to write

$$\Delta_n Z_n = \pi k_B T \sum_m \frac{\Delta_m}{|\omega_m|} \left(\frac{\lambda \Omega_0^2}{\Omega_0^2 + (\omega_n - \omega_m)^2} - \frac{\mu \Omega_F^2}{\Omega_F^2 + (\omega_n - \omega_m)^2} \right) \tag{9.60}$$

where we have assumed for simplicity an Einstein phonon spectrum at frequency Ω_0. The positive coupling constant μ describes the effect of Coulomb repulsion. For this Coulomb term, the summation over ω_m converges for ω_m large compared to Ω_F. Since our approach is anyway approximate, it is easier to replace this convergent Coulomb term with a constant μ, with a cut-off at Ω_F. Since the phonon term has already converged by far when this cut-off is reached, we can apply this cut-off to the whole summation and write instead

$$\Delta_n Z_n = \pi k_B T \sum_m^{\Omega_F} \frac{\Delta_m}{|\omega_m|} \left(\frac{\lambda \Omega_0^2}{\Omega_0^2 + (\omega_n - \omega_m)^2} - \mu \right) \tag{9.61}$$

It is easily seen from Eq. (9.61) that in the range $\Omega_0 \ll |\omega_n| \ll \Omega_F$, Δ_n is essentially a constant, independent of ω_n, which we note as Δ_∞. Indeed in this range Z_n goes to 1, and the phonon contribution goes to zero, as is fairly clear from its expression. Hence the remaining expression for Δ_n, which is the μ term on the right-hand side, is explicitly independent of ω_n.

Let us see these two points in more detail. Indeed in Eq. (9.43) for Z_n, the sum over the λ_p terms has, for $|\omega_n| \gg \Omega_0$, fully converged to its high frequency limiting value, which is of order $\lambda \Omega_0$. (The sum can be evaluated approximately by replacing it by an integral.) So the first term ω_n on the right-hand side is dominant, which gives $Z_n \simeq 1$. This is the physically expected result since at high frequency phonons cannot follow the fast motion of an electron, and accordingly this electron behaves as a bare electron so that its normal state self-energy is zero, corresponding to $Z_n = 1$.

To examine the behavior of the phonon contribution on the right-hand side, we introduce an additional cut-off Ω_c, large enough so that the summation over $|\omega_m|$ has fully converged, but small compared to Ω_F, so that $\Omega_0 \ll \Omega_c \ll \Omega_F$. Accordingly, we split the sum in Eq. (9.61) into two sums. The first sum goes up to Ω_c, and the second one from Ω_c to Ω_F. In this phonon term, for the contribution to the sum from $|\omega_m| \leq \Omega_c$, we have, for large ω_n, $\Omega_0^2 + (\omega_n - \omega_m)^2 \simeq \omega_n^2$ since all the other quantities are bounded. Hence this contribution behaves as Ω_0^2 / ω_n^2 for large ω_n, and it is accordingly negligible. Considering now the contribution from $\Omega_c \leq \omega_m \leq \Omega_F$, the dominant contribution comes from ω_m in the vicinity of ω_n. Hence we can approximate $\Delta_m/|\omega_m|$ by $\Delta_\infty/|\omega_n|$, making use of our hypothesis $\Delta_m \simeq \Delta_\infty$ in this range. For such large values of ω_m the remaining summation can be replaced by an integral over frequency, which leads to evaluate the contribution of this sum to the right-hand side of Eq. (9.61) as $(\pi/2)\lambda\Omega_0\Delta_\infty/|\omega_n|$. This is again negligible in the large ω_n domain. Hence Eq. (9.61) leads indeed to the conclusion that Δ_n is independent of ω_n for large enough ω_n.

Since we have found the phonon contribution negligible in this regime, we can easily find the value of Δ_∞ from Eq. (9.61). We have merely to consider the contributions from the Coulomb repulsion. This gives

$$\Delta_\infty = -\mu\pi k_B T \sum_m^{\Omega_c} \frac{\Delta_m}{|\omega_m|} - \mu\pi k_B T \sum_{\Omega_c}^{\Omega_F} \frac{\Delta_m}{|\omega_m|} \qquad (9.62)$$

With the property $\Delta_m = \Delta_\infty$ in the range $[\Omega_c, \Omega_F]$, the last summation is easily performed

$$\pi k_B T \sum_{\Omega_c}^{\Omega_F} \frac{\Delta_m}{|\omega_m|} = 2\pi k_B T \Delta_\infty \sum_{\omega_m=\Omega_c}^{\Omega_F} \frac{1}{\omega_m} \simeq \Delta_\infty \int_{\Omega_c}^{\Omega_F} d\omega \frac{1}{\omega} = \Delta_\infty \ln \frac{\Omega_F}{\Omega_c} \qquad (9.63)$$

where, in the first equality, the factor 2 takes into account that we have to sum over negative, as well as positive, values of ω_m, and in the second one we may replace the summation over the large variable $\omega_m = (2m+1)\pi k_B T$ with an integral. This leads to

$$\Delta_\infty = -\mu^* \pi k_B T \sum_m^{\Omega_c} \frac{\Delta_m}{|\omega_m|} \qquad \mu^* = \frac{\mu}{1 + \mu \ln \frac{\Omega_F}{\Omega_c}} \qquad (9.64)$$

Coming back to the equation for Δ_n Eq. (9.61), we see from Eq. (9.62) that the μ term on the right-hand side is just equal to Δ_∞. Hence making use of Eq. (9.64) for Δ_∞, this equation can be rewritten for frequencies $|\omega_n| \leq \Omega_c$

$$\Delta_n Z_n = \pi k_B T \sum_m^{\Omega_c} \frac{\Delta_m}{|\omega_m|} \left(\frac{\lambda\Omega_0^2}{\Omega_0^2 + (\omega_n - \omega_m)^2} - \mu^* \right) \qquad (9.65)$$

where we have neglected the contribution to the phonon sum of the frequencies $\Omega_c \leq |\omega_m| \leq \Omega_F$, since in our range of interest for $|\omega_n|$ the phonon sum is essentially fully converged for $|\omega_m| = \Omega_c$ so that the contribution from Ω_c to Ω_F is indeed negligible. It is easy to summarize our steps from Eq. (9.61) to Eq. (9.65): the cut-off Ω_F has merely been replaced by the cut-off Ω_c, at the price of replacing μ by μ^*.

The relation Eq. (9.64) between μ and μ^* is conveniently rewritten as

$$\frac{1}{\mu^*} = \frac{1}{\mu} + \ln \frac{\Omega_F}{\Omega_c} \qquad (9.66)$$

As we have mentioned, we have no precise theory of the effect of Coulomb repulsion because there is no small parameter, and all the energy scales are expected to be of the same order as the Fermi energy E_F. Accordingly, although we have no way to calculate it precisely, it is reasonable to estimate that the dimensionless parameter μ is of order unity. On the other hand, taking Ω_c of the order of a few phonon energy, and Ω_F of the order of the Fermi energy, we have $\Omega_F/\Omega_c \simeq E_F/\Omega_0 \sim 10^2 - 10^3$, where we have taken for this ratio the typical value in a standard metal. Fortunately, this rough estimate enters a logarithm, which is fairly insensitive to the chosen value. Indeed $\ln 10^2 \simeq 4.6$, whereas $\ln 10^3 \simeq 6.9$. Taking $\mu = 1$, this leads to μ^* equal to 0.18 or 0.13, which are not so different. In practice one takes often $\mu^* \simeq 0.1$, which frequently gives results in fair agreement with experiments. The essential point is that this parameter is fairly small, because its expression Eq. (9.66) is mostly dominated by the logarithmic term, which is the important one because the energy scales for electron–phonon interaction and Coulomb interaction are very different. We notice that μ^* depends on the chosen cut-off Ω_c (and is an increasing function of this cut-off). This is not so satisfactory for a physical parameter, but in practice this is fairly unimportant since this dependence is quite weak.

Let us finally come to a specific result from Eq. (9.65) for the critical temperature. If we consider the weak coupling regime, our treatment in the preceding section is equivalent to writing $\lambda \Omega_0^2/(\Omega_0^2 + (\omega_n - \omega_m)^2) \simeq \lambda$. Hence taking the Coulomb repulsion into account is equivalent to replacing λ with $\lambda - \mu^*$. Following otherwise the same argument as at the end of the preceding section, leading to our writing $f_n \simeq (1 + \lambda)|2n + 1|$, we obtain

$$k_B T_c = 0.688\,\hbar\Omega_0\,\exp\left[-\frac{1 + \lambda}{\lambda - \mu^*}\right] \tag{9.67}$$

where we have taken the numerical prefactor in such a way that for $\mu^* = 0$, we recover the proper result Eq. (9.56) with $0.688\,e^{-1} = 0.253$. This result is quite similar to the McMillan formula, Eq. (9.59). Naturally we see that the presence of the Coulomb repulsion μ^* lowers the critical temperature with respect to its value in the absence of this repulsion.

This formula is also fairly similar to the result we have obtained in Section 3.6, where we have handled the Coulomb repulsion by an extension of the BCS weak-coupling approach. In this treatment, it was the dependence of the order parameter on wavevector $\mathbf{k}$, more specifically on $\xi_{\mathbf{k}}$, which was relevant, whereas in the present more correct treatment within Eliashberg theory, it is rather the dependence on the frequency ω that is coming in. Nevertheless the two treatments follow similar steps, and both lead to the conclusion that the negative effect of Coulomb repulsion is strongly attenuated due to the large difference between the characteristic frequencies of phonons and electrons. This appears explicitly in both cases through the logarithm of the ratio between these two frequencies. We could naturally make use of Eq. (9.67), or better of McMillan formula Eq. (9.59), to discuss the isotope effect. However, all the qualitative conclusions would be the same as those of Section 3.6, and so there is no need to repeat them here.

9.6 Density of States

Let us finally investigate the reduced density of states

$$n(\omega) = \mathrm{Re}\left(\frac{\omega}{\sqrt{\omega^2 - \Delta^2(\omega)}}\right) \tag{9.68}$$

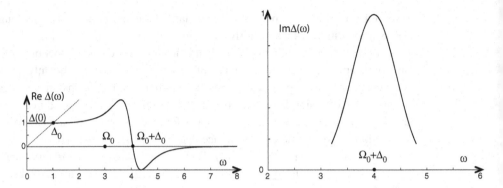

Fig. 9.7   Qualitative behavior of $\mathrm{Re}\,\Delta(\omega)$ and $\mathrm{Im}\,\Delta(\omega)$, as it results from the solution of Eliashberg equations, for the phonon spectrum and the parameters indicated in the text.

we have seen appearing in Eq. (9.28). This is indeed a very important quantity, and just as in BCS theory, we expect it to appear in many important physical results. Moreover, it can be directly measured by tunneling experiments, just as we have shown in Section 3.2. As we will see, it bears the mark of Eliashberg theory, and so tunneling experiments can give detailed support to this theory.

In order to examine the behavior of $n(\omega)$, we need to know the dependence of $\Delta(\omega)$ on frequency. We give here its qualitative features as they result from Eliashberg equations, providing some justification. We restrict ourselves first to the simplest $T = 0$ situation. Naturally the answer depends on the phonon spectrum, on the coupling strength, and on Coulomb repulsion. We take the simplest case where the small effect of Coulomb repulsion can be omitted, that is, $\mu^* = 0$. We consider a situation with a moderate coupling strength $\lambda \sim 1$, so that we are not so far from a weak coupling case. Finally we choose the phonon spectrum to be essentially an Einstein spectrum at frequency Ω_0, but slightly broadened in order to avoid singularities. So the corresponding normalized weight $g(\Omega) = 2\alpha^2 F(\Omega)/(\lambda\,\Omega)$ has the shape of a broadened peak.

The resulting qualitative behavior of $\mathrm{Re}\,\Delta(\omega)$ is shown in Fig. (9.7). It starts at some finite value $\Delta(0)$ and rises at first gently as a function of ω. In this domain, $\mathrm{Im}\,\Delta(\omega)$ is zero. Hence in Eq. (9.68), as long as $\omega < \Delta(\omega)$, $\sqrt{\omega^2 - \Delta^2(\omega)}$ is purely imaginary and the density of states is zero $n(\omega) = 0$. This inequality implies that $\omega < \Delta_0$, where Δ_0 is solution of $\Delta_0 = \Delta(\Delta_0)$ (this corresponds in Fig. (9.7) to the crossing between $\mathrm{Re}\,\Delta(\omega)$ and the straight line ω). On the other hand, as soon as $\omega > \Delta_0$, we have $\omega > \Delta(\omega)$, the square root is real and we have $n(\omega) > 0$. Moreover, we find the same divergent square root behavior as we have in BCS theory just above the gap. Accordingly, Δ_0 is the value of the gap in Eliashberg strong coupling theory. Since we have taken a coupling strength $\lambda \sim 1$, we expect this gap Δ_0 to be fairly small compared to the phonon frequency Ω_0. Actually, in this case Δ_0 is in practice not so different from $\Delta(0)$, as is the case for example in Fig. (9.7).

On the other hand, the large ω behavior comes also naturally from Eliashberg equations Eq. (9.26) and Eq. (9.27). Indeed, in our case, only the first term remains in the brackets on the right-hand side, which is merely $1/(\omega - \omega' - \Omega)$. This clearly makes the integral go to zero at large ω. Hence $Z(\omega)$ goes to 1 in this limit, as physically expected. And from Eq. (9.27), we see that $\Delta(\omega)$ goes also to zero, in agreement with the sketch in Fig. (9.7).

The final noticeable qualitative structure for $\text{Re}\Delta(\omega)$ is the one where it goes across zero, with a dispersive-like behavior. This structure arises at a frequency $\omega \simeq \Omega_0 + \Delta_0$. Its origin is again merely the pole $1/(\omega - \omega' - \Omega)$ appearing in Eliashberg equations. When the integration over ω' is performed, the presence of $1/\sqrt{\omega'^2 - \Delta^2(\omega')}$ in the integrand, that is essentially $n(\omega')$, forces to have $\omega' \gtrsim \Delta_0$, with actually a strong weight around $\omega' = \Delta_0$ due to the BCS like divergent behavior of $n(\omega')$. Hence the pole contribution is dominantly $1/(\omega - \Delta_0 - \Omega)$, giving rise to a structure around $\Omega_0 + \Delta_0$. Actually, the situation is quite similar to the one we have seen in Eq. (9.16) for the self-energy in the normal state, where we had a logarithmic structure at the phonon frequency Ω. If we average the logarithmic term appearing in Eq. (9.16) for a broadened Einstein spectrum like the one we consider here, we will obtain a dispersive-like structure analogous to the one in Fig. (9.7), instead of a logarithmic divergence. In the present case, this structure is shifted by Δ_0, due to the presence of the gap.

Similarly we consider now $\text{Im}\Delta(\omega)$. In the normal state in Eq. (9.16), the existence of an imaginary part for the self-energy required a frequency larger than a phonon frequency, so that an excitation at frequency ω could decay by phonon emission. Here, in the same way, one needs a frequency larger than Ω_0 to have dissipative phenomena. But in the same way as for $\text{Re}\Delta(\omega)$, the dominant processes occur at a frequency shifted by Δ_0, due to the strong density of states in the vicinity of the gap. This turns out to give for $\text{Im}\Delta(\omega)$ a behavior as sketched in Fig. (9.7), with a peak around $\Omega_0 + \Delta_0$.

The structures we have seen appearing in $\Delta(\omega)$ lead to corresponding structures in $n(\omega)$. This can be seen more explicitly in our case by taking advantage of the fact that they arise for ω around $\Omega_0 + \Delta_0$, which is fairly high compared to $|\Delta(\omega)|$. So we can expand the square root and write

$$n(\omega) = \text{Re}\left(1 - \frac{\Delta^2(\omega)}{\omega^2}\right)^{-1/2} \simeq 1 + \frac{1}{2\omega^2}\,\text{Re}\,\Delta^2(\omega) \tag{9.69}$$

$$= 1 + \frac{1}{2\omega^2}\left([\text{Re }\Delta(\omega)]^2 - [\text{Im }\Delta(\omega)]^2\right)$$

which should be compared, for example, to the corresponding expansion $n(\omega) = 1 + \Delta^2(0)/2\omega^2$ for the BCS density of states. The difference between these two results is

$$\delta n(\omega) = \frac{1}{2\omega^2}\left([\text{Re }\Delta(\omega)]^2 - \Delta^2(0) - [\text{Im }\Delta(\omega)]^2\right) \tag{9.70}$$

It starts from 0 and rises to positive values since at first $\text{Im }\Delta(\omega) = 0$ and $\Delta(\omega) > \Delta(0)$. Then for $\omega \simeq \Omega_0 + \Delta_0$, it goes to negative values since $\text{Re }\Delta(\omega)$ goes to zero, while $\Delta^2(0) + [\text{Im }\Delta(\omega)]^2 > 0$. Finally it goes naturally back to zero for larger values of ω. The corresponding behavior of $n(\omega)$ is clearly seen, around 7 meV, in the experimental result shown in Fig. (9.8a).

This density of states $n(\omega)$ is, in principle, directly observable in tunneling experiments. Hence this provides a qualitative check for the whole theoretical description. Indeed the dominant phonon contributions, which ar essentially the peaks in the phonon density of

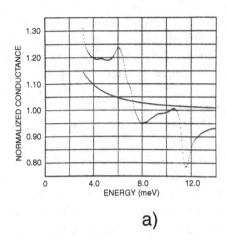
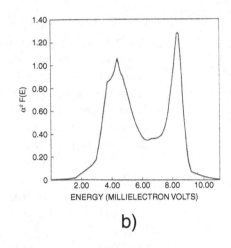

a)   b)

Fig. 9.8 (a) Experimental data of McMillan and Rowell for the reduced conductance of their Pb-I-Pb tunneling junction, as a function of voltage. (b) Corresponding $\alpha^2 F(\omega)$ obtained by fitting the conductance shown in (a).

states, must appear as we have just seen in $n(\omega)$, with their position shifted by the gap Δ_0. On the other hand, the phonon spectrum can be explored completely independently, for example by neutron scattering experiments. The coincidence from two independent experiments between the position of the peaks in the phonon spectrum provides a very strong indication on the attractive mechanism responsible for superconductivity, in the present case the electron–phonon interaction. Unfortunately, these interesting structures in $n(\omega)$ appear at fairly high frequency. So, as is clear, for example, from Eq. (9.70), they provide experimentally fairly small signatures that are not so easy to exploit.

Nevertheless McMillan and Rowell [55] have taken the density of states $n(\omega)$ obtained from their tunneling experiments on lead (which has $T_c = 7.2$ K), where the deviations from the BCS density of states are fairly large, of the order of 5%. They have been able to invert Eliashberg equations in order to find the corresponding $\alpha^2 F(\omega)$ and μ^* for lead. This was done by an iterative numerical procedure, allowing to find the best fit for the experimental data. Their starting conductance data for their Pb-I-Pb tunneling junction is shown in Fig. (9.8a) and their result for $\alpha^2 F(\omega)$ in Fig(9.8b). They found $\mu^* = 0.12$ for the Coulomb repulsion. They could then use these results to calculate various physical quantities for lead, such as its superconducting condensation energy, and find good agreement with known experimental results.

Finally they used the fact that in their fitting procedure, they did not use the high energy data corresponding to the domain $\omega - \Delta_0 > 11$ meV. This domain still contains a fair deal of structure, corresponding in particular to multiple phonon processes. In this domain, the comparison between theory and experiment has no adjustable parameters. The result is shown in Fig. (9.9). The agreement between theory and experiment is quite impressive, taking into account the fairly strange shape of the data in this range. The quality of this test of Eliashberg theory is certainly one of the most remarkable and striking successes of our refined understanding of standard superconductors.

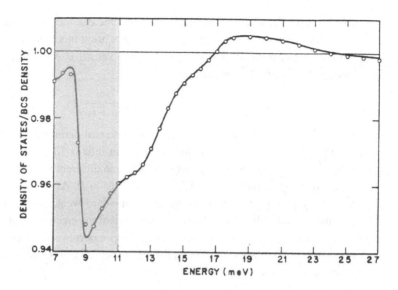

Fig. 9.9 Comparison between experimental data (small circles) and theory (full line), for the density of states compared to the BCS density of states. The adjustable parameters were obtained from the energy domain below 11 meV (shaded area), where $\alpha^2 F(\omega)$ is nonzero (see Fig. (9.8b). In the energy domain beyond 11 meV, there are no adjustable parameters.

-Nonzero Temperature

Let us finally consider the situation at nonzero temperature. It is immediately more complicated than at $T = 0$ because there is no true gap in this case. One has only something similar to a gap, which one might call a "pseudogap." The origin of this problem can be found by considering $\mathrm{Im}\, Z(\omega)$. In contrast with the situation we had at $T = 0$, it is nonzero, whatever the frequency, because an excitation with any ω can always decay by absorbing a phonon, which is always present due to thermal excitation. This situation is analogous to the one we have seen in the normal state with the last term on the right-hand side of Eq. (9.8). Since Eliashberg equation Eq. (9.27) gives $\Delta(\omega)Z(\omega)$, we have as a result $\mathrm{Im}\,\Delta(\omega) \neq 0$, since no coincidental cancellation occurs in this respect with the right-hand side of this equation. Accordingly, from Eq. (9.68), we have $n(\omega) \neq 0$ regardless of the value of ω, and consequently there is no gap.

One can get some more insight into the physical situation by making use of a simple approximation. It actually corresponds to a low-frequency expansion, which is reasonably valid in the frequency domain we are interested in. Just as in the normal state from Eq. (9.16), one can see from Eq. (9.26) that $\mathrm{Re}\,\Sigma(\omega)$ is an odd function of ω, while $\mathrm{Im}\,\Sigma(\omega)$ is an even function. This leads us to the expansion $\omega[1 - Z(\omega)] = \Sigma(\omega) \simeq \omega(1 - A_1) - i/\tau_1$, that is, $\omega Z(\omega) = \omega A_1 + i/\tau_1$. In the normal state, A_1 would merely be the mass renormalization coefficient $1 + \lambda$. We have used the notation $1/\tau_1$ in the imaginary part of $Z(\omega)$ to stress that it has the physical signification of an inverse lifetime, as we have seen in Section 9.2.2. On the other hand, from Eq. (9.27) the real part of $\Delta(\omega)Z(\omega)$ is an even function of ω, and if we stay at the lowest level in our expansion, we merely take

$\Delta(\omega)Z(\omega) \simeq B_1$, where B_1 is a real constant. One can see that including in this expansion the next order term, proportional to ω, does not bring much change to the peak position in $n(\omega)$ we consider below (but the peak height is sensitive to the presence of this term). This leads us to the simple approximation

$$\Delta(\omega) = \delta \, \frac{\omega}{\omega + i v_0} \tag{9.71}$$

where $\delta = B_1/A_1$ and $v_0 = 1/(\tau_1 A_1)$. All these real parameters are understood as having a temperature dependence. We note that for an infinite lifetime $v_0 = 0$, we merely have $\Delta(\omega) = \delta$, so we expect it in general to be not so different from the standard BCS gap.

A surprising feature of Eq. (9.71) is that for $\omega \to 0$, $\Delta(\omega)$ goes to zero while becoming essentially purely imaginary. However, in standard weak coupling superconductors, the lifetime (which is directly linked to the strength of the electron–phonon interaction) is very long, so that v_0 is quite small. Hence this behavior occurs only at very small frequency, and in practice it is inobservable. When Eq. (9.71) is used in Eq. (9.68) to obtain the density of states, the result is

$$n(\omega) = \mathrm{Re}\left(\frac{\omega + i v_0}{\sqrt{(\omega + i v_0)^2 - \delta^2}} \right) \tag{9.72}$$

This expression looks quite reasonable, since it merely corresponds to the replacement of ω with $\omega + i v_0$ to take into account the existence of a finite lifetime of excitations.

Physically the parameter v_0/δ grows when temperature increases. Indeed at $T = 0$ there are no phonons present to produce excitations decay, but when the temperature increases, the number of thermally excited phonons increases, so the excitation decay rate grows, which means that v_0 increases. On the other hand, we expect a decrease of δ with temperature, in a way that is analogous to the one occurring in BCS theory. As a result, we see that v_0/δ grows with temperature. A few examples of $n(\omega)$ are given in Fig. (9.10) for increasing values of the parameter v_0/δ, that is increasing values of T. For $v_0 = 0$, the result coincides with the standard BCS expression, and it is not qualitatively much different for $v_0/\delta = 0.1$. However, one notices that $n(\omega) \neq 0$ for all values of ω. This corresponds to the absence of a true gap, as mentioned above. As v_0/δ increases to 0.3 and 0.5, one sees that there is a progressive "filling" of the gap, that is there is, roughly speaking, an increase of the values of $n(\omega)$ for ω below the value ω_m corresponding to the maximum of $n(\omega)$.

At $T = 0$, this maximum (which corresponds to the divergence) occurs for $\omega_m = \delta(T = 0)$. However, one sees that when temperature grows, the reduced position of this maximum increases regularly beyond the value $\delta(T)$, that is the ratio ω_m/δ increases. One can see[3] that the position of this maximum is given by $\omega_m = [(3\delta^2 + 4v_0^2)^{1/2} + v_0]/\sqrt{3}$. So, for $T=0$ where $v_0(T=0)=0$, one has indeed $\omega_m=\delta(T=0)$. However, when one goes toward T_c where $\delta(T)$ goes to zero, the position of the maximum goes to $\sqrt{3}\, v_0(T = T_c)$, instead of going to zero as in the BCS case. In general, one expects $\sqrt{3}\, v_0(T = T_c) < \delta(T = 0)$, so that when temperature goes from 0 to T_c, the location $\omega_m(T)$ of the maximum is actually a

---

[3] One has, from Eq. (9.72) $dn(\omega)/d\omega = -\delta^2 \, \mathrm{Re}[(\omega + i v_0)^2 - \delta^2]^{-3/2}$. So $dn(\omega)/d\omega = 0$ implies that $[(\omega_m + i v_0)^2 - \delta^2]^{3/2}$ is purely imaginary, proportional to $e^{i\pi/2}$. Hence $(\omega_m + i v_0)^2 - \delta^2 = \omega_m^2 - v_0^2 - \delta^2 + 2i v_0 \omega_m$ is proportional to $e^{i\pi/3}$, so that $\omega_m^2 - v_0^2 - \delta^2 = 2 v_0 \omega_m/\sqrt{3}$, which leads to the result in the text.

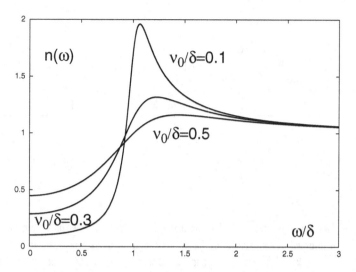

Fig. 9.10 Reduced density of states $n(\omega)$ from Eq. (9.72) as a function of the reduced frequency ω/δ, for increasing values of the parameter v_0/δ, corresponding to increasing values of the temperature. With this increase of T, the peak gets broader, its height decreases, and its reduced position ω_m/δ shifts to higher values, respectively 1.06, 1.23, and 1.44.

decreasing function of temperature, as in the BCS situation. However, instead of going to zero at T_c, $\omega_m(T)$ extrapolates to a nonzero value related to the excitation decay rate in the normal state. Now, if one obtains $n(\omega)$ experimentally, a natural quantity to characterize it is the location $\omega_m(T)$ of its maximum, which is tempting to consider as the "gap" value. However, as we have seen, such an identification does not work within strong coupling theory. Actually, this kind of behavior with a gap decreasing, but not extrapolating to zero at T_c, may be seen in some experimental results.

9.7 Further Reading: Migdal's Theorem

Despite this common reference name, Migdal's theorem is not really a theorem but rather the result of a close examination, calculation, and understanding of the terms coming in the perturbative expansion of the self-energy. The lowest-order contribution is the one pictured in Fig. (9.2). All the involved quantities should actually be understood as being renormalized by "high-energy" processes (of the order of the Fermi energy) due to Coulomb interactions that are responsible for the build-up of the quasi-electron (see Sections 2.2 and 9.5). But, for our purposes, this does not bring any change in practice. We have calculated this contribution in Section 9.2.2.

At the next order in perturbation, we compare the contributions from the two processes pictured in Fig. (9.11). In the first process a first phonon is emitted, then a second one, which is then reabsorbed by the electron before the final reabsorption of the first one. In the second process, the order of phonon reabsorption is inverted, so that the first emitted phonon is reabsorbed before the second one. Apparently the contributions from these two processes should be of similar order. However, upon closer examination, one sees that

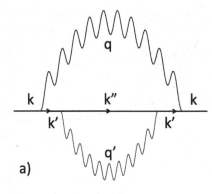
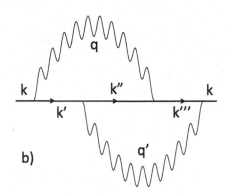

a)   b)

Fig. 9.11   Two phonon processes' contributions to the self-energy. (a) The two phonons $\mathbf{q}$ and $\mathbf{q}'$ are emitted and then reabsorbed in reverse order. (b) The two phonons $\mathbf{q}$ and $\mathbf{q}'$ are emitted and then reabsorbed in the same order. Note that the $\mathbf{k}$ lines at the beginning and the end of these diagrams do not actually enter the expression for the self-energy. Here, and in the following figures, they are only drawn to make it easier to read the processes involved in the diagrams.

this is not correct. Indeed, in all these perturbative calculations, the results contain energy denominators, involving energy differences between the various states involved in the processes. At lowest order this is well known, as it appears, for example, in Eq. (9.9). But the same happens in the higher orders.

This can be seen from the Brillouin–Wigner formulation of standard perturbative theory. When a perturbative term V is added to Hamiltonian H_0, the energy E_0 of the ground state $|0\rangle$ is shifted to E, which satisfies in this formulation the implicit equation

$$E = E_0 + \langle 0| V \sum_{n=0}^{\infty} \left(\frac{P_\perp}{E - H_0} V \right)^n |0\rangle \qquad (9.73)$$

where $P_\perp$ is the projection operator on the subspace perpendicular to the ground state $|0\rangle$. When $1/(E - H_0)$ is evaluated by inserting the closure relation $\sum_i |i\rangle\langle i| = 1$ for the eigenstates $|i\rangle$ (with energy E_i) of H_0, and the lowest-order approximation $E \simeq E_0$ is used, one sees the appearance of energy denominators $1/(E_0 - E_i)$. These denominators involve the difference between the energy E_0 of the initial state and the energy E_i of any of the intermediate states occurring in the considered process.

When we compare the contributions of various processes, the dominant ones, giving the larger contributions, are those that contain the larger number of small energy denominators. In our case we are interested, as we have already seen, in an initial state corresponding to the electron $\mathbf{k}$ close to the Fermi surface, so that $\xi_\mathbf{k}$ is small. More precisely, "small" means to be of the order of a typical phonon energy ω_D. When the first phonon $\mathbf{q}$ is emitted, the electron goes by momentum conservation to wavevector $\mathbf{k}' = \mathbf{k} - \mathbf{q}$. The phonon wavevector $\mathbf{q}$ can be chosen in such a way that $\mathbf{k}'$ is also close to the Fermi surface. In this way, the energy difference $\xi_\mathbf{k} - \xi_{\mathbf{k}'}$ can be small. (We forget to take into account in this energy evaluation the phonon energy $\Omega_\mathbf{q}$, which is anyway always small.) Then, in a

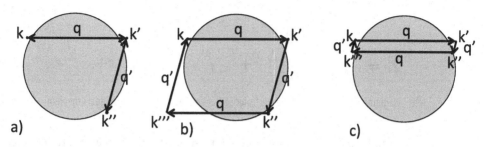

Fig. 9.12   Geometry with respect to the Fermi surface of the wavevectors involved in the two phonons processes considered in Fig. (9.11). In case (a), corresponding to Fig. (9.11a), the electron is, at the third step, automatically back at $\mathbf{k}'$, which is near the Fermi surface. On the contrary, if this is phonon $\mathbf{q}$ which is first reabsorbed, the electron goes to $\mathbf{k}'''$, which has no reason to be near the Fermi surface, as shown in (b). An exception shown in (c) occurs if phonon wavevector $\mathbf{q}'$ is small, in which case $\mathbf{k}'''$ is also near the Fermi surface.

similar way, when the second phonon $\mathbf{q}'$ is emitted, the electron goes to $\mathbf{k}'' = \mathbf{k}' - \mathbf{q}'$, which can be also close to the Fermi surface, so that the energy difference $\xi_{\mathbf{k}} - \xi_{\mathbf{k}''}$ is also small. The geometry of these wavevectors with respect to the Fermi surface is shown in Fig. (9.12).

Now, for the first process shown in Fig. (9.11a), when the second phonon is reabsorbed, the electron goes back to $\mathbf{k}' = \mathbf{k}'' + \mathbf{q}'$, which is automatically close to the Fermi surface, so that the energy difference $\xi_{\mathbf{k}} - \xi_{\mathbf{k}'}$ for the third stage in this process is automatically small. This is shown in Fig. (9.12a). In this way, the three energy difference denominators are small for this process. However, consider now the second process shown in Fig. (9.11b), with exactly the same involved phonons $\mathbf{q}$ and $\mathbf{q}'$, but reabsorbed in reverse order. Since the electron at $\mathbf{k}''$ absorbs first phonon $\mathbf{q}$, it goes to $\mathbf{k}''' = \mathbf{k}'' + \mathbf{q}$, which has no specific reason to be near the Fermi surface, as shown in Fig. (9.12b). Hence the energy difference $\xi_{\mathbf{k}} - \xi_{\mathbf{k}'''}$ is not small; it is typically of order E_F instead of ω_D. So, compared to the first one, this process contains two instead of three small denominators of order ω_D. As a result, its contribution is smaller by a factor of order ω_D/E_F.

Naturally there are cases where, for this second process, $\xi_{\mathbf{k}} - \xi_{\mathbf{k}'''}$ happens to be small and the contribution of the second process is in such cases of the same order as the contribution of the first one. However, these are restricted cases, corresponding to a specific arrangement of wavevectors. This implies that there is a phase space restriction associated with these particular arrangements. Hence the overall contribution to the self-energy of this second kind of process will be small anyway. An example of such an exception is the case where, for example, $\mathbf{q}'$ is small (with respect to the size of the Fermi surface), as is shown in Fig. (9.12c). In this case, since $\mathbf{k}''' = \mathbf{k}'' + \mathbf{q} = \mathbf{k} - \mathbf{q}'$, $\mathbf{k}'''$ is near $\mathbf{k}$ and the energy difference $\xi_{\mathbf{k}} - \xi_{\mathbf{k}'''}$ is small. Hence this is the situation where "center of the Brillouin zone" phonons are involved (and actually this case happens to be important only for the corresponding optical phonons). However, the number of such small wavevector phonons is small; there is indeed a phase space restriction, so that their overall contribution to the self-energy is negligible compared to the overall contribution of our first kind of processes.

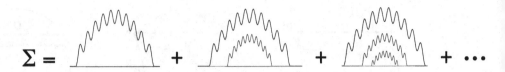

$$\Sigma = \quad + \quad + \quad + \cdots$$

Fig. 9.13   Diagrammatic expression of the electronic self-energy, as it results from Migdal's theorem.

To establish Migdal's theorem a number of other processes are investigated, with two or more phonons involved. In most cases, one comes to the conclusion that their contribution to the self-energy is negligible, by some power of the small ratio ω_D/E_F. The reasons are analogous to the ones we have seen for our second kind of two phonon process. The obvious exception is for contributions similar to our first kind of two-phonon process, but in higher order. For example, one can see that the contribution of the third-order process, shown in Fig. (9.13), where three phonons are successively emitted and then reabsorbed in reverse order, is in principle of the same order as the corresponding first- and second-order contributions. The basic reason is the following. We have seen by our explicit calculation in Section 9.2.2 that for a coupling constant λ of order unity, taking into account the second-order processes represented in Fig. (9.2) does not bring small corrections but rather modifications of order unity through the self-energy. It is accordingly reasonable that repeating this process Fig. (9.2) also brings non-negligible contributions that must be taken into account. Hence Migdal's theorem actually states that the correct self-energy is obtained, to lowest order in powers of ω_D/E_F, by taking into account all similar contributions at any order (that is, involving any number of emitted and then reabsorbed phonons). This means that to obtain the self-energy, one has to sum up the corresponding perturbative series, as shown in Fig. (9.13).

One can also reformulate this result by noticing that taking into account Migdal's theorem, the sum of all the contributions coming after the first phonon emission is equal to the propagator taking fully into account the self-energy (since one can clearly see that all the corresponding diagrams are present). This propagator is the full electronic propagator. Hence Migdal's theorem states that the self-energy is merely given by the diagram in Fig. (9.2), provided that the free $\mathbf{k}'$ propagator is replaced by the full $\mathbf{k}'$ propagator. Naturally this makes the equation for the self-energy $\Sigma(\omega)$ an implicit equation, since the self-energy is expressed in terms of a quantity that itself contains the self-energy. Actually, in the normal state, one can show that taking into account the fact that the self-energy is to a very good approximation independent of the wavevector $\mathbf{k}$, this substitution brings essentially no modification. So the simple diagram in Fig. (9.2) turns out to be enough to calculate the self-energy according to Migdal's theorem, and this justifies the calculation we have performed in Section 9.2.2. However, this is no longer true in the superconducting state, and we have to use in the self-energies the full propagators. This leads to the fact that in Eliashberg equations Eq. (9.26) and Eq. (9.27), $\Delta(\omega)$ appears on the right-hand side of the equations, so that these equations are integral equations.

It is interesting to note that the small parameter $\omega_D/E_F \sim (m/M)^{1/2}$, which is used to establish Migdal's theorem, is the same as the one that comes in the Born–Oppenheimer approximation, for the calculation of the phonons dispersion relation $\Omega_{\mathbf{q}}$. In this

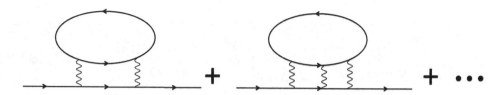

Fig. 9.14 Diagrammatic infinite series leading to the superconducting instability.

approximation, one considers first the ionic mass M as infinite so that when ions are in specific positions, they do not move and one can calculate (in principle) electronic energies for these ionic positions. In the next step, one considers the slow ionic motions. One assumes that the electrons adjust their wave functions and their energies instantaneously, because they have very fast motion compared to ionic motion, due to the light electronic mass m. Hence electrons adiabatically follow the ionic motion. In this way, the energy of any ionic geometry can be evaluated, and this leads to the phonon frequencies $\Omega_{\mathbf{q}}$. One can then evaluate the corrections to this adiabatic approximation, and one finds that these corrections correspond to an expansion in powers of $(m/M)^{1/2}$. As a result of this link with the Born–Oppenheimer approximation, Migdal's theorem is also often called the "adiabatic theorem."

Finally, let us add a last interesting remark on the validity of Migdal's theorem. If one considers the diagrammatic series represented in Fig. (9.14) (where only the terms with two and three phonons have been represented, but the series contains naturally corresponding diagrams with any number of phonons), one can check that all the terms are indeed negligible by some power of ω_D/E_F, in agreement with Migdal's theorem. However, the sum of this series diverges when, lowering temperature, one reaches $T = T_c$. Clearly the existence of this divergence shows that Migdal's theorem is not valid strictly speaking, since an infinite contribution can clearly not be neglected. However, physically, this divergence is just the manifestation of the superconducting instability. It signals the appearance of a bound state, corresponding to the formation of the Cooper pairs. The divergence in this series is analogous to the one we would find if, in Chapter 8, we wanted to calculate, by a series expansion in powers of the interaction potential, the scattering amplitude and the scattering length. We would find a divergent series and accordingly an infinite scattering length $a = \infty$, when the bound state due to the attractive potential appears. In the present case, the problem raised by these terms and the corresponding divergence is taken into account explicitly by introducing the superconducting order parameter with correspondingly the anomalous propagators and self-energy, the other "normal" terms being handled following Migdal's theorem.

Appendix Second Quantization

We give here a short summary of the second quantization formalism. For more detail, we refer the reader to standard textbooks, such as Landau and Lifschitz [56].

Second quantization is equivalent to the standard formalism of quantum mechanism, but it is by far much more convenient as soon as one deals with many particle systems. The operators involved in this formalism appear quite naturally when one deals with the standard one-dimensional harmonic oscillator, with Hamiltonian

$$H = \frac{1}{2m} p^2 + \frac{m\omega^2}{2} x^2 \tag{A.1}$$

the operator x and p satisfying the standard commutation relation $[x, p] = i\hbar$. One introduces the annihilation operator a, and its Hermitian conjugate the creation operator $a^\dagger$, defined by

$$a = \sqrt{\frac{m\omega}{2\hbar}} x + i \sqrt{\frac{1}{2\hbar m\omega}} p \qquad a^\dagger = \sqrt{\frac{m\omega}{2\hbar}} x - i \sqrt{\frac{1}{2\hbar m\omega}} p \tag{A.2}$$

which satisfy

$$[a, a^\dagger] = 1 \tag{A.3}$$

and allow one to write the Hamiltonian Eq. (A.1) as

$$H = \hbar\omega \left(a^\dagger a + \frac{1}{2} \right) \tag{A.4}$$

The normalized eigenstates $|n\rangle$, with $n = 0, 1, 2, \cdots$, satisfy $a^\dagger a |n\rangle = n|n\rangle$. They are obtained from the ground state $|0\rangle$ by the repeated action of the creation operator $a^\dagger$, since it results from the commutation relation that $a^\dagger a (a^\dagger |n\rangle) = (n + 1)(a^\dagger |n\rangle)$. Hence $a^\dagger |n\rangle = \sqrt{n+1} |n+1\rangle$, the coefficient being obtained by calculating the norm $(\langle n|a)(a^\dagger |n\rangle) = n+1$ from the commutation relation. Conversely the repeated action of the annihilation operator a lets go down the ladder formed by the eigenstates, since $a^\dagger a (a|n\rangle) = (n - 1)(a|n\rangle)$, so that $a|n\rangle = \sqrt{n}|n - 1\rangle$. This goes on until one reaches $n = 0$ for which one finds from the preceding relation $a|0\rangle = 0$.

If one switches to the standard language in terms of bosons, the eigenstates $|n\rangle$ are considered as corresponding to the presence of $n = 0, 1, 2, \cdots$ bosons, with the creation and annihilation operator adding or removing one boson at a time. In particular, when no boson is present, that is $n = 0$, which corresponds to vacuum, the annihilation operator acting on this vacuum gives 0. The number n of bosons is given by the operator $a^\dagger a$. All these properties are directly linked to the commutation relation Eq. (A.3) satisfied by a and $a^\dagger$.

This means of recording the number of bosons present in a specific state can also be used when one deals with a system made of N identical bosons. To represent any state, one first needs a complete orthonormal basis set for single particles. This may be the set $\phi_\mathbf{k}(\mathbf{r}) = e^{i\mathbf{k}\cdot\mathbf{r}}$ of plane waves when the bosons are in a box of unit volume. We use here this specific basis for simplicity and clarity, but the results below apply as well for any appropriate basis $\phi_\mathbf{k}(\mathbf{r})$. We want now to use this basis for the case where there are N bosons present. To be definite and simple, we will consider explicitly the case $N = 2$ and give the results for the general case. If the two bosons, with position $\mathbf{r}$ and $\mathbf{r}'$, are in two different states $\mathbf{k}_1$ and $\mathbf{k}_2$, we cannot write their wave function as $e^{i(\mathbf{k}_1\cdot\mathbf{r}+\mathbf{k}_2\cdot\mathbf{r}')}$, since it must be symmetric with respect to the exchange of $\mathbf{r}$ and $\mathbf{r}'$. Hence we must write the symmetrized version of it, namely

$$\Psi_{1,1}(\mathbf{r}, \mathbf{r}') = \frac{1}{\sqrt{2}}\left(e^{i(\mathbf{k}_1\cdot\mathbf{r}+\mathbf{k}_2\cdot\mathbf{r}')} + e^{i(\mathbf{k}_1\cdot\mathbf{r}'+\mathbf{k}_2\cdot\mathbf{r})}\right) \tag{A.5}$$

The index $1, 1$ for Ψ indicates that there is one boson in state $\mathbf{k}_1$ and another one in a different state $\mathbf{k}_2$. The prefactor is for a proper normalization. In the general situation of N bosons in N different states, the sum in Eq. (A.5) would contain $N!$ terms, and the normalization coefficient would be accordingly $1/\sqrt{N!}$. On the other hand, if the two bosons are in the same state $\mathbf{k}_1$, the symmetrized and normalized wave function is

$$\Psi_{2,0}(\mathbf{r}, \mathbf{r}') = e^{i\mathbf{k}_1\cdot(\mathbf{r}+\mathbf{r}')} \tag{A.6}$$

In general the presence of N_1 bosons in some state $\mathbf{k}_1$ would produce a corresponding further factor $\sqrt{N_1!}$ in the wave function, for proper normalization.

Let us now consider a single particle operator $\hat{U} = \sum_i U(\mathbf{r}_i)$ for these bosons. In the present case of two bosons, this is $U(\mathbf{r}) + U(\mathbf{r}')$. Since $U(\mathbf{r}_i)$ acts on a single boson variable, $\hat{U}$ can only connect states (that is, have nonzero matrix elements) that differ by the occupation of a single plane wave. For example, the final state may have $\mathbf{k}_3$ occupied by a boson, instead of $\mathbf{k}_1$ in the initial state. In the calculation of the matrix element of $\hat{U}$ between this final state and the initial state, making use of the above expression Eq. (A.5) for the corresponding wave functions, this will produce, for example, a factor

$$U_{\mathbf{k}_3,\mathbf{k}_1} = \int d\mathbf{r}\, e^{-i\mathbf{k}_3\cdot\mathbf{r}} U(\mathbf{r}) e^{i\mathbf{k}_1\cdot\mathbf{r}} \tag{A.7}$$

which depends only on the emptied state $\mathbf{k}_1$ and the filled state $\mathbf{k}_3$. Otherwise one can easily see that the other wavevectors and variables behave as "spectators" in this calculation, and contribute in the same way as in the calculation of the normalization of the wave function. As a result, this part of the calculation gives just a factor equal to 1, so that the matrix element of $\hat{U}$ is just $U_{\mathbf{k}_3,\mathbf{k}_1}$.

If the initial state is Eq. (A.6) instead of Eq. (A.5), the calculation of the matrix element of $\hat{U}$ works in the same way. The only difference is that there is no $1/\sqrt{2}$ factor in Eq. (A.6), so that the result is $\sqrt{2}\, U_{\mathbf{k}_3,\mathbf{k}_1}$. In the general case with N_1 bosons present in the initial state, the factor $\sqrt{N_1!}$ present in the wave function leads in a similar way to a factor $\sqrt{N_1}$ in the matrix element. This factor is just equal to the matrix element $\langle N_1 - 1|a_{\mathbf{k}_1}|N_1\rangle$

of an annihilation operator $a_{\mathbf{k}_1}$ satisfying the same commutation relation as the annihilation operator a in the above harmonic oscillator. Here N_1 and $N_1 - 1$ are the number of bosons in state $\mathbf{k}_1$ in the initial state and in the final state. In a symmetrical way, if the final state contains N_3 bosons, the matrix element has a factor $\sqrt{N_3}$, which is equal to the matrix element $\langle N_3 | a_{\mathbf{k}_3}^\dagger | N_3 - 1 \rangle$ of a creation operator $a_{\mathbf{k}_3}^\dagger$. Finally, if $\mathbf{k}_3 = \mathbf{k}_1$, so that one looks at its contribution to the diagonal matrix element of $\hat{U}$, one merely finds multiplying $U_{\mathbf{k}_1,\mathbf{k}_1}$ a factor N_1, which is equal to $\langle N_1 | a_{\mathbf{k}_1}^\dagger a_{\mathbf{k}_1} | N_1 \rangle$. Hence, for these involved wavevectors $\mathbf{k}_3$ and $\mathbf{k}_1$, the bosonic factors are in all cases the same as those produced by $a_{\mathbf{k}_3}^\dagger a_{\mathbf{k}_1}$. Extending these considerations to all wavevectors, we can write as a result the single particle operator $\hat{U}$ as

$$\hat{U} = \sum_{\mathbf{k}_i,\mathbf{k}_j} U_{\mathbf{k}_i,\mathbf{k}_j} \, a_{\mathbf{k}_i}^\dagger a_{\mathbf{k}_j} \tag{A.8}$$

since all its matrix elements are the same as those of $\hat{U} = \sum_i U(\mathbf{r}_i)$. Naturally the creation and annihilation operators corresponding to different wavevectors commute, since they act on different variables, namely the occupation numbers of different wavevectors.

One can proceed to similar evaluations for a two-particle operators, such as the interaction between all the bosons. For $\hat{V} = \sum_{ij} V(\mathbf{r}_i, \mathbf{r}_j)$, one finds in this way the generalization of Eq. (A.8) as

$$\hat{V} = \sum_{\mathbf{k}_i,\mathbf{k}_j,\mathbf{k}_m,\mathbf{k}_n} V_{\mathbf{k}_i,\mathbf{k}_j,\mathbf{k}_m,\mathbf{k}_n} \, a_{\mathbf{k}_i}^\dagger a_{\mathbf{k}_j}^\dagger a_{\mathbf{k}_m} a_{\mathbf{k}_n} \tag{A.9}$$

with

$$V_{\mathbf{k}_i,\mathbf{k}_j,\mathbf{k}_m,\mathbf{k}_n} = \int d\mathbf{r} \, d\mathbf{r}' e^{-i(\mathbf{k}_i \cdot \mathbf{r} + \mathbf{k}_j \cdot \mathbf{r}')} V(\mathbf{r}, \mathbf{r}') \, e^{i(\mathbf{k}_m \cdot \mathbf{r}' + \mathbf{k}_n \cdot \mathbf{r})} \tag{A.10}$$

All the above considerations are for bosons, for which the wave function must be symmetric under the exchange of the variables of any two bosons. We consider now the case of fermions, for which any such exchange must change the sign of the wave function. We consider again explicitly the case of two fermions, and then give the generalization for any number of fermions. If their positions are $\mathbf{r}$ and $\mathbf{r}'$, and they are in two different states $\mathbf{k}_1$ and $\mathbf{k}_2$, again we cannot write the wave function of two fermions as $e^{i(\mathbf{k}_1 \cdot \mathbf{r} + \mathbf{k}_2 \cdot \mathbf{r}')}$ since it must be antisymmetric with respect to the exchange of $\mathbf{r}$ and $\mathbf{r}'$. The properly antisymmetrized and normalized version of it is

$$\Psi_{\mathbf{k}_1,\mathbf{k}_2}(\mathbf{r}, \mathbf{r}') = \frac{1}{\sqrt{2}} \left(e^{i(\mathbf{k}_1 \cdot \mathbf{r} + \mathbf{k}_2 \cdot \mathbf{r}')} - e^{i(\mathbf{k}_2 \cdot \mathbf{r} + \mathbf{k}_1 \cdot \mathbf{r}')} \right) \tag{A.11}$$

On the other hand, since two fermions cannot be in the same state, we do not have to consider the case where they have the same wavevector.

In contrast with Eq. (A.5), we have to choose a specific order in the basis functions, here the wavevectors $\mathbf{k}_1$ and $\mathbf{k}_2$, to write the specific wave function Eq. (A.11). This is indicated by the indices $\mathbf{k}_1, \mathbf{k}_2$ of the wave function Ψ. The opposite ordering (corresponding to the exchange of $\mathbf{k}_1$ and $\mathbf{k}_2$) leads to the opposite sign for the wave function $\Psi_{\mathbf{k}_2,\mathbf{k}_1}(\mathbf{r}, \mathbf{r}') = -\Psi_{\mathbf{k}_1,\mathbf{k}_2}(\mathbf{r}, \mathbf{r}')$. In the general situation of N fermions with variables $\mathbf{r}_n$ in N different states $\mathbf{k}_i$, the sum in Eq. (A.11) would contain $N!$ terms, with alternating signs, and the normalization coefficient would be again $1/\sqrt{N!}$. This sum may be written as the determinant of the $e^{i\mathbf{k}_j \cdot \mathbf{r}_n}$.

Just as for bosons, we want to make use of creation and annihilation operators $c_{\mathbf{k}}^\dagger$ and $c_{\mathbf{k}}$ to take into account the number of fermions present in specific states. However, one sees immediately that the commutation rules must be different if we want to be able to represent $\Psi_{\mathbf{k}_1,\mathbf{k}_2}(\mathbf{r}, \mathbf{r}')$ as the result of the two creation operators $c_{\mathbf{k}_1}^\dagger$ and $c_{\mathbf{k}_2}^\dagger$ acting on the vacuum $|0\rangle$ of fermions, that is to write $\Psi_{\mathbf{k}_1,\mathbf{k}_2}(\mathbf{r}, \mathbf{r}') = \langle \mathbf{r}, \mathbf{r}' | c_{\mathbf{k}_2}^\dagger c_{\mathbf{k}_1}^\dagger |0\rangle$. Indeed, since $\Psi_{\mathbf{k}_2,\mathbf{k}_1}(\mathbf{r}, \mathbf{r}') = -\Psi_{\mathbf{k}_1,\mathbf{k}_2}(\mathbf{r}, \mathbf{r}')$, this implies that the operators $c_{\mathbf{k}_1}^\dagger$ and $c_{\mathbf{k}_2}^\dagger$ must anticommute, that is

$$c_{\mathbf{k}_2}^\dagger c_{\mathbf{k}_1}^\dagger = -c_{\mathbf{k}_1}^\dagger c_{\mathbf{k}_2}^\dagger \tag{A.12}$$

and similarly for the annihilation operators, which are their Hermitian conjugates. Introducing the anticommutator of two operators $\{A, B\} \equiv AB + BA$, this can be written as

$$\{c_{\mathbf{k}_1}^\dagger, c_{\mathbf{k}_2}^\dagger\} = 0 \qquad \{c_{\mathbf{k}_1}, c_{\mathbf{k}_2}\} = 0 \tag{A.13}$$

In particular, setting $\mathbf{k}_1 = \mathbf{k}_2$, these relations imply $c_{\mathbf{k}} c_{\mathbf{k}} = 0 = c_{\mathbf{k}}^\dagger c_{\mathbf{k}}^\dagger$. These relations are obvious since any basis state can contain zero or one fermion, so that it is impossible to create or annihilate two fermions in the same state $\mathbf{k}$.

Moreover, for a given wavevector $\mathbf{k}$, with respect to occupation number, one can only have $|0\rangle$ or $|1\rangle$, corresponding to zero or one fermion present. Hence these two states form a basis with respect to particle number. Since $c_{\mathbf{k}}|0\rangle = 0$, $c_{\mathbf{k}}^\dagger|1\rangle = 0$, $c_{\mathbf{k}}|1\rangle = |0\rangle$ and $c_{\mathbf{k}}^\dagger|0\rangle = |1\rangle$, one finds that $c_{\mathbf{k}} c_{\mathbf{k}}^\dagger + c_{\mathbf{k}}^\dagger c_{\mathbf{k}}$, acting on $|0\rangle$ or $|1\rangle$, gives $|0\rangle$ or $|1\rangle$, that is it behaves as the unit operator.[1] Hence we have for the anticommutator of $c_{\mathbf{k}}$ and $c_{\mathbf{k}}^\dagger$

$$\{c_{\mathbf{k}}, c_{\mathbf{k}}^\dagger\} = 1 \tag{A.14}$$

Finally, for different wavevectors, we have naturally

$$\{c_{\mathbf{k}_1}, c_{\mathbf{k}_2}^\dagger\} = 0 \tag{A.15}$$

since this is again a matter of basis state ordering. This is seen explicitly below.

Now we consider again a single particle operator $\hat{U} = \sum_i U(\mathbf{r}_i)$ for these fermions, and see that in a way similar to Eq. (A.8), it can be written as

$$\hat{U} = \sum_{\mathbf{k}_i,\mathbf{k}_j} U_{\mathbf{k}_i,\mathbf{k}_j} c_{\mathbf{k}_i}^\dagger c_{\mathbf{k}_j} \tag{A.16}$$

in terms of the fermionic creation and annihilation operators satisfying the above anti-commutation rules. Indeed let us start with state Eq. (A.11) and consider the case where, in the final state, wavevector $\mathbf{k}_3$ is occupied instead of $\mathbf{k}_2$. If we choose to order the plane wave states in the order $\mathbf{k}_1$, $\mathbf{k}_2$, and $\mathbf{k}_3$, the wave function corresponding to the final state is $\Psi_{\mathbf{k}_1,\mathbf{k}_3}(\mathbf{r}, \mathbf{r}') = (1/\sqrt{2})(e^{i(\mathbf{k}_1 \cdot \mathbf{r} + \mathbf{k}_3 \cdot \mathbf{r}')} - e^{i(\mathbf{k}_3 \cdot \mathbf{r} + \mathbf{k}_1 \cdot \mathbf{r}')})$. The calculation of the matrix element of $\hat{U}$ between these initial and final states runs actually exactly as for

---

[1] Note that, since as a result $c_{\mathbf{k}}^\dagger c_{\mathbf{k}}|1\rangle = |1\rangle$ and $c_{\mathbf{k}}^\dagger c_{\mathbf{k}}|0\rangle = 0$, $n_{\mathbf{k}} = c_{\mathbf{k}}^\dagger c_{\mathbf{k}}$ is the operator giving the fermion number in state $\mathbf{k}$.

the boson case, and one gets merely in the same way $U_{\mathbf{k}_3,\mathbf{k}_2}$ for this matrix element. On the other hand, if we consider a different final state where $\mathbf{k}_3$ is occupied instead of $\mathbf{k}_1$, because of our chosen ordering the wave function corresponding to this final state is $\Psi_{\mathbf{k}_2,\mathbf{k}_3}(\mathbf{r},\mathbf{r}') = (1/\sqrt{2})(e^{i(\mathbf{k}_2\cdot\mathbf{r}+\mathbf{k}_3\cdot\mathbf{r}')} - e^{i(\mathbf{k}_3\cdot\mathbf{r}+\mathbf{k}_2\cdot\mathbf{r}')})$. As a result, proceeding to the same calculation of the matrix element of $\hat{U}$ between the initial state and this new final state, we find now $-U_{\mathbf{k}_3,\mathbf{k}_1}$, which has a sign opposite to the preceding case.

We check now that Eq. (A.16) gives the correct result for both these cases. In the first case, the only involved term in Eq. (A.16) is $U_{\mathbf{k}_3,\mathbf{k}_2}\, c^{\dagger}_{\mathbf{k}_3} c_{\mathbf{k}_2}$. With our ordering conventions, the initial state is $c^{\dagger}_{\mathbf{k}_2} c^{\dagger}_{\mathbf{k}_1}|0\rangle$ and the final state is $c^{\dagger}_{\mathbf{k}_3} c^{\dagger}_{\mathbf{k}_1}|0\rangle$. Hence we have to calculate

$$\langle 0|c_{\mathbf{k}_1} c_{\mathbf{k}_3}(c^{\dagger}_{\mathbf{k}_3} c_{\mathbf{k}_2})c^{\dagger}_{\mathbf{k}_2} c^{\dagger}_{\mathbf{k}_1}|0\rangle = 1 \tag{A.17}$$

The result is obtained by noticing that creating a $\mathbf{k}_2$ fermion by $c^{\dagger}_{\mathbf{k}_2}$, then immediately annihilating it by $c_{\mathbf{k}_2}$, is the same as doing nothing. So we can replace $c_{\mathbf{k}_2} c^{\dagger}_{\mathbf{k}_2}$ with 1. More formally we may use the anticommutation rule Eq. (A.14) to write $c_{\mathbf{k}_2} c^{\dagger}_{\mathbf{k}_2} = 1 - c^{\dagger}_{\mathbf{k}_2} c_{\mathbf{k}_2}$. By anticommuting $c_{\mathbf{k}_2}$ with $c^{\dagger}_{\mathbf{k}_1}$, and making use of $c_{\mathbf{k}_2}|0\rangle = 0$, the second term gives indeed 0. Then, in the same way, we can replace $c_{\mathbf{k}_3} c^{\dagger}_{\mathbf{k}_3}$ and $c_{\mathbf{k}_1} c^{\dagger}_{\mathbf{k}_1}$ with 1, so we are left for this matrix element with the result $\langle 0|0\rangle = 1$.

In our second case, the only involved term in Eq. (A.16) is $U_{\mathbf{k}_3,\mathbf{k}_1}\, c^{\dagger}_{\mathbf{k}_3} c_{\mathbf{k}_1}$. With our ordering conventions the initial state is $c^{\dagger}_{\mathbf{k}_2} c^{\dagger}_{\mathbf{k}_1}|0\rangle$ and the final state is $c^{\dagger}_{\mathbf{k}_3} c^{\dagger}_{\mathbf{k}_2}|0\rangle$, so we have to calculate

$$\langle 0|c_{\mathbf{k}_2} c_{\mathbf{k}_3}(c^{\dagger}_{\mathbf{k}_3} c_{\mathbf{k}_1})c^{\dagger}_{\mathbf{k}_2} c^{\dagger}_{\mathbf{k}_1}|0\rangle = -1 \tag{A.18}$$

We obtain this result by first making use of the anticommutation rule, Eq. (A.15), to write $c_{\mathbf{k}_1} c^{\dagger}_{\mathbf{k}_2} = -c^{\dagger}_{\mathbf{k}_2} c_{\mathbf{k}_1}$. Then we proceed with the operators $c_{\mathbf{k}} c^{\dagger}_{\mathbf{k}}$, as we have done for Eq. (A.17). Hence the matrix element of $\hat{U}$ is indeed $-U_{\mathbf{k}_3,\mathbf{k}_1}$, as we have found above by our direct calculation with the wave functions. We note that here we indeed require the anticommutation rule Eq. (A.15) to be valid in order to obtain the proper result.

In conclusion, we see that thanks to the anticommutation rules satisfied by the creation and annihilation operators $c^{\dagger}_{\mathbf{k}}$ and $c_{\mathbf{k}}$, we find the proper signs resulting from the antisymmetry of the fermionic wave functions. The manipulation of these operators is much more convenient than working with huge determinants to keep track of all the fermions, in case we have many of them. Just as for bosons, our considerations can be extended to the case of two-particle operators. For $\hat{V} = \sum_{ij} V(\mathbf{r}_i,\mathbf{r}_j)$, the generalization of Eq. (A.16) is

$$\hat{V} = \sum_{\mathbf{k}_i,\mathbf{k}_j,\mathbf{k}_m,\mathbf{k}_n} V_{\mathbf{k}_i,\mathbf{k}_j,\mathbf{k}_m,\mathbf{k}_n}\, c^{\dagger}_{\mathbf{k}_i} c^{\dagger}_{\mathbf{k}_j} c_{\mathbf{k}_m} c_{\mathbf{k}_n} \tag{A.19}$$

with

$$V_{\mathbf{k}_i,\mathbf{k}_j,\mathbf{k}_m,\mathbf{k}_n} = \int d\mathbf{r}\, d\mathbf{r}'\, e^{-i(\mathbf{k}_i\cdot\mathbf{r}+\mathbf{k}_j\cdot\mathbf{r}')} V(\mathbf{r},\mathbf{r}')\, e^{i(\mathbf{k}_m\cdot\mathbf{r}'+\mathbf{k}_n\cdot\mathbf{r})} \tag{A.20}$$

In Eq. (A.19) one has to take care of the proper order of the operators because of the anticommutation relations.

References

[1] H. Kammerlingh Onnes, *Leiden Comm.* **122b, 124c** (1911).

[2] W. Meissner and R. Ochsenfeld, *Naturwissenschaften* **21**, 787 (1933).

[3] F. London and H. London, *Proc. Roy. Soc.* (London) A **149**, 71 (1935).

[4] A.B. Pippard, *Proc. Roy. Soc.* (London) A **216**, 547 (1953).

[5] M. Gell-Mann and K. Brueckner, *Phys. Rev.* **106**, 364 (1957).

[6] L.D. Landau, *Sov. Phys. JETP* **3**, 920 (1956).

[7] D. Pines and P. Nozières, *The Theory of Quantum Liquids*, Benjamin, New York (1966).

[8] P. Kapitza, *Nature* **141**, 74 (1938).

[9] J.F. Allen and A.D. Misener, *Nature* **141**, 75 (1938).

[10] F. London, *Nature* **141**, 643 (1938).

[11] A. Einstein, *Sitzber. Preuss. Akad. Wiss.* **1**, 3 (1925).

[12] L.D. Landau, *J. Phys. USSR*, **5**, 71 (1941).

[13] D.W. Osborne, B. Weinstock, and B.M. Abraham, *Phys. Rev.* **75**, 988 (1949).

[14] H. Fröhlich, *Phys. Rev.* **79**, 845 (1950).

[15] E. Maxwell, *Phys. Rev.* **78**, 477 (1950); C.A. Reynolds, B. Serin, W.H. Wright, and L.B. Nesbitt, *Phys. Rev.* **78**, 487 (1950).

[16] M.R. Schafroth, *Phys. Rev.* **96**, 1149 (1954).

[17] L.N. Cooper, *Phys. Rev.* **104**, 1189 (1956).

[18] J. Bardeen, L.N. Cooper and J.R. Schrieffer, *Phys. Rev.* **108**, 1175 (1957).

[19] L.S. Gradstein and I.M. Ryzhik, *Tables of integrals, series and products*, Academic Press (2000).

[20] D. Mattis and E. Lieb, *J. Math. Phys.* **2**, 602 (1961).

[21] R.W. Richardson and N. Sherman, *Nucl. Phys.* **52**, 221 (1964).

[22] J. Dukelsky, S. Pittel, and G. Sierra, *Rev. Mod. Phys.* **76**, 643 (2004).

[23] M. Combescot and S.Y. Shiau, *Excitons and Cooper Pairs*, Oxford University Press (2015)

[24] N.N. Bogoliubov, *Nuovo Cimento* **7**, 794 (1958); *Usp. Fiz. Nauk* **67**, 549 (1959).

[25] J.G. Valatin, *Nuovo Cimento* **7**, 843 (1958).

[26] I. Giaever, *Phys. Rev. Lett.* **5**, 147 (1960).

[27] L.P. Pitaevskii, *Sov. Phys. JETP* **10**, 1267 (1960).

[28] D.D. Osheroff, R.C. Richardson and D.M.Lee, *Phys. Rev. Lett.* **28**, 885 (1972).

[29] W. Kohn and J.M. Luttinger, *Phys. Rev. Lett.* **15**, 524 (1965).

[30] P.W. Anderson, *J. Phys. Chem. Solids* **11**, 26 (1959).

[31] D.C. Mattis and J. Bardeen, *Phys. Rev.* **111**, 412 (1958).

[32] R.E. Glover and M. Tinkham, *Phys. Rev.* **104**, 844 (1956).

[33] L.C. Hebel and C.P. Slichter, *Phys. Rev.* **107**, 901 (1957).

[34] F. London, *Superfluids*, Vol. I, p. 152, John Wiley, New York (1950).

[35] B.S. Deaver and W.M. Fairbank, *Phys. Rev. Lett.* **7**, 43 (1961).

[36] B.D. Josephson, *Phys. Lett.* **1**, 251 (1962).

[37] L.P. Gor'kov, *Sov. Phys. JETP* **9**, 1364 (1959).

[38] V.L. Ginzburg and L.D. Landau, *J. Exptl. Theoret. Phys. (U.S.S.R.)* **20**, 1064 (1950).

[39] L.D. Landau, *J. Exptl. Theoret. Phys. (U.S.S.R.)* **7**, 19 (1937); *J. Exptl. Theoret. Phys. (U.S.S.R.)* **7**, 627 (1937).

[40] A.A. Abrikosov, *Sov. Phys. JETP* **5**, 1174 (1957).

[41] R.D. Parks and W.A. Little, *Phys. Rev.* **133**, A97 (1964).

[42] S.N. Bose, *Zeitschr. für Physik* **26**, 178 (1924).

[43] N.N. Bogoliubov, *J. Phys. USSR*, **11**, 23 (1947).

[44] R.P. Feynman, *Phys. Rev.* **94**, 262 (1954).

[45] L.D. Landau, *J. Phys. USSR*, **11**, 91 (1947).

[46] L. Tisza, *Nature* **141**, 913 (1938).

[47] A.M. Clogston, *Phys. Rev. Lett.* **9**, 266 (1962); B.S. Chandrasekhar, *Phys. Rev. Lett.* **1**, 7 (1962).

[48] P. Fulde and R.A. Ferrell, *Phys. Rev.* **135**, A550 (1964).

[49] A.I. Larkin and Y.N. Ovchinnikov, *Sov. Phys. JETP* **20**, 762(1964).

[50] G.M. Eliashberg *Sov. Phys. JETP* **11**, 696 (1960); *Sov. Phys. JETP* **12**, 1000 (1961).

[51] A.B. Migdal, *Sov. Phys. JETP* **7**, 996 (1958).

[52] V.L. Ginzburg and D.A. Kirzhnits, *Phys. Rep.* **4**, 343 (1972); P.B. Allen and R.C. Dynes, *Phys. Rev. B* **12**, 905 (1975).

[53] B.T. Geilikman, V.Z. Krezin, and N.F. Masharov, *J. Low Temp. Phys.* **18**, 241 (1975); R. Combescot, *Phys. Rev. B* **42**, 7810 (1990).

[54] W.L. McMillan, *Phys. Rev.* **167**, 331 (1968).

[55] W.L. McMillan and J.M. Rowell, *Tunneling and strong-coupling superconductivity*, in *Superconductivity* ed. by R.D. Parks, Taylor and Francis (1969).

[56] L.D. Landau and E. Lifschitz, *Quantum Mechanics*, Ch. 9, Pergamon Press (1977).

Index

Printed in the United States
by Baker & Taylor Publisher Services